BIOLOGY FOR THE HEALTH SCIENCES

Biology is central to our understanding of health and disease and to the development of effective treatments, and thus it is critical that health professionals have a solid grounding and knowledge comfort in the pathogenesis and mechanisms of disease processes. This innovative new textbook draws these topics together, providing an accessible introduction across four central disciplines: basic biology, biotechnology, non-infectious disease, and infectious disease.

Key Features:

- Provides students of biology and those going into health-care professions with a strong foundation to understand the pathogenesis of disease at the molecular and cellular levels

- Focuses on the etiology and pathophysiology of the major human diseases by body system, including diabetes and nutritional disorders, cardiovascular disease, neurodegenerative diseases, and cancer, aligned to medicine and health science course structure

- Covers mechanisms of infectious disease transmission, as well as disease pathophysiology, and considers the impact of antibiotic resistance

- Reviews the applications of biotechnology and genomics to human health in diagnosis and treatment, as well as to our understanding of disease and disease surveillance

- Each chapter contains a mini-glossary of key terms, and review questions for students to assess how much of the chapter they have assimilated.

Enhanced throughout with plentiful illustrations, *Biology for the Health Sciences* is an essential companion for any student of the health sciences and for biological science students studying the causes of disease as part of a wider course.

Mark F. Wiser is Associate Professor and Vice Chair, School of Public Health and Tropical Medicine, Tulane University, New Orleans, Louisiana, USA.

BIOLOGY FOR THE HEALTH SCIENCES

MECHANISMS OF DISEASE

Mark F. Wiser

Associate Professor and Vice Chair
School of Public Health and Tropical Medicine
Tulane University, New Orleans, USA

CRC Press
Taylor & Francis Group
Boca Raton London New York

CRC Press is an imprint of the
Taylor & Francis Group, an **informa** business

Designed cover image: From Paulista/Shutterstock.com, with permission.

First edition published 2023
by CRC Press
6000 Broken Sound Parkway NW, Suite 300, Boca Raton, FL 33487-2742

and by CRC Press
4 Park Square, Milton Park, Abingdon, Oxon, OX14 4RN

CRC Press is an imprint of Taylor & Francis Group, LLC

ISBN: 978-1-032-35726-3 (hbk)
ISBN: 978-0-8153-4586-2 (pbk)
ISBN: 978-1-003-32820-9 (ebk)

DOI: 10.1201/9781003328209

Typeset in ITC Leawood Std
by Evolution Design & Digital

Access the student resources: https://routledge.com/cw/wiser

Contents

Preface

Biology plays a central role in our understanding of health and disease. It is therefore critical that health professionals have a solid foundation in biology and especially the biological basis of disease. This is also true for public health professionals in that a solid understanding of biological principles will become increasingly important in disease management, prevention, and control. Furthermore, biotechnology is being rapidly applied to medicine as well as public health. So going forward, health professionals will need a strong grounding in the fundamentals of the biological basis of disease, as well as an understanding of biotechnologies such as stem cells, vaccines and other immunotherapies, genetic engineering, and genomics.

The origins of this book are from a Biological Basis of Disease course that was initially taught at the graduate level and then transformed into an advanced undergraduate course within a school of public health. However, the content of the book is also of interest to pre-medical students as well as students pursuing other health-related professions, such as pharmacy, veterinary medicine, or nursing. The book has three major sections: Basic Biology and Biotechnology, Non-infectious Disease, and Infectious Disease. The combining of these three topics into a single book is unique and allows for a merger of the basic sciences with the more applied biomedical sciences.

Section 1 consists of nine chapters that cover basic cell and molecular biology, genetics, and immunology and vaccines, as well as recombinant DNA technology and gene therapy, genomics, and stem cells. In advanced courses the students may already be sufficiently knowledgeable about the basic aspects of cell structure and function and genetics, and these chapters could serve as a review or reference within the same book. In the later chapters there are numerous signposts referring the reader back to the specific pages in the earlier chapters where the basic concepts are explained. In this way, readers who are unsure about the fundamental concepts can quickly review basic biology as they pursue the biological basis of disease. Conversely, there may be courses in which students do not have a strong background in basic biology and the beginning of the course needs to cover these topics. The book could even be used as part of two sequential courses in which the first course focused on basic biology and biotechnology followed by a second course focusing on the biological basis of disease.

The major non-infectious diseases of humans are covered in the nine chapters of Section 2. The first chapter of this section gives an overview of disease and discusses aging, senescence, and degenerative diseases.

This is followed by a chapter on monogenic genetic diseases. Sickle cell disease, cystic fibrosis, and Huntington's disease are described in detail as examples of how genetic defects lead to pathogenesis at the molecular and cellular levels. The next chapter deals with the role of nutrition in disease with a focus on overnutrition and obesity since obesity is a factor in many diseases. Thereafter, there are chapters on diabetes, cardiovascular disease, chronic kidney disease, cancer, immunological disorders, and addictive disorders. Each of the disease-specific chapters gives an overview of the disease and describes disease etiology and pathogenesis. The molecular and cellular aspects of etiology and pathogenesis are also discussed for each of the diseases. Clinical manifestations, as well as diagnosis, treatment, and disease management, are also described for these various diseases. The clinical aspects of the disease are described at a level suitable for non-physicians and pre-professional students.

The final six chapters of the book make up Section 3 and are devoted to infectious diseases. Infectious diseases tend to have superficial coverage in most biology of disease books. However, the COVID-19 pandemic has certainly awakened the world to the importance of infectious diseases in both resource-rich regions and resource-poor regions. Section 3 starts with an overview of pathogens and infectious disease. Included in the first chapter of this section is a discussion of drug resistance and the mechanisms by which pathogens become resistant to drugs. The subsequent chapters focus on a particular type of infectious disease, which includes chapters on respiratory tract infections, gastrointestinal infections and diarrheal diseases, HIV and other sexually transmitted infections, vector-borne disease with a focus on malaria, and parasitic worms. In each of these chapters the biologies of the various pathogens are extensively covered, as well as the biology of the host–pathogen interaction, including immunity. As in the chapters on non-infectious diseases, the pathogenesis associated with infectious diseases is described, as well as clinical manifestations, diagnosis, and treatment. Examples of interactions between the pathogen and human host at the molecular and cellular levels are also described. In regard to disease etiology, modes of pathogen transmission are thoroughly discussed including methods to prevent transmission.

In some respects, this book is three books in one. And that was some of the impetus for writing the book since there was not a single book that comprehensively covered both non-infectious and infectious diseases as well as the applications of modern biotechnology. This book could serve as the primary textbook in many public health biology courses or courses in pre-medical or pre-professional curricula. In addition, many pre-medical or pre-professional students may find this book helpful in preparing for qualifying exams and interviews. At the end of each chapter are a summary of the key points, definitions of the key terms, and review questions that can assist students in studying for exams.

Mark F. Wiser

Student and Instructor Resources

STUDENT RESOURCES

In addition to the review questions found at the end of each chapter in the book, *Biology for the Health Sciences* also comes with a number of interactive questions, hosted online, which the student can use to test their understanding of the chapter material.

This interactive resource makes use of several different question types: multiple choice, true or false, fill-in-the-blanks, and matching-type questions. Where necessary, feedback is also provided to help the student to understand why a particular answer is correct.

The students can access these resources by visiting https://routledge.com/cw/wiser

INSTRUCTOR RESOURCES

For instructors, all the figures found in the textbook have been compiled into Figure Slides, which the instructors may wish to use in their lectures. These are available in two convenient formats: PowerPoint and PDF. *To access these, please visit* https://routledge.com/cw/wiser

Basic Biology and Biotechnology

As we better understand disease at the molecular, cellular, genetic, and immunological levels, it is important to have a strong foundation in basic molecular and cellular biology, genetics and evolution, and immunology. Furthermore, recombinant DNA technology, genomics, stem cells, and other technologies are increasingly being applied to the diagnosis, treatment, and management of disease. The first four chapters of the first section provide an overview of cellular and molecular biology. Genetics also plays a role in all disease and, accordingly, a chapter on genetics that covers the basic principles of inheritance follows the chapters on basic cellular and molecular biology. Also included in this chapter is a brief discussion of evolution and our current understanding of human evolution. These opening chapters can serve as a review or a reference for students who have prior knowledge of these topics. Students who do not have a sufficient background can use these chapters to acquire basic knowledge that will be needed later in the book.

Understanding basic molecular biology and genetics is also important to understanding the biotechnologies that are being applied at a rapid rate to medicine and related fields. Following the chapter on genetics is a chapter on genomics, which describes genomics and its derivatives, such as transcriptomics, proteomics, and metabolomics. Also included in this chapter are discussions on how genomics can be used to improve human health through the development of better diagnostics and therapies and the prospect of personalized medicine. Genome-wide association studies are also described. The chapter following genomics describes recombinant DNA technology including gene therapy and how recombinant DNA technology leads to the more efficient production of biopharmaceuticals and its applications to agriculture. Chapter 8 discusses the organization of cells into tissues and organs and gives an overview of the cell differentiation process. Also included in this chapter is a discussion of stem cells and cell-based therapies. The final chapter in this section provides an overview of the immune system and immunology. Discussions of serology and vaccines are also included.

Chemical Basis of Life

1

All living organisms, at the most fundamental level, are composed of basic chemical elements. Chemical elements form the molecules that make up cellular structures and participate in chemical reactions necessary to sustain life. These molecules and chemical reactions occur in an aqueous environment and the unique properties of water play an essential role in the existence and sustenance of life. Biologically important molecules consist of a carbon backbone and fall into four chemically distinct classes of molecules: carbohydrates, lipids, proteins, and nucleic acids.

ATOMIC STRUCTURE

Atoms are composed of subatomic particles called electrons, protons, and neutrons. Atoms are typically depicted as a central nucleus, composed of protons and neutrons, being orbited by electrons (FIGURE 1.1). Protons (p+) are positively charged and the attraction between negative and positive charges maintains the negatively charged electrons (e-) in this orbit. Neutrons have no charge since they are composed of a proton and an electron. Atoms containing the same number of protons and electrons have no net charge. A gain or loss of electrons orbiting the nucleus results in the ionization of the atom that imparts either a negative or a positive charge on that atom.

Atoms define the basic chemical elements that make up all of nature

Atoms are defined as substances that cannot be broken down into another substance by ordinary means. The atomic nuclei provide stability to atoms and ordinary forces, such as heat, electricity, and light, do not adversely affect or change the atomic structure. There are 92 elements that occur naturally and each element corresponds to a distinct type of atom. These basic elements are defined by the number of protons making up the atom. For example, hydrogen, the simplest atom, is composed of one proton and its corresponding electron. The next element, which is helium (He), is composed of two protons and the corresponding electrons, as well as two neutrons. All 92 naturally occurring elements are defined by the number of protons ranging from 1 to 92 and the number of protons is referred to as the atomic number.

Protons and neutrons have mass and atoms can also be defined by the total number of neutrons and protons. The mass of the electron is negligible compared with the masses of the protons and neutrons and is therefore not included in the atomic mass. So, in addition to defining

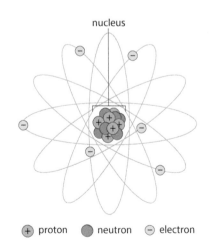

nucleus

(+) proton   (•) neutron   (−) electron

FIGURE 1.1 Model of atomic structure. Atoms are typically depicted as a core nucleus composed of neutrons and protons being circled by electrons. The number of protons defines an element.

elements by the atomic number corresponding to the number of protons, elements can also be defined by an atomic mass, which is the combined number of protons and neutrons. In the case of normal hydrogen, both the atomic number and the atomic mass are equal to one since there are no neutrons, whereas for helium the atomic number is two and the atomic mass is four since the helium nucleus also has two neutrons. The primary elements making up greater than 95% of the mass of living matter are oxygen, carbon, hydrogen, and nitrogen (TABLE 1.1).

Table 1.1	Common Elements Making Up Living Things					
Element	Symbol	Atomic number	Common atomic mass	Percentage in human body	Percentage in Earth's crust	Percentage in Universe
Hydrogen	H	1	1	9.5	0.14	91
Carbon	C	6	12	18.5	0.03	0.02
Nitrogen	N	7	14	3.3	Trace	0.04
Oxygen	O	8	16	65	47	0.06
Sodium	Na	11	23	0.2	2.8	Trace
Phosphorous	P	15	31	1.0	0.07	Trace
Chlorine	Cl	17	35	0.2	0.01	Trace
Potassium	K	19	39	0.4	2.6	Trace
Calcium	Ca	20	40	1.5	3.6	Trace
Iron	Fe	26	56	Trace	5.0	Trace

Atomic numbers and masses of selected elements and their approximate percentage by weight in the human body, the Earth's crust, and the Universe.

Variations in the number of neutrons result in different forms of an element

Elements can exist in different forms called isotopes that are defined by a different number of neutrons. All isotopes of a particular element have the same atomic number but have different atomic masses. For example, a form of hydrogen with one proton and one neutron in the nucleus is called deuterium and has an atomic mass of two. The atomic number, however, is still one. Similarly, tritium is another form of hydrogen with an atomic mass of three due to one proton and two neutrons. The various isotopes of an element are chemically the same in regard to their interactions with other elements. In nature elements are generally represented by a predominant isotope and a mixture of the other possible isotopes.

Some isotopes are unstable and tend to break apart or decay. These are referred to as radioactive isotopes or radioisotopes. Energy is released when these unstable isotopes break apart and this energy is called radioactivity. The released energy can be in the form of subatomic particles or electromagnetic radiation similar to X-rays. For example, carbon-14 (^{14}C) is an isotope of carbon containing six protons and eight neutrons. Carbon-14 is not as stable as the more common carbon-12 (^{12}C) consisting of six protons and six neutrons. To increase stability, one of the neutrons of carbon-14 converts into a proton by ejecting an electron from the nucleus. This ejected electron is the radioactivity that is detectable and can be measured. Since one of the neutrons converts to a proton, the atomic number is now seven and the atom is now nitrogen-14 (^{14}N), a common and stable isotope of nitrogen. Radioactivity and radioisotopes are widely used in biomedical research and medicine (TABLE 1.2) due to the well-defined properties of radioactive decay and the ease at which radioactivity can be detected and measured. Various imaging techniques, such as positron emission tomography (PET) scans (BOX 1.1), can be used to evaluate the integrity and function of internal organs.

Table 1.2 **Examples of Radioactivity in Diagnosis and Treatment of Disease**
• Imaging techniques such as X-rays and magnetic resonance imaging (MRI) to evaluate internal organs
• Use of radioactive substances to evaluate physiology such as using radioactive iodine to evaluate thyroid function
• Positron emission tomography (PET) scans to evaluate organ function
• Radiation therapy to treat cancer

Positron Emission Tomography

Positron emission tomography, commonly known as PET, is used to observe metabolic processes in the body as a means to diagnose disease. Radionuclides are incorporated either into compounds normally used by the body, such as glucose, or into molecules that bind to receptors or other sites of drug action. Such labeled compounds are often called radiotracers. Following injection, ingestion, or inhalation of the radiotracer, PET technology determines the distribution of the radiotracers in specific organs or tissues. The most often used radiotracer is a radioactive analog of glucose. Glucose is the central metabolite for cellular energy production (Chapter 3) and tissues or organs using excessive glucose will show up as 'hot spots' on the PET scan. PET scans are widely used in the diagnosis of cancer and neurological disorders.

BOX 1.1

Electrons orbit the nucleus at defined distances, forming shells

The positive charge of the nucleus attracts the orbiting electrons and thus the electrons are drawn close to the nucleus. However, the negatively charged electrons repel each other and therefore only a limited number of electrons can occupy the space close to the nucleus. In the case of large atoms with many electrons, the electrons need to be distributed at increasing distances from the nucleus. These increasing distances are called electron shells and only a fixed number of electrons can occupy a particular shell. For example, only two electrons can occupy the space closest to the nucleus and the next shell can accommodate up to eight electrons. The electrons of an atom fill in the shells closest to the nucleus first and then expand outward. For example, the carbon atom with six electrons has two electrons in the first shell and four electrons in the second shell, whereas oxygen with eight electrons has two electrons in the first shell and six in the second shell (FIGURE 1.2).

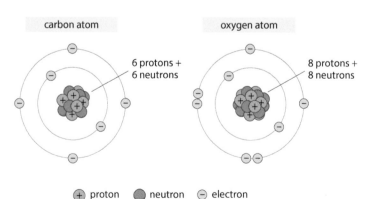

FIGURE 1.2 Electron shells in carbon and oxygen. The electrons in an atom distribute in 'shells' defined by the distance from the nucleus. The first shell can only hold two electrons and the next shell can hold up to eight electrons. Carbon has four electrons in its outer shell and oxygen has six.

The electrons orbiting the nucleus can be gained, lost, or shared with other atoms

The electrons in the outermost shells tend to be rather dynamic and especially in atoms with a partially filled outermost electron shell. Atoms with partially filled outer shells have the propensity to gain or lose electrons in order to fill or empty the outermost shell. Having a completely full outermost electron shell increases the stability of the atom. For example, atoms in which the outer shell is completely full in its normal state do not react with other atoms and are called inert. Conversely, atoms with partially full outer shells are more likely to interact with other atoms by forming chemical bonds.

INTERACTIONS BETWEEN ATOMS

Atoms can combine with other atoms to form molecules. For example, water – defined chemically as H_2O – is composed of two hydrogen atoms (H) and one oxygen atom (O). The interaction between atoms to form molecules involves gaining, losing, or sharing electrons in the outermost shell. Losing, gaining, or sharing electrons between two or more atoms results in the formation of **chemical bonds**. Chemical bonds are attractive forces that hold the atoms close together. Each element has a characteristic set of possible chemical bonds that can be formed, which depends largely on the configuration of its electrons in the outer shell. These chemical bonds can be either weak forces or strong forces (TABLE 1.3).

Table 1.3	**Types of Bonds between Atoms or Molecules**	
	Bond type	**Nature of the bond/interaction**
Strong forces	Covalent	A sharing of electrons between atoms
Weak forces	Hydrogen	Interactions between a hydrogen atom involved in a polar covalent bond and another atom involved in a polar covalent bond
	Ionic	Interactions between positive and negatively charged atoms
	Hydrophobic	Due to the exclusion of hydrophobic molecules by water

Atoms share electrons in a covalent bond

An atom with a partially full outer electron shell can become more stable by sharing electrons with another atom and thereby form a covalent bond. For example, hydrogen has a single electron in a shell designed for two electrons. By combining with another hydrogen atom, the two atoms can share the electrons (FIGURE 1.3) to form molecular hydrogen (H_2). In this case, each electron contributed by each hydrogen atom orbits both nuclei and, thus, each atom now appears to have a full electron shell and the hydrogen molecule is more stable than the two individual hydrogen atoms. Therefore, the natural state of hydrogen is predominantly two hydrogen atoms bound by a covalent bond and free hydrogen atoms are relatively rare. The covalent bond is depicted as a single line between the atoms (H—H).

Two oxygen atoms also interact to form molecular oxygen (O_2) in a similar manner (see Figure 1.3). However, in this case each oxygen atom has six electrons in its outer shell and needs two electrons to complete the shell. Therefore, two electrons are shared to fill the shell and this results in a double covalent bond that is denoted by a double line between the atoms (O=O). And if three electrons are shared by two

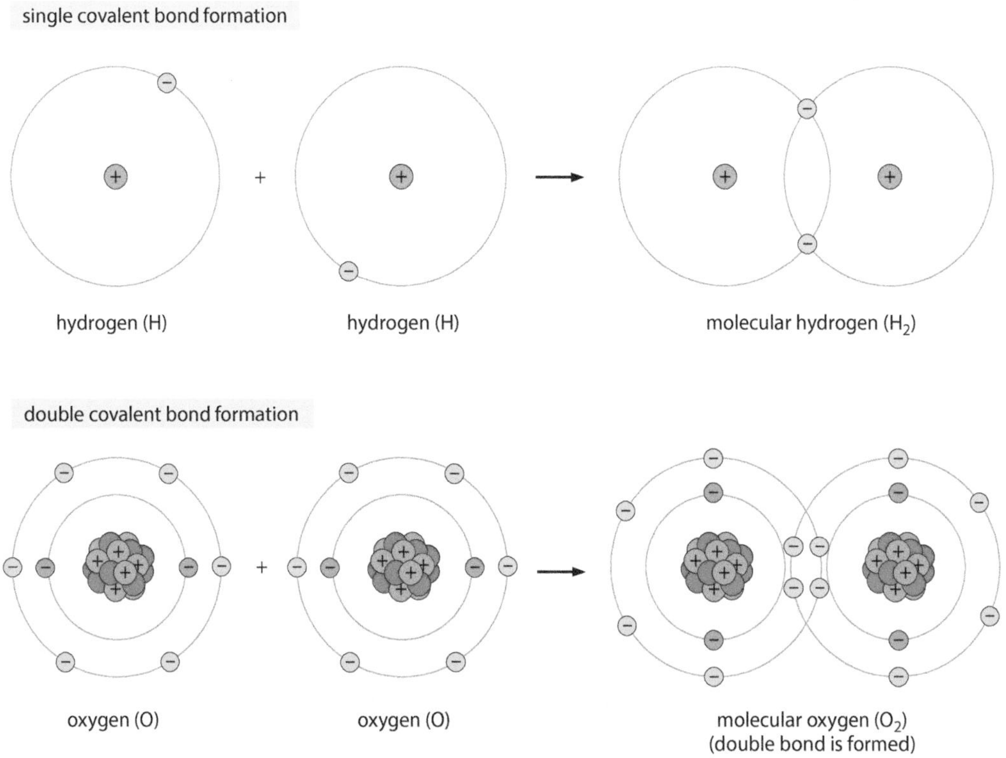

single covalent bond formation

hydrogen (H)   hydrogen (H)   molecular hydrogen (H₂)

double covalent bond formation

oxygen (O)   oxygen (O)   molecular oxygen (O₂)
                                                 (double bond is formed)

FIGURE 1.3 **Covalent bonds.** In molecular hydrogen one electron from each atom is shared to form a single covalent bond. The two electrons are shared equally between the two nuclei and therefore the first electron shell is filled for both nuclei. In molecular oxygen two electrons from each atom are shared to form a double covalent bond. By sharing two electrons there are now eight apparent electrons in the outermost electron shell instead of six, and thus the shell is filled.

atoms a triple bond is formed as in the case of nitrogen gas ($N\equiv N$) which has five electrons in the outer shell.

In the cases of molecular hydrogen (H_2), molecular nitrogen (N_2), and molecular oxygen (O_2), both nuclei are identical and the electrons spend an equal amount of time with each nucleus. This equal sharing of electrons results in symmetrical molecules that are electrically neutral. However, not all covalent bonds exhibit such equality in electron sharing between atoms. In covalent bonds between different atoms, the nucleus with the most protons has a greater positive charge and therefore attracts the electrons more strongly. In other words, the electrons spend more time over the more positively charged nucleus. This results in a molecule that is asymmetric in terms of the distribution of electrons and the molecule exhibits a polarity. For example, water is formed between an oxygen atom and two hydrogen atoms (FIGURE 1.4). The electrons spend more time over the oxygen atom, resulting in a slight negative pole over the oxygen atom and slight positive poles over the hydrogen atoms.

Chemical reactions involve the breaking of covalent bonds

Covalent bonds are significantly stronger than the weak forces between atoms and molecules. However, not all covalent bonds are of equal strength since some are more stable than others. For example, the bond forming molecular hydrogen (H_2) is extremely stable and difficult to break. Other covalent bonds are not as strong and can more easily be broken.

Energy is released during chemical reactions in which weaker bonds are broken and more stable bonds are formed. For example, fuels such as

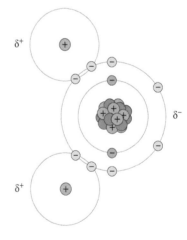

FIGURE 1.4 **Polarity of water.** Water is formed via covalent bonds between two hydrogen (H) atoms and one oxygen atom (O). In each of the two bonds electrons are shared between hydrogen and oxygen. Since the oxygen nucleus has more protons and a greater positive charge than the hydrogen nucleus the electrons spend more time orbiting the oxygen nucleus. This unequal sharing of the electrons results in a slight negative charge (designated by δ^-) near the oxygen atom and slight positive charges (designated by δ^+) near the hydrogen atoms.

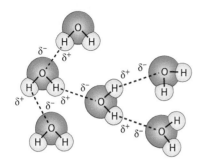

FIGURE 1.5 Hydrogen bonds between water molecules. The partial positive charge (δ^+) of hydrogen (H) interacts with the partial negative charge (δ^-) of oxygen (O) as depicted by the dashed lines. This placement of hydrogen between two atoms with partial negative charges is called a hydrogen bond.

gasoline and molecular oxygen (O_2) are composed of relatively unstable bonds. Burning fuels in the presence of oxygen is a chemical reaction that results in the formation of carbon dioxide (CO_2) and water (H_2O) – both of which contain highly stable bonds. This results in a release of energy in the form of heat. A similar process occurs in living organisms during the metabolism of carbohydrates to generate energy (Chapter 3). In this case the energy is captured and used to maintain the living organism.

Hydrogen bonds play important roles in biology

Another type of chemical bond between atoms and molecules is the hydrogen bond. This type of bond is weaker than a covalent bond, but nevertheless, it plays an important role in the structure and properties of water as well as being important in the structure and function of many biological molecules. The basis of the hydrogen bond is the slight polarity of the covalent bond formed between a hydrogen atom and atoms such as oxygen and nitrogen. The hydrogen atom involved in such a polar covalent bond has a slight positive charge and is attracted to another atom with a slight negative charge. For example, oxygen has a slight negative charge when it forms a covalent bond with hydrogen. Therefore, water molecules interact with each other via hydrogen bonds (FIGURE 1.5).

Ionic bonds are formed between atoms of opposite charges

Atoms can also gain or lose electrons to stabilize the outermost electron shell. The loss of an electron results in a positively charged atom, whereas the gain of an electron results in a negatively charged atom. Charged atoms or molecules are called ions. Positively charged ions are called cations and negatively charged ions are called anions. Atoms that have an almost empty outer shell tend to lose those electrons and become cations. Conversely, atoms with an almost full outer shell tend to gain electrons and become anions. Therefore, atoms with a nearly empty outer electron shell tend to donate electron(s) to atoms with a nearly full outer electron shell. For example, the formation of sodium chloride (NaCl) is due to sodium donating an electron to chlorine (FIGURE 1.6). This donor/acceptor relationship results in sodium becoming positively charged (Na$^+$) and chlorine becoming a negatively charged anion called chloride (Cl$^-$). These two oppositely charged ions now exhibit a mutual attraction and this electrical attraction is an ionic bond.

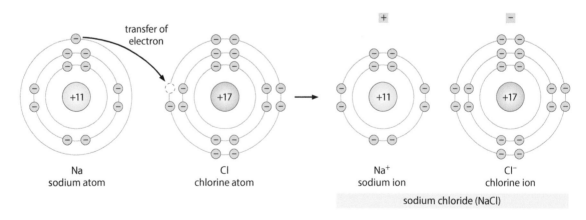

FIGURE 1.6 Ionic bond formation in sodium chloride. The sodium atom (Na) donates its lone electron in the outermost shell to chlorine (Cl) which accepts the electron to fill its outermost shell. This results in sodium having 10 electrons and a net positive charge of +1 and chlorine having 18 electrons and a net negative charge of −1. The resulting sodium cation (Na$^+$) and chloride anion (Cl$^-$) are attracted to each other due to their opposite charges to form an ionic bond.

Electrostatic interactions between oppositely charged ions can result in the formation of ionic crystals which are generally called salts. The atoms of ionic crystals are held together in an orderly array of alternating anions and cations. For example, sodium chloride, common dietary salt, forms a cubic array (FIGURE 1.7). In the crystal form these ionic interactions are quite strong and difficult to break. However, in an aqueous environment ionic bonds are easily broken. The polar water molecules interact with the salt ions to break the ionic bonds and dissolve the salt in water.

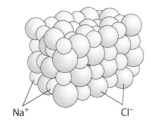

FIGURE 1.7 Crystal structure of sodium chloride. Anions and cations can interact to form ionic crystals in which the anions and cations alternate to form a lattice.

Many important biological processes involve the gain and loss of electrons

Oxidation is the loss of electrons and reduction is the gain of electrons. These oxidation and reduction reactions occur together in pairs with one substance being oxidized by donating electrons to another substance. The substance receiving the electrons is reduced. In the example of sodium chloride, sodium is oxidized and chlorine is reduced to chloride. These combined reactions are often called **redox reactions** as a combination of reduction and oxidation. Many important biological processes involve redox reactions. For example, the aerobic metabolism of glucose to produce energy is a series of redox reactions in which glucose is ultimately oxidized to carbon dioxide (CO_2) and oxygen is reduced to water (Chapter 3).

Substances with the ability to cause other substances to lose electrons are called oxidizing agents. Oxygen is the quintessential oxidizing agent, hence the term oxidation is used to describe the loss of electrons in general. In contrast, reducing agents are substances that cause other substances to gain electrons.

Inappropriate oxidation causes molecular damage

Although redox reactions are a normal part of cellular metabolism, oxidation potentially presents problems for organisms. For example, a possible by-product of aerobic metabolism is the generation of **reactive oxygen species** (ROS). The aberrant generation of ROS represents a situation in which the reduction of oxygen was not immediately coupled with an oxidation reaction. Due to their excess electrons, ROS are highly unstable and readily react with a lot of other molecules. This high reactivity of ROS causes cellular damage by inappropriately oxidizing important cellular molecules (FIGURE 1.8). Just as rust is viewed as an undesirable oxidation of iron, inappropriate oxidation of biological molecules causes cellular damage and disease.

Oxidative damage is involved in many diseases. Thus, cells have mechanisms to remove and detoxify ROS and minimize the undesirable consequences of redox reactions. In addition, some chemicals called antioxidants function as strong reducing agents and can remove ROS by reducing them to less reactive compounds. For this reason, antioxidants are generally considered to be healthy supplements to our diet (BOX 1.2). It should also be noted, though, that ROS are also exploited by the immune system to kill pathogens (Chapter 9).

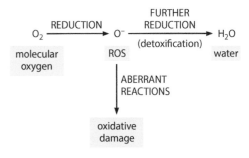

FIGURE 1.8 Reactive oxygen species. Reactive oxygen species (ROS) are formed by the reduction of molecular oxygen. ROS can cause oxidative damage by oxidizing other molecules and destroying their normal function. ROS can be further reduced to non-toxic forms such as water to prevent aberrant oxidation.

BOX 1.2

Antioxidants and Health

Reactive oxygen species (ROS) are produced as a by-product of metabolism and inflammation. Environmental sources of ROS include pollution, certain foods, smoking, and radiation. ROS cause oxidative stress and this may result in cellular damage that can increase the risk of developing certain diseases, such as cancer (Chapter 16) or atherosclerosis (Chapter 14), as well as being part of the aging process (Chapter 10). The body has several mechanisms to deal with oxidative stress and remove ROS. It may also be possible to supplement the diet with antioxidants as a means of reducing disease risks and further slowing cellular damage. In particular, vitamins A, C, and E have strong antioxidant properties (Chapter 12) and fruits and vegetables tend to be good sources of antioxidants. However, there is a lack of evidence clearly demonstrating that a higher intake of specific antioxidants reduces the risk of disease. In fact, high-dose supplements of antioxidants, such as vitamin E, may be linked to health risks in some cases.

LIFE-AFFIRMING PROPERTIES OF WATER

Living organisms contain 60–90% water and the evolution of life was facilitated by water. Several properties of water contribute to its essentialness to life. In particular, water serves as a buffer against temperature fluctuations, functions as a solvent, and promotes hydrophobic interactions. All these attributes are due in part to the polarity of the water molecule and the hydrogen bonding that occurs between water molecules. Like most chemical matter, water can exist in gaseous, liquid, or solid states. Life depends on water being in the liquid state and hydrogen bonding keeps much of the water on Earth in the liquid state.

Water can serve as a buffer against changes in temperature

Compared with other fluids, water requires a greater input of energy in the form of heat to raise its temperature. This is because the hydrogen bonds between water molecules absorb much of the applied energy. The absorption of heat is important since living organisms can only survive in a limited temperature range. Temperatures that are too high damage biological molecules. Temperatures below freezing result in the formation of ice crystals that damage the membranes of cells.

Water also requires a relatively high temperature to convert the liquid form of water into a gaseous state. Enough energy in the form of heat must be absorbed to break all the hydrogen bonds between water molecules for water molecules to evaporate as water vapor. This absorption of energy during evaporation also has a cooling effect that is exploited by human physiology to moderate body temperature through perspiration. For example, a person can absorb a lot of heat energy from the sun through an activity like sunbathing without the body temperature increasing to a deadly level. Heating causes perspiration and the subsequent evaporation of the sweat has a cooling effect on the body without a tremendous loss of water. Water also freezes more slowly than other liquids and thus moderates the effects of lower temperatures.

Water is also unique in that the solid form of water is less dense than liquid water. The solid forms of most molecules are denser than the liquid forms. This lower density of solid water results in ice formation on the surface of a body of water, such as a lake, when it freezes. The ice forming on the surface results in an insulating layer that delays freezing of the remaining liquid water. Therefore, fish and other aquatic organisms can survive below the ice layer. If the solid form of water had been denser than the liquid form, then ice would sink to the bottom of the lake and, if the temperature remained below the freezing point for sufficient time – as is the case for most of North America in the winter – the lakes would freeze into a solid mass and not permit the survival of aquatic organisms. Furthermore, liquid water would be less available to terrestrial organisms.

Water is a solvent for many substances

The **polar** nature of water allows it to interact with other polar substances and ions by forming spheres of hydration. Therefore, polar substances are readily dissolved in water. For this reason ions and polar molecules are also referred to as being hydrophilic (Greek for 'water loving'). Biological molecules, such as sugars, amino acids, and nucleotides, are also hydrophilic and therefore readily dissolve in water. Water also dissolves gases such as oxygen and carbon dioxide. The large variety of concentrated solutes in water provides an ideal environment for chemical reactions that are necessary to sustain life. Furthermore, water also participates in many of these chemical reactions.

Salts readily dissolve in water due to a replacement of ionic bonds by interactions with water. For example, sodium chloride consists of positively charged sodium atoms and negatively charged chloride atoms. The partially positive pole of the hydrogen atom of the water molecule interacts with the negatively charged chloride and, conversely, the partially negative pole of the oxygen atom of water interacts with the positively charged sodium ion. Therefore, when salt is mixed with water, the ionic bonds holding the sodium and chloride ions together are broken and the individual sodium and chloride ions are surrounded by water molecules (FIGURE 1.9). These spheres of hydration around the ions

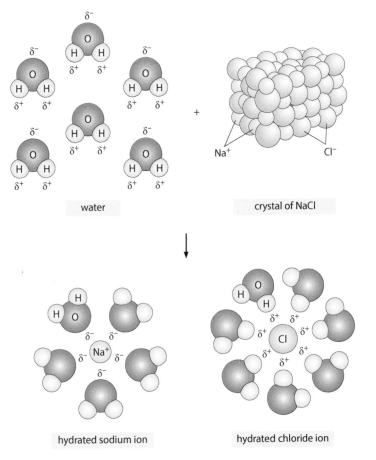

FIGURE 1.9 Dissolution of ions in water. Water disrupts the ionic interactions of atoms in a salt crystal, which results in the ions becoming dissolved in water. The partial negative charge (δ^-) of oxygen (O) interacts with sodium ions (Na^+) and the partial positive charge (δ^+) of hydrogen (H) interacts with chloride ions (Cl^-). These interactions result in a sphere of hydration around the ions.

result in a dispersal of the individual ions throughout the water solution. In other words, sodium chloride (NaCl) no longer exists as a crystal, but rather as a solute that is uniformly dispersed throughout the water solution. A solute is a general term to describe a substance in solution.

As in the case of ions, the polarity of water can disrupt polar interactions between other molecules and form spheres of hydration. Furthermore, hydrogen bonds can form between the polar solute molecules and the water molecules. For example, most sugars are not charged. However, sugars are polar molecules due to polar covalent bonds between oxygen and hydrogen that form hydroxyl (OH) groups. Water molecules interact with the hydroxyl groups in sugar molecules via hydrogen bonds to form spheres of hydration as the sugar dissolves in the water (FIGURE 1.10).

FIGURE 1.10 Hydrogen bonding between water and polar solute. Hydration of uncharged polar molecules, such as the sugar glucose, involves hydrogen bonds (dashed lines) between water molecules and the solute. The hydroxyl (OH) groups of the glucose are polar bonds that can participate in hydrogen bonds with other polar bonds. As with ions, a sphere of hydration forms around polar molecules when dissolved in water. These interactions result in a sphere of hydration around the ions.

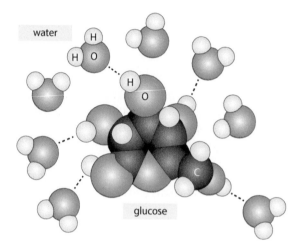

Non-polar substances interact via hydrophobic bonds

Non-polar substances, such as fats and oils, do not readily dissolve in water and are also called hydrophobic (Greek for 'water fearing'). If water and oil, such as an oil and vinegar salad dressing, are mixed and vigorously shaken, the oil forms globules dispersed throughout the water. However, the oil is not dissolved in the water and, if this mixture is left alone, the water and oil will separate into distinct layers. Thus, one needs to periodically re-shake oil and vinegar salad dressings to maintain a dispersed mixture. The limited ability of non-polar substances to dissolve in water is due to limited interactions between water molecules and oil molecules.

Since hydrophobic molecules do not interact well with water, they have a strong tendency to interact with each other when placed in an aqueous environment. In some sense, the hydrogen bonding that occurs between water molecules squeezes the non-polar molecules out of solution. This tendency for non-polar substances to interact with each other in an aqueous environment is referred to as hydrophobic bonding and represents another type of weak interaction between molecules (see Table 1.2). Hydrophobic interactions are important in living organisms because these hydrophobic interactions are a major factor in the formation of biological membranes that define cells (Chapter 2) and hydrophobic bonds are important in the structure of proteins (Chapter 3).

IONIZATION AND pH

The water molecule can also become ionized. This occurs when one of the bonds between a hydrogen atom and the oxygen atom is broken and the shared electron remains associated with the oxygen atom to form a negatively charged hydroxide ion (OH⁻). The free hydrogen now lacks an electron and forms a positively charged hydrogen ion (H⁺). The hydrogen ion is equivalent to a proton since the standard hydrogen nucleus consists solely of a single proton. This gain and loss of hydrogen ions happen spontaneously and result in an equilibrium reaction consisting of water (H_2O), the hydroxide ion (OH⁻), and the hydrogen ion (H⁺) as depicted in the following formula:

$$H_2O \leftrightarrow OH^- + H^+.$$

Solutions with a high concentration of H⁺ ions are acidic and solutions with a low concentration of H⁺ ions are alkali or basic. The concentration of H⁺ ions in a solution is expressed as pH, which is defined as the negative log of the hydrogen ion concentration expressed in moles per liter. For example, a solution that has a concentration of one mole per liter of hydrogen ions has a pH of 1.0. This represents an extremely acidic condition and the beginning of the pH scale, which has a range of 1–14. As the hydrogen ion concentration decreases, the pH value increases. A pH of 14 is an extremely basic condition and corresponds to a hydrogen ion concentration of 10^{-14} moles per liter. Pure water has a pH of 7.0 (hydrogen ion concentration of 10^{-7} moles per liter) and pH 7.0 is defined as neutral. This neutral pH represents the natural level of dissociation of pure water into hydrogen and hydroxide ions. However, pH is strongly influenced and determined by the nature of the substances dissolved in water.

Acids and bases are different states of ionizable molecules

Many substances are defined as being **acids** or **bases** depending on how they influence the pH when dissolved in water. Acids are substances that release, or donate, H⁺ ions. For example, when hydrochloric acid (HCl) is mixed with water it completely dissociates into H⁺ and Cl⁻ resulting in a high concentration of H⁺ and therefore a low pH. A base is the opposite of an acid and is a substance that accepts H⁺, and thus lowers the H⁺ ion concentration. Sodium hydroxide (NaOH) is a strong base and completely dissociates into sodium ion (Na⁺) and hydroxide (OH⁻) when dissolved in water. The OH⁻ ion combines with, or accepts, the H⁺ ion to form water (H_2O). This results in a lower H⁺ ion concentration and therefore a higher pH.

In some respects, acids and bases are opposites. However, they interact with each other to determine the equilibrium of the dissociation of water into H⁺ and OH⁻ ions. For example, if HCl and NaOH are mixed in water, the H⁺ released from HCl combines with the OH⁻ released from NaOH to form water. A NaCl solution with neutral pH of 7 results if equal amounts of the two are mixed. In other words, acids and bases can neutralize each other and mixing acids and bases results in the formation of salts. In terms of practical applications, an overly acidic stomach can be neutralized with basic substances called antacids (BOX 1.3).

The Acidity of the Stomach

In contrast with most other organs, the stomach is highly acidic. This acidity facilitates the digestion of food by breaking down large molecules via hydrolysis into smaller molecules that can be absorbed. Gastric acid is produced by specialized cells in the lining of the stomach that secrete hydrochloric acid. Other cells in the stomach produce bicarbonate, a base, that neutralizes the acid and prevents the stomach from becoming too acidic. Over-the-counter antacids are also bases that function by neutralizing acid to relieve indigestion, also known as heartburn. These antacids relieve pain and reduce damage, but only work transiently. Drugs that inhibit the secretion of hydrogen ions from the specialized cells of the stomach are a longer-lasting treatment for indigestion.

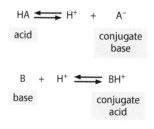

FIGURE 1.11 Acids and bases. An acid (A) donates a proton (H⁺) by dissociating into the conjugate base (A⁻) and H⁺. A base (B) accepts a proton to become a conjugate acid (BH⁺). Weak acids and bases form an equilibrium with their respective conjugate bases and conjugate acids.

Although acids and bases are opposites, they are also different manifestations of the same compound. For example, when an acid dissociates into an anion plus the H^+ ion, the anion becomes a base since it can accept a proton (FIGURE 1.11). This anion form of the acid is referred to as the conjugate base. Similarly, when a base accepts a H^+ ion it becomes a conjugate acid.

Buffers maintain solutions at relatively constant pH

The pH of the interior of cells and the extracellular fluids that bathe most cells and tissues is generally between 7.3 and 7.4. Deviations from this nearly neutral pH can affect the structure and function of many biological molecules and result in cellular death. However, many of the metabolites and chemical reactions occurring within cells either generate hydrogen ions, and thus raise the pH, or remove hydrogen ions, and thus lower the pH. Maintaining a constant pH is mediated by substances called buffers.

Buffers are ionizable compounds that do not completely dissociate into ions and exhibit an equilibrium as exemplified by water and its dissociation into H^+ and OH^- ions. Such compounds are called weak acids or weak bases. HCl and NaOH are a strong acid and base, respectively, since they completely dissociate into ions when mixed with water. The amount of dissociation of a specific weak acid or base is an inherent property of that compound. This means that dissolving a weak acid or base in water produces a specific pH value that is determined by the equilibrium between the acid and conjugate base forms of the compound. Furthermore, a relatively constant pH is maintained even with the addition or removal of H^+ ions. For example, blood pH is maintained at a constant value through an equilibrium established between bicarbonate and carbonic acid (BOX 1.4).

Maintaining Blood pH

Blood and body fluids typically exhibit a pH of 7.3–7.4 and this pH is maintained with a buffer based on dissolved carbon dioxide. Carbon dioxide (CO_2) dissolved in water combines with a water molecule to become carbonic acid (H_2CO_3). Carbonic acid is a weak acid that dissociates into bicarbonate (HCO_3^-) and H^+ ions and exhibits the following equilibrium:

$$H_2CO_3 \leftrightarrow HCO_3^- + H^+$$

If a metabolic process produces H^+ ions, these excess H^+ ions combine with the bicarbonate to form carbonic acid. Conversely, if H^+ ions are removed, the carbonic acid dissociates into bicarbonate and H^+ ions. In this way the H^+ remains approximately constant and the pH does not drastically change with the addition or removal of hydrogen ions. Therefore, small changes in the H^+ ion concentrations only have a minor effect on the pH of blood. This buffering capacity can be overwhelmed in some disease states. The process of blood or tissues becoming too acidic is called acidosis.

THE CARBON ATOM

Carbon is a major component of biologically important molecules. Carbon has four electrons in its outer shell which can hold up to eight electrons (see Figure 1.2). This means that a carbon atom can form up to four single covalent bonds with other atoms (FIGURE 1.12), as well as forming double and triple bonds. In biological molecules, carbon atoms are often linked together in linear chains or rings with hydrogen atoms making up the remaining bonds. These hydrocarbon chains and rings form relatively stable structures since the carbon and hydrogen atoms share electrons equally.

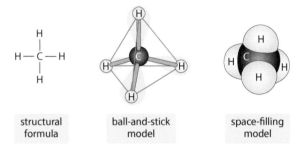

| structural formula | ball-and-stick model | space-filling model |

FIGURE 1.12 Carbon bonds. Carbon atoms can form four bonds with other molecules as illustrated by methane (CH_4). The angles of the bonds form a tetrahedron shape as depicted by the ball-and-stick model.

Other atoms such as oxygen and nitrogen can also form covalent bonds with the carbon atom. Such atoms are functional groups that impart specific chemical properties to the hydrocarbon (TABLE 1.4). For example, oxygen and nitrogen form covalent polar bonds and impart polarity on molecules and can participate in hydrogen bonding. Other functional groups can become ionized and impart either a negative or a positive charge on the molecule. These various carbon-based molecules can be classified into distinct classes of molecules based on their chemical properties.

Table 1.4	Common Functional Groups in Biological Molecules	
Group	**Structure**	**Properties**
Hydrogen	—H	Polar or non-polar depending on a partner atom
Hydroxyl	—O—H	Polar
Carboxyl	—C(=O)—OH	Acidic (negative charge)
Amino	—N(H)(H)	Basic (positive charge)
Phosphate	—O—P(=O)(OH)(OH)	Acidic (negative charge)
Methyl (hydrocarbon)	—C(H)(H)—H	Non-polar; tends to make molecules hydrophobic

The functional groups of these carbon-based molecules may also participate in chemical reactions. Two particularly common chemical reactions, which are involved in many reactions involving biological molecules, are condensation and hydrolysis. **Condensation** reactions involve the formation of water when two molecules are joined together. Typically, one of the molecules donates an hydroxyl group and the other molecule donates hydrogen to form water as the molecules are joined together (FIGURE 1.13). **Hydrolysis** is the reverse reaction in which water breaks the bond holding together the molecules by donating hydrogen to one molecule and an hydroxyl to the other molecule. So, in addition to being a solvent for biologically important molecules and participating in the ionization of molecules, water molecules can also directly participate in chemical reactions.

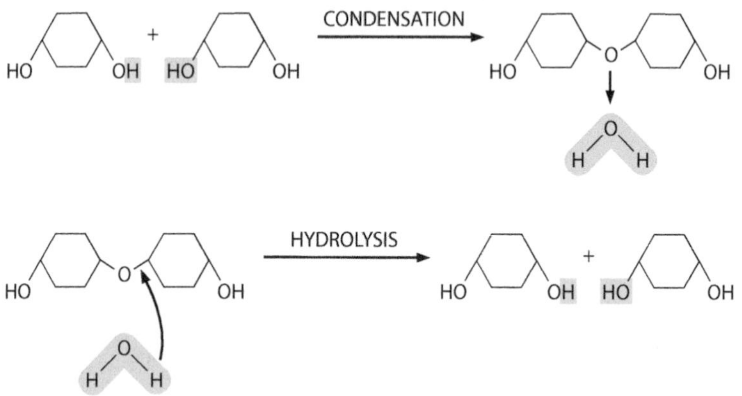

FIGURE 1.13 Condensation and hydrolysis reactions. During condensation hydrogen from one molecule joins with an hydroxyl in the other molecule to form water (blue shading) as the two molecules are joined together. In the reverse reaction, or hydrolysis, a water molecule breaks the bond between the molecules by adding an hydroxyl to the carbon and a hydrogen to the oxygen (blue shading).

CLASSES OF BIOLOGICAL MOLECULES

Biologically important molecules can be divided into four major classes according to their chemistry: carbohydrates, lipids, proteins, and nucleic acids (RNA and DNA). All these molecules are hydrocarbons but have different chemical properties due to different combinations of functional groups. In addition to their different chemistries, these major classes of molecules also have distinct major functions within an organism (TABLE 1.5). These biologically important molecules also tend to be large and complex and are formed from smaller molecules. The small molecules that make up larger molecules can be viewed as building blocks. In the case of proteins and nucleic acids the molecules can be very large and are often referred to as macromolecules. These macromolecules are long polymers of building blocks that can be difficult to study by conventional chemical methods. Gel electrophoresis is a method widely used in the study of proteins and nucleic acids (BOX 1.5).

Table 1.5	Chemical Basis of Life	
Molecule	**Building block**	**Major functions**
Carbohydrates	Sugars	Energy production and storage
Lipids	Fatty acids	Membrane component and energy storage
Protein	Amino acids	Cell structure and function
RNA	Nucleotides	Protein synthesis
DNA	Nucleotides	Genetic material

BOX 1.5

Gel Electrophoresis

Gel electrophoresis is a powerful technique for the analysis of proteins and nucleic acids. It allows individual macromolecules from complex mixtures to be separated from one another for their identification and analysis. The gel is a three-dimensional matrix of cross-linked polymers in an aqueous medium. Macromolecules can be forced through this porous matrix by applying an electric field under conditions in which the macromolecules have a net charge. In most situations, macromolecules have a net negative charge, and thus migrate toward the positive pole in the electric field. Smaller molecules more easily pass through the spaces between the polymers and therefore migrate faster than larger molecules (**BOX 1.5 FIGURE 1**). This molecular sieving separates the macromolecules according to size. This technology allows for the identification and characterization of individual DNA, RNA, or protein molecules.

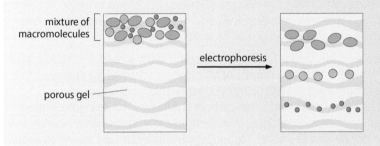

BOX 1.5 FIGURE 1 Gel electrophoresis. Subjecting a mixture of macromolecules to gel electrophoresis separates molecules primarily by their size. Smaller molecules migrate faster through the gel than larger molecules. Following electrophoresis, the molecules will be located at distinct positions in the gel, which allows for their identification. Calibration of the gel also allows an approximate size of the molecule to be determined.

Carbohydrates include various sugars and complex saccharides

Carbohydrates are composed primarily of carbon, hydrogen, and oxygen and most of the carbon atoms of a carbohydrate have both a hydrogen atom and an hydroxyl group attached. This reflects the name carbohydrate, which means carbon plus water. Carbohydrates include simple molecules commonly called sugars, as well as more complex molecules composed of simple sugars. The primary function of carbohydrates is energy production and glucose plays a central role in the generation of energy (Chapter 3). Most simple sugars are composed of five and six carbons that form a ring and are referred to as monosaccharides.

Monosaccharides can also react with each other to form larger molecules. For example, the common table sugar, called sucrose, is a disaccharide composed of glucose and fructose (FIGURE 1.14). Another common disaccharide is lactose, or milk sugar, formed from glucose and galactose. Numerous monosaccharides can be joined together to form polysaccharides. The most common polysaccharides are glycogen, starch, and cellulose. Glycogen is a storage form of glucose in animal cells

FIGURE 1.14 Formation of disaccharides from monosaccharides. The monosaccharides glucose and fructose react via a condensation reaction, as described in Figure 1.13, to form the disaccharide sucrose and water (H_2O).

and starch is a storage form of glucose in plant cells, whereas cellulose forms the cell walls of plant cells.

Lipids are hydrophobic molecules that make up the fats and oils

There are many types of **lipids** with a wide range of functions. The fundamental property of lipids is that they are composed primarily of hydrogen and carbon with non-polar carbon–carbon and hydrogen–carbon bonds. Thus, these molecules are hydrophobic and do not readily dissolve in water. Lipids function as a key component in biological membranes, in energy storage, as coatings, and as hormones. Regarding health, excess levels of lipids in the blood are a major risk factor for developing cardiovascular disease (Chapter 14). In terms of their chemical structure, lipids are grouped into two major types: those composed of fatty acids and those composed of steroid rings.

Fatty acids consist of a hydrocarbon chain with a carboxyl group at one end (FIGURE 1.15). The length of the hydrocarbon chain ranges from 4 to 28 carbon atoms. Hydrocarbon chains that contain no double bonds are called saturated since the carbon chain has the maximum possible number of hydrogen atoms. Unsaturated fatty acids have one or more double bonds and therefore can accept more hydrogen atoms to replace the double bonds, and hence the term unsaturated.

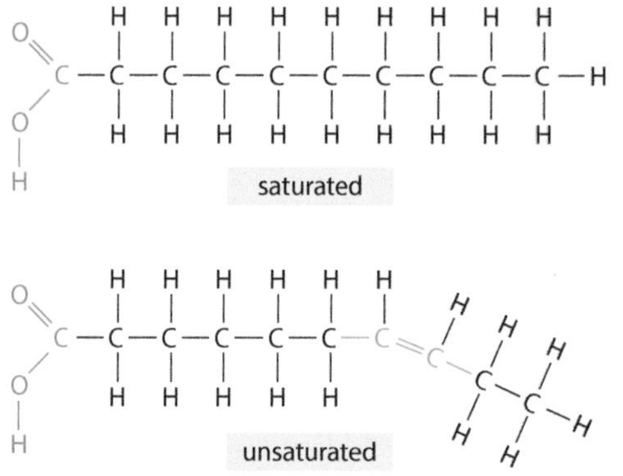

FIGURE 1.15 Fatty acid structure. Fatty acids consist of a carboxyl group (red) and a chain of hydrocarbons (black). Hydrocarbon chains without double bonds are called saturated, whereas hydrocarbon tails with double bonds are called unsaturated. Unsaturated fatty acids have bends that result in poorer packing between the fatty acids and therefore tend to be liquid at room temperature.

Fats formed from saturated fatty acids tend to be solid at room temperature and unsaturated fats tend to be oils, or liquid, at room temperature. This is due to the double bonds creating kinks in the long-chain hydrocarbons and disrupting the packing between the fatty acid molecules. In contrast, saturated fatty acids stack together better and form a solid at room temperature. Artificial *trans*-fats are produced in a process that adds hydrogen to the double bonds in liquid vegetable oils to make them solid. These *trans*-fats are widely used in processed and prepared foods and are associated with developing heart disease (BOX 1.6).

BOX 1.6

Trans Fats and Heart Disease

Trans fats are made from vegetable oils through a process called hydrogenation in which hydrogens are added to the double bonds in unsaturated fatty acids. These trans fats, also called partially hydrogenated oils, have been widely used in processed foods. Foods prepared with trans fats generally have a longer shelf-life and are often tastier and have a better texture than foods prepared with natural oils or fats. Beginning in the 1990s, studies indicated that diets high in trans fats greatly increased the risk of developing coronary heart disease (Chapter 14). Recently there has been a strong push to eliminate or minimize the use of artificially prepared trans fats in the food-processing industry and some countries, states, or cities have passed laws to restrict the use of trans fats.

Glycerol forms a backbone for many lipids

Fatty acids form lipids by reacting with a three-carbon linear carbohydrate called glycerol. The carboxyl groups of the fatty acids react with the hydroxyls of the glycerol in a condensation reaction (FIGURE 1.16). If fatty acids react with all three carbons of the glycerol, the resulting lipid is called a triglyceride. Triglyceride is a major storage form of energy that is found in fat tissue (Chapter 12).

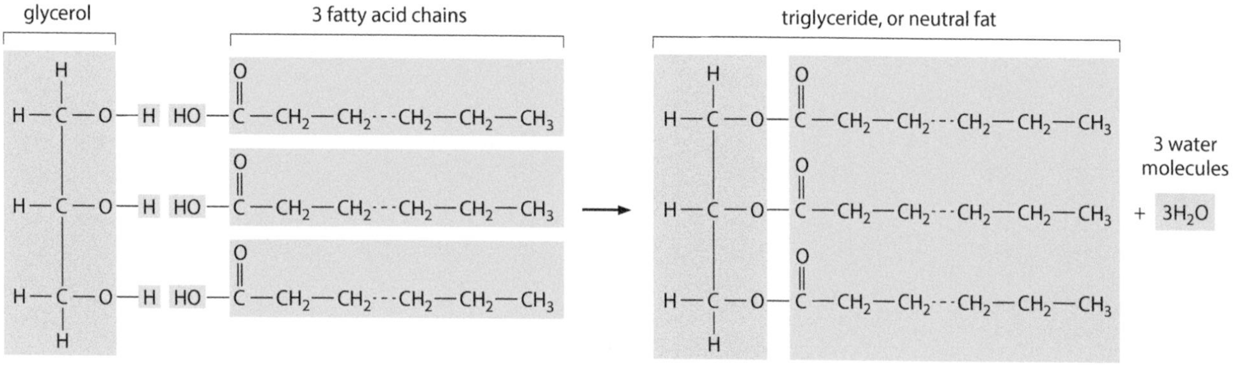

FIGURE 1.16 Triglyceride formation. The hydroxyls of the carboxyl groups of the fatty acids react with the hydrogen groups of the glycerol backbone (blue shading) in a condensation reaction to form lipids and water (blue shading). Lipids that have fatty acids attached to all three carbon atoms of the glycerol are called triglycerides.

Some lipids only have two fatty acid molecules bound to the glycerol backbone and the third position is occupied by a polar group. In particular, the lipids making up membranes have a polar group and two non-polar fatty acid tails (FIGURE 1.17). The polar functional groups in the third position include phosphate (phospholipids), carbohydrates (glycolipids), amino acid derivatives (sphingolipids), or combinations of the three. This polar portion of the molecule is called the polar head group and the non-polar portion is called the fatty acid tail. Thus, such lipids have both polar and non-polar properties. Molecules with both polar and non-polar moieties are called amphipathic, and this amphipathic property of lipids is important in the formation of biological membranes (Chapter 2).

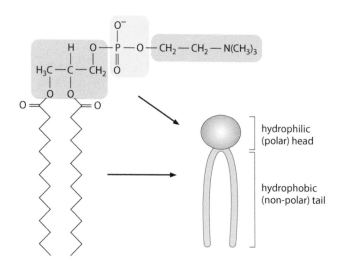

FIGURE 1.17 Lipid structure. Many lipids contain two fatty acid chains on the glycerol backbone (pink shading) and a polar group in the third position on the glycerol backbone. Possible polar groups include phosphate (green shading), amino (blue shading), or carbohydrate (not shown) groups. Such lipids have hydrophilic (polar) heads and hydrophobic (non-polar) tails.

Many steroids function as hormones

Steroids are a class of lipids that consist of a backbone of four carbon rings. The various steroids differ in the various functional groups that are attached to the carbon rings making up the steroid backbone (FIGURE 1.18) and have distinct roles in cellular function and physiology. For example, the steroid cholesterol is a component of cell membranes in animal cells. Cholesterol is also a component of bile, which facilitates the absorption of fat by the intestines. Many steroids function as hormones controlling a wide array of metabolic and physiological functions. For example, estrogen and testosterone are steroids that control sexual development. Other steroid hormones regulate various physiological functions like metabolism (cortisol) and blood pressure (aldosterone).

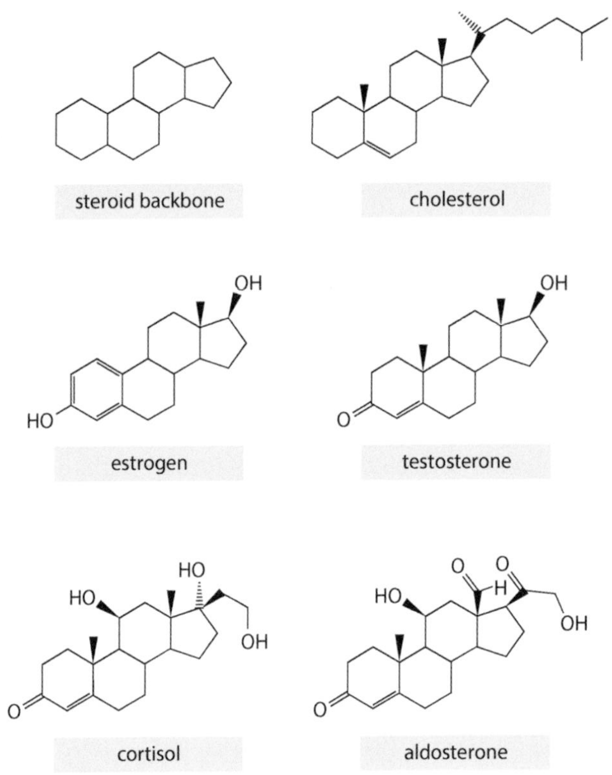

FIGURE 1.18 Steroid structure. Steroids consist of a four-ring backbone with various functional groups attached. Cholesterol is a membrane lipid found in animal cells and a major component of bile. The other compounds are hormones that play various roles in human physiology.

steroid backbone

cholesterol

estrogen

testosterone

cortisol

aldosterone

Proteins and amino acids

Proteins are responsible – at least in part – for all structural elements of cells, as well as all cellular functions (Chapter 3). They are complex polymers made of monomers called amino acids. As the name implies, the molecular structure of an amino acid includes a basic amino group and an acidic carboxyl group. These two functional groups are bound to a carbon atom called the α-carbon (FIGURE 1.19). The other two positions on the α-carbon are occupied by hydrogen and something called the R group. A wide variety of functional groups can serve as the R group. The various R groups found within amino acids have unique chemical properties and thus each individual amino acid has chemical properties reflective of its R group.

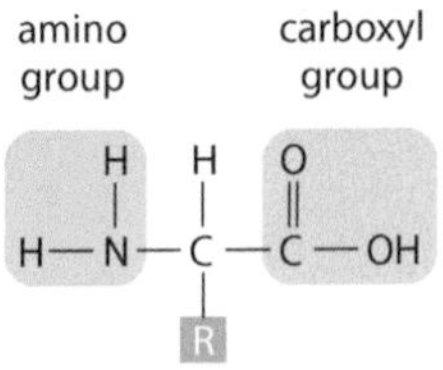

FIGURE 1.19 Amino acid structure. Amino acids consist of a central carbon atom called the α-carbon. Three of the carbon bonds are occupied by a hydrogen atom (H), an amino group (blue shading), and a carboxyl group (pink shading). The fourth bond can be occupied by a variety of functional groups and is called the R group (green shading).

amino group   carboxyl group

There are 20 distinct amino acids defined by their R groups that are naturally found in proteins. The R groups range from the simplest being a hydrogen atom, to various hydrocarbons that may also contain other functional groups such as hydroxyls, amines, carboxylic acids, and sulfur atoms. A convenient way to classify the amino acids is by whether they are polar or non-polar (TABLE 1.6). Among the polar amino acids, some are uncharged, whereas others can be positively or negatively charged. Since each individual protein is composed of a unique combination of amino acids, each protein has unique chemical and physical properties as determined by the overall amino acid composition, as well as the order of the amino acids. For example, some proteins can be more hydrophobic than other proteins due to having a greater number of non-polar amino acid residues. Similarly, some proteins could have an overall negative charge or an overall positive charge depending on the amino acid composition of that specific protein.

Table 1.6 Amino Acids	
Non-polar	**Polar**
Alanine (Ala, A)	Arginine (Arg, R)
Glycine (Gly, G)	Asparagine (Asn, N)
Isoleucine (Ile, I)	Aspartic acid (Asp, D)
Leucine (Leu, L)	Cysteine (Cys, C)
Methionine (Met, M)	Glutamic acid (Glu, E)
Phenylalanine (Phe, F)	Glutamine (Gln, Q)
Proline (Pro, P)	Histidine (His, H)
Tryptophan (Trp, W)	Lysine (Lys, K)
Valine (Val, V)	Serine (Ser, S)
	Threonine (Thr, T)
	Tyrosine (Tyr, Y)

The 20 naturally occurring amino acids found in proteins including the standard three letter and one letter abbreviations and group according to whether non-polar or polar.

The amino acids are joined together in long polymers via a condensation reaction between the carboxyl group of one amino acid and the amino group of a second amino acid (FIGURE 1.20). The resulting bond between the consecutive amino acids is called a peptide bond. The linear chain of amino acids is called a protein or a polypeptide. Relatively short polypeptides are often simply called peptides. Polypeptides are linear strings of amino acids that fold into unique three-dimensional structures. The structure of the polypeptide is determined by complex interactions between the amino acids and other polypeptides (Chapter 3).

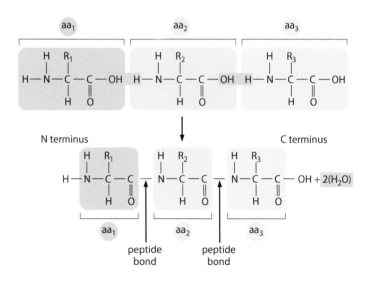

FIGURE 1.20 Peptide bond formation. The carboxyl and amino groups of amino acids undergo a condensation reaction in which a water molecule is generated (blue shading) as the peptide bond is formed. Polypeptides have a free amino group on one end, often called the N terminus, and a carboxyl group on the other end called the C terminus.

Nucleic acids are long polymers that function in protein synthesis and as genetic material

Nucleic acids are long polymers composed of nucleotides. Each nucleotide has three parts: a five-carbon sugar called ribose, a phosphate group, and a nitrogen-containing base (FIGURE 1.21). The base is analogous to the R group of proteins in that there are five distinct bases that make up five different nucleotides. The five bases found in nucleic acids are: adenine, guanine, cytosine, thymine, and uracil. The base plus the ribose sugar group without the phosphate group are nucleosides with the corresponding names of adenosine, guanosine, cytidine, thymidine, and uridine, which are often abbreviated using their first letters (A, G, C, T, or U). The nucleosides can have one, two, or three phosphates bound to the 5'-carbon to form the nucleotides and are designated by the nucleoside and the number of phosphates. For example, adenosine with one phosphate is called adenosine monophosphate, which is abbreviated as AMP. Similarly, adenosine with two phosphates is adenosine diphosphate, or ADP, and with three phosphates is adenosine triphosphate, or ATP.

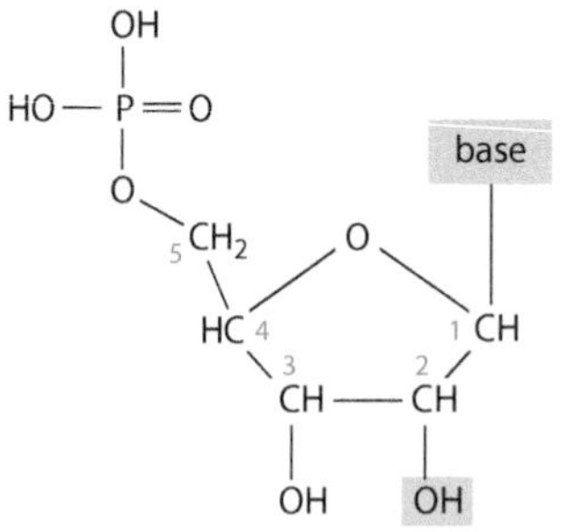

base	nucleoside
adenine	adenosine
guanine	guanosine
cytosine	cytidine
thymine	thymidine
uracil	uridine

FIGURE 1.21 Nucleotide structure. Nucleotides consist of a five-carbon sugar called ribose with a nitrogen-containing base bound to the number one carbon and a phosphate bound to the number five carbon. Nucleosides lack the phosphate group. Two classes of nucleotides are defined by the presence or absence of an hydroxyl group (blue shading) at the number two carbon. Nucleotides with the hydroxyl group are ribonucleotides and make up RNA and nucleotides without the hydroxyl group are deoxyribonucleotides and make up DNA. Five distinct bases define the five nucleotides that make up nucleic acids.

The nucleotides are joined together in linear chains via a condensation reaction of the phosphate group found in the number five position of the ribose sugar with the hydroxyl group found in the number three position (FIGURE 1.22). The order of the nucleotides defines a sequence starting with the nucleotide with the free 5'-phosphate and ending with the nucleotide with the free 3'-hydroxyl.

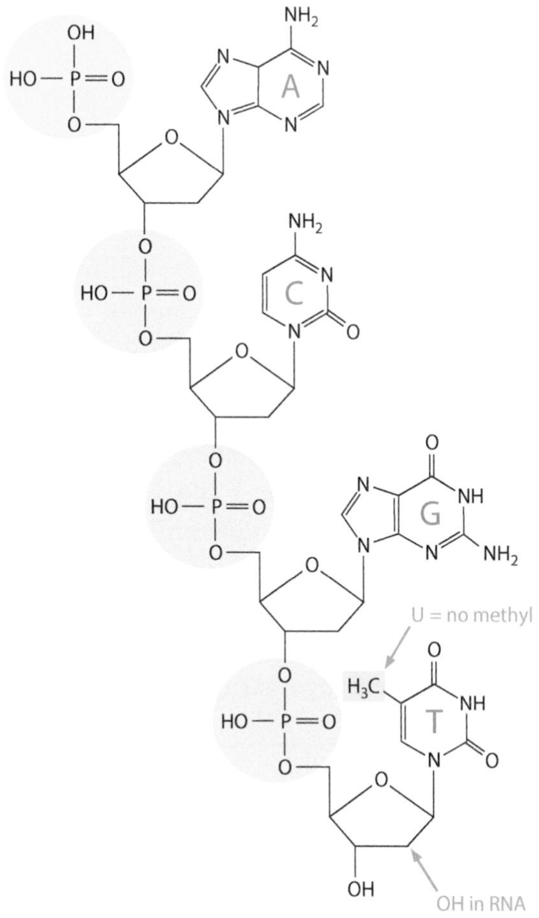

FIGURE 1.22 Nucleic acid structure. Nucleic acids are formed by a condensation reaction between the 5′-phosphate of one nucleotide and the 3′-hydroxyl of another nucleotide to form a phosphate bond (orange shading) between the nucleotides. Nucleic acids have a free phosphate at the 5′-end of the molecule and a free hydroxyl group at the 3′-end. DNA and RNA have similar structures and only differ in that RNA has hydroxyl groups in the 2′-carbon (green arrow) and RNA has uridine (U) residues instead of thymidine (T) residues. Uridine and thymidine only differ by a single methyl group (green shading and green arrow) which is lacking in uridine.

Two major types of nucleic acids with different functions are associated with cells (see Table 1.6). In a chemical sense these two molecules, DNA and RNA, are quite similar despite their distinct roles in protein synthesis and as genetic material. One difference between DNA and RNA is the exact nature of the nucleotides making up the long polymers. In the case of DNA, the nucleotides lack an hydroxyl group in the number two position of the ribose sugar and are called deoxyribonucleotides and the resulting nucleic acid is deoxyribonucleic acid, or DNA. Similarly, RNA stands for ribonucleic acid and is composed of ribonucleotides that have an hydroxyl in the number two position. Another difference concerns the specific bases that make up DNA and RNA. Adenine, guanine, and cytosine are found in both DNA and RNA, whereas thymine is only found in DNA and uracil is only found in RNA. Thus, nucleic acids only consist of 4 different nucleotides compared with the 20 amino acids that make up proteins. The other difference between DNA and RNA is that within a cell DNA is usually a double-stranded molecule, whereas RNA is usually a single-stranded molecule (Chapter 4).

Nucleotides have additional roles within cells besides serving as the building blocks for DNA and RNA. For example, adenosine triphosphate, or ATP, plays a central role in the energy metabolism of cells (Chapter 3). Other specialized nucleotides bind to proteins and function as co-factors by assisting those proteins in carrying out chemical reactions. Another nucleotide that plays a key role in cellular function is cyclic adenosine monophosphate (cAMP) which functions as a second messenger to relay signals from hormone receptors on the cell surface to the interior of the cell (Chapter 3).

SUMMARY OF KEY CONCEPTS

- Atoms define the key chemical elements that make up the universe
- The primary components of the atom are positively charged protons, negatively charged electrons, and neutrons with no charge
- Atoms can interact in various ways to form bonds leading to the formation of molecules
- Forces holding atoms together include covalent bonds, ionic bonds, and hydrogen bonds
- Water plays a key role in life due to its ability to function as a solvent of polar substances, its ability to stabilize temperature fluctuations, its participation in chemical reactions, and its role in hydrophobic interactions
- Hydrogen, oxygen, carbon, and nitrogen are the most abundant elements in organisms and form molecules that are important in the physiology of organisms
- Carbon forms a backbone for the formation of biologically important molecules due to its ability to form several chemical bonds with different functional groups
- The major classes of biologically important molecules are carbohydrates, lipids, proteins, and nucleic acids
- Macromolecules, such as proteins and nucleic acids, are large complex molecules composed of smaller molecules held together by covalent chemical bonds

KEY TERMS

acid	A molecule that is able to donate a hydrogen ion (proton) to another substance or lower the pH when dissolve in water.
base	A molecule that is able to accept a hydrogen ion (proton) from another substance or raise the pH when dissolve in water.
carbohydrate	Molecules composed primarily of carbon, hydrogen, and oxygen that are the primary fuel of organisms to produce energy.
chemical bond	An interaction between atoms to form molecules.
condensation	A chemical reaction involving the combination of two molecules which also produces a water molecule as a byproduct.
electron shell	A region surrounding the atomic nucleus containing a specific number of orbiting electrons and defined by a specific energy range.
hydrolysis (reaction)	A chemical reaction in which a water molecule breaks a covalent bond to split one molecule into two molecules.
isotopes	Variants of chemical elements (atoms) that have the same number of protons and electrons but different numbers of neutrons.
lipid	A non-polar molecule that plays roles in membrane formation, energy storage, and cellular communication.
monosaccharide	A simple sugar (carbohydrate) that cannot be hydrolyzed into smaller carbohydrates.
non-polar	Refers to hydrophobic substances with minimal charge that do not readily dissolve in water.
nucleic acid	A long polymer composed of nucleotides that functions as the hereditary material and directs the synthesis of proteins.
nucleotide	The molecular unit of nucleic acids (DNA and RNA) which consists of a ribose sugar, a purine or pyrimidine base, and a phosphate group.

polar | Refers to hydrophilic substances that readily interact with water and easily dissolve in water due to being charged or partially charged.

protein | A long polymer composed of amino acids that play a central role in the structure and function of cells.

reactive oxygen species | Oxygen atoms that have been reduced due to the gain of electron(s) that are highly reactive and can damage biological molecules.

redox reaction | A chemical reaction between an oxidizing substance and a reducing substance in which the oxidizing substance loses electrons and the reducing substance gains electrons.

REVIEW QUESTIONS

1. Briefly describe the four types of chemical bonds involved in the formation of molecules. Indicate which are strong or weak forces.

2. What are redox reactions? How are reactive oxygen species formed?

3. Discuss the properties of water that are life affirming. Why are hydrogen bonds important in the structure of liquid water?

4. Identify and briefly describe the four major classes of biological molecules based on chemical characteristics.

5. Briefly discuss the condensation reactions that are involved in the formation of the biological molecules (i.e., carbohydrates, lipids, proteins, and nucleic acids). What is the reverse of condensation reactions?

ADDITIONAL QUESTIONS

In addition to the Review Questions provided above, there is a range of free online questions designed to enable students to further test their understanding of the chapter material. In order to access these interactive questions, please visit the book's website: https://routledge.com/cw/wiser

Cell Structure and Function

2

Cells are described as the basic units of life. The reason for this somewhat grandiose definition is due to the cell being the smallest unit that has the two most fundamental properties of living organisms: the ability to maintain itself, or homeostasis, and the ability to replicate itself. Living organisms are capable of maintaining themselves in the presence of a changing environment, as are cells. Both cells and living organisms take up nutrients and convert those nutrients to energy and to biomolecules. In this way organisms and cells maintain themselves by continuously synthesizing the biomolecules that make up their structures and carry out their functions. Living organisms can reproduce themselves and generate offspring. Similarly, cells also replicate themselves, and thus produce more cells. This ability of cells to reproduce is the basis for the continuity of life.

In some cases, such as bacteria, protozoa, and some fungi, an organism consists solely of a single cell and is referred to as being unicellular. Plants, animals, and some fungi consist of numerous different cells of specialized functions and are thus multicellular. Most cells, with a few exceptions, are quite small and require a microscope to be seen. Whether a unicellular organism or part of a multicellular organism, all cells have the same basic structural features and are composed primarily of nucleic acids, proteins, lipids, and carbohydrates (Chapter 1). Viruses confuse the issue a bit in that they somewhat resemble cells in their biomolecular makeup and can direct their replication utilizing the machinery of the host cell (BOX 2.1).

BASIC CELLULAR FEATURES

MEMBRANE STRUCTURE AND FUNCTION

TRANSPORT OF SUBSTANCES ACROSS MEMBRANES

CELLULAR ORGANELLES

CYTOSKELETON

CELL REPLICATION AND MITOSIS

CELLULAR DAMAGE AND DISEASE

Are Viruses Organisms?

Whether viruses are alive, or not, has been the subject of a long-standing debate. Initially viruses were viewed as poisons. But later it was observed that they behave like bacteria – albeit extremely small – in their transmissibility and ability to increase their numbers. However, the ability to crystallize viruses and the demonstration that they were complex molecular particles consisting of nucleic acid surrounded by a protein coat again raised issues with whether they met the definition of life – namely, exhibiting homeostasis and autonomous replication. Viruses are clearly biological entities since they are composed of the same molecules that make up other organisms and contain genes that are passed down through subsequent generations. Some viruses are even surrounded by a lipid-bilayer membrane and, thus, even look like miniature and rather simple cells. This, combined with the ability to pass from person to person and cause disease

due to their proliferation, certainly makes viruses look like living organisms.

However, a virus cannot proliferate independently of its host and viruses outside of cells are rather inert. In other words, the purified virus is a super-molecular complex that is rather chemical in nature and lacks metabolic functions that are generally associated with life. But upon entry into an appropriate cell, viral genes become active, and the virus takes over the machinery of the cell and redirects cellular functions toward replicating the virus. Therefore, it could be argued that viruses do have metabolic activity and the ability to reproduce, even though it is borrowed in a sense. Perhaps a compromise to the debate can be to claim the virus exists in both a living and a non-living state. Outside of the host cell the virus is a non-living biochemical entity, whereas within the cell it is alive.

BOX 2.1

Since cells are the fundamental unit of life, disease is often described at the cellular level. The observation that pathogenesis has a cellular basis was recognized by early pathologists and disease is often associated with damage to cells or cellular death. Thus, having a basic knowledge of cell biology is important to understand disease.

BASIC CELLULAR FEATURES

Cells exist in a variety of shapes and sizes and exhibit a wide range of functions. Even though animals, plants, fungi, protozoa, and bacteria are all quite distinct, they are all composed of cells that exhibit a similar overall composition and structure. A major defining feature of the cell is the plasma membrane, which marks the outermost boundary of the cell. The plasma membrane separates the internal components of the cell from the outside environment and allows metabolic and other activities of the cell to occur in an organized fashion. Enclosed within the plasma membrane is the cytoplasm of the cell. The cytoplasm is a semifluid substance in which biomolecules and other subcellular structures are organized. It is composed of water, ions, and a wide variety of molecules.

There are two basic types of cells: prokaryote and eukaryote

The major distinguishing difference between prokaryotic cells and eukaryotic cells is the presence of a nucleus. The word eukaryote can be translated from Greek to mean 'true nucleus' whereas prokaryote means 'before the nucleus.' Prokaryote cells have a simpler structure and evolutionarily pre-date the eukaryotic cells. The nucleus is a membrane-bound structure within eukaryotic cells and has a central role in the biology of the cell since DNA is localized to the nucleus. In prokaryotes the DNA is not localized in a membrane-bound compartment, but rather, is localized to a special region of the cytoplasm sometimes called the nucleoid. Another major distinguishing feature of prokaryotes and eukaryotes is the presence of membrane-bound organelles and an internal cytoskeleton in eukaryotic cells (TABLE 2.1).

Table 2.1   Major Distinctions between Prokaryotes and Eukaryotes

Feature	Prokaryotes	Eukaryotes
DNA arrangement	Circular molecules located in nucleoid within the cytoplasm	Linear molecules packaged into chromosomes within the nucleus
Reproduction	Binary fission	Mitosis and meiosis
Organelles	Absent	Present
Plasma membrane	No carbohydrates or steroids	Contains steroids and glycoproteins
Cell wall	Composed of peptidoglycan	Lacking or composed of cellulose (plants) or chitin (fungi, protozoa)
Internal cytoskeleton	Lacking	Actin filaments and microtubules and intermediate filaments in animal cells
Size	Typically 0.2–2 µm	Typically 5–100 µm

Prokaryotic cells are essentially bacteria, whereas other organisms, such as animals, plants, fungi, and protozoa, fall into the eukaryotic group. Generally prokaryotic cells are smaller than eukaryotic cells. The larger size of the eukaryotic cells necessitates an internal cytoskeleton and the membrane-bound organelles found within the cytoplasm. For example, a 10-fold difference in a linear dimension translates into a

100-fold difference in surface area and a 1000-fold difference in volume. Prokaryotic cells lack an internal cytoskeleton, and they have an external cell wall to provide mechanical support. Many eukaryotic cells, such as fungi and plants, also have cell walls, but these cell walls are of a different chemical composition than prokaryotic cells. These fundamental differences between eukaryotic cells and prokaryotic cells are often exploited in the design of anti-bacterial drugs, as well as by our immune system (Chapter 9) to eliminate bacterial infections.

MEMBRANE STRUCTURE AND FUNCTION

The external membrane forming the boundary of the cell is called the **plasma membrane**. Membranes found within the cytoplasm define organelles. Cellular membranes form a barrier that is impermeable to most polar solutes dissolved in water. However, the exchange of substances across the membrane is essential and the membrane regulates the movement of ions and molecules across the membrane. In addition, the plasma membrane must interact with the environment of the cell. This could include binding to surfaces or to other cells, as well as communicating with other cells. Membranes are complex structures composed primarily of lipids and proteins. The lipids are primarily responsible for the barrier function of the membrane, whereas proteins regulate the exchange of substances across the membrane and the communication and interaction of the cell with its environment.

Membranes are composed of lipids organized into a bilayer

Many lipids consist of a polar head group and two hydrophobic tails (see Figure 1.17). This means that part of the lipid molecule (the polar head group) interacts well with water, whereas the hydrophobic tails do not. Thus, lipids in an aqueous environment create a bit of a dilemma. This dilemma is resolved by the lipids aggregating into spherical structures called micelles. Micelles are arranged so that the polar head groups are all facing outward, and are in contact with the water, whereas the hydrophobic tails are internal and are not in contact with the water (FIGURE 2.1). This is an example of hydrophobic bonds (see Table 1.3). Micelles are soluble in water and the formation of micelles is the basis of detergent action (BOX 2.2).

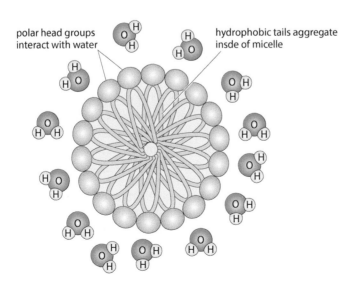

polar head groups interact with water

hydrophobic tails aggregate insde of micelle

FIGURE 2.1 **Lipids form micelles in aqueous environments.** The interaction of the polar head groups of lipid molecules with water (H_2O) forces the aggregation of the non-polar (hydrophobic) tails in the center of spherical structures called micelles. This hydration means that micelles are soluble in water.

Detergents

BOX 2.2

Detergents have a similar structure to lipids and are composed of polar and non-polar groups that are also capable of forming micelles. This ability to form micelles is the basis for how detergents 'clean.' Dirt is generally hydrophobic and thus is not soluble in water. However, the dirt from your clothes or dishes can be incorporated into the inner hydrophobic core of the detergent micelle. Since the micelles are soluble by virtue of the polar head groups, the mixed micelles of detergent and dirt remain soluble in water and are rinsed away. Detergents are also widely used in biochemical research. Detergents can extract lipids from membranes and help solubilize membrane-associated and other hydrophobic proteins.

Another possible configuration for lipids in an aqueous environment is to form a bilayer. In this arrangement the lipids assemble into a sheet in which the polar head groups are on the outside and the hydrophobic tails are in the middle (FIGURE 2.2). In contrast with the micelles, which are relatively small, this bilayer is a continuous sheet that forms cellular membranes. This sheet is not a planar structure with edges but wraps around it to form a continuous enclosed structure, and thus has an internal aqueous environment. These lipid-bilayer sheets form the plasma membrane and internal membranes that make up the organelles. Both the inside of the cell or organelle and the outside of the cell or organelle are aqueous environments. The hydrophobic fatty acid tails are not in contact with either aqueous environment due to the outwardly facing polar heads of the lipid. This configuration provides a barrier between the inside and outside aqueous environments.

FIGURE 2.2 Organization of lipids in a bilayer. The most fundamental unit of a biological membrane is a lipid bilayer in which the polar head groups of lipids are oriented on the outside and the hydrophobic tails are in the middle.

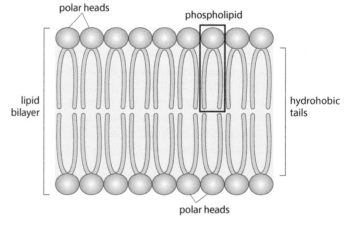

polar heads

phospholipid

lipid bilayer

hydrohobic tails

polar heads

Proteins are integrated into or associated with biological membranes

On average approximately half of the mass of a biological membrane is composed of protein. Some of these proteins pass through the lipid bilayer and are referred to as integral membrane proteins (FIGURE 2.3). These proteins have hydrophobic domains that interact with the hydrophobic tails of the membrane lipids and therefore are anchored into the bilayer. Quite often, the domains of the proteins that are exposed on the external surface of the membrane have attached carbohydrate groups and are called glycoproteins. Other membrane-associated proteins interact with the polar head groups or the exposed regions of the integral membrane proteins, and do not pass through the bilayer. These proteins are referred to as peripheral proteins. Some peripheral proteins form a meshwork of filaments, or a submembrane cytoskeleton, that supports and strengthens the membrane. Other peripheral proteins interact with and participate in the functions of the integral membrane proteins.

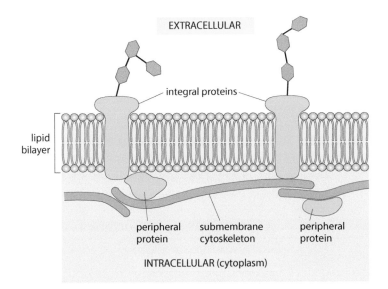

FIGURE 2.3 **Membrane structure.** A diagram showing the interactions between lipids and proteins to form the plasma membrane. Proteins that pass through the lipid bilayer are called integral membrane proteins. Other proteins are associated with the polar head groups of lipids or integral membrane proteins on the cytoplasmic side of the membrane. These proteins are called peripheral membrane proteins. Some peripheral membrane proteins form a two-dimensional filamentous network called the submembrane cytoskeleton.

Many different proteins with a variety of functions are associated with membranes. For example, some membrane proteins function in the transport of substances across membranes. Furthermore, these proteins can regulate which substances are moved across membranes. In other words, membranes exhibit a selective permeability. Other membrane proteins participate in the adherence of cells to surfaces or other cells. This is particularly important in multicellular organisms in which similar cells tightly attach to each other to form tissues (Chapter 8). Similarly, some membrane proteins function as receptors and bind to other molecules. In general, molecules that bind to proteins are called ligands. Hormones are an example of ligands used in cell–cell communication (Chapter 3). Binding to ligands may be a means by which cells sense their environment.

TRANSPORT OF SUBSTANCES ACROSS MEMBRANES

The movement of substances across membranes is critical for the function of cells. For example, cells need to import nutrients from the extracellular environment and export metabolic waste products, as well as maintain the correct osmotic balance of solutes. Generally, only water, dissolved gases, and some lipid molecules can readily cross membranes. Therefore, solutes can exist at different concentrations inside and outside of cells. Isotonic refers to conditions in which solute concentrations are equal on both sides of the membrane. A solution with a higher solute concentration than the cytoplasm of the cell is called hypertonic and, conversely, a lower solute concentration is hypotonic.

Cells need to maintain the correct solute balance between the cytoplasm and the external environment of the cell. Since water can pass through the bilayer, water molecules tend to move from a hypotonic solution to a hypertonic solution. For example, cells in a hypotonic environment can swell and possibly rupture due to water moving into the cell. In the reverse situation in which cells are in a hypertonic environment, water moves out of the cell leading to cell shrinkage and dehydration. This movement of water across membranes in response to concentration gradients is called osmosis.

The cell has several mechanisms by which it moves substances, such as nutrients and ions, across membranes (TABLE 2.2). Some of the movement across membranes is due to simple diffusion in which water and solutes that readily cross the membrane move from areas of higher concentration to areas of lower concentration. Most solutes, though, do not easily cross the bilayer and membrane proteins facilitate the movement of ions and molecules across membranes.

Table 2.2 Transport Mechanisms across Membranes	
Passive transport	Non-energy-requiring process involving the movement of substances across a membrane down a concentration, pressure, or electrical charge gradient
Simple diffusion	Movement of water, dissolved gases, or lipid-soluble molecules across the lipid bilayer
Facilitated diffusion	Movement of solutes across a membrane involving a channel or transport protein
Osmosis	Movement of water across a membrane in response to a concentration or pressure gradient
Energy-requiring transport	Transport processes requiring the expenditure of energy by the cell
Active transport	Energy-dependent movement of small molecules or ions across a membrane utilizing specific transport proteins called pumps
Endocytosis	Uptake of large particles or molecules by engulfing extracellular material with the plasma membrane and enclosing the material in a membrane-bound vesicle
Exocytosis	Excretion of materials from the cell via fusion of vesicles with the plasma membrane

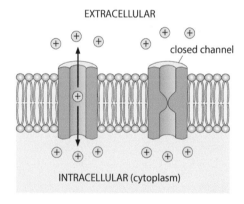

FIGURE 2.4 **Membrane channels.** Proteins can form channels through membranes that allow for the passage of some small molecules, particularly ions, across the membrane. Ion channels are specific for either anions or cations due to the type of charge on the inside of the channel. For example, a negatively charged channel only allows the passage of cations. The cell can also open and close the channels to regulate the passage of ions.

Facilitated diffusion utilizes channels and transport proteins

Membrane proteins in the form of channels or carriers mediate the movement of substances across membranes in a process called facilitated diffusion. Channel proteins form a pore in the lipid bilayer (FIGURE 2.4). Most channel proteins have a pore of a defined diameter and electrical charges may line the pore. These properties only allow certain ions to cross the membrane through a particular pore. For example, in many cells there are separate channels for the transport of sodium (Na^+), potassium (K^+), and calcium ($Ca2^+$) ions due to their size differences. Similarly, negatively or positively charged ions utilize distinct channels that are oppositely charged. Channel proteins allow molecules that normally could not cross the membrane to readily move across membranes in response to concentration gradients. This passive diffusion maintains ion concentrations in an equilibrium state. In some cases, the channel can be opened or closed and this allows for some regulation of the movement of substances across membranes. For example, the opening of ion channels is associated with nerve impulses (see Box 18.1).

Transport proteins bind to specific molecules or a class of similar molecules. Binding of the solute to the transport protein triggers a change in the conformation of the protein so that it allows the molecule to cross the membrane (FIGURE 2.5). These transport proteins function as carriers in that they carry the solute from one side of the membrane to the other. Like channels, transport proteins only move molecules down the concentration gradient from higher to lower concentration. Facilitated diffusion with transport proteins is generally a slower process than simple diffusion across membranes or facilitated diffusion through channels.

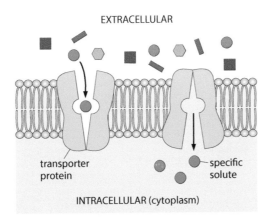

EXTRACELLULAR

transporter protein

specific solute

INTRACELLULAR (cytoplasm)

FIGURE 2.5 Transport proteins. Transport proteins are carrier proteins that move solutes across membranes. Any specific transporter binds to a limited range of solutes. Binding of the solute to the transporter causes a change in the structure of the protein so that the solute is released on the other side of the membrane. Transporters can move solutes from the outside to the inside or vice versa (not shown). The direction of movement is determined by the concentration gradient so that the solute moves from a higher concentration to a lower concentration.

Active transport uses energy to move substances against concentration gradients

Cells often need to move substances against a concentration gradient. For example, many nutrients are maintained at higher concentrations inside cells than in the external environment. To avoid losing these nutrients, cells actively pump such molecules into cells. Similarly, cells maintain ionic compositions distinct from their external environment. For example, in animals the level of sodium in the extracellular fluid is higher than in the cell's cytoplasm, whereas potassium levels are reversed. A membrane protein called the sodium–potassium pump maintains the correct distribution of these two ions. Cystic fibrosis is a disease in which the correct ion balance of chloride is not maintained (see Box 11.2).

Membrane proteins use cellular energy to move molecules and ions against a concentration gradient to maintain concentration differences between the inside and outside of cells. Active-transport proteins, like the facilitated diffusion carriers, bind to specific substrates and actively move the substrate across the membrane. The transporter proteins, often called pumps, generally utilize ATP as the source of energy. This involves an associated ATPase activity that converts ATP to ADP and phosphate. The energy released from the hydrolysis of ATP to ADP is captured by the pump and promotes a change in the protein's conformation so that the substrate is translocated to the other side of the membrane (FIGURE 2.6).

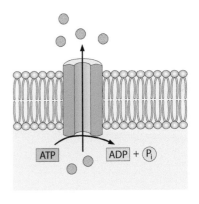

ATP   ADP + P_i

FIGURE 2.6 Active transport. Solutes can be moved against a concentration gradient with membrane pumps. These proteins use the energy released through the hydrolysis of ATP to ADP and phosphate (P_i) to move solutes across a membrane. These membrane pumps exhibit specificity in regard to the solutes that can be moved across the membrane.

Endocytosis and exocytosis involve a vesicle-mediated movement of substances into and out of cells

Large molecules and particles cannot cross membranes via channels, transporters, or pumps and require a vesicle-mediated process. Vesicles are small membrane-bound sacs that pinch off from the plasma membrane or an internal membrane within the cytoplasm of the cell. In the case of endocytosis, a patch on the plasma membrane indents and forms a vesicle that encloses material from the extracellular milieu (FIGURE 2.7). This vesicle then transports the material to other subcellular compartments. Exocytosis is the opposite process of endocytosis. In this case the vesicle fuses with the plasma membrane and thereby releases its contents to the outside of the cell.

Pinocytosis, receptor-mediated endocytosis, and phagocytosis are distinct types of endocytosis. Pinocytosis involves the uptake of extracellular fluid and is sometimes described as 'cell drinking.' Extracellular solutes including macromolecules such as proteins are taken up in a non-specific fashion. In contrast, receptor-mediated endocytosis is a process by which cells take up specific molecules. For

FIGURE 2.7 Endocytosis and exocytosis.
During endocytosis cells take up extracellular fluid or particles by enclosing the material in a vesicle. During exocytosis secretory vesicles fuse with the plasma membrane to release their contents into the extracellular environment.

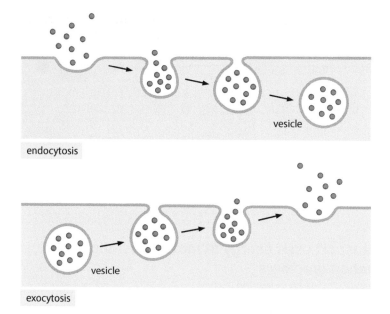

endocytosis

vesicle

vesicle

exocytosis

example, iron binds to a carrier protein called transferrin and transferrin receptors on the cell surface facilitate the uptake of transferrin and the bound iron. Phagocytosis, or cell eating, is the uptake of large particles such as bacteria by macrophages as part of immunity against pathogens (Chapter 9). In this case, the plasma membrane extends outward to surround the bacterium, which is subsequently engulfed and destroyed by the macrophage.

CELLULAR ORGANELLES

In addition to having a distinct and well-defined nucleus, eukaryotic cells contain cytoplasmic **organelles**. Organelles are membrane-bound compartments within eukaryotic cells that have specialized functions (TABLE 2.3).

Table 2.3	Major Membrane-bound Organelles
Organelle	**Major Function**
Nucleus	Location of DNA in the form of chromosomes. Site of transcription (RNA synthesis)
Endoplasmic reticulum	Synthesis of lipids and proteins destined for secretion or localization to other membranous compartments
Golgi apparatus	Modification, transport, and sorting of proteins within the secretory pathway
Lysosome	Digestion of substances within cells and killing pathogens within macrophages
Mitochondrion	Synthesis of ATP
Chloroplast	Found in plants and some algae and functions in the capture of light energy

The nucleus is the most prominent among the organelles and is visible with light microscopy

The **nucleus** was the first organelle identified due to its large size and the ability to stain the nucleus with biological dyes that bind DNA. In fact, the discovery of the nucleus led to the concept of eukaryotic and prokaryotic cells. The nucleus is also the site of RNA synthesis, or transcription. Transcription is the first step in gene expression leading to the synthesis of proteins (Chapter 4). The double membrane surrounding the nucleus, called the nuclear envelope, has relatively large pores so that RNA and proteins can enter or exit the nucleus in a regulated manner.

The mitochondrion plays a key role in energy generation

Mitochondria function in energy metabolism and contain enzymes needed for the generation of ATP (Chapter 3). ATP functions as the energy currency of the cell and most energy-requiring processes utilize ATP through the action of an ATPase activity associated with proteins that are involved in the energy-requiring processes. The mitochondrion has a double membrane and contains its own DNA. Chloroplasts are organelles of plant cells that function to capture energy from light.

Many organelles are part of a complex endomembrane system

The intracellular endomembrane system includes the nuclear envelope, endoplasmic reticulum, Golgi apparatus, lysosomes, and the vesicles that transport material between these subcellular compartments. The various parts of this endomembrane system exchange membrane material with one another via vesicles that move between the compartments. This membrane trafficking between the membrane compartments also intersects with endocytic and exocytic pathways. This flow of membranes within the cell is important for the overall cell homeostasis and function and several diseases have been described involving deficits in membrane flow and trafficking (BOX 2.3).

Diseases Associated with Membrane Flow

Problems with the flow of material between the components of the endomembrane system can result in the improper accumulation of proteins and other substances within the ER, Golgi, or lysosomes. As the first step in intracellular transport, the ER also functions in quality control and ensures that proteins are properly folded. Misfolded or unfolded proteins aggregate in the ER and induce an ER stress response to remove the aggregated proteins. If not removed, these aggregated proteins are toxic to cells, with neurons being especially sensitive to protein aggregates. For example, Parkinson's disease is characterized by the accumulation of aggregated proteins into intracellular structures called Lewy bodies. Interestingly, mutations in a gene named *parkin* may be associated with Parkinson's disease. The parkin protein may function as part of the ER stress response, suggesting that deficiency in the removal of aggregated proteins may contribute to the disease. Other neurological diseases due to the accumulation of protein aggregates in neurons include Huntington's disease (see Box 11.3) and prion diseases (Chapter 19).

Lysosomal storage disorders are a group of approximately 50 rare inherited metabolic disorders that are characterized by an accumulation of waste products in the lysosome. Lysosomes function somewhat like a cell recycling center in that they contain hydrolytic enzymes that digest or break down macromolecules into molecules that can be utilized by the cell. If the cell has a defective hydrolase or lacks a specific hydrolase, the cell is unable to break down specific molecules and this leads to an accumulation of those substrates within the lysosome. Similarly, if a particular transport protein on the lysosomal membrane is defective or missing, the substrate accumulates in the lysosome. The accumulation of these substrates within the lysosome results in large vacuoles filled with potentially toxic substances. The symptoms of lysosomal storage disease vary and range from mild to severe, depending on the specific disorder and other variables like the age of onset. Pathogenesis can include delayed development, heart or lung problems, movement disorders, seizures, dementia, deafness, and blindness.

BOX 2.3

The endoplasmic reticulum and Golgi apparatus function in the targeting of proteins to their correct intracellular locations

The endoplasmic reticulum (ER) is a network of membrane-bound tubes and flattened sacs that is continuous with the nuclear envelope. Some regions of the ER contain ribosomes and are referred to as the rough ER, whereas regions without ribosomes are the smooth ER. The smooth ER primarily functions in lipid synthesis and the rough ER is the site of protein synthesis for proteins destined for secretion or localization to other membrane-bound compartments. Thus, a major role of the ER is the synthesis of cellular membranes.

Newly synthesized membrane material exits the ER by forming vesicles (FIGURE 2.8). These vesicles are transported to other membrane-bound compartments and fuse with those compartments. Not only does vesicle fusion add membrane to the other compartment, but any soluble material within the vesicle, called cargo, is also delivered to that compartment. Vesicles derived from the ER deliver cargo and membranes to the Golgi apparatus by fusing with the Golgi.

FIGURE 2.8 The endomembrane system and trafficking. Newly synthesized lipids and membrane-associated proteins are synthesized in the endoplasmic reticulum (ER) and then transported to the Golgi via vesicles. In the Golgi proteins are further processed and sorted according to their final destination. Endocytosed material is processed through a compartment called the endosome and subsequently moved to the lysosome.

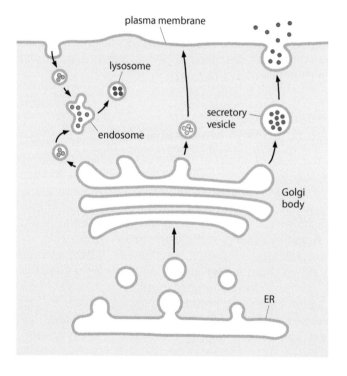

The Golgi apparatus is also called the Golgi body, Golgi complex, or simply Golgi after its discoverer. It is generally a stack of flattened membrane stacks located near the ER. A major function of the Golgi is sorting proteins and membranes so that proteins are delivered to their correct and final destinations. For example, proteins destined for secretion are accumulated in specialized vesicles called secretory vesicles (see Figure 2.8). These secretory vesicles fuse with the plasma membrane and the cargo proteins are delivered to the external environment by exocytosis. Similarly, proteins localized to the plasma membrane are pinched off into vesicles that are delivered to the plasma membrane. Fusion of the vesicles with the plasma membrane results in the proteins being incorporated into the plasma membrane. Likewise, hydrolytic enzymes are delivered to the lysosomes.

The lysosomes are part of the endocytic pathway

Vesicles generated by endocytosis generally fuse with **lysosomes**. Lysosomes are acidic compartments containing hydrolytic enzymes that break down macromolecules into smaller molecules. These smaller molecules can then be used for the biosynthesis of other macromolecules. Thus, endocytosis is involved in the acquisition of nutrients that are too large for carriers and channels. In addition, lysosomes in macrophages and related cells play a key role in the destruction of pathogens during the immune response (Chapter 9).

Some substances are taken up by a process called receptor-mediated endocytosis (FIGURE 2.9). For example, lipids are associated with large complexes containing both lipids and proteins called lipoproteins (see Box 14.1). These lipoproteins bind to receptors on the cell surface and are internalized into vesicles that fuse into a compartment called the endosome. The pH of the endosome is slightly acidic and this acidity results in the lipoprotein separating from the receptor. Receptors are then returned to the plasma membrane via vesicles where they can bind and internalize more lipoprotein. Endosomes fuse with lysosomes, which are even more acidic, and the resulting decrease in pH causes a separation of the lipid and protein components of the lipoprotein. Hydrolytic enzymes within the lysosome break down the lipoproteins into cholesterol, fatty acids, and amino acids that are used in cellular metabolism.

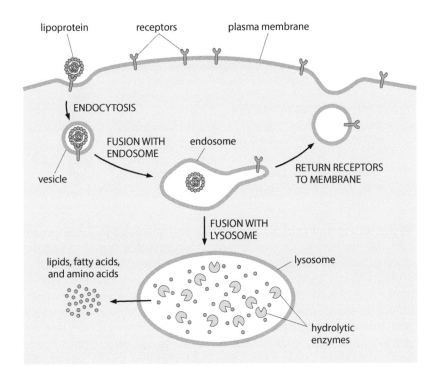

FIGURE 2.9 Receptor-mediated endocytosis. Lipoproteins bind to receptors and are endocytosed. Upon fusion with the endosome the lipoprotein and receptor separate. The receptor is returned to the plasma membrane for recycling and the endosome with the lipoproteins is fused with the lysosome. Hydrolytic enzymes in the lysosome break down the proteins and lipids to release lipids, fatty acids, and amino acids that can be utilized by the cell.

CYTOSKELETON

Diagrams of eukaryotic cells lead to the impression that the various organelles are freely floating about in a relatively fluid cytoplasm. This is not true and the cytoplasm is actually gel-like due to a highly organized network of fibrous elements (FIGURE 2.10). These fibrous networks are composed of proteins and are known as the cytoskeleton since they help determine cell shape and provide mechanical support to the cell. Three distinct filamentous elements make up the cytoskeleton: microtubules, microfilaments, and intermediate filaments (TABLE 2.4). Microtubules and microfilaments are found in all eukaryotic cells, whereas intermediate filaments are only found in animal cells. Although these cytoskeletal elements provide mechanical strength, they are also quite dynamic and are continuously assembled and dissembled in response to changing cellular physiology.

FIGURE 2.10 Network of fibrous structures makes up the cytoskeleton. Cytoskeletal proteins assemble into filaments and tubules, providing mechanical support for the cell. The cytoskeleton provides mechanical support and gives the cytoplasm a gel-like quality.

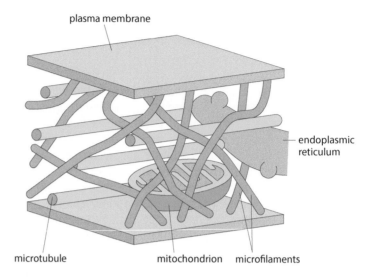

plasma membrane

endoplasmic reticulum

microtubule   mitochondrion   microfilaments

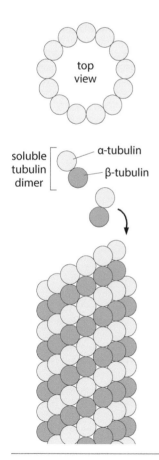

top view

soluble tubulin dimer

α-tubulin
β-tubulin

FIGURE 2.11 Microtubule structure. Microtubules are composed primarily of tubulin. The dimers consisting of α- and β-tubulin are arranged in a spiral staircase fashion to form a hollow tube.

Table 2.4	Cytoskeletal Elements	
Element	**Description**	**Composition**
Microtubules	Rigid tubular structures 25 nm (nm = 10^{-9} m) in diameter that plays a key role in cell shape, the spindle apparatus, and cellular movement involving cilia and flagella	Tubulin and associated proteins
Microfilaments	Thin filaments 6–7 nm in diameter that are associated with many subcellular structures and are involved in force generation	Actin and associated proteins
Intermediate filaments	Rope-like fibers approximately 10 nm in diameter that provide mechanical strength to cells and tissues that experience high levels of stress (e.g., muscle, epithelium)	Intermediate filament proteins

Microtubules and microfilaments are also involved in force generation and cell movement. In this regard, the cytoskeleton is more analogous to a subcellular musculoskeletal system. Force is generated through proteins that interact with the filamentous cytoskeletal elements called motor proteins. Motor proteins generate force by converting the chemical energy associated with ATP hydrolysis into mechanical energy. Kinesins and dynein are motor proteins associated with microtubules and myosin is associated with microfilaments. Kinesins, dyneins, and myosins are protein families, with each individual member of the family having a specific role in motility. The force produced by these motor proteins and cytoskeletal elements can move cells as well as move subcellular structures within cells.

Microtubules are important for cell structure, cell division, and cell motility

Microtubules are hollow tubes primarily composed of the protein tubulin. Tubulin is a dimer consisting of α-tubulin and β-tubulin and the dimer is stacked in a spiral staircase fashion to form the tubes (**FIGURE 2.11**). These tubes are rather rigid and provide mechanical support for the cell. In animal cells the microtubules usually originate near the center of the cell at a structure called the centrosome and project outward (FIGURE 2.12). This arrangement provides substantial support to maintain cell shape. In addition, the microtubules serve as tracks by which vesicles can move within the cell. The microtubules also reorganize during cell division to form the mitotic spindle (BOX 2.4).

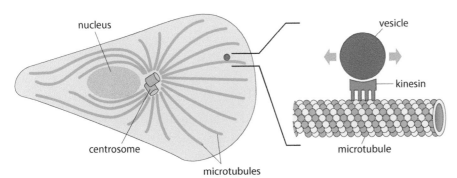

FIGURE 2.12 Microtubules. In animal cells microtubules usually emanate from a structure called the centrosome. The centrosome is generally near the center of the cell and the microtubules project toward the outer edges of the cell and provide support for the three-dimensional shape of the cell. Vesicles can also move along the microtubules via motor proteins such as kinesin that are attached to the vesicles.

Microtubules are also a key component of cilia and flagella, which are specialized organelles for cellular movement. Both cilia and flagella are hair-like projections composed of parallel rows of microtubules arranged in a ring (FIGURE 2.13). Flagella are generally long structures projecting from one end of the cell with the tail of the sperm being the classic example. Cilia are generally shorter, more numerous, arranged in rows, and beat in unison like oars on ancient ships. *Paramecium* species and related protozoa use cilia for swimming. Cilia also line the respiratory tract and function to move particles and mucus from the lungs to be expelled.

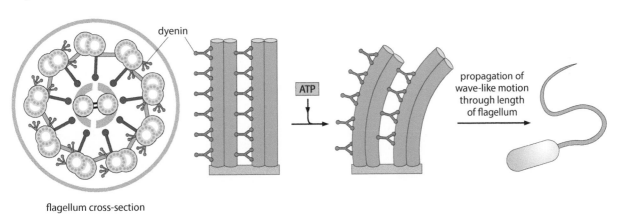

FIGURE 2.13 Flagellar structure and function. Cilia and flagella are made up of a bundle of microtubules consisting of nine doublets that form a ring and two microtubules in the center (flagellum cross-section). This 9 + 2 arrangement of microtubules is surrounded by the plasma membrane and projects from the cell as a long tail. Proteins, such as the motor protein dynein, hold the microtubule bundle together. Dynein is permanently embedded into one microtubule doublet and the other end of dynein associates with the neighboring microtubule doublet. Hydrolysis of ATP is associated with a movement of dynein on the neighboring microtubule doublet which creates a bending of the microtubule. This bending is propagated as a wave through the length of the flagellum due to the coordinate binding and release of dynein.

Motility associated with cilia and flagella is due to a whip-like motion that is propagated along the length of this cellular appendage. The whip-like motion is generated by the motor protein dynein that causes the microtubules to slide past each other (see Figure 2.13). As the microtubules are fixed at the base of the flagellum, this sliding results in a bending of the flagellum that is sequentially propagated along the length of the flagellum.

BOX 2.4

The Cytoskeleton and Cell Division

Both microtubules and microfilaments are involved in the process of cell replication. As cells begin to divide the microtubules disassemble into the tubulin subunits. The centrosome duplicates and the two centrosomes move to opposite sides of the cell and direct the reassembly of the microtubules into the mitotic spindle (**BOX 2.4 FIGURE 1**). The mitotic spindle and motor proteins called kinesins supply the force to separate the chromosomes. Chromosomes move along the microtubules toward opposite poles via their attachment to a kinesin. At the same time the microtubules slide past each other leading to further separation of the poles. Other kinesins bind to both microtubules and generate the force involved in this sliding movement.

In animal cells, cytokinesis is mediated by actin and myosin. A ring of actin–myosin filaments forms on the inner surface of the plasma membrane in the center of the cell. This ring then contracts and produces a narrowing at the cell equator. This constriction continues until a membrane fusion can occur that results in the two newly formed daughter cells being pinched apart, somewhat analogous to the pulling of a purse drawstring.

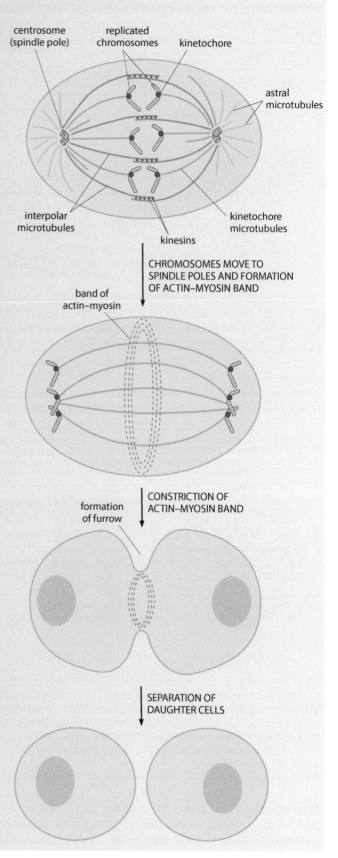

BOX 2.4 FIGURE 1 The mitotic spindle and cytokinesis. The centrosomes of the spindle poles function as microtubule organizing centers from which microtubules originate and expand. The astral microtubules connect to the cell membrane and help orient the cell division plane. Kinetochore microtubules connect to the chromosomes and the chromosomes migrate along these microtubules toward the centrosomes. The kinetochore is a structure on the chromosomes that contains kinesin, which pulls the chromosomes along the microtubules. These microtubules shorten as the chromosomes move. The interpolar microtubules originate from both centrosomes and interact with each other via kinesins. The kinesins force the microtubules to slide away from each other and these microtubules grow in length during the mitotic process. This results in further separation of the spindle poles. The division of the cytoplasm in animal cells involves microfilaments composed of actin and myosin contracting around the center of the cell to pinch the cell in two. A band of actin–myosin forms around the center of the cell and the myosin mediates a constriction that forms a furrow. At some point the constriction becomes small enough that the membranes can fuse and form the two daughter cells.

Microfilaments are associated with many cellular structures and processes

Microfilaments are thin filaments composed of actin plus associated proteins (FIGURE 2.14). Microfilaments can be cross-linked to form a three-dimensional meshwork (see Figure 2.10). Like microtubules, the cross-linked microfilaments also provide support for the cell and influence the overall cell shape. Microfilaments are also quite dynamic and assemble and reassemble with physiological and morphological changes in the cell. For example, the microfilaments rearrange during processes such as endocytosis and cell division. Microfilaments can also be rearranged into bundles that form various subcellular structures.

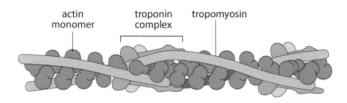

actin
monomer
troponin
complex
tropomyosin

FIGURE 2.14 Microfilament structure. Microfilaments are composed primarily of a protein called actin. Actin monomers associate into long thin filaments. Other proteins such as troponin and tropomyosin bind to actin filaments to provide more support.

Actin filaments also participate in cellular processes involving force generation with a motor protein called myosin. Myosin moves along microfilaments utilizing energy generated by the hydrolysis of ATP. The best known example of actin–myosin-dependent force generation is muscle contraction. In this case, actin filaments within a muscle cell are pulled toward each other, resulting in a contraction of the muscle cell. Contraction of numerous muscle cells in unison results in muscle contraction. Many other forms of cell movement also depend on the actin–myosin cytoskeleton. For example, cells crawling along a surface rearrange their microfilaments and generate force through myosin. This type of crawling motility is called amoeboid movement.

CELL REPLICATION AND MITOSIS

A fundamental property of cells is their ability to replicate. Cellular replication involves a doubling of all the cellular components and copying DNA (Chapter 4). The resulting progeny, or daughter cells, contain the necessary machinery to carry out cellular functions and the ability to synthesize new cellular components such as structural proteins, enzymes to carry out metabolism, membranes, and cellular organelles. In the case of single-celled organisms, cellular replication is how they reproduce. This is primarily a clonal process involving asexual reproduction in which the progeny cells are complete organisms and are essentially identical to each other.

In multicellular organisms, reproduction involves a specialized type of cell division called meiosis and the production of gametes in a sexual process. Growth and development of multicellular organisms involve a type of asexual replication called **mitosis**. These asexually replicating cells are called somatic cells and will develop into specialized tissues and organs that carry out distinct functions within the organism (Chapter 8). Cell division is often defined by a cell cycle (BOX 2.5).

Details of meiosis are described in Chapter 5, pages 91-94.

SIGNPOST

BOX 2.5

The Cell Cycle

After cell division, the cell undergoes a period of growth before beginning another round of cell division. This period before starting another cell division cycle is called the G1 phase (short for gap 1). Most cells are characterized by a defined size range. Therefore, before dividing, cells need to take up nutrients and increase to approximately double their normal size. Furthermore, the process of cell division is highly regulated and the G1 phase serves as a checkpoint to determine whether a cell will ultimately divide or not (**BOX 2.5 FIGURE 1**). Once committed to cell division the S phase (short for synthesis) begins and DNA is replicated. Cells start the G2 phase (short for gap 2) following the completion of DNA synthesis. During the G2 phase the cell carries out quality control by checking for DNA damage and chromosome integrity. If the DNA has been accurately replicated, the cell enters the mitotic, or M, phase of the cell cycle. The M phase consists of chromosome separation and cytokinesis resulting in two daughter cells. Some cells are terminally differentiated and are no longer able to divide. Such cells can be viewed as being in a permanent G1 phase, often referred to as G0.

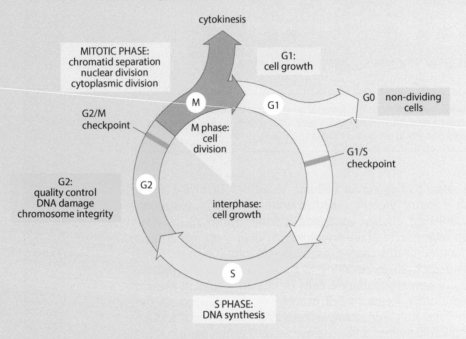

BOX 2.5 FIGURE 1 Phases of the cell cycle. Distinct steps in the process of cell division are recognized, including two pauses (G1 and G2) for quality control.

Bacteria replicate by binary fission

The DNA of prokaryotic cells is not associated with a nucleus and is found within the cytoplasm. In addition, the chromosomal DNA of prokaryotes is usually a single circular DNA molecule. Therefore, the replication of prokaryotes is somewhat simpler than the mitosis of eukaryotic cells. And thus, under favorable conditions most bacteria can rapidly replicate. For example, the common intestinal bacterium *Escherichia coli* can replicate in 20–30 minutes and under favorable conditions a single bacterium can increase to approximately 300 trillion bacteria in 24 hours. This ability to rapidly replicate contributes to the virulence of some bacteria. Inoculation of even a few pathogenic organisms can lead to severe disease in a relatively short period if the immune system is unable to control their replication.

The first step in **binary fission** is a duplication of the bacterial chromosome involving DNA synthesis in a template-mediated fashion (Chapter 4), resulting in two identical copies of the circular DNA molecule. Each copy of the circular DNA molecule is attached to the plasma membrane and as the bacterial cell grows via elongation the two copies of the bacterial chromosome are pulled apart (FIGURE 2.15). As the duplicated DNA is being moved to opposite sides of the cell, the plasma membrane in the middle of the cell grows inward and will eventually fuse, resulting in a double membrane separating the cell into two halves. The cell wall is synthesized on these newly formed membranes and the two newly formed cells separate, resulting in two essentially identical daughter cells.

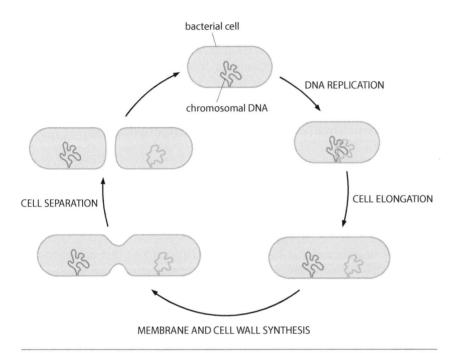

FIGURE 2.15 Binary fission in prokaryotic cells. Most bacterial cells have a single circular DNA molecule as their chromosomal DNA. After replication, the chromosomal DNA remains attached to the cell membrane and the two chromosomes are separated as the cell elongates. Cell separation is accomplished by the synthesis of the plasma membrane and cell wall material in the middle of the cell.

Eukaryotic cells replicate by mitosis to generate two genetically identical cells

Eukaryotic cells also replicate by a process similar to binary fission in that one cell divides into two nearly identical cells. Eukaryotic cells contain significantly more DNA than prokaryotic cells and this DNA is organized into multiple chromosomes which are located within the nucleus of the cell. As the first step in replication, the individual chromosomes are duplicated. The duplicated chromosomes remain attached to each other at a specialized region of the chromosome called the centromere (FIGURE 2.16). These two attached chromosomes are called sister chromatids.

As the chromosomes are being duplicated, the centrosome also duplicates, and the microtubules reorganize into the mitotic spindle (see Box 2.4). In many eukaryotes, the nuclear envelope disappears as the mitotic spindle is formed (FIGURE 2.17). The duplicated chromosomes line up in the middle of the cell and attach to microtubules of the mitotic spindle at their centromeres. The individual sister chromatids are then pulled apart and move in opposite directions along the microtubules. Once the chromosomes are completely separated, the nuclear envelope

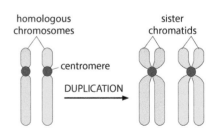

FIGURE 2.16 Chromosome duplication.
After duplication eukaryotic chromosomes remain attached at the centromere and the newly replicated chromosomes are called sister chromatids. Many eukaryotes, such as humans, contain two copies of each of their chromosomes which are called homologous chromosomes.

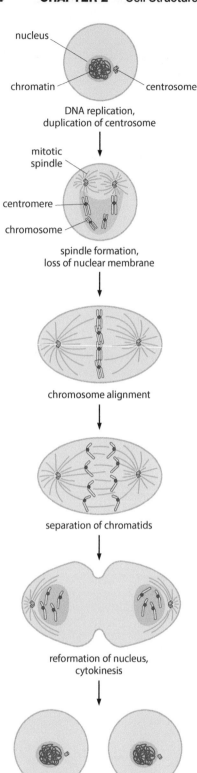

nucleus

chromatin   centrosome

DNA replication,
duplication of centrosome

mitotic
spindle

centromere

chromosome

spindle formation,
loss of nuclear membrane

chromosome alignment

separation of chromatids

reformation of nucleus,
cytokinesis

two identical daughter cells

FIGURE 2.17 Mitosis. The process of mitosis involves the replication of chromosomes and their separation into daughter cells. The centrosome duplicates and forms the mitotic spindle that pulls apart the sister chromatids which are segregated into two identical daughter cells.

re-forms around the chromosomes and the cell undergoes cytokinesis (the division of the cytoplasm), resulting in two daughter cells with the same genetic makeup. Cytokinesis is facilitated by microfilaments and myosin (see Box 2.4).

The cytoplasm and organelles are partitioned between the newly formed daughter cells

Some organelles, such as the mitochondria and chloroplasts in plant cells, replicate autonomously. These organelles contain their own DNA, which is replicated, and these organelles divide by binary fission. The mitochondria are distributed throughout the cytoplasm and each daughter cell incorporates approximately equal numbers during cytokinesis. The various membranous compartments that make up the secretory and endocytic pathways are all derived from the ER. During cell division, the Golgi apparatus and various vesicular compartments are in essence reabsorbed by the ER. The ER is then divided between the two daughter cells and regenerates the secretory and endocytic pathways.

CELLULAR DAMAGE AND DISEASE

Similar cells associate to form tissues and various tissues will form organs (Chapter 8). It is widely recognized that tissue and organ pathology can be ascribed to cellular damage. Thus, disease has a cellular basis. Damaged cells are unable to fulfill their normal functions and this contributes to disease progression. Severe cellular damage or injury can result in cell death, called necrosis. Necrosis is distinct from a natural physiological cell death known as apoptosis (see Box 16.1).

A possible consequence of cell death is the release of cellular contents such as enzymes. Therefore, the measurement of specific enzymes associated with a particular cell type or organ in the blood can be used to assess the health of a particular organ or to assist in disease diagnosis. For example, high levels of aspartate aminotransferase (AST) and alanine aminotransferase (ALT) in the blood are indicative of liver damage. Similarly, proteins associated with heart muscle are biomarkers for damage to the heart, such as a heart attack.

Chemical, physical, and biological agents can damage cells

Cells can be injured in several ways (TABLE 2.5). A common cause of cellular injury is inadequate oxygen, called hypoxia, due to insufficient blood flow. Oxygen is important for normal metabolism and initiates chemical and acid–base imbalances that result in cell death if oxygenation is not restored. Many chemicals are toxins that cause cellular damage. Similarly, direct physical actions due to mechanical force, extremes in temperature, ionizing radiation, strong acids, or strong bases directly damage or kill cells. Pathogens, such as bacteria and viruses, directly damage cells or produce toxins that affect cells. Excesses or deficiencies in nutrients also affect cellular physiology and result in cellular damage. Some cellular damage may have a genetic etiology or just simply be due to normal cellular aging. In some cases, though, disease is exacerbated by repair processes that follow injury or stress. For example, cellular injury or infection produces an inflammatory response from the immune system that often causes collateral damage (Chapter 9).

Details of cellular senescence can be found in Chapter 10, page 210-212.

SIGNPOST

Table 2.5	**Factors Causing Cellular Injury**
Hypoxia	Lack of oxygen causes chemical imbalances within the cell and ultimately results in cell death
Toxins	A diverse array of substances that are toxic for cells by interfering with normal cellular processes
Mechanical force	Direct physical trauma can disrupt cells, tissues, or organs
Temperature extremes	Extreme heat or freezing directly kills cells
Strong acids or bases	Ingestion profoundly alters blood pH and contact with epithelium results in necrosis
Radiation	Directly damages DNA and affects cell reproduction
Pathogens	Bacteria, viruses, and other disease-causing organisms directly or indirectly damage cells
Nutritional imbalance	Deficiencies or excesses in nutrients can affect cellular metabolism and cause damage
Genetic defects	Mutant genes can have profound effects on cell structure and function
Inflammation	Immune responses to injury or infection can cause collateral damage to cells
Aging	Cell aging is progressive and renders cells less able to respond to injury

Changes in cell morphology are often associated with mild injury or stress

Mild injury to cells can affect cellular physiology without resulting in cell death. Injury or stress, such as unusual physiological demands, result in cellular changes or adaptations of the cell (**FIGURE 2.18**). If the injury or stress is not too great the cell can return to normal once the injurious or stressful conditions have been removed. These cellular alterations

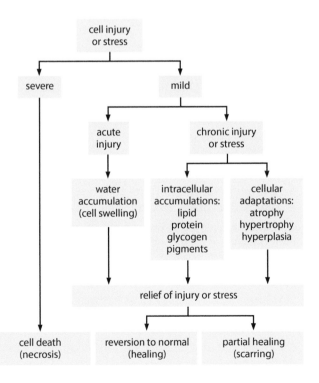

FIGURE 2.18 Cellular adaptations to stress or injury. Severe injury or stress can result in cell death. Cell death due to injury is called necrosis. Cells change or adapt to milder injury or stress and can completely or partially heal if the stress or injurious condition is removed.

may be observed microscopically and thus can be diagnostic. A common cellular alteration is the accumulation and retention of water resulting in cell swelling. This is particularly common with acute injury. Long-term or chronic injuries or stress can lead to the accumulation of other substances such as lipids, protein, glycogen, or pigments.

Cellular injury or stress can also result in a change in cell size or cell number. Atrophy is a decrease in cell (or organ) size or function. It is an adaptive response to decreased demand or increased stress. The opposite of atrophy is hypertrophy, which is an increase in the size or functional capacity of a cell or organ. Increased demand or hormonal stimulation is a possible cause of hypertrophy. Hyperplasia is an increase in tissue or organ size due to an abnormal increase in cell number (Chapter 16).

SUMMARY OF KEY CONCEPTS

- Cells are composed of molecules with distinct functional roles within the cell
- Cells exhibit metabolism and can replicate and, thus, are considered the basic unit of life
- Membranes are composed of a lipid bilayer with associated proteins, which encloses the contents of the cell and regulates the movement of substances in and out of the cell
- Prokaryotic cells (bacteria) and eukaryotic cells have a distinct structural organization, with eukaryotic cells having subcellular compartments called organelles and a cytoskeleton
- Membrane-bound organelles of eukaryotic cells carry out specialized functions
- Cellular replication involves passing genetic instructions plus the cytoplasmic machinery necessary to carry out cellular functions to two daughter cells
- Eukaryotic cell replication (mitosis) involves the duplication of the nucleus and chromosomes followed by cytoplasmic division (cytokinesis)
- Cellular injury is the basis of disease

KEY TERMS

binary fission	The asexual replication of prokaryotic cells (bacteria) yielding two identical daughter cells.
centromere	The region of a chromosome at which the sister chromatids are held together until they separate during cell division. It is also the position where the spindle fibers attach during chromosome and chromatid separation.
cytoskeleton	System of protein filaments in eukaryotic cells that determines cell shape and participates in cellular movements.
lysosome	An intracellular membrane-bound organelle containing hydrolytic enzymes that break down molecules taken up by endocytosis.
microfilaments	Cytoskeletal element composed of actin and associated proteins which play a role in cell shape and motility. Motility is associated with the motor protein myosin.
microtubules	Cytoskeletal element composed of tubulin and associated proteins. Plays a role in cell shape, motility via flagella or cilia, and forms the mitotic spindle. Motility is associated with the motor proteins dynein or kinesin.
mitosis	A type of asexual cell division in eukaryotic cells in which duplicated chromosomes are distributed equally between two daughter cells.

necrosis	Cell death due to irreparable damage.
nucleus	The double-membrane-bound organelle of eukaryotic cells that contains DNA and chromosomes.
organelle	A structure found within eukaryotic cells that performs a specific function. Examples of membrane-bound organelles include the nucleus, mitochondria, lysosomes, ER, and Golgi.
plasma membrane	The outer membrane of a cell that contains the cytoplasm and defines the boundary of the cell.

REVIEW QUESTIONS

1. Briefly describe the various mechanisms of passive transport and energy-requiring transport for the movement of molecules and substances across biological membranes.

2. What are lysosomes and the endocytic pathway? Briefly describe their primary function.

3. Describe how elements of the cytoskeleton are involved in cell motility.

4. Describe and discuss the differences between binary fission and mitosis.

5. Describe how cells change in response to stress or damage.

ADDITIONAL QUESTIONS

In addition to the Review Questions provided above, there is a range of free online questions designed to enable students to further test their understanding of the chapter material. In order to access these interactive questions, please visit the book's website: https://routledge.com/cw/wiser

Protein Structure and Function

3

Proteins are long polymers of amino acids that fold into unique three-dimensional shapes. Some proteins interact and bind to other proteins or molecules to form cellular structures that are visible with the microscope or electron microscope. Other proteins bind to various molecules with a high degree of specificity as part of their cellular function. For example, enzymes are proteins that catalyze chemical reactions associated with cellular metabolism, whereas transporters and channels move substances across cellular membranes. Proteins also function in cellular signaling and regulatory processes. Thus, proteins play a central role in cellular structure and function and having a solid understanding of protein structure and function is key to the basic workings of cells and organisms.

In that proteins are central to the well-being and normal function of organisms, perturbations in protein structure and function result in disease. For example, many chemicals and processes can damage proteins and these damaged proteins are likely to affect cellular physiology and homeostasis. Therefore, not only does a better understanding of protein structure and function increase our understanding of basic biology, but also understanding proteins contributes to our knowledge about disease and pathogenesis. In addition, proteins are often the target for disease treatment and intervention. Accordingly, the mechanisms of action of many drugs involve binding to specific proteins and affecting the function of those protein targets.

PROTEIN STRUCTURE

FUNCTIONAL ROLES OF PROTEINS

CELLULAR METABOLISM

HORMONES, RECEPTORS, AND CELLULAR COMMUNICATION

MECHANISM OF DRUG ACTION

PROTEIN STRUCTURE

Amino acids react with each other to form a peptide bond between the amine group of one amino acid and the carboxylic acid of another amino acid (see Figure 1.20). In this fashion, a linear array of amino acids is joined together to form a polypeptide that is characterized by a free amine group at one end of the molecule, called the amino or N terminus, and a free carboxylic acid on the other end, called the carboxy or C terminus. In this respect, polypeptides can be viewed as a string of beads with each bead being an amino acid (FIGURE 3.1). In general, the words polypeptide and protein are synonymous.

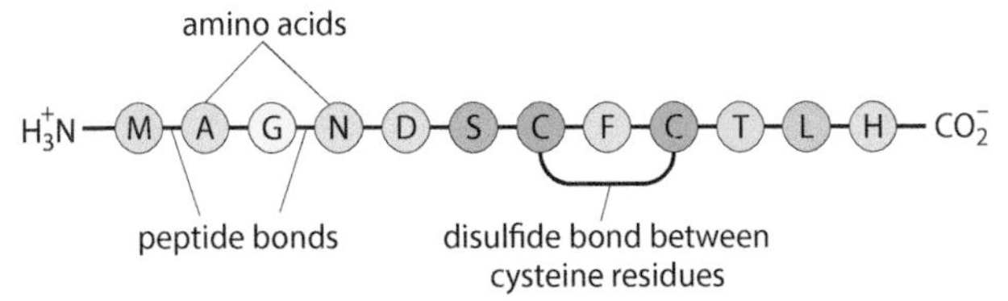

amino acids

H_3^+N —M—A—G—N—D—S—C—F—C—T—L—H— CO_2^-

peptide bonds   disulfide bond between
                       cysteine residues

FIGURE 3.1 Polypeptide sequence. Amino acids are joined together via peptide bonds resulting in a linear sequence of amino acids (designated with single-letter codes) characterized by an amino group on the N terminus and a carboxylic acid group on the C terminus. Covalent disulfide bonds (see Figure 3.2) can also form between cysteine residues.

The order of the amino acids is called the sequence and each protein is defined by its sequence. The sequence plays a major role in the properties and structure of a particular protein. For example, the amino acid composition of a protein determines its overall chemical properties as the various R groups in the amino acids define the overall polarity and hydrophobicity of a particular protein. Proteins with a larger percentage of non-polar amino acids are more hydrophobic than proteins with a larger percentage of polar amino acids. Similarly, proteins composed of a larger percentage of negatively charged R groups than positively charged R groups have an overall negative charge, and *vice versa*. Therefore, each individual protein has a distinct set of chemical and physical properties. These properties include hydrophobicity, polarity, charge, size, and ligand-binding activity.

Covalent bonds determine the primary structure of a protein

The amino acid sequence of a protein is also referred to as the primary structure of the protein. Sometimes included in the primary protein structure is the interaction between cysteines. Cysteine is an amino acid that contains a reactive sulfur group, called a thiol (–SH), as the R group. Thiols can undergo redox reactions. Oxidation results in the two sulfur atoms forming a covalent bond called a disulfide bond and reduction breaks the disulfide bond to form the two thiols (FIGURE 3.2). Since the disulfide bonds are covalent – like the peptide bonds holding together the amino acids – the locations of the disulfide bonds between cysteines are often included as part of the primary structure. Furthermore, disulfide bonds play a key role in protein conformation and, thus, their positions are often noted as part of a protein's sequence. For example, mature insulin has three disulfide bonds that play an integral role in the molecule's overall structure (BOX 3.1).

SIGNPOST

Redox reactions are described in Chapter 1, pages 9-10

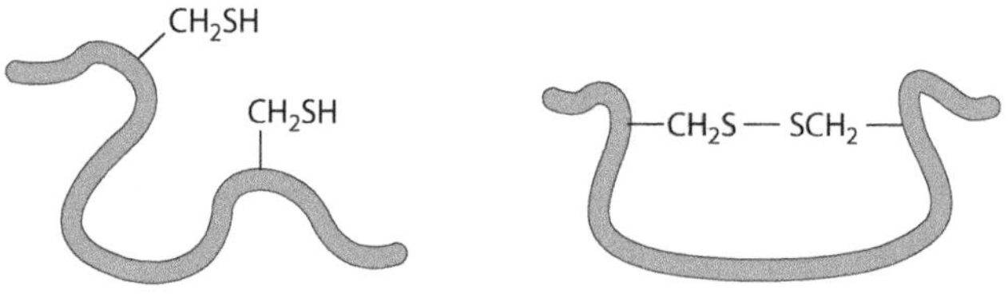

FIGURE 3.2 Formation of disulfide bonds between cysteine residues. The R group of cysteine contains a free thiol group (–SH) that can be oxidized to form a covalent disulfide bond (–S–S–). Disulfide bonds can form between cysteines within the same polypeptide chain and contribute to tertiary structure or form between cysteines on different polypeptides and contribute to the quaternary structure.

BOX 3.1

Insulin Structure

Insulin is a hormone that plays a key role in regulating blood glucose levels and dysfunction of this hormone is the cause of diabetes (Chapter 13). Of historical note, insulin was the first polypeptide hormone to be described. Insulin is produced by β-cells found in specialized regions of the pancreas called the islets of Langerhans. Following synthesis, insulin undergoes two major post-translational processing events to produce mature insulin. The first step is the removal of the signal sequence from the immature insulin, called preproinsulin, as the polypeptide is translocated into the endoplasmic reticulum (ER) during its translation. This type of processing is typical of many proteins that are secreted (Chapter 4). The resulting proinsulin is further processed in the Golgi by removal of the central region of the polypeptide called the C-chain (**BOX 3.1 FIGURE 1**). The C-chain is necessary for the correct folding of mature insulin and the formation of the correct disulfide bonds. The remaining polypeptides are called the A-chain, composed of 21 amino acids, and the B-chain, composed of 30 amino acids. Mature insulin is secreted from β-cells in response to rising levels of glucose in the blood. Insulin receptors are found on most types of cells and are enzyme-linked receptors.

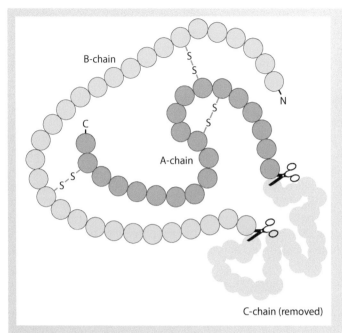

BOX 3.1 FIGURE 1 Structure of insulin. Mature insulin is composed of two polypeptides called the A-chain and the B-chain. Insulin is synthesized as one polypeptide and a central region called the C-chain is removed by proteolysis. Specific proteases recognize amino acid residues at the boundaries (scissors). Two disulfide (S–S) bonds hold the A-chain and B-chain together and a third disulfide bond is found within the A-chain.

Polypeptide chains fold into unique three-dimensional shapes as determined by the amino sequence

Hydrogen bonds, ionic interactions, hydrophobic interactions, and disulfide bonds dictate a distinct conformation for a particular polypeptide sequence. These various interactions between amino acid residues and the polypeptide backbone cause the string of amino acids to fold into a unique shape determined by a combination of bends, loops, and turns. Different levels of protein structure have been defined based on the type of interaction between amino acids and the distance between the interacting amino acids (TABLE 3.1). For example, interactions between amino acids that are relatively close together within a polypeptide generally determine the secondary structure of a protein and interactions between amino acids further apart in the sequence dictate the tertiary structure. Quaternary structure refers to interactions between distinct polypeptide chains making up a protein.

Table 3.1	Levels of Protein Structure
Primary	Refers to the amino acid sequence and the location of disulfide bonds (covalent bonds)
Secondary	Refers to H-bonds between peptide bonds of amino acids that are close together (α-helix, β-sheet) or lack of interactions (random coil)
Tertiary	Refers to interactions between R groups of amino acids that can be far apart (domains)
Quaternary	Refers to interactions between two or more polypeptide chains (protein subunits)

Hydrogen bonds between peptide bonds of neighboring amino acids determine the secondary structure

The two major types of secondary protein structures are the **α-helix** and the **β-sheet**. Both structures are due to hydrogen bonds between the NH group of the peptide bond and the carbonyl group (C=O) of a neighboring peptide bond. The α-helix forms spontaneously during protein synthesis

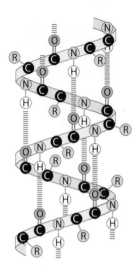

and resembles a spiral staircase or coiled spring. Hydrogen bonds hold the polypeptide backbone in this coiled conformation with the R groups facing toward the outside of the helix (FIGURE 3.3). Some R groups promote the formation of this helix, whereas other R groups can prevent the formation of a helix. Therefore, the local amino acid sequence determines the propensity of α-helix formations within a protein.

The β-sheet also has hydrogen bonding between the amide and carbonyl groups of the peptide bonds of two or more β-strands. The polypeptide backbone of a β-strand is extended and interacts with another β-strand to form a somewhat flat β-sheet (FIGURE 3.4). Every other R group either projects above or below the β-sheet. β-sheets can exist in either parallel or antiparallel configurations depending on the orientation relative to the N and C termini. The regions between β-strands consist of turns or loops of varying lengths. As with the α-helix, the propensity to form a β-sheet is determined by the local amino acid sequence.

FIGURE 3.3 α-Helix structure. Shown is the polypeptide backbone (yellow ribbon) in the α-helix conformation and the hydrogen bonds (purple dashes) between NH groups (blue) and carbonyl groups (C=O in orange). These hydrogen bonds hold the polypeptide in a coiled structure with the R groups (green) projecting away from the outer edge of the coil.

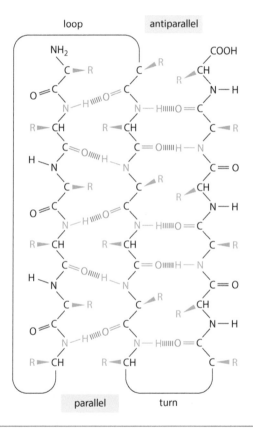

FIGURE 3.4 β-Strands and β-sheets. β-Strands are in an extended conformation and interact with other β-strands to form a β-sheet. Hydrogen bonds (purple dashes) between the NH (blue) and carbonyl (C=O in red) groups hold the β-strands together in the β-sheet. β-Sheets are somewhat flat with the R groups (green) projecting downward and upward in an alternating pattern. β-Sheets in which the β-strands have the same N- and C-terminus orientation are parallel and require a loop of flexible sequence between the β-strands. Antiparallel β-sheets have β-strands in the opposite N- and C-terminus orientation and only require a few amino acids to make a turn.

Different proteins have varying compositions of α-helices and β-sheets. For example, some proteins may contain predominantly α-helices and no β-sheet or *vice versa*, or some contain a mixture of the two types of secondary structure. Between the various secondary structural elements are loops that do not have a defined structure. These loops are referred to as a random coil, extended conformation, or flexible sequence. In contrast with the α-helices and β-sheets, which are somewhat rigid, the loops are more flexible and do not exhibit a defined conformation. The positions and interactions of the α-helices, β-sheets, and flexible loops within a polypeptide chain determine the tertiary structure of the protein (FIGURE 3.5).

Each polypeptide has a unique shape that is determined by secondary and tertiary structures

Generally, proteins exhibit either a globular or a fibrous shape. Globular proteins have an overall rounded or irregular shape with grooves and crevices for binding other molecules (FIGURE 3.6). These grooves and crevices exhibit a high degree of specificity in regard to binding other molecules, often called ligands. Protein–ligand interactions are often described as a lock-and-key fit. Fibrous proteins are generally long and cable-like and often composed predominantly of α-helices. The final shape of any polypeptide is unique to that polypeptide and is determined in part by its sequence through the formation of secondary and tertiary structures. As with secondary structures, interactions between amino acids within the polypeptide determine the tertiary structure. However, the tertiary structure is determined by interactions between the R groups and these interactions are often between amino acids that are relatively far apart in the linear polypeptide sequence. Many polypeptides also consist of distinct domains that are also determined by secondary and tertiary structures.

The interactions between R groups responsible for tertiary structure include covalent disulfide bonds, ionic bonds, hydrogen bonds, and hydrophobic interactions. The type of interaction is dictated by the chemical nature of the specific amino acids. For example, covalent bonds relevant for tertiary structure are only between two cysteine residues. Ionic bonds can form between a positively charged R group, such as lysine, and a negatively charged R group, such as aspartate. Similarly, hydrogen bonds are formed by sharing hydrogen between oxygen and nitrogen atoms.

Hydrophobic interactions are also a major component in the overall shape of a protein. Amino acids with non-polar R groups tend to be found in the middle of globular proteins to avoid interacting with water, whereas polar residues are more likely to be on the outer surface of the protein molecule (FIGURE 3.7). The combinations of these various interactions hold the polypeptide chain in a distinct conformation of folds, loops, and secondary structures that determines a protein's overall three-dimensional shape. The tendency for polar amino acids to be exposed on the outer surface of a protein means that many proteins are soluble in an aqueous environment. Exceptions to this tendency are integral membrane proteins that are embedded within the lipid bilayer of the membrane (Chapter 2). In this case, hydrophobic amino acids on the surface of the protein interact with the hydrophobic tails of the membrane lipids.

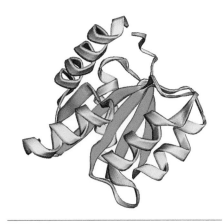

FIGURE 3.5 Interactions between secondary structures within a protein. A ribbon diagram showing the interactions of α-helices (yellow), β-strands (green), and flexible loops (gray) that make up the tertiary structure of a protein. This example has an overall globular shape.

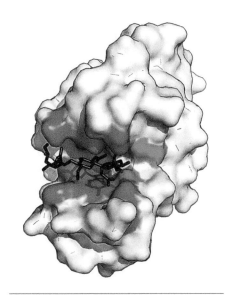

FIGURE 3.6 Three-dimensional shapes of proteins. A space-filling model of an enzyme showing the substrate-binding site (highlighted in blue) and catalytic site (highlighted in red). The substrate (black), or any ligand in general, fits into a groove of the protein. The shape of the binding site is highly specific for the ligand that binds to a particular protein. (By Thomas Shafee, published under CC BY 4.0.)

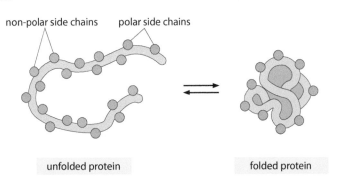

FIGURE 3.7 Hydrophobic interactions promote protein folding. Proteins tend to fold up into a conformation so that non-polar R groups (orange) are in the center of the folded protein and polar R groups (blue) are on the outer surface of the protein. The polar R groups interact with water (hydration) and allow the protein to be soluble in aqueous environments.

Many proteins are composed of multiple polypeptides held together in a complex

In some cases, a single polypeptide chain can serve as a functional protein. In many cases, though, multiple polypeptides interact to form a complex of polypeptides that is the functional protein. The individual polypeptides making up the complete protein are referred to as subunits. The subunits can be identical polypeptides or a mixture of different polypeptides. Interactions between polypeptide subunits represent the quaternary protein structure. The forces involved in the quaternary structure are the same forces that are involved in tertiary protein structure: hydrophobic interactions, ionic interactions, hydrogen bonding, and covalent disulfide bonds. Proteins that interact with each other have complementary surfaces to promote binding and the various interactions between R groups hold the polypeptides firmly together.

Except for covalent disulfide bonds, the forces responsible for secondary, tertiary, and quaternary structures are relatively weak (see Table 1.3) and more easily disrupted than covalent bonds. Thus, many proteins are somewhat fragile and sensitive to conditions that disrupt hydrogen bonding, ionic interactions, and hydrophobic interactions. For example, heating proteins or exposing them to extreme pH conditions results in most proteins partially or completely unfolding. This unfolding, called denaturation, results in a loss of protein function. The analysis of proteins can be difficult due to their lability and tendency to denature.

Many proteins also require a non-protein co-factor to be functional

A wide variety of chemical compounds can serve as co-factors and these co-factors can either be carbon-based organic compounds or inorganic metallic ions. Many trace elements that are required for proper nutrition, like zinc, function as co-factors, and this explains their dietary requirement. Similarly, many of the organic co-factors are vitamins or are synthesized from vitamins. Co-factors are tightly associated with proteins and are considered an integral part of that protein. A co-factor performs some essential aspects of a protein's function. For example, many co-factors play a role in chemical reactions catalyzed by enzymes.

The structure of hemoglobin illustrates the concepts of quaternary structure and co-factors (**FIGURE 3.8**). Hemoglobin is the major protein found in red blood cells and hemoglobin functions to carry oxygen from the lungs to the tissues. In the tissues, oxygen is released and replaced with carbon dioxide (CO_2), which is a waste product of cellular metabolism. The carbon dioxide is then carried from the tissues to the lungs where it is expelled and replaced with oxygen. Hemoglobin is composed of four polypeptide chains: two identical α-chains and two identical β-chains. Each of the four polypeptides has a tightly associated heme group as the co-factor. Heme is composed of an organic component called the porphyrin ring with an atom of iron in the center. It is the ionized iron that binds the oxygen and carbon dioxide. Thus, iron in the form of heme plays an essential role in the function of hemoglobin.

Protein folding is assisted by chaperones

The proper function of a protein depends upon the protein folding into its correct three-dimensional shape. Some proteins can spontaneously fold into their correct conformations. This folding process is driven initially by the hydrogen bonding that forms the secondary structures and this is followed by the interactions that form tertiary structures. The protein-

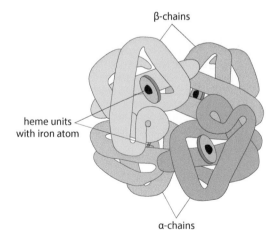

β-chains

heme units
with iron atom

α-chains

FIGURE 3.8 Structure of hemoglobin.
Hemoglobin is composed of four polypeptide subunits called the α- and β-chains. Each of the polypeptides has a bound co-factor consisting of a heme group with an iron atom.

folding process, however, can be relatively slow and thus the hydrophobic regions of the polypeptide are exposed to the aqueous environment of the cell during the folding process. Due to the high protein content within cells, the partially folded proteins can interact with each other via non-specific hydrophobic interactions. This can lead to the formation of non-functional protein aggregates that are detrimental to the cell (BOX 3.2).

Protein Misfolding and Neurodegenerative Disease

Alzheimer's disease, Huntington's disease (see Box 11.3), Parkinson's disease (see Box 2.3), and diseases related to prions (Chapter 19) are examples of neurodegenerative diseases in which the pathology is due to misfolded proteins that aggregate in the cytoplasm of cells. In healthy cells misfolded proteins are either degraded or refolded by molecular chaperones. In the disease state, genetic defects in proteins combined with cellular stressors lead to the misfolded proteins

aggregating into toxic clumps (**BOX 3.2 FIGURE 1**). This overwhelms the cell's normal ability to remove misfolded protein aggregates and, over time, these toxic clumps accumulate and grow larger. Furthermore, once formed the aggregates are highly resistant to degradation and they propagate by recruiting normally folded proteins to unfold. The accumulation of the protein aggregates leads to cellular dysfunction and ultimately death of the neuron.

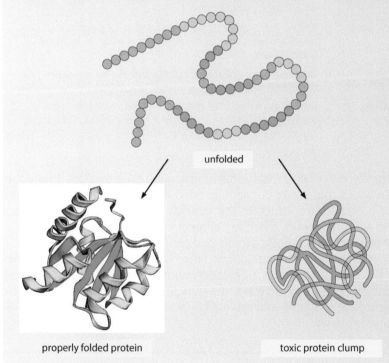

unfolded

properly folded protein

toxic protein clump

BOX 3.2 FIGURE 1 Protein misfolding produces cytotoxic clumps. Properly folded proteins exhibit the correct secondary, tertiary, and quaternary structures, whereas misfolded proteins aggregate into intracellular clumps that are toxic to cells.

BOX 3.2

To prevent protein aggregation, folding is assisted by a class of proteins called molecular **chaperones**. Chaperones bind the partially folded protein during the folding process and prevent non-specific protein aggregation. They also function in response to the accumulation of unfolded denatured proteins caused by excesses in temperature, pH, or oxidative stress. In fact, chaperones were first discovered based on this stress-response activity and originally named heat-shock proteins, a name that is still widely used to describe chaperones. An increase in misfolded proteins within the cell increases the expression of chaperones and these chaperones prevent the aggregation of misfolded proteins and promote the destruction and recycling of the misfolded proteins. The endoplasmic reticulum (ER) stress response (see Box 2.3) is also mediated by chaperones.

Post-translational modifications affect protein structure and function

Many proteins are also modified following their synthesis via the covalent attachment of chemical groups. These post-translational modifications are generally carried out by enzymes that add the chemical groups to R groups of specific amino acids or to either the N or C terminus. Many different modifications have been described and these modifications play many distinct roles in terms of protein structure and function (TABLE 3.2). Such modifications can affect the overall structure of a protein. For example, amino acids with charged or polar R groups can be made non-polar, or vice versa. Post-translational modifications can affect the folding of the protein or how the protein interacts with other proteins or cellular structures. For example, the addition of lipid molecules to a protein can promote an association with membranes. In many cases the post-translational modification regulates the function of the protein.

Table 3.2 Some Common Post-translational Modifications of Proteins	
Modification	**Description**
Phosphorylation	Reversible addition of a phosphate group by a protein kinase that regulates the activity of the target protein
Glycosylation	Addition of a carbohydrate group to a protein to stabilize protein structure. Common in plasma membrane proteins expressed on the surface of eukaryotic cells
Lipidation	Addition of a lipid group to a protein that promotes the association of the protein with membranes
Proteolysis	Removal of a portion of the protein by breaking a peptide bond with an enzyme called a protease. Sometimes converts a precursor form of the protein into a functional protein (e.g., insulin)

Another type of post-translational modification is proteolysis. **Proteases** are enzymes that break the peptide bond between amino acid residues. Some proteases are used for the general degradation of proteins into amino acids. However, there are some proteases that break the peptide bond between two specific amino acid residues within a specific protein, resulting in smaller polypeptides. In some cases, such as insulin (see Box 3.1), this is part of a post-translation maturation of a protein. Specific proteolysis can also activate the function of some proteins. Many physiological functions are regulated by proteases that activate hormones or other proteins.

FUNCTIONAL ROLES OF PROTEINS

Proteins play a central role in the structural aspects of the cell as well as in carrying out cellular functions (TABLE 3.3). Some proteins are found in all types of cells and carry out the basic functions necessary for cell survival. Some cells, however, have specialized functions. For example, the functions of muscle cells and nerve cells are quite distinct, and this distinction is reflected by the types of proteins found in those cells. Both types of cells have the same set of proteins that carry out the basic functions or the so-called housekeeping activities. However, muscle cells have proteins involved in contraction that are not found in nerve cells. Similarly, nerve cells have proteins that are important for the specialized functions of nerve cells which are not found in other types of cells. The protein composition of a cell is determined by differential gene expression (Chapter 4).

Table 3.3	Protein Functions
Structural and mechanical	Provides mechanical support to the cell or its structures (e.g., cytoskeleton, chromosomes, extracellular matrix) and participates in cellular movements
Enzymatic and biosynthetic	Catalyzes the various chemical reactions associated with cellular metabolism and the synthesis of macromolecules such as DNA, RNA, and proteins
Carriers, transporters, and channels	Bind ligands for distribution throughout body, especially regarding moving substances across membranes
Environmental perception and cell communication	Receptors, signal transduction proteins, and some hormones are made of proteins that facilitate communication between cells and monitor the cellular environment
Regulatory and accessory	Control of cellular functions by binding to other proteins or macromolecules to regulate enzyme activity, gene expression, protein folding (i.e., chaperones), or assembly of supramolecular complexes
Storage proteins	Proteins functioning primarily as amino acid reserves such as casein in mammalian milk, ovalbumin in egg whites, and various proteins found in plant seeds

Some proteins play structural and mechanical roles within cells

Proteins make up the various structures that can be seen within cells or provide support for cells or subcellular structures. The cytoskeleton is an example of cellular structures made up of proteins. The various filamentous systems making up the cytoskeleton are composed of proteins that provide mechanical support for the cell and are a major determinant of cell shape. In addition, cytoskeletal proteins are involved in force generation and cell movement. The extracellular matrix (ECM), which provides external support for cells and is composed of fibrous proteins secreted by cells, is another example of proteins that primarily play a structural or mechanical role. The ECM also holds cells together and gives support to tissues. This support is further strengthened by interactions between the ECM and the cytoskeleton. Other examples of proteins with structural roles are proteins that interact with DNA to form chromosomes and proteins that interact with RNA to form ribosomes (Chapter 4).

The cytoskeleton and motor proteins are described in Chapter 2, pages 37–41.

SIGNPOST

Enzymes are proteins that catalyze chemical reactions

During metabolism nutrients are taken up by cells and converted to the various molecules that are needed by cells to carry out cellular functions. The various chemical reactions associated with metabolism are facilitated by proteins called **enzymes**. Enzymes are not permanently altered by chemical reactions and, thus, are catalysts. Catalysts are substances that speed up chemical reactions without being consumed by the chemical reaction. As an enzyme is not consumed by the reaction, as soon as one reaction is completed another reaction can occur.

Enzymes bind to specific molecule(s) and hold these substrates in a conformation that promotes a chemical reaction and results in the formation of a new molecule called the product. For example, an enzyme may bind two molecules and hold them adjacent to each other so that a chemical reaction is more likely to occur between the substrates (e.g., condensation reaction). Such a chemical reaction results in the formation of a covalent bond between the two molecules and the formation of a single molecule (FIGURE 3.9). Conversely, some enzymes may break chemical bonds and generate two products from a single substrate (e.g., hydrolysis reaction). Many enzymes catalyze reactions that are reversible in that the direction of the reaction is determined by the relative abundance of the substrates and products. In the example of two substrates being combined into a single product, the reverse reaction would be a substrate converted into two products.

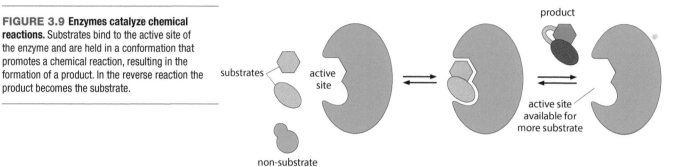

FIGURE 3.9 Enzymes catalyze chemical reactions. Substrates bind to the active site of the enzyme and are held in a conformation that promotes a chemical reaction, resulting in the formation of a product. In the reverse reaction the product becomes the substrate.

The site on the protein that binds the substrate is called the active site. The shape of the active site matches the shape of the substrate, resulting in a highly specific interaction between the substrate and the enzyme. The R groups of amino acids that form the active site interact with the substrate and hold it in the correct position and orientation for the chemical reaction to occur (see Figure 3.6). In some cases, the R groups in the active sites may also participate in reactions. Alternatively, co-factors may participate in the chemical reaction. For example, in redox reactions (Chapter 1) a co-factor may temporarily accept an electron that will be donated to the substrate.

Some proteins carry substances from one location to another

Most proteins that function as carriers are involved in moving substances across membranes. This includes channels, transport proteins, and pumps, as well as receptor-mediated endocytosis. Substances taken up by receptor-mediated endocytosis involve the carrier protein and the receptor that binds to the carrier protein. For example, transferrin is a carrier protein that binds iron and the transferrin receptor binds to transferrin and facilitates its entry into cells. Similarly, cells have receptors for the apolipoproteins that are a constituent of lipoproteins (see Box 14.1). Transferrin and lipoproteins are found in the blood and function to distribute their ligands, iron and lipids, respectively, throughout the body, as well as participating in the uptake of their ligands by cells. Albumins are another type of blood protein that function as carriers.

Receptor and many other proteins function in cellular signaling

Organisms sense and respond to their environment. Sensing the environment is largely mediated by proteins called **receptors**. Like enzymes and carrier proteins, receptors have an active site that binds to a specific molecule called the **ligand** (FIGURE 3.10). Receptor–ligand binding is instrumental in cell–cell communication and hormones are examples of ligands involved in cellular communication and signaling. Some hormones, like insulin, are proteins (see Box 3.1). Many hormones, though, are small molecules that are synthesized by enzymes. Furthermore, there are other proteins that play accessory roles in cellular signaling.

SIGNPOST

See the section on 'Transport of Substances across Membranes' in Chapter 2 for details, pages 31–34.

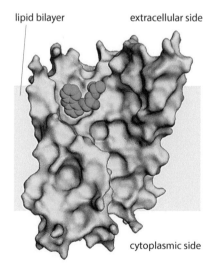

lipid bilayer extracellular side

cytoplasmic side

FIGURE 3.10 Receptor–ligand binding. A three-dimensional model of an epinephrine receptor with a bound agonist is shown. The epinephrine binding site is highlighted in yellow and matches the shape of the agonist depicted as a space-filling model. Many receptors are located on the plasma side with the ligand-binding site facing toward the extracellular side. Binding of the ligand sends a signal to the cytoplasmic side of the receptor.

Other roles of proteins include accessory functions and for storage of amino acids

Some proteins participate in cellular processes by binding to other proteins and assisting in their functions. For example, chaperones assist in the proper folding of proteins. Similarly, proteins can assist in the assembly of supramolecular structures or the directing of proteins to their correct locations within cells. A few proteins are essentially reservoirs for amino acids. Casein is the primary protein in milk and is broken down into amino acids by proteases in the developing infant. Similarly, ovalbumin is the major protein found in egg white and it too functions as a reservoir of amino acids for developing birds. Up to 25% of the dry weight of some plant seeds can be storage proteins that serve as the source of amino acids for the developing plant embryos. An example of a plant storage protein is the glutens of wheat that can trigger celiac disease (Chapter 17).

CELLULAR METABOLISM

Cells take up nutrients and convert those nutrients into energy and molecules that are needed by the cell to carry out cellular functions. This metabolism consists of catabolic processes in which molecules are broken down into smaller molecules and energy is released and captured. These smaller molecules can then be used in biosynthetic processes to synthesize larger molecules, such as lipids, amino acids, and nucleotides. Enzymes catalyze chemical reactions of metabolism by converting one molecule to another and function in metabolic pathways in which the product of one enzyme serves as the substrate of another enzyme (FIGURE 3.11). These various metabolic pathways carry out the catabolic and biosynthetic processes necessary to maintain the cell.

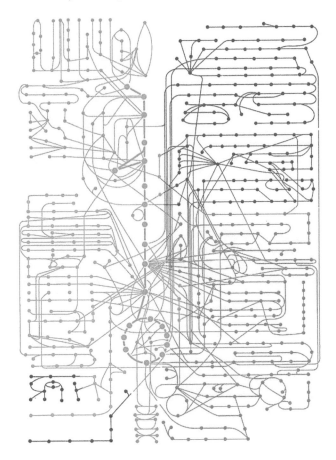

FIGURE 3.11 Metabolic pathways. Depiction of cellular metabolism in which each dot represents a molecule and the lines between the dots represent the chemical reaction catalyzed by an enzyme. Carbohydrate metabolism is represented in green, lipid metabolism in blue, amino acid metabolism in red, and nucleic acid metabolism in purple. Various protein co-factors are shown in brown.

The various types of cells within a multicellular organism exhibit different metabolism depending on the function of that specific cell. Each specific cell type has its own unique metabolism and phenotype that reflect the overall function of that specific cell. This phenotype is determined by the specific genes that are expressed (Chapter 4). However, since all cells require energy, all cells have the same basic metabolism for the generation of energy and other basic metabolic housekeeping functions. These housekeeping enzymes are expressed in all cells, albeit at different levels depending on the energy requirements of any specific cell.

ATP is the major energy carrier of the cell

Some organisms, such as plants, can capture the energy from light and convert it into chemical energy in the form of carbohydrates by a process called photosynthesis. Non-photosynthetic organisms obtain energy through the ingestion of food and capture that energy via metabolism. In either case, the energy captured needs to be transferred to an energy carrier. The predominant energy carrier of all organisms is the nucleotide adenosine triphosphate, or ATP (FIGURE 3.12). Cellular processes that require energy generally involve proteins that bind ATP and remove the third, or gamma (γ), phosphate from the ATP resulting in the production of adenosine diphosphate (ADP) and phosphate (P_i).

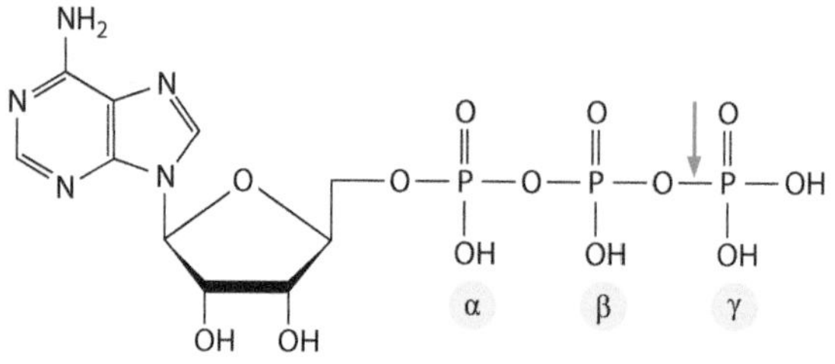

FIGURE 3.12 Structure of adenosine triphosphate (ATP). The three phosphate groups of ATP are designated alpha (α), beta (β), and gamma (γ). The red arrow shows the phosphate bond that is normally hydrolyzed to release energy.

The energy released from the hydrolysis of the γ-phosphate bond is captured or utilized in some fashion. For example, enzymes that carry out energy-requiring biosynthetic processes often transfer the γ-phosphate and its associated energy to another substrate. Such enzymes that transfer the γ-phosphate to another substrate are called kinases. In the case of motor proteins, which interact with the cytoskeletal proteins to produce movement (Chapter 2), the energy released from the hydrolysis of ATP results in a conformational change of the protein. This conformational change produces movement. The hydrolysis of ATP into ADP and phosphate is called ATPase activity. Similarly, many membrane pumps (see Figure 2.6) also have an ATPase activity that is associated with a conformational change, and this conformational change is part of the mechanism that moves solutes across biological membranes.

ATP is often compared with currency in that cells 'spend' ATP to grow and maintain homeostasis. This currency is renewed by metabolic processes in which ATP is reformed from ADP via a phosphorylation reaction in which the γ-phosphate is added to the β-phosphate (FIGURE 3.13). The synthesis of ATP requires energy and various metabolic pathways capture energy through catabolism of other molecules to produce ATP. The ATP is then utilized in energy-requiring activities of the cell resulting in an ADP/ATP cycle.

Metabolism of glucose is the primary source of ATP

The primary function of carbohydrates is to provide energy for the cell. Energy released from the metabolism of carbohydrates is used to synthesize ATP. Complex carbohydrates are converted to glucose,

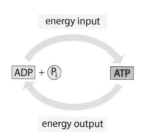

FIGURE 3.13 ATP as the cellular energy carrier. Energy captured from catabolic processes is used in the synthesis of ATP from ADP and phosphate (P_i). Energy-requiring processes release the energy from ATP by hydrolysis of the γ-phosphate group, producing ADP and phosphate.

which is then metabolized to carbon dioxide (CO_2) and water (H_2O) by a series of enzymatic reactions encompassing three major stages: glycolysis, tricarboxylic acid (TCA) cycle, and oxidative phosphorylation (FIGURE 3.14). In eukaryotic cells, glycolysis occurs within the cytoplasm, whereas the TCA cycle and oxidative phosphorylation take place in the mitochondria. For this reason, mitochondria are viewed as the energy factories of the cell. Dysfunction of the mitochondria can cause a wide range of disease symptoms, but the most prominent among them are poor growth and development and muscle weakness.

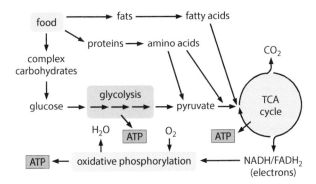

FIGURE 3.14 Synthesis of ATP. Food is metabolized to produce energy through glycolysis, the tricarboxylic acid (TCA) cycle, and oxidative phosphorylation. ATP is generated at various steps in this process with two molecules of ATP generated during glycolysis, two molecules or more during the TCA cycle, and more than 30 molecules during oxidative phosphorylation.

During **glycolysis**, the glucose molecule, which contains six carbon atoms, is converted into two molecules of pyruvate, which contains three carbon atoms. Nine distinct enzymatic steps are involved in the conversion of glucose to pyruvate and some energy is captured resulting in the net synthesis of two ATP molecules per glucose molecule. Pyruvate then enters the next stage, or the **TCA cycle**, previously called the Krebs' cycle in honor of its discoverer. The TCA cycle consists of another nine reactions in which the pyruvate is broken down into carbon dioxide through a series of reactions with intermediates. It is referred to as a cycle since the substrate of the first reaction is the product of the last reaction of the cycle. In other words, there is no net gain or loss of these metabolic intermediates. The catabolism of the pyruvate molecule during the TCA cycle results in the net production of one ATP molecule. As two pyruvates are formed from glucose, this yields another two ATP molecules per glucose molecule.

Glycolysis and the TCA cycle also result in the capture of electrons by co-factors known as nicotinamide adenine dinucleotide (NADH) and reduced flavin adenine dinucleotide ($FADH_2$). These co-factors function as electron carriers that deliver the electrons to another series of chemical reactions called **oxidative phosphorylation**. In the last step of oxidative phosphorylation, the captured electrons are donated to molecular oxygen to form water. As a by-product of this series of electron transfers, hydrogen ions (H^+) accumulate on one side of the inner mitochondrial membrane resulting in a hydrogen ion gradient. The osmotic force of this gradient causes the hydrogen ions to flow across the membrane through the ATP-synthesizing enzyme and provides the energy to produce ATP. This is analogous to producing electricity by hydroelectric dams. The hydrogen ions produced from a single glucose molecule are sufficient to synthesize 32–34 ATP molecules. Thus, in total, a single glucose molecule can yield 36–38 ATP molecules.

Metabolic pathways are interconnected and regulated

Lipids and amino acids can also be utilized to produce energy by entering the TCA cycle (FIGURE 3.14). And metabolites from glucose metabolism can be utilized for the biosynthesis of nucleotides, amino acids, and lipids (see Figure 3.11). This complexity of metabolism requires some regulation, and the regulation of metabolic processes is on three levels. One level is at the level of gene expression which reflects the specific metabolism and phenotype of any specific cell type (TABLE 3.4). Different types of cells express different proteins depending on the function of that cell. Or cells can express specific genes in response to external stimuli. The second level of regulation is through the modification of key enzymes in metabolic pathways in which hormones can turn enzymatic activity on or off.

Table 3.4	Regulation of Metabolic Activity
Cellular phenotype	The enzymes expressed by a specific cell type determine the metabolism of that cell
Hormone action	Key enzymes in metabolic pathways can be 'turned on' or 'turned off' depending on metabolic needs
Supply and demand	The availability of metabolites determines the direction of reversible enzymes and metabolic pathways

The third level of control is not necessarily directed at specific enzymes in the various metabolic pathways but reflects the immediate needs of the cell or organism. Since many of the enzymes involved in metabolic pathways are reversible, the direction of the reaction is determined by the concentrations of the various metabolites in the pathways. For example, if there is an excess of amino acids, the amino acids can be shunted into the energy production pathways. Similarly, an actively growing cell has a high demand for nucleotides, amino acids, and lipids. Therefore, if the cell has sufficient energy, metabolites of glucose metabolism can be utilized for biosynthetic purposes. Some metabolic pathways can be controlled by feedback inhibition in which the end-product of the pathway binds to and inhibits a key enzyme in the pathway. In that way the cell does not synthesize excess metabolites that are not needed.

HORMONES, RECEPTORS, AND CELLULAR COMMUNICATION

The individual cells of multicellular organisms must communicate with one another and function in a coordinated manner. At the molecular and cellular levels, this cell–cell communication involves receptors and ligands. The interaction between ligands and receptors is generally highly specific in that a receptor typically only recognizes a single ligand (see Figure 3.10). Binding of the ligand causes some change in the receptor that then has some effect on cell physiology or behavior. Ligands involved in cell–cell communication are generally called **hormones**. Other signaling ligands involved in cell–cell communication are neurotransmitters of the nervous system (Chapter 18) and cytokines and chemokines of the immune system (Chapter 9).

Cell–cell communication is mediated by hormones

Like any form of communication, there is a sender and a receiver. In the case of cell–cell communication the senders synthesize and secrete hormones, or other signaling molecules, and receivers have receptors for those molecules. Cells without receptors do not hear the message. Secreted hormones can affect neighboring cells and this is referred to as paracrine signaling (FIGURE 3.15). If the cell secreting the hormone also has receptors for that hormone, autocrine signaling is possible. Cells from distal parts of the body can also communicate by releasing hormones into the blood. Cells that secrete hormones or other substances into the blood are called endocrine cells. Exocrine cells secrete substances in epithelial tissues or ducts composed of epithelial cells and are considered to be outside of the body.

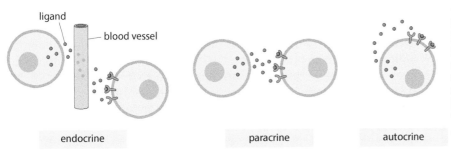

endocrine   paracrine   autocrine

FIGURE 3.15 Cell–cell communication. Hormones (ligands) are synthesized and secreted by one cell and affect another cell(s). Hormones diffusing to nearby cells are paracrine interactions and hormones being carried by the blood to distal cells are endocrine interactions. An autocrine interaction is the case in which the cell that secretes the hormone also has receptors for the hormone.

Various molecules derived either from amino acids or lipids function as hormones (TABLE 3.5). Peptide hormones, hormones derived from amino acids, and hormones derived from oxidized lipids (e.g., eicosanoids) are generally water soluble and thus are readily transported throughout the body by the circulatory system or can diffuse in the tissues in the space between cells. Such polar hormones are unable to cross the plasma membrane of the cell and bind to receptors found on the surface of the target cell. In contrast, steroid hormones can cross membranes. The receptors for steroid hormones and other non-polar hormones are proteins found within the nucleus of the cell. These nuclear receptors regulate the expression of genes.

> The mechanism of action of nuclear receptors is discussed in detail in Chapter 4, pages 79–81.
>
> SIGNPOST

Table 3.5	Chemical Classes of Hormones	
Type	**Chemical composition**	**Examples**
Peptides/proteins	Linear chains of amino acids	Insulin
Amino acid	Derivatives of tryptophan or tyrosine	Epinephrine
Eicosanoids	Oxidized fatty acids	Prostaglandin
Steroids	Derivatives of cholesterol	Testosterone

Hormones binding to cell surface receptors induce a signal transduction event

Binding of a ligand to its receptor results in some change in the receptor, and this change then relays a signal to the cytoplasmic side of the cell membrane. In other words, ligand binding can be viewed as an activation of the receptor and the activated receptor 'transduces' the signal across the membrane. There are three primary mechanisms by which the signal is transduced across the membrane (TABLE 3.6).

Table 3.6	Classes of Cell Surface Receptors
Class	**Mode of action**
Ion channel linked	Ligand binding opens an ion channel resulting in an electrical signal
Enzyme linked	Ligand binding activates an enzyme activity on the cytoplasmic side of the plasma membrane
G-protein linked	Ligand binding activates a G-protein and second messenger production on the cytoplasmic side of the plasma membrane

One mechanism of signal transduction involves receptors that are linked to ion channels. Binding of the ligand – often a neurotransmitter – promotes a shape change in the receptor that opens the ion channel and allows ions to pass through the channel (see Figure 2.4). Opening this gated channel creates an electrical potential across the membrane that elicits an electrical signal. Hence communication between nerve cells is often akin to electricity passing through wires.

Some receptors, such as the insulin receptor, are associated with enzyme activity. These enzyme-linked receptors have a large extracellular domain with ligand-binding activity and a large domain on the cytoplasmic face of the membrane with an enzyme activity (FIGURE 3.16). The two domains are usually connected by a single α-helix that spans the lipid bilayer. Binding of the ligand generally induces a conformational change in the receptor that then activates the enzymatic activity. Many enzyme-linked receptors have an associated protein kinase or protein phosphatase activity that directly activates or inactivates other proteins. Another example of an enzyme linked to a receptor is guanylate cyclase which produces the second messenger cyclic GMP.

FIGURE 3.16 Enzyme-linked receptors.
Binding of a ligand to an enzyme-linked receptor activates the enzymatic domain of the protein on the cytoplasmic side of the membrane. Activation of most enzyme-linked receptors involves two inactive receptor monomers forming a dimer that is active.

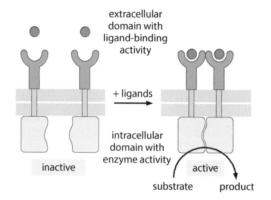

G-proteins are a major mediator of signal transduction

G-protein-linked receptors are the largest class of cell surface receptors and are associated with a wide variety of cellular functions. The cytoplasmic domains of these receptors bind to G-proteins. They are called G-proteins because they bind to the nucleotides GDP and GTP. Hormone binding to the receptor causes a change in the receptor and this change activates the G-protein. Activation of the G-protein involves replacing the bound GDP with GTP. The activated G-protein with the bound GTP then associates with another protein and either activates or inhibits that protein (FIGURE 3.17). The interaction of the G-protein with another protein converts the GTP to GDP and resets the G-protein. G-proteins interact either with enzymes that produce second messengers or with gated ion channels.

FIGURE 3.17 G-protein linked receptor.
Hormone binding to the receptor activates the G-protein by inducing an exchange of bound GDP for GTP. The activated G-protein binds to the target enzyme and activates its enzyme activity. However, some G-proteins inactivate the target protein, and these are called inhibitory G-proteins. G-proteins activate the production of second messengers and during this process the bound GTP is converted to GDP. Second messengers activate protein kinases that mediate the cellular response by regulating other proteins.

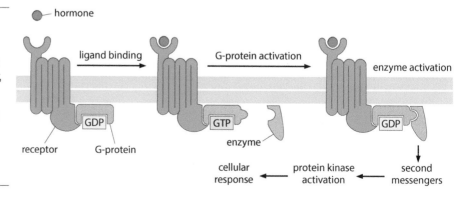

Second messengers are small molecules produced in response to hormone activation of receptors

Hormones can be viewed as primary messengers between cells. The small molecules produced after binding of hormones to their receptors are called **second messengers**. The second messengers then interact with other cellular proteins to alter the physiology or function of that cell in response to the hormone. Examples of second messengers include cyclic adenosine monophosphate (cAMP), inositol triphosphate (IP$_3$) plus diacylglycerol (DAG), calcium ions (Ca^{2+}), and cGMP (TABLE 3.7). Second messengers can be produced from all three classes of cell surface receptors. For example, Ca^{2+} is a second messenger produced by an ion-channel-linked receptor, cGMP is a second messenger produced by an enzyme-linked receptor, and cAMP is a common second messenger activated by G-proteins. Intracellular concentrations of second messengers rapidly rise in response to hormones binding to receptors.

Table 3.7 Second Messengers and Protein Kinases		
Second messenger	**Production/Enzyme (reaction)**	**Protein kinase**
cAMP	G-protein-linked adenylate cyclase (ATP $\rightarrow$ cAMP)	Protein kinase A
Inositol triphosphate (IP$_3$) + diacylglycerol (DAG)	G-protein-linked phospholipase C (phosphatidylinositol $\rightarrow$ IP$_3$ + DAG)	Protein kinase C
Calcium ion (Ca^{2+})	Receptor-linked Ca^{2+} channel on the plasma membrane or release from ER by IP$_3$-gated Ca^{2+} channel	Calmodulin-dependent protein kinase
cGMP	Receptor-linked guanylate (or guanylyl) cyclase (GTP $\rightarrow$ cGMP)	Protein kinase G

Second messengers affect cellular function by activating protein kinases

The various second messengers bind to specific **protein kinases** (see Table 3.7). Protein kinases are enzymes that transfer a phosphate group from ATP to a target protein (FIGURE 3.18). The phosphorylation of the target protein (i.e., substrate) results in a conformational change in the protein that alters its function. For example, a conformational change associated with phosphorylation could make a ligand-binding site more assessable. In the case of an enzyme, exposure of the catalytic site could activate, or turn on, the enzyme. Conversely, phosphorylation could block a binding site or turn the protein off. In other cases, phosphorylation may affect the assembly or disassembly of protein complexes. In other words, protein kinases regulate the activities of other proteins and this regulation is in response to hormones acting via second messengers (see Figure 3.17). Another enzyme called phosphoprotein phosphatase removes the phosphate group and returns the target protein back to its original state.

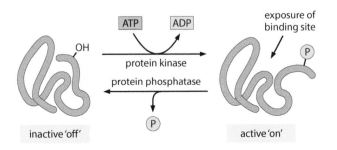

FIGURE 3.18 **Mechanism of protein kinase action.** Protein kinases transfer phosphate (P) groups from ATP to a protein substrate. This results in a conformational change in the protein that can turn its function on or off. Protein phosphatases remove the phosphate group to return the protein to its previous state.

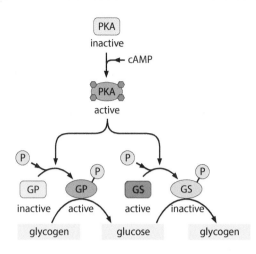

FIGURE 3.19 Regulation of glucose production by protein phosphorylation. Hormones such as epinephrine and glucagon result in the production of the second messenger cAMP, which activates cAMP-dependent protein kinase (PKA). PKA phosphorylates glycogen phosphorylase (GP) and glycogen synthase (GS). The simultaneous activation of glycogen phosphorylase and inactivation of glycogen synthase result in the net production of glucose.

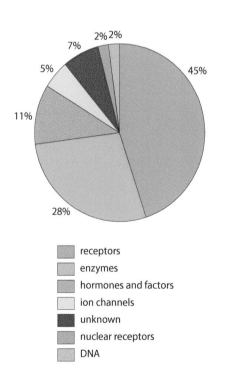

receptors
enzymes
hormones and factors
ion channels
unknown
nuclear receptors
DNA

FIGURE 3.20 Common drug targets. Pie chart showing the distributions of types of proteins targeted by drugs. (Courtesy of Creative Biomart.)

A protein kinase may also phosphorylate several different proteins in response to hormone activation. This allows for coordinated actions within a cell. For example, two enzymes involved in opposite aspects of glucose metabolism are regulated differently by phosphorylation (FIGURE 3.19). Glycogen synthase converts glucose into glycogen and is inactivated by phosphorylation. Conversely, glycogen phosphorylase converts glycogen into glucose and is activated by phosphorylation. The net result is the production of glucose due to the simultaneous phosphorylation of both proteins. This means that a hormone such as epinephrine, which stimulates the production of glucose during the fight-or-flight response, efficiently and rapidly leads to the production of glucose.

MECHANISM OF DRUG ACTION

Drugs are chemical substances that can be used to treat and manage disease. In general, drugs stimulate or suppress a physiological function to bring about a more normal state. For example, high cholesterol levels are associated with cardiovascular disease and statins are a class of drugs that lower cholesterol levels. Similarly, there are a wide variety of drugs that can lower blood pressure in people with hypertension. In the case of infectious diseases, drugs serve as 'magic bullets' by specifically targeting pathogens such as bacteria and viruses. Drugs carry out these various actions by interacting with specific molecules, called targets, that affect cellular functions.

Most drugs bind to specific protein targets

Although there are a few drugs that target DNA or other molecules, the majority of drug targets are proteins, such as enzymes, receptors, and ion channels (FIGURE 3.20). In some respects, drugs can be considered ligands that bind to specific proteins. For example, some drugs that target enzymes resemble the natural substrates of the targeted enzyme. Similarly, drugs that target receptors may structurally resemble the natural ligands of those receptors. This means that drugs bind to specific protein targets and affect the function of those targets.

Drugs that bind to the active or ligand-binding site of the protein and prevent the binding of the natural substrate or ligand are called competitive inhibitors (FIGURE 3.21). In many cases, the drug has a higher affinity for the target protein than the natural substrate or ligand. Therefore, the normal ligand or substrate is prevented from binding to the target, and this inhibits the normal activity of that protein. There are also non-competitive inhibitors that bind to sites on the target proteins distinct from the active sites. Quite often these sites are involved in the regulation of the protein function and are called allosteric sites. Binding of an inhibitor to an allosteric site results in a change in protein conformation that affects the binding domain or active site of the protein.

Drugs can function as either antagonists or agonists

Drugs that inhibit the normal activity of a protein or block the normal physiological activity associated with that drug target are called **antagonists**. For example, drugs that function as hormone antagonists bind to the receptor, but do not activate the signal transduction pathway. Such drugs are often called 'blockers.' For example, a class of drugs called beta-blockers is used to manage abnormal heart rhythms by competing with epinephrine and related hormones for their receptors.

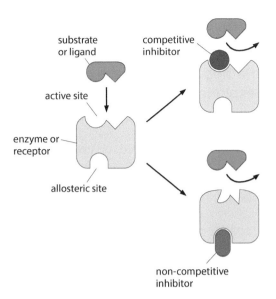

FIGURE 3.21 Competitive and non-competitive inhibitors. Some drugs function by inhibiting the normal activity of a protein by interfering with the binding of the normal substrate or ligand. Competitive inhibitors compete with the normal substrate or ligand and block it from binding to the active site. Non-competitive inhibitors bind to allosteric sites on the protein and cause a conformational change, which prevents binding of the normal substrate or ligand.

Other drugs function as **agonists** and stimulate the normal physiological properties associated with the target protein (FIGURE 3.22). For example, some drugs mimic hormones by binding to the receptor and promoting signal transduction and second messengers in the same manner as the natural ligand. In other words, the drug is functioning as a hormone and can be used to regulate cellular metabolism and physiology. Progesterone agonists are used in birth control pills to prevent ovulation by mimicking the normal role progesterone plays in the maintenance of pregnancy. In summary, drugs can be used to stimulate or inhibit physiological or metabolic processes through their interactions with target proteins.

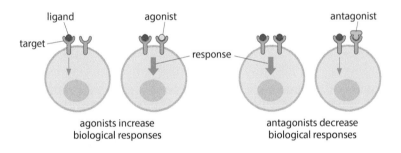

FIGURE 3.22 Agonists and antagonists. Agonists mimic the natural ligand and increase biological responses. Antagonists inhibit the natural ligand and decrease the biological responses.

Drugs exhibit selective toxicity

All drugs can be considered toxic and can cause disease or mortality at high doses. The therapeutic use of drugs strives to exploit differences between the disease state and the normal state. This means that ideally drugs are selectively toxic for the diseased state and have minimal effect on the normal state. Selective toxicity is often expressed as a **therapeutic index**, which is the ratio of the amount of drug that has a therapeutic effect and the amount of drug that is toxic. Drugs with high therapeutic indices are relatively safe in that the dose needed to have a therapeutic effect is well below the doses which cause adverse reactions. Conversely, the therapeutic and toxic doses are relatively close in drugs with low therapeutic indices.

Another aspect of selective toxicity is that many drugs function by shifting the diseased state to a more normal state. For example, hypercholesteremia and high blood pressure are major risk factors for developing cardiovascular disease (Chapter 14). Statins are drugs that lower cholesterol levels and there are many drugs that can lower blood pressure. Similarly, the high blood sugar levels associated with diabetes (Chapter 13) can be managed with drugs that lower blood glucose levels.

Pathogens often have unique drug targets that can be exploited for their elimination

The concept and mechanisms of selective toxicity are probably best exemplified with pathogens that cause infectious diseases (Chapter 19). In this case, selective toxicity means that the drug is more toxic for the pathogen than for the host. In other words, drugs can be given at a dose that kills the pathogen but has minimal toxicity to humans. Several possible mechanisms are associated with selective toxicity directed against pathogens (TABLE 3.8). Many pathogens have unique biosynthetic or metabolic pathways that are not found in human cells. For example, many antibiotics target the synthesis of the cell wall found in bacteria (Chapter 19). Human cells do not synthesize a peptidoglycan cell wall and drugs that specifically inhibit the synthesis of the bacterial cell wall are highly toxic for bacteria, while exhibiting minimal toxicity to the human host. Penicillin, one of the first antibiotics identified, is an example of a drug that interferes with bacterial cell wall synthesis.

Table 3.8 Basis of Selective Toxicity	
Unique target	The pathogen or disease state has a unique target not found in the host or normal state
Target discrimination	The drug has a higher affinity for the target in the pathogen than for the target in the host
Target importance	The target is more important for the pathogen or disease state than the target in the host or normal state
Drug accumulation	The drug accumulates in the pathogen to higher concentrations than in host cells
Drug activation	The drug is metabolized to an active form in the pathogen, but not in the host

Many drug targets are parts of essential metabolic pathways found in all organisms. However, the drug target may exhibit some species-dependent differences regarding drug binding. For example, the drug may have a higher affinity for the drug target in the pathogen compared with the homologous protein in the human host. Homologous proteins from different species exhibit differences in their sequence and therefore have slightly different structures, which may influence the binding of drugs. Therefore, some drugs may bind preferentially to the protein target from the pathogen and, thus, are toxic for the pathogen at lower concentrations than for the host.

The target may be more important for the pathogen than the host and therefore the pathogen is more sensitive to the drug. For example, a rapid rate of cell replication is important for the survival and virulence of most pathogens. However, most human cells are not replicating. Therefore, pathogens tend to be more sensitive to drugs that inhibit processes involved in DNA synthesis and cell replication. Similarly, many anti-cancer therapies also target cells that are replicating rapidly (Chapter 16).

Selective toxicity can also be due to greater accumulation of the drug by the pathogen or activation of the drug by the pathogen. Cells exhibit selective permeability and drugs need to be taken up by cells in sufficient amounts to affect the target. Thus, selective toxicity is due to the pathogen accumulating toxic levels of the drug, whereas host cells do not significantly accumulate the drug. In other cases, the drug per se is not active and needs to be converted to the active form of the drug that is toxic for the pathogen. Pathogens activate some drugs due to their unique metabolic pathways compared with humans. Drugs that need to be activated are called prodrugs.

SUMMARY OF KEY CONCEPTS

- Proteins are polymers of amino acids that fold into defined shapes which are largely determined by their amino acid sequences

- Proteins are responsible for the structural elements of a cell and carry out cellular functions

- Cellular functions are carried out by proteins through their activities as enzymes, receptors, transporters, signaling molecules, and accessory proteins

- Enzymes act together in metabolic and biosynthetic pathways that convert nutrients into energy, metabolites, and macromolecules that are needed by the cell

- Receptors play a key role in environmental sensing and cell–cell communication by binding to ligands, such as hormones, and regulating cellular functions

- Many enzymes, hormone receptors, and transporters are affected by drugs and thus serve as drug targets

KEY TERMS

agonist	A drug that binds to a ligand-binding site and triggers an active response that mimics the natural ligand (e.g., hormone or neurotransmitter).
alpha-helix	A common secondary structure of proteins in which the polypeptide backbone is in a tight coil like a spiral staircase.
antagonist	A drug that binds to a target, such as a receptor, and blocks the normal function of the target.
beta-sheet	A common secondary structure of proteins in which two or more polypeptide strands are held together laterally by hydrogen bonds.
chaperone (protein)	Proteins that assist in the proper folding of protein substrates or protect proteins in the process of folding from aggregation.
enzyme	A protein that functions as a biological catalyst by promoting chemical reactions between molecules. Enzymes convert substrates to products.
glycolysis	Metabolic pathway involving sequential enzymes in which glucose is converted into pyruvate and energy is captured in the form of ATP.
hormone	A messenger molecule synthesized and secreted by one cell that affects other cell(s). The hormone functions as a ligand that binds to a receptor on the target cell.
ligand	Any molecule that binds to a specific site on a protein or other molecule (e.g., substrate or hormone).
oxidative phosphorylation	Metabolic pathway in bacteria and mitochondria producing ATP via the transfer of captured electrons from the TCA cycle to molecular oxygen.
protease	An enzyme that hydrolyzes the peptide bond between amino acids in proteins resulting in smaller polypeptides, peptides, or amino acids.

protein kinase	An enzyme that transfers a phosphate group from ATP to a protein substrate. Protein phosphorylation generally affects the function of the substrate protein.
receptor	A protein that binds a specific ligand such as a hormone or neurotransmitter.
second messenger	A molecule (e.g., cAMP) that is produced in response to hormone (first messenger) binding to a cell surface receptor. The second messenger generally regulates cellular physiology via protein kinases.
TCA cycle	Tricarboxylic acid cycle. Circular metabolic pathway in which the products of glycolysis (pyruvate) are converted into carbon dioxide and water and the electrons are captured for oxidative phosphorylation.
therapeutic index	The ratio of the drug dose that produces toxicity to the dose that produces a clinically desired or effective response. A measure of drug safety.

REVIEW QUESTIONS

1. What are the four levels of protein structure and how do these four elements affect the overall structure of a protein?

2. Identify and briefly describe the six major functional classes of proteins.

3. Define ligand and describe the ligands associated with enzymes, membrane transporters, and receptors.

4. Describe how glucose is metabolized into energy.

5. What are second messengers and what role do they play in cellular physiology?

6. How do protein kinases regulate cellular functions?

7. What is meant by 'selective toxicity' and how does that relate to the drug target?

ADDITIONAL QUESTIONS

In addition to the Review Questions provided above, there is a range of free online questions designed to enable students to further test their understanding of the chapter material. In order to access these interactive questions, please visit the book's website:
https://routledge.com/cw/wiser

Molecular Biology

4

Cells are the fundamental unit of living organisms and have the capacity to maintain themselves in a changing environment, as well as the ability to replicate (Chapter 2). Proteins, either directly or indirectly, are responsible for cellular structures and functions (Chapter 3). The information about the sequence of proteins is encoded in DNA in the form of genes. Decoding this information is a two-step process consisting of **transcription** and **translation** (FIGURE 4.1). During transcription DNA directs the synthesis of RNA molecules, which are then translated into proteins. The combination of transcription and translation is also referred to as gene expression. Following gene expression proteins are modified and directed to their correct locations within cells and either make up structural elements of cells or carry out cellular functions.

DNA REPLICATION

RNA SYNTHESIS AND TRANSCRIPTION

PROTEIN SYNTHESIS AND TRANSLATION

PROTEIN TRAFFICKING

FIGURE 4.1 Central dogma of molecular biology. Cells and organisms can replicate their DNA and pass the DNA to their progeny. DNA directs the synthesis of RNA by transcription and RNA directs the synthesis of proteins by translation.

The information encoded by DNA is also passed from generation to generation and DNA is the hereditary material. At the cellular level, DNA is replicated before cell division and each daughter cell obtains a complete copy of DNA from the parental cell. In this way newly formed cells have the complete instructions for synthesizing proteins, which are also the instructions for maintaining cellular function. Reproduction in multicellular organisms involves sexual reproduction in which DNA from each parent is combined to form an offspring. DNA replication, transcription of RNA, and translation of proteins are complex biochemical processes that are at the heart of molecular biology. Understanding these processes is essential to the understanding of the basic biology of cells and organisms, as well as understanding disease etiology and pathophysiology. For example, malfunctions in gene expression and cellular replication result in disease. Therefore, having fundamental knowledge of molecular biology is essential to understanding disease.

DNA REPLICATION

SIGNPOST

The structure of nucleic acids is described in Chapter 1, pages 22–23.

DNA is a long polymer composed of four deoxyribonucleotides called adenosine (A), guanosine (G), cytidine (C), and thymidine (T). The bases (adenine, guanine, cytosine, and thymine) within the polynucleotide chains can interact with each other through hydrogen bonding. However, this hydrogen bonding is specific in that adenine forms hydrogen bonds with thymine and guanine forms hydrogen bonds with cytosine (FIGURE 4.2). The specificity of these hydrogen bonds between the nucleotides is the basis of DNA replication and RNA transcription.

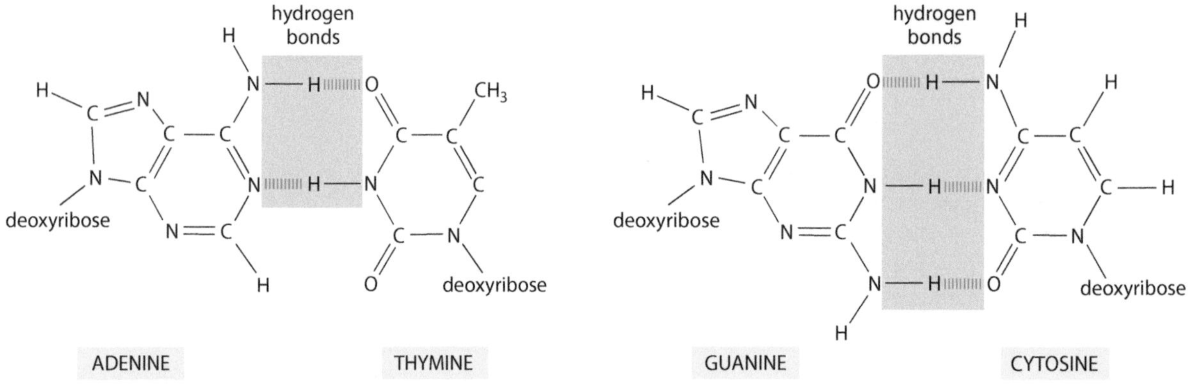

FIGURE 4.2 **Hydrogen bonding between nucleotides.** Adenine and thymine interact via the formation of two hydrogen bonds (red dashes) and guanine and cytosine interact via the formation of three hydrogen bonds.

Two complementary DNA molecules interact to form a double-stranded molecule

DNA most often exists as a double-stranded molecule formed from two molecules of single-stranded DNA. The two DNA molecules (i.e., strands) are held together through hydrogen bonds between A/T pairs and G/C pairs. This means that not just any two DNA molecules can form a double-stranded molecule, but only two DNA molecules in which the complementary bases align in both sequences. Thus, these two strands are called **complementary** strands. This ability to form double-stranded molecules between complementary polynucleotides is also the basis for the detection of specific nucleic acids (BOX 4.1).

BOX 4.1

Detection of Specific DNA and RNA Molecules

Nucleic acids can exist in either single-stranded or double-stranded states and it is relatively easy to convert from one state to the other. For example, heating DNA breaks the hydrogen bonds that hold the two strands together without breaking the covalent bonds between the nucleotides. Conversely, under appropriate conditions two single-stranded molecules can form a double-stranded molecule if the two strands are complementary (**BOX 4.1 FIGURE 1**).

This ability to form double-stranded molecules in vitro is exploited to detect and measure specific nucleic acid molecules. A short single-stranded DNA molecule is synthesized or generated via recombinant DNA methods (Chapter 7). A label that can be easily measured, such as fluorescence, radioactivity, or an enzyme, is attached to this molecule. This labeled DNA molecule functions as a 'probe' to evaluate the presence of a specific DNA or RNA molecule in a sample. The probe specifically binds to nucleic acid molecules that are complementary to the probe, often called the 'target,' and does not bind to other nucleic acids in the sample. Typically, the sample containing the target nucleic acid is immobilized on a solid support and then incubated with the probe. Any unbound probe is washed away, and the amount of bound probe can be determined by measuring the fluorescence, radioactivity, or enzyme activity (**BOX 4.1 FIGURE 2**). This allows for precise detection and measurement of the target DNA or RNA in a complex mixture containing other DNA or RNA molecules.

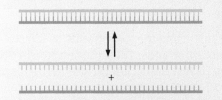

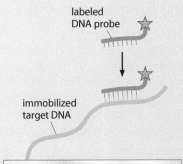

BOX 4.1 FIGURE 1 Separation and reformation of double strands. Double-stranded DNA can be separated into single-stranded molecules in the test tube by heating or with chemicals that weaken hydrogen bonds. The double-stranded molecule can be reformed by removing the condition that interferes with hydrogen bonding.

BOX 4.1 FIGURE 2 Complementary DNA probes detect specific nucleic acid targets. DNA probes complementary to the target DNA (or RNA) are used to evaluate the presence of a specific nucleic acid. Generally, the target is immobilized and incubated with a probe that is conjugated with a fluorescent molecule, radioactivity, or an enzyme. The amount of bound label (star) is measured after washing away any unbound probe.

Polynucleotides have a polarity with a phosphate group on the 5'-end of the molecule and an hydroxyl group on the 3'-end of the molecule. The two strands of the double-stranded DNA orient in an antiparallel arrangement in that two strands of DNA are running in opposite directions according to their 5'-ends and 3'-ends (FIGURE 4.3). Because of the antiparallel orientation and the A/T and G/C base-pairing rules, the sequences of the two strands are quite distinct. Nonetheless, if you

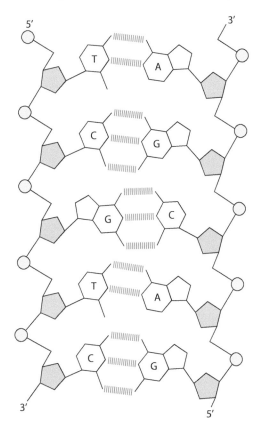

FIGURE 4.3 Complementary nature of double-stranded DNA. Double-stranded DNA is composed of two complementary polynucleotide chains in which the two strands have an opposite orientation of their 5'/3' directionality and are held together by hydrogen bonds (red dashes) between A and T and G and C. The strand on the left has the sequence 5'-TCGTC-3' and the complementary strand is 5'-GACGA-3.'

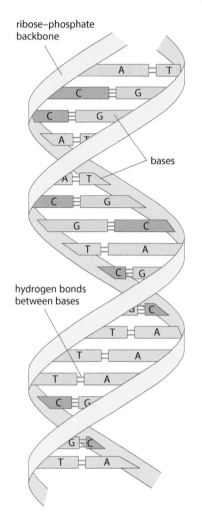

know the sequence of one of the DNA strands you can easily fill in the sequence of the other DNA strand. These two DNA strands twist into a helical structure widely known as the double helix (FIGURE 4.4). The double helix looks somewhat like a twisted ladder with the phosphate-ribose backbone forming the outside edges and the rungs are composed of the base pairs.

One DNA strand serves as the template for synthesis of the complementary strand

The cell synthesizes a DNA molecule using the two strands as templates for the synthesis of the complementary strands. Thus, the first step in DNA replication is separation of the two DNA strands. Since hydrogen bonds are weaker than covalent bonds (see Table 1.3), the complementary strands are separated without breaking the covalent bonds of the phosphate-ribose backbone. Each of the strands then serves as a template for the synthesis of the complementary strands (FIGURE 4.5). The base-pairing rules dictate which of the four possible nucleotides complements the nucleotides on the template strand. Using the two separated strands as templates results in the synthesis of two double-stranded molecules with identical sequences to the original parental DNA molecule. Each of the two newly formed molecules is composed of one strand from the original parental molecule and one newly synthesized strand. This configuration of one original nascent strand and one nascent strand is referred to as semi-conservative replication.

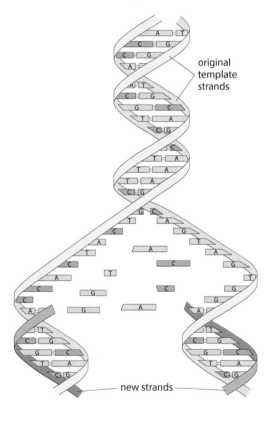

FIGURE 4.4 Double helix structure of DNA. Two strands of DNA interact through hydrogen bonds between bases (A/T and G/C) and the double-stranded DNA twists into a helical conformation.

FIGURE 4.5 Template-mediated DNA replication. Following strand separation each DNA strand can serve as a template for the synthesis of the complementary strand. For each nucleotide on the template strand there is a pairing with the complementary nucleotide.

Although the concept of semi-conservative replication is easy to understand, the actual process of replicating DNA is rather complex. Numerous biochemical activities are involved in the DNA synthesis process. First, the double helix needs to be unwound and the hydrogen bonds holding together the DNA strands need to be broken. These two activities are carried out by an enzyme called helicase. DNA synthesis is initiated by the synthesis of a complementary RNA primer and an enzyme called primase carries out this process. And, finally, the nucleotides need to be joined together by the formation of covalent bonds between the 5′-phosphate of the nucleotide and the 3′-hydroxyl of the growing polynucleotide chain via a condensation reaction. These various biochemical activities associated with DNA replication are carried out by a large protein complex called **DNA polymerase** which includes helicase, primase, and other proteins.

DNA polymerase copies both strands simultaneously

DNA molecules are extremely long and, to efficiently copy the entire DNA of a cell or organism, numerous DNA polymerase complexes act simultaneously. DNA polymerase starts replicating DNA at positions called the origins of replication. These origins of replication have some sequence conservation and are spread out along the DNA molecule. The DNA polymerase complexes start at the origins of replication and move in both directions (FIGURE 4.6). Nucleotides are added to the 3′-hydroxyl of the growing DNA strand and, thus, the DNA polymerase moves along the template DNA strand toward the 5′-end. This transfer of nucleotides to the growing polynucleotide is repeated one nucleotide at a time and eventually the DNA polymerase complexes run into each other. The fragments of newly synthesized DNA are joined together by an enzyme called DNA ligase, resulting in a complete copy of the DNA molecule.

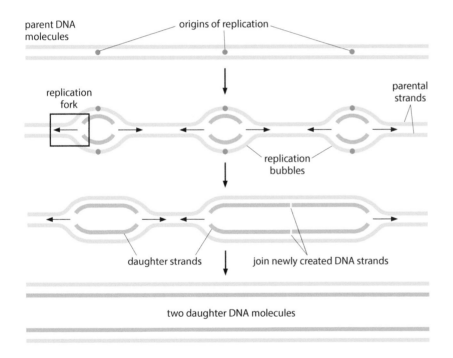

FIGURE 4.6 Origins of replication. DNA synthesis starts at the origin of replication and proceeds in both directions forming replication bubbles. The fragments of newly synthesized DNA are joined together to form two new DNA molecules composed of one parental strand and one daughter strand. Details about the replication fork (box) are in Figure 4.7.

The region of the DNA molecule with the bound DNA polymerase and the separated DNA strands is called the replication fork. One of the template strands is in the correct orientation in regard to the direction of movement of DNA polymerase and the nascent strand is synthesized continuously as the DNA polymerase progresses along the template. However, the antiparallel nature of the DNA presents somewhat of a logistic problem in that the directions of DNA synthesis are opposite for the two template strands. This problem is solved by one of the strands being synthesized in a discontinuous manner by synthesizing short fragments that are later joined together. The strand synthesized in a continuous manner is called the leading strand and the strand synthesized in a discontinuous manner is the lagging strand.

As the replication fork advances, the lagging strand loops around so that the template is now oriented in the same direction as the leading strand (FIGURE 4.7). DNA synthesis starts at a primer and proceeds until running into the beginning of the previously synthesized segment. The ribonucleotides of the RNA primer are replaced with deoxyribonucleotides by the DNA polymerase. A DNA ligase activity associated with DNA polymerase joins these two newly synthesized DNA segments. After ligating the segments, the DNA polymerase temporarily releases the lagging strand, re-attaches to the lagging strand at the position of the next primer, and starts DNA synthesis again. Coincident with this the primase adds another primer further upstream on the template strand. This stepwise action of synthesizing short DNA segments on the lagging strand is repeated until the DNA molecule is completely copied.

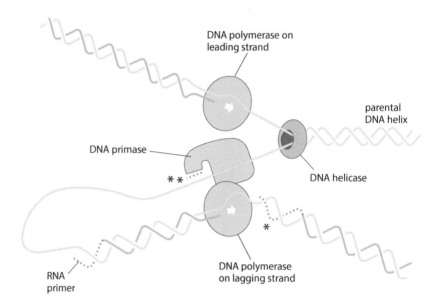

FIGURE 4.7 DNA polymerase and the replication fork. DNA polymerase moves along the parental DNA molecule as it is being copied. At the replication fork DNA helicase unwinds the DNA and separates the strands and primase periodically adds RNA primers to the lagging strand. DNA polymerase uses the parental DNA strands (light yellow) as templates to synthesize the complementary strands (dark yellow). The leading strand is synthesized continuously as the DNA polymerase moves along the DNA template. The lagging strand is synthesized in segments as the replication fork progresses along the parental DNA molecule. When the DNA polymerase catches up to the previous segment (*) the ribonucleotides in the RNA primer are replaced with deoxyribonucleotides and then the DNA polymerase detaches from the template and re-attaches to the template at the next RNA primer (**).

Replication of linear DNA molecules does present a problem since the ends of the lagging strands are not copied due to not enough of the template strand being left to form a loop. Thus, chromosomes would become shorter with every round of DNA replication. However, this problem is solved through the addition of telomeres to the ends of chromosomes (see Box 10.2). The telomeres are added in a non-template-mediated fashion through the action of telomerase, which protects the ends of the chromosomes.

RNA SYNTHESIS AND TRANSCRIPTION

The chemistries of DNA and RNA are quite similar with only minor differences (Chapter 1) and, accordingly, the synthesis of RNA is mechanistically similar to DNA synthesis in that both utilize a DNA template. For example, ribonucleotides are used in RNA synthesis and deoxyribonucleotides are used in DNA synthesis, and RNA has uracil as a base instead of thymine (TABLE 4.1). However, uracil and thymine are nearly identical in chemical structure, and uridine forms base-pairs with adenine in the same manner as thymine.

The synthesis of RNA is mediated by a complex of proteins called **RNA polymerase**. RNA polymerase, like DNA polymerase, moves along the DNA molecule, separates the DNA strands, and uses one of those strands

Table 4.1	DNA vs. RNA Synthesis	
	DNA	**RNA**
Nucleotides*	dA, dG, dC, dT	A, G, C, U
Enzyme	DNA polymerase	RNA polymerase
Stands copied	Both strands	One strand
Amount copied	Entire chromosome	Define sequences (= genes)

*The small 'd' is the designation for deoxyribonucleotides. The nucleotides are adenosine (A), guanosine (G), cytidine (C), thymidine (T), and uridine (U). The corresponding bases are adenine, guanine, cytosine, thymine, and uracil.

as a template for the synthesis of RNA (FIGURE 4.8). Nucleotides that are complementary to the template strand are added to the 3′-end of the growing RNA strand one base at a time. In contrast with DNA synthesis, RNA polymerase only copies one DNA strand. The newly synthesized RNA molecule has the same sequence as one of the DNA strands except the thymidines (T) in DNA are replaced with uridines (U) in RNA. The DNA strand with the same sequence as the RNA strand is called the sense strand and the DNA strand that is used as the template is the antisense strand.

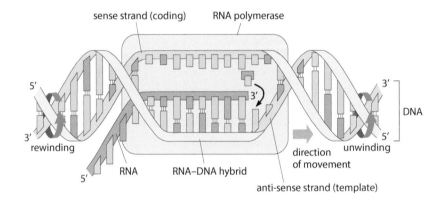

FIGURE 4.8 RNA transcription. RNA polymerase moves along DNA unwinding the helix, separates the strands, and adds nucleotides to the 3′-end of the growing RNA molecule. The antisense strand serves as the template for the synthesis of an RNA molecule corresponding to the sense strand.

Regions corresponding to genes are transcribed during gene expression

Another difference between DNA and RNA synthesis is that cells replicate their entire DNA content as a prelude to cell division, whereas only certain regions of the DNA molecule are transcribed into RNA. The regions of DNA that are transcribed are **genes**. Most genes encode the information for the amino acid sequence of proteins and the RNA synthesized from these genes is called **messenger RNA** (mRNA). The mRNA is subsequently translated into protein. There are also other types of RNA that are not translated into protein, but, nonetheless, play important roles in gene expression (TABLE 4.2). Some of these RNA molecules are in large complexes with proteins such as ribosomes, spliceosomes, and telomerase. Other RNA molecules can regulate the transcription or degradation of mRNA molecules and thus regulate gene expression.

Table 4.2	Types of RNA
Type	**Function**
Messenger RNA (mRNA)	Encodes the amino acid sequence of a protein
Transfer RNA (tRNA)	Link between nucleotide sequence (codon) and amino acid sequence
Ribosomal RNA (rRNA)	Component of ribosomes and functions in protein synthesis
Small nuclear RNA (snRNA)	Component of the spliceosome and functions in RNA splicing
Telomere RNA	Component of telomerase that serves as a template for telomere synthesis
Micro-RNA (miRNA)	Small RNA molecules that regulate gene expression
Small interfering RNA (siRNA)	Small double-stranded RNA that targets specific mRNA molecules for degradation

Genes have specified start points and end-points on the DNA molecule (FIGURE 4.9). The beginning of the gene is the position on the 5'-end of the sense strand where transcription begins and, likewise, the end of the gene at the 3'-position where transcription stops. In addition, genes have directionality in that one DNA strand is the sense strand and the other strand is the antisense strand. However, the various genes located on a DNA molecule can be oriented in either direction. This means that the sense and antisense strands are a feature of each individual gene and not a feature of the entire DNA molecule.

FIGURE 4.9 Relation of RNA transcripts to genes. Coding regions of DNA (yellow) define RNA transcripts (blue block arrows) and the gene. Regulatory regions (orange) precede the coding regions of genes and DNA between genes (light blue) are non-coding sequences. Each gene has a specific orientation in regard to the DNA strands that serve as sense and template strands.

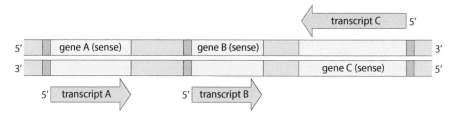

Transcription of genes is highly regulated

Cells can activate or repress the transcription of specific genes as part of normal cellular function, or in response to environmental factors. Quite often activating transcription is called turning genes on and repressing transcription is turning genes off. The specific genes that are expressed, and the relative amounts of the various transcripts determine the overall properties of a cell or organism and contribute to the phenotype. Phenotype is often defined as the observable traits of a cell or organism. For example, muscle cells express a different milieu of genes than nerve cells and, accordingly, the two types of cells have different phenotypes. The differences in the specific genes expressed by different types of cells account for the phenotypic differences between cells, despite the fact that all the cells in an individual have the same DNA sequence.

A key element in the regulation of transcription is a region of DNA that immediately precedes the start site of transcription, called the **promoter**. RNA polymerase binds to the promoter and initiates transcription from the promoter. Other proteins that regulate transcription also bind to the promoter and RNA polymerase (FIGURE 4.10). There are other regulatory regions in the DNA sequence near genes that bind proteins and regulate transcription. These various proteins that regulate transcription are generally called **transcription factors**. Some transcription factors activate RNA polymerase and turn on transcription, whereas other transcription factors function as repressors and turn off transcription. Thus, the initiation of RNA transcription is a complex process involving many transcription factors to ensure that the appropriate genes are expressed at the appropriate times. It is the combination of these various transcription factors that determine which genes are expressed and how many RNA transcripts are synthesized from a specific gene.

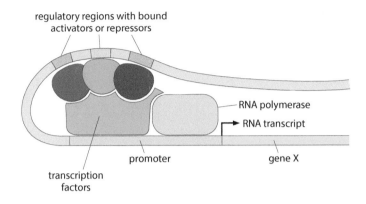

regulatory regions with bound
activators or repressors

RNA polymerase

RNA transcript

promoter

gene X

transcription
factors

FIGURE 4.10 Regulation of transcription. RNA polymerase binds to the promoter region, which precedes the coding region of a gene. Transcription factors bind the promoter and other regulatory regions on the DNA. These various transcription factors control the activity of RNA polymerase and regulate the initiation of transcription. Some transcription factors activate gene expression, whereas other transcription factors repress gene expression.

Environmental factors can also regulate transcription and gene expression so that organisms can respond to changes in their surroundings by turning genes on or off. Similarly, many hormones affect transcription by directly or indirectly affecting transcription factors. For example, many hormones activate protein kinases and protein phosphorylation as part of signal transduction. Many transcription factors are regulated by protein phosphorylation and can be activated or repressed by the addition of a phosphate group. Some transcription factors bind to ligands, such as hormones, and this ligand binding activates or represses transcription. These transcription factors that are directly regulated by ligands are called nuclear receptors.

Hormones, signal transduction, and protein phosphorylation are described in Chapter 3, see pages 62–66.

SIGNPOST

Nuclear receptors are transcription factors regulated by hormones or other ligands

Nuclear receptors function in the cell nucleus, instead of on the cell surface, and regulate gene transcription. Forty-eight different genes for nuclear receptors that bind to a wide range of ligands have been identified in humans. In many cases the ligand is a hormone or vitamin. However, some of the receptors may function as sensors of metabolites or foreign substances called xenobiotics. The ligands that bind to nuclear receptors are small non-polar molecules that can cross biological membranes and enter cells. Many of these ligands are derived from cholesterol-like lipids called steroids and these hormones were originally called steroid hormones, a term that is still widely used. Some receptors are located in the cytoplasm and binding the hormone results in the hormone–receptor complex being translocated to the nucleus (FIGURE 4.11). Other nuclear receptors reside in the nucleus and the hormone diffuses to the nucleus. In either case, the hormone–receptor complex binds to regulatory regions of DNA near the gene and either activates or represses the expression of specific genes.

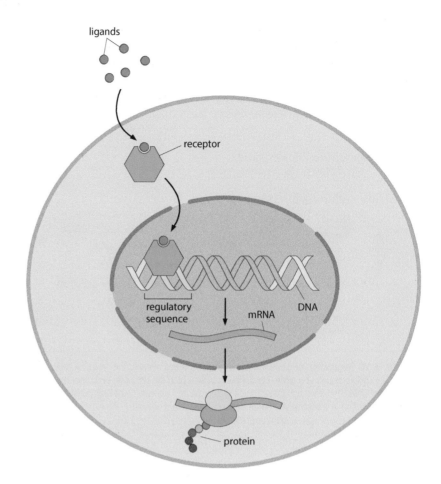

FIGURE 4.11 Mechanism of nuclear receptor action. A ligand, such as a hormone, diffuses across the plasma membrane and binds to the intracellular receptor. The receptor–ligand complex translocates to the nucleus, binds to a regulatory region of DNA, and activates or represses transcription of those genes. The resulting mRNA is translated into protein and affects cellular functions and properties.

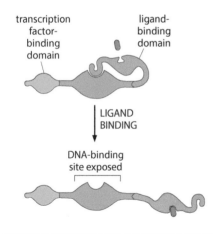

FIGURE 4.12 Activation of nuclear receptors. Nuclear receptors have three domains: transcription factor binding, DNA binding, and ligand binding. Binding of a ligand, such as a hormone, to the ligand-binding site causes a change in shape of the receptor resulting in exposure of the DNA-binding site. The ligand–receptor complex then binds to regulatory regions near genes and either activates or represses gene transcription.

The various nuclear receptors all have a similar amino acid sequence and a similar structure consisting of three distinct domains. These domains are the ligand-binding domain, a DNA-binding domain, and a domain that binds to other transcription factors (FIGURE 4.12). Differences in the sequence of the ligand-binding domain account for the different ligands that are recognized by the various nuclear receptors. Binding of the ligand to the receptor induces a conformational change in the receptor that exposes the DNA-binding domain. The receptor with the bound ligand can now bind to regulatory regions associated with the regulated genes. Binding of the receptor–hormone complex to these regulatory regions, plus interactions with the other transcription factors, either activates or represses the transcription of those genes. Thus, the hormone or ligand turns on or turns off gene expression and results in a change in the cellular phenotype.

In accordance with their ability to regulate gene expression and to change cellular phenotypes, many of the steroids and related hormones play key roles in cellular growth, differentiation, and metabolism (TABLE 4.3). For example, testosterone and estrogen increase during puberty in boys and girls, respectively, leading to sexual maturation of the organs and tissues associated with sexual reproduction. This represents a change in the phenotype of the cells associated with those organs due to a change in the composition of the genes that are expressed. Other nuclear receptors, such as the glucocorticoid receptor and thyroid hormone receptor help to regulate glucose and lipid metabolism. These key roles that nuclear receptors play in development and metabolism make them attractive drug targets and many drugs target nuclear receptors (BOX 4.2).

Table 4.3 Examples of Nuclear Receptors and Their Ligands

Ligand	Precursor	Receptor	Major functions
Estrogen	Steroid	Estrogen	Development of female sexual characteristics Regulation of menstrual cycle
Testosterone	Steroid	Androgen	Development of male sexual characteristics Anabolic effects
Progesterone	Steroid	Progesterone	Regulates pregnancy and menstrual cycle
Cortisol	Steroid	Glucocorticoid	Stimulates gluconeogenesis Inhibits inflammatory response
Aldosterone	Steroid	Mineralocorticoid	Promotes sodium and water retention Regulates blood pressure
Retinoids	Vitamin A	Retinoid	Roles in vision, immune function, epithelial cell development
Calcitriol	Vitamin D	Vitamin D	Regulation of calcium homeostasis
Thyroid hormone	Tyrosine	Thyroid hormone	Increases basal metabolic rate
Xenobiotics	Various	Pregnane X	Upregulates the expression of proteins involved in detoxification of foreign substances

Nuclear Receptors as Drug Targets

Many drugs target nuclear receptors. In fact, approximately 13% of the drugs approved by the United States Food and Drug Administration (US FDA) target nuclear receptors. These drugs can function as an agonist by mimicking the hormone, or function as antagonists by blocking the action of the hormone. One of the earliest examples of targeting nuclear receptors was the development of oral contraceptives based on progesterone. Drugs targeting nuclear receptors are available for a wide array of diseases and conditions ranging from cancer to inflammatory diseases to metabolic disorders. For example, derivatives of estrogen can be used in the treatment of certain types of breast cancer. Dexamethasone and related drugs that bind to the glucocorticoid receptor have long been used as anti-inflammatories to treat allergies and autoimmune-type disorders (Chapter 17). Such drugs function as agonists of the glucocorticoid receptor and suppress an overactive inflammatory response. Nuclear receptors are also involved in lipid and glucose metabolism and many diseases, such as diabetes and cardiovascular disease, are associated with imbalances in the metabolism of glucose and lipids, which can be corrected with drugs.

BOX 4.2

Eukaryotic mRNA undergoes post-transcriptional processing

After transcription mRNA is enzymatically modified to produce mature mRNA. One modification is the addition of a guanosine derivative to the 5'-end of the mRNA called a cap (FIGURE 4.13). On the 3'-end numerous adenosines are added in a non-template-mediated fashion to create a poly(A) tail. Both modifications are carried out by enzymes and these modifications stabilize the mRNA molecule and prevent degradation.

Many eukaryotic genes are also interspersed with regions that do not code for protein. These regions that do not encode protein are called introns and the protein-encoding regions are called exons. Before the mRNA can be correctly translated, the introns need to be removed by a process often called splicing. Removal of the introns is carried out by a complex of proteins and RNA called the spliceosome. In some cases, some introns may not be removed, resulting in a longer mRNA molecule encoding a longer protein. This is referred to as alternate splicing and results in the possibility of a single gene encoding different forms of the same protein.

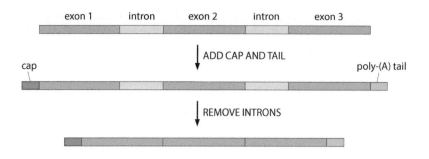

PROTEIN SYNTHESIS AND TRANSLATION

Many genes directly encode the information needed for the synthesis of proteins and, in general, one gene contains the instructions for the synthesis of one protein. An exception to this is the possible production of different proteins through alternate splicing. Protein-encoding genes are transcribed to produce mRNA molecules. The mRNA is then translated into proteins on subcellular structures called ribosomes. Other types of RNA, such as ribosomal RNA (rRNA) and transfer RNA (tRNA) are also involved in protein synthesis. Thus, DNA provides the information needed for protein synthesis through RNA intermediates. Specifically, during translation the information in the nucleotide sequence of mRNA is converted into the amino acid sequence of the protein by ribosomes and tRNA.

The genetic code uses three nucleotides to specify an amino acid

The linear nucleotide sequence of a gene is converted to a linear nucleotide sequence of mRNA via transcription. This linear nucleotide sequence of mRNA needs to be converted into a linear amino acid sequence of the protein. The chemistries of nucleotides and amino acids are distinct and a direct template-mediated synthesis, as in transcription, is not possible. Instead, three sequential nucleotides encode for a single amino acid. These three sequential nucleotides are called a **codon**. This genetic code is essentially the same in all organisms with very few exceptions. The codons are read from the 5′-end of the mRNA to the 3′-end and translated into a protein starting at the N terminus and proceeding to the C terminus (FIGURE 4.14). The triplet genetic code also means that there are three possible reading frames depending upon the nucleotides making up the first codon. However, only one of the reading frames is correct for any given gene.

FIGURE 4.14 Information flow from gene to protein. Regions corresponding to genes that encode proteins are transcribed into mRNA. The sense strand (upper) of DNA has the same sequence as the RNA except that thymidines (T) are replaced by uridines (U). Untranslated regions (UTRs) are included with the transcript on the 5′- and 3′-sides of the mRNA. Start and stop codons define the coding region of the gene that is translated into protein. Codons in the incorrect reading frames are indicated with short red lines above and below the mRNA. The N terminus (NH_3) and C-terminus (CO_2) of the protein correspond to the 5′-end and 3′-end of the mRNA, respectively.

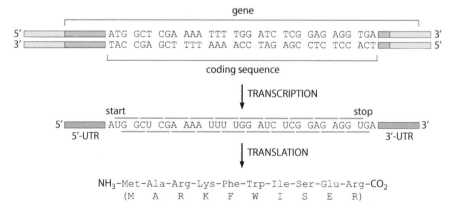

The four nucleotides making up either DNA or RNA generate 64 (= 4^3) distinct codons known as the genetic code (FIGURE 4.15). Thus, there is a surplus of codons compared with the 20 amino acids that make up proteins. Three of the codons do not encode amino acids, but rather serve as stop signals, indicating the end of the protein-coding region of mRNA. In addition, there is redundancy in the genetic code as multiple codons can encode the same amino acid. In particular, the first two positions in many of the codons determine the amino acid, and the third position can be any nucleotide.

1st base	2nd base								3rd base
	U		C		A		G		
U	UUU	Phe	UCU	Ser	UAU	Tyr	UGU	Cys	U
	UUC	Phe	UCC		UAC	Tyr	UGC	Cys	C
	UUA	Leu	UCA	Ser	UAA	stop	UGA	stop	A
	UUG		UCG		UAG	stop	UGG	Trp	G
C	CUU	Leu	CCU	Pro	CAU	His	CGU	Arg	U
	CUC		CCC		CAC	His	CGC		C
	CUA		CCA		CAA	Gln	CGA		A
	CUG		CCG		CAG	Gln	CGG		G
A	AUU	Ile	ACU	Thr	AAU	Asn	AGU	Ser	U
	AUC	Ile	ACC		AAC	Asn	AGC		C
	AUA		ACA		AAA	Lys	AGA	Arg	A
	AUG	Met	ACG		AAG	Lys	AGG		G
G	GUU	Val	GCU	Ala	GAU	Asp	GGU	Gly	U
	GUC		GCC		GAC	Asp	GGC		C
	GUA		GCA		GAA	Glu	GGA		A
	AUA		GCG		GAG	Glu	GGG		G

non-polar   polar   basic

acidic   stop codon

FIGURE 4.15 The standard genetic code. Shown are the amino acids encoded by the 64 potential codons of mRNA. Three codons do not encode amino acids, but function as 'stop' codons, indicating the end of the protein-coding sequence. The AUG-encoding methionine is a 'start' codon as it is the first codon of nearly all mRNA molecules. Colors indicate the chemical properties of the encoded amino acids.

On paper the simplicity of the genetic code allows easy conversion from a nucleotide sequence to an amino acid sequence. However, the molecular mechanisms involved in decoding nucleic acid sequences into amino acid sequences are complex and involve several distinct types of RNA molecules and subcellular structures called **ribosomes**. A key component of the decoding process is a type of RNA called **transfer RNA** (tRNA). Transfer RNA has internal complementarity and folds into a structure consisting of stems and loops (FIGURE 4.16). One of the loops contains a region with the anticodon. The anticodon is the complement of the codon, and there are up to 61 different tRNA molecules corresponding to the 61 codons that encode amino acids. On the stem that opposes the anticodon loop is an amino acid attachment site. Amino acids corresponding to the sequence of the anticodon are attached to this site by enzymes called amino-acyl-tRNA synthases. Thus, if the anticodon sequence is CAU, which is the complement of the codon AUG, a methionine is attached to the stem of that tRNA molecule.

Translation is carried out by ribosomes

Protein synthesis, as does DNA replication and RNA transcription, involves many complex biochemical processes. These complex processes are carried out on large molecular structures called **ribosomes**. The ribosomes consist of two subunits, designated large and small subunits, and each subunit contains numerous proteins and RNA molecules

anticodon

amino acid attachment site

FIGURE 4.16 Structure of transfer RNA. Internal hydrogen bonds (dashes) between complementary bases in the tRNA molecule result in three loops and a stem. The middle loop has the anticodon (complement of the codon) and the stem is the attachment site for an amino acid that corresponds to the sequence of the anticodon.

called ribosomal RNA (rRNA). Ribosomes can be viewed as a subcellular machine that synthesizes proteins by bringing together all the components necessary for protein synthesis and facilitating chemical reactions that are part of protein synthesis. The main activity of the ribosome is to hold the mRNA and tRNA in a configuration that promotes the formation of the peptide bond between two amino acids (FIGURE 4.17).

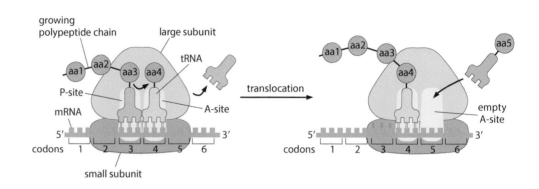

FIGURE 4.17 Ribosomes and translation. Ribosomes, composed of large and small subunits, hold the mRNA and two tRNA molecules in a conformation that allows for a condensation reaction to form a peptide bond between two amino acids. The P-site contains a tRNA with the growing polypeptide chain and the A-site contains the tRNA with the amino acid (aa) to be added next as determined by the codons of the mRNA. The polypeptide chain is transferred to the amino acid on the tRNA in the A-site and the resulting tRNA without an amino acid is released from the ribosome, freeing up the P-site. Translocation of the mRNA and the bound tRNA from the A-site to the P-site empties the A-site and the tRNA corresponding to the next codon is recruited into the A-site. This process is repeated codon-by-codon until a stop codon is reached.

The ribosome has sites that can bind two tRNA molecules with attached amino acids. One of these sites binds the tRNA with the growing polypeptide chain and is called P-site (polypeptide). Next to the P-site is the A-site (amino acid) that binds a tRNA with a single amino acid. The growing polypeptide chain in the P-site is removed from the tRNA and transferred to the amino acid in the A-site. The juxtaposition of the two tRNA molecules allows a peptide bond to be formed between the amino acid corresponding to the codon in the A-site and the growing polypeptide chain. This change results in the tRNA in the P-site no longer having an associated amino acid. This empty tRNA then exits the ribosome and is subsequently recharged by the addition of the appropriate amino acid that corresponds to its anticodon.

The mRNA/tRNA/growing polypeptide complex is then shifted so that the tRNA with the growing polypeptide chain moves from the A-site to the P-site. This leaves the A-site empty and a tRNA corresponding to the codon in the A-site is now recruited and fills the A-site. The process of transferring the growing polypeptide to the adjacent amino acid is repeated and again the mRNA/tRNA/polypeptide complex is shifted by one codon. This stepwise addition of amino acids to the growing polypeptide chain is continuously repeated until reaching a stop codon. During termination the ribosomal subunits are separated, and the newly synthesized polypeptide is released (FIGURE 4.18). The released ribosomal subunits and mRNA are then available for another round of translation. This action means that a single mRNA molecule is translated into numerous protein molecules. In fact, several ribosomes line up along a single mRNA and sequentially translate the mRNA into protein resulting in the production of numerous protein molecules per mRNA transcript.

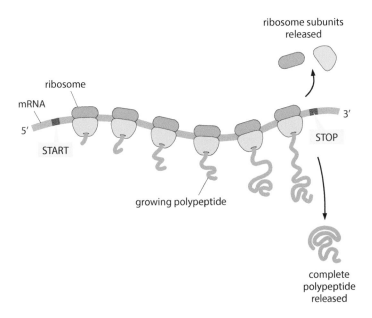

FIGURE 4.18 Overview of translation. Ribosomes move along an mRNA molecule and synthesize a protein. The start and stop codons determine the beginning and end of the protein being synthesized. When the process is completed the large and small subunits of the ribosome dissociate and can begin the process again. The mature protein is released and folds into its proper structure.

PROTEIN TRAFFICKING

After synthesis, proteins need to fold into their correct conformations and many proteins are post-translationally modified. Eukaryotic cells are further complicated by the various organelles that have defined subcellular functions and protein compositions. Thus, proteins also need to be correctly targeted to their correct organelle after synthesis in the cytoplasm (FIGURE 4.19). This correct targeting is accomplished by accessory proteins that guide proteins to their final destinations and assist in the translocation of the proteins from the cytoplasm into the organelle.

Signal sequences direct proteins to the correct location within cells

Within the sequence of a protein destined for a particular organelle is a stretch of amino acids with defined characteristics. This stretch of amino acids functions somewhat like an address in that it directs the protein to the correct organelle (FIGURE 4.20). For example, proteins that function within the nucleus of the cell have a stretch of basic amino acids somewhere within the sequence that serves as a signal for that protein to be imported into the nucleus. Proteins with the nuclear import signal bind to another protein called importin and this protein complex moves through the nuclear pores. The complex then dissociates, and the imported protein remains in the nucleus. Similarly, proteins destined for the mitochondria have signal sequences. In this case, the signal sequence is found at the N terminus of the protein and consists of basic residues plus serines and threonines. The signal sequence binds to a receptor protein on the surface of the mitochondrion and the protein is imported into the mitochondrion.

Chaperones, protein folding, and protein modifications are discussed in Chapter 3, pages 54–56.

SIGNPOST

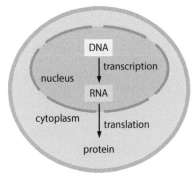

post-translational events:
proper folding
modifications
targeting to organelle

FIGURE 4.19 Gene expression in eukaryotes. Transcription occurs in the nucleus and the mRNA is exported to the cytoplasm where it is translated. After translation, proteins are folded into their correct conformation. Some proteins are enzymatically modified or moved to their final destinations.

FIGURE 4.20 Signal sequences. Proteins are targeted to their correct subcellular destinations by short stretches of amino acids or modifications on the protein. Signals directing proteins to the mitochondria or endoplasmic reticulum are generally found at the N terminus (red). Nuclear targeting sequences are positively charged internal sequences. Proteins destined for the lysosomes have a specific carbohydrate added in the Golgi at a specific internal site (green).

organelle	signal	location
nucleus	5 sequential basic (+) amino acids	internal
mitochondrion	3–5 basic amino acids and serines and threonines	N terminus
endoplasmic reticulum	6–12 hydrophobic amino acids preceded by 1–3 basic amino acids	N terminus
lysosomes	mannose-6-phosphate (glycosylation in the Golgi)	internal

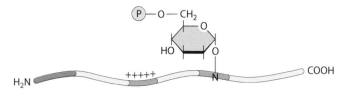

The secretory pathway directs proteins to lysosomes, the plasma membrane, and export from the cell

Eukaryotic cells have an endomembrane system composed of the endoplasmic reticulum (ER), Golgi apparatus, lysosomes, and the vesicles that move proteins between these compartments (see Figure 2.8). Proteins that are part of these endomembranes, as well as plasma membrane proteins and secreted proteins, have hydrophobic signal sequences that are recognized by the ER and thus imported into the ER. In many cases, these proteins are synthesized on ribosomes bound to the ER and are imported directly into the ER during translation. These imported proteins are then incorporated into vesicles and carried to the Golgi apparatus. Many plasma membrane proteins and secreted proteins are glycoproteins and carbohydrate groups are added to these proteins by enzymes in the Golgi. Proteins destined for lysosomes have a specific carbohydrate called mannose-6-phosphate that directs the protein to the lysosomes. These various signal sequences ensure that the proteins are delivered to their correct locations in the cell.

SUMMARY OF KEY CONCEPTS

- DNA functions as the hereditary material and can be replicated by cells in a template-mediated fashion by DNA polymerase

- DNA is an information storage molecule that encodes the basic directions for the synthesis of proteins

- Decoding the information stored in DNA involves processes called transcription and translation

- Genes are transcribed into RNA from a DNA template by RNA polymerase

- Transcription factors, such as nuclear receptors, regulate RNA synthesis

- Nucleotide sequences are translated into amino acid sequences via the genetic code in which three nucleotides (codon) represent one amino acid

- Ribosomes, messenger RNA, and transfer RNA direct protein synthesis through the process of translation

- After translation, proteins must be properly folded, modified, and targeted to the correct location within the cell

KEY TERMS

codon	Three sequential nucleotides in mRNA that specify the amino acid to be incorporated into the polypeptide chain. The collection of all codons is called the genetic code.
complementary	Describes the relationship between nucleotides or nucleic acids in which adenine pairs with thymine (or uracil) and guanine pairs with cytosine via hydrogen bonds.
DNA polymerase	A complex of proteins that synthesizes DNA by adding nucleotides to the 3'-hydroxyl groups of DNA polymers using the complementary DNA strands as templates.
gene	A functional segment of DNA located at a particular place on the chromosome which corresponds to a region that is transcribed. A unit of heredity that encodes the amino acid sequence of a protein.
messenger RNA	An RNA molecule transcribed from a protein-encoding gene that directs the synthesis of a protein by providing a sequence of codons.
nuclear receptor	A class of transcription factors that bind non-polar hormones or other ligands. Ligand binding exposes a DNA-binding domain that binds to regulatory regions of genes and regulates transcription.
promoter	A region of DNA on the 5'-side of a gene that regulates transcription by binding to RNA polymerase and transcription factors.
ribosome	Subcellular structures composed of proteins and ribosomal RNA that function in protein translation by positioning mRNA and tRNA so that peptide bonds can form between amino acids.
RNA polymerase	A complex of proteins that synthesizes RNA from a DNA template by adding nucleotides to the 3'-hydroxyl groups of the RNA polymer.
transcription	The synthesis of RNA from a DNA template that corresponds to a gene.
transcription factor	Proteins that bind to promoters and other regulatory regions of genes and activate or repress gene transcription.
transfer RNA	Small RNA molecules that facilitate the transfer of amino acids to growing polypeptide during translation by bringing the amino acid that corresponds to a codon to the ribosome.
translation	The process by which a messenger RNA sequence is converted into an amino acid sequence of a protein using ribosomes, messenger RNA, and transfer RNA.

REVIEW QUESTIONS

1. Describe the various activities associated with the replication fork including the leading and lagging strands.

2. What are the differences between DNA synthesis and RNA synthesis?

3. What are nuclear receptors and how do they regulate transcription?

4. What is the genetic code? Discuss its redundancy.

5. Describe the process of translation including the roles of mRNA, tRNA, and ribosomes.

6. How are proteins targeted to specific organelles in eukaryotic cells?

ADDITIONAL QUESTIONS

In addition to the Review Questions provided above, there is a range of free online questions designed to enable students to further test their understanding of the chapter material. In order to access these interactive questions, please visit the book's website:
https://routledge.com/cw/wiser

Genetics and Evolution

5

The molecular aspects of DNA replication and gene expression (Chapter 4) are the fundamental basis of inheritance and genetics. The information stored in the DNA sequence is passed down through generations from parents to offspring in the form of genes. In multicellular eukaryotes this flow of genetic information through the generations entails sexual reproduction in which genetic information is shuffled in the gametes. This results in individuals having different versions of their genes, and therefore having unique traits despite having similar genes. Traits are observable features that are determined by the expression of genes and variations in genes between individuals can produce differences in traits. These various traits can be passed on to progeny and the inheritance of traits exhibits certain qualities and patterns. Genetics is the study of how this genetic information in the form of genes flows through the generations and how genes determine traits.

Genes also play an important role in health and disease. For example, some diseases are due to defects in a particular gene and thus are inherited as a genetic trait (Chapter 11). In other cases, variations in genes may not directly lead to the development of a disease but affect the predisposition to develop disease in combination with environmental or behavioral factors. In this regard, the various versions of genes can be regarded as favorable or unfavorable and may affect the reproductive potential of some individuals. Thus, the gene pool of a population can change over the generations and such changes are the basis of evolution, including human evolution. The analysis of changes in the DNA sequence over time is called molecular phylogenetics. Molecular phylogenetic studies can provide a means to elucidate evolutionary relationships and insight into many aspects of human health and disease.

CHROMOSOME STRUCTURE

The word **chromosome** means 'colored body' and refers to the observation that chromosomes can be stained with biological dyes and are visible during cell division under a microscope. This visibility allowed for the observation that chromosomes duplicate and segregate into the daughter cells during cell replication. Because of these observations, chromosomes have long been recognized as having a role in heredity. Even before the discovery of DNA, traits could be mapped to locations on particular chromosomes.

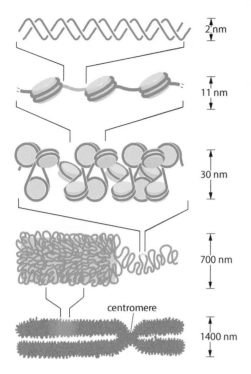

2 nm

11 nm

30 nm

700 nm

centromere

1400 nm

FIGURE 5.1 DNA is associated with proteins to form chromosomes. DNA (red) interacts with proteins called histones (yellow disks) to form a shorter wider fiber. This fiber is then coiled into yet another shorter wider fiber that is looped, coiled, and folded until condensed into a chromosome.

Histones and other proteins package DNA into chromosomes

Chromosomes are composed of both DNA and protein, with DNA being the component responsible for inheritance and the proteins playing a structural role. DNA is an extremely long molecule that needs to fit inside the cell. In fact, the total length of the human genome is approximately 1.8 meters and it is packaged into cells that are often smaller than 50 micrometers (1/20th of a millimeter) in diameter. One of the functions of the proteins found associated with DNA in chromosomes is to fold the long DNA molecule into a convenient package that fits within the nucleus of the cell. Furthermore, this compact package is more easily segregated into the daughter cells during cell division.

Histones are proteins that bind to DNA and play a key role in the formation of eukaryotic chromosomes. A DNA molecule is initially wrapped around histones, resulting in a DNA/protein fiber that is shorter and wider than the naked DNA molecule (**FIGURE 5.1**). Other proteins interact with this DNA/protein fiber and subsequent coiling, folding, and looping produce a chromosome. During cell division the chromosomes become even more compact and are called condensed chromosomes.

Chromosomes are distinguished by size and banding patterns

The condensed chromosomes also exhibit bands when stained with biological dyes. The banding pattern is due to regional differences in precisely how the DNA is packaged within a chromosome and is relatively constant for each individual chromosome. The complete set of condensed chromosomes is referred to as the karyotype. For example, the human karyotype consists of 23 pairs of chromosomes (BOX 5.1) with each chromosome representing a single DNA molecule. There are two copies of each chromosome with one copy originating from the mother and the other from the father. This state of having two copies of the chromosomes is called **diploid** and the two members of the chromosome pair are called homologous chromosomes. Cytogenetics refers to a study of the karyotype to look for abnormalities in chromosome numbers or banding patterns and the association of these abnormalities with genetic disease. The best known example is Down's syndrome due to three copies of chromosome 21, called trisomy 21.

BOX 5.1

The Human Karyotype

The genomic DNA of human cells consists of 46 linear DNA molecules that are packaged into 23 pairs of chromosomes. Chromosomes within a cell are different sizes, corresponding to the length of the DNA molecule making up that chromosome. The human chromosomes are numbered 1–22 with chromosome 1 being the largest and chromosome 22 being the smallest. The other two chromosomes are the X and Y sex chromosomes. Females have two copies of the X chromosome and males have one X chromosome and one Y chromosome. A chromosomal karyotype is produced by staining cells during cell division when the chromosomes are in the most condensed state. Pictures of the individual chromosomes are then arranged in descending order (**FIGURE BOX 5.1 FIGURE 1**). The chromosomes are also characterized by a banding pattern. The study of karyotypes is referred to as cytogenetics. Aberrant chromosomes or chromosome number can serve as markers of certain genetic diseases.

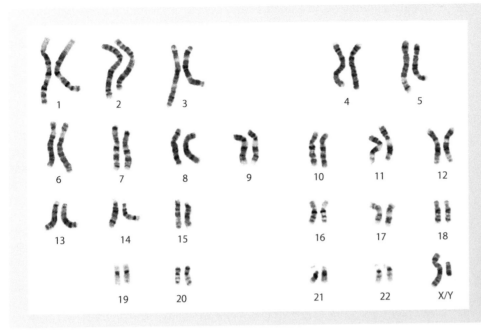

BOX 5.1 FIGURE 1 Karyotype of normal human male. Chromosomes are characterized by size and banding patterns. (With permission from Shutterstock/Zuzanae.)

Genes can be mapped to specific locations on chromosomes

The unique banding pattern of each chromosome allowed early geneticists to demonstrate that certain traits were associated with distinct regions of a specific chromosome. Before the description of genes in terms of DNA sequence (Chapter 4), the word locus (loci plural) was used to describe the inheritance of traits in reference to the association of traits with specific locations on chromosomes. The designation of the location of genes on specific chromosomes is generally referred to as chromosome mapping. Thus, genes can be defined by their location on a chromosome, the specific region of DNA that is transcribed, and the protein that is encoded.

Chromosomes are divided into a p-arm (short) and a q-arm (long) based on the position of the centromere. The centromere is a constriction of the chromosome that the mitotic spindle attaches to during cell division. The bands, called cytogenetic coordinates, are numbered going outward from the centromere and designated as either p or q (FIGURE 5.2). Gene mapping studies associate a gene with a specific cytogenetic coordinate. For example, the gene responsible for cystic fibrosis (see Box 11.2) is located near band 31.2 of the q-arm of chromosome 7 and is designated 7q31.2. Since a chromosome corresponds to a linear DNA molecule, genes can also be mapped in reference to the chromosome length as measured in base-pairs. In this regard, the gene for cystic fibrosis is encoded by DNA starting at base-pair 116,907,253 and ending at base-pair 117,095,955.

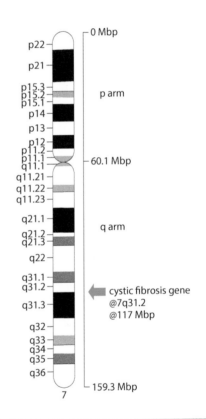

FIGURE 5.2 Map of chromosome 7. Shown is a depiction of chromosome 7 with cytogenetic coordinates based on the banding pattern and location of the centromere (pink) on the left side and a scale based on a million basepairs (Mbp) on the right. The blue arrow indicates the position of the gene responsible for cystic fibrosis with coordinates based on banding pattern and Mbp. (From the National Center for Biotechnology Information, United States National Library of Medicine.)

SEXUAL REPRODUCTION AND MEIOSIS

Unicellular organisms can reproduce through cell division. Eukaryotic cells replicate via mitosis and prokaryotes replicate via binary fission. These asexual processes involve copying the DNA and dividing it between two daughter cells. All the progeny of this asexual replication are clonal in that the progeny all have the same DNA sequence except for errors made during the DNA replication process. Reproduction in multicellular organisms is more complex and involves specialized reproductive organs and the formation of eggs and sperm. Collectively, eggs and sperm are

SIGNPOST

For a detailed description of cellular replication see Chapter 2, pages 41–44.

called gametes and gametes carry genetic information from generation to generation. This sexual reproduction, in contrast to asexual reproduction, is generally not a clonal process as the progeny resulting from sexual reproduction all have unique DNA sequences.

Sexual reproduction involves the formation of gametes via meiosis

Gametes are produced from progenitor cells via gametogenesis and a specialized type of cell division called **meiosis**. Meiosis consists of two sequential rounds of cell division called meiosis 1 and meiosis 2 and, therefore, four cells are generated. In the case of sperm formation, all four cells can develop into sperm (FIGURE 5.3). An egg, however, needs to be able to support the early development of the zygote and thus is larger. Oogenesis, or egg formation, proceeds with unequal cytoplasmic divisions resulting in the formation of a single egg and three other smaller cells called polar bodies.

During meiosis, diploid cells are converted to **haploid** cells that only contain one set of chromosomes. For example, the normal human karyotype consists of 23 pairs of chromosomes (see Box 5.1) and each egg or sperm produced via meiosis only has one of each of the 23 pairs. This means that even though gametes have half as much total DNA as somatic

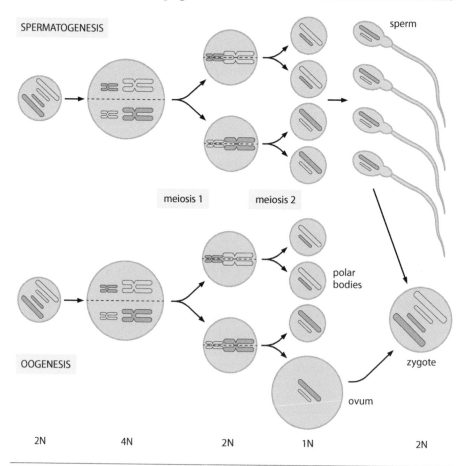

FIGURE 5.3 Meiosis and the formation of gametes. The gamete progenitor cells replicate their DNA and duplicate their chromosomes as they go from a diploid state (2N) to a tetraploid state (4N). During the first meiosis the homologous pairs of chromosomes line up and segregate, resulting in a return to the cells to a diploid state (2N). In the second meiotic division the chromatids are separated resulting in four haploid cells (1N). In spermatogenesis all four cells develop into sperm. In oogenesis there is an unequal cytoplasmic division during meiosis resulting in a single egg and three polar bodies. The diploid state (2N) is regenerated upon fertilization of the ovum by a sperm.

cells, gametes all have a complete set of chromosomes. However, since males contain one X chromosome and one Y chromosome, individual sperm have slightly different karyotypes as half of the sperm have an X chromosome and half have a Y chromosome. Thus, the sex of the progeny is determined by whether a sperm has an X (female) or Y (male) chromosome. The diploid state is regenerated at fertilization with the fusion of an egg and a sperm that combines the maternal and paternal chromosomes. In other words, half of the chromosomes in the zygote are derived from the mother and half are derived from the father resulting in 23 pairs of chromosomes.

Meiosis occurs in two steps to produce haploid gametes

The two rounds of cell division associated with meiosis exhibit some similarities to mitosis. Meiosis 1 starts the same as mitosis with the DNA being copied and the chromosomes being duplicated. In contrast with mitosis, the duplicated homologous chromosomes associate with each other and line up in the center of the cell. In addition, during the first cell division the chromosome pairs are segregated (FIGURE 5.4), instead of the sister chromatids, as is the case during mitosis. This results in a diploid cell containing one copy of each chromosome that is composed of the joined sister chromatids. These newly formed cells undergo a second round of cell division called meiosis 2. During meiosis 2 there is no DNA synthesis and no chromosome duplication. The chromosomes now line up in the middle of the cell and the sister chromatids are pulled apart, as in conventional mitosis. This results in progeny cells that are haploid with each cell containing only one copy of each chromosome.

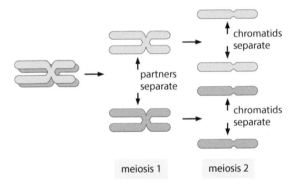

FIGURE 5.4 **Chromosome segregation during meiosis.** During meiosis 1 the homologous chromosomes pair and then segregate into the daughter cells. During meiosis 2 the individual chromatids separate and segregate into the daughter cells.

Maternally and paternally derived genes are randomly shuffled during meiosis

The individual chromosomes are segregated independently into the daughter cells during meiosis 1. This means that each individual gamete contains a unique mixture of maternally and paternally derived chromosomes. Since the chromosomes segregate independently, the number of unique combinations of chromosomes in a gamete can be defined mathematically as 2^n where n equals the number of chromosomes. In the case of humans this would be 2^{23}, equal to 8.4 million, different possible chromosome combinations found in the gametes. Thus, siblings from the same parents can have relatively different mixtures of chromosomes derived from their parents, and this accounts for the

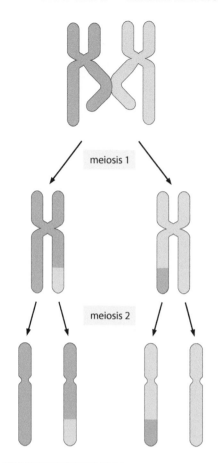

FIGURE 5.5 Crossing over during meiosis.
During the first meiosis the homologous chromosomes associate and exchange DNA via a genetic recombination called crossing over. This recombination results in the chromosomes becoming hybrids of the original paternal and maternal chromosomes.

differences between siblings that are often observed. Nonetheless, siblings are still more similar to each other than to non-relatives as they do share the same genes that originated from their parents.

There is a further mixing of the maternally and paternally derived genes during meiosis by a genetic recombination process called crossing over. During meiosis 1 when the homologous chromosomes are associated with each other there are recombination events between the chromosomes in which the DNA is broken in the same position on two of the chromatids and then rejoined to form hybrid chromosomes (FIGURE 5.5). These recombination events are an integral and necessary part of meiosis and occur at least once on every chromosome. Crossing over tends to occur in hotspots along the chromosome, but the hotspots are randomly dispersed along the chromosome. This recombination results in a further mixing of the maternally and paternally derived genes beyond the mixing that occurs during independent chromosome segregation.

ALLELES AND INHERITANCE

Although all humans essentially have the same genes, each individual gene can exist in multiple forms. Different forms of the same genes are called **alleles**. The alleles are due to differences in the DNA sequence of a particular gene. DNA sequence differences can result in a different amino acid sequence for the protein encoded by that gene. In most cases, such differences in amino acid sequence do not have a major effect on the structure and function of that protein. In some cases, the polymorphism can result in a protein that does not function correctly which results in a genetic disease (Chapter 11). Some sequence differences, even though they do not adversely affect a protein's structure or function, can result in a different trait. This partially explains the variation in traits we see between individuals, despite all humans having essentially the same genes. Even though individuals all have the same genes, different alleles of these genes result in different traits and variations between the individuals.

Changes in the DNA sequence introduce polymorphisms leading to the formation of new alleles

During DNA replication, mistakes are occasionally made resulting in changes in the DNA sequence. This can include the substitution of an incorrect nucleotide or the insertion or deletion of nucleotides. In addition, radiation or some chemicals can damage DNA and cause errors in the DNA sequence. The cell does have DNA repair mechanisms, but errors persist, nonetheless. If these errors are in the coding region of a gene the amino acid sequence of that protein can also be affected and, in turn, such a change could affect the function of that protein. Changes in non-coding regions may influence the levels of gene expression if the promoter or other regulatory sequence elements are affected. A change that has an adverse effect on the function of a protein or its expression might cause a genetic disease. In general, changes in the DNA sequence that have adverse effects are called mutations. However, most changes in a person's DNA sequence do not have an adverse effect and simply account for differences that are observed between individuals. In such cases the differences in the DNA sequence are referred to as **polymorphisms**.

Alleles can be homozygous or heterozygous

The diploid state means that every individual has two copies of every gene with the exception of males, who have one X chromosome and one Y chromosome. If both gene copies are the same on both chromosomes this is called **homozygous** (FIGURE 5.6). **Heterozygous** alleles refer to the situation in which the two chromosomes have different alleles of that gene. During meiosis only one allele for each gene is segregated into the gametes. For genes with heterozygous alleles, this means that half of the gametes will have one allele and half will have the other allele. Therefore, one would expect approximately half of the progeny of an individual to have one of the two possible alleles. This explains why siblings with the same parents can express numerous different traits as well as numerous similar traits. Substantial diversity is seen at the population level as humans are estimated to have 20,000–25,000 genes and in the human population each gene has an average of 250 alleles.

Possible allele combinations of single genes from the same parents can be illustrated by a Punnett square (FIGURE 5.7). The Punnett square is named after a geneticist from the early 1900s and provides a means to predict the frequency of alleles in the offspring. In the case of two parents who are heterozygous for the same two alleles of a particular gene, half of the offspring are expected to be heterozygous like the parents. However, one-quarter of the offspring would be expected to be homozygous for one allele and another quarter homozygous for the other allele. Thus, potentially half of the children have different allele combinations than either parent. The Punnett square is a useful tool for looking at the inheritance of one or a few genes in which each gene has a few possible alleles. The situation becomes more complex for traits determined by multiple genes and for genes that have a large number of alleles in the human population.

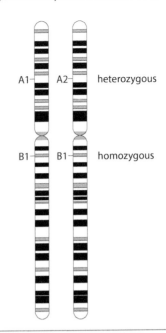

FIGURE 5.6 Interrelationship of genes, alleles, and chromosomes. Genes are located at defined positions on chromosomes as depicted by gene A and gene B. The gene is homozygous if the same allele is found on both chromosomes (gene B, allele 1), and heterozygous if different alleles are found on the two chromosomes (gene A, allele 1 and gene A, allele 2).

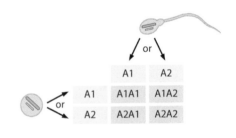

FIGURE 5.7 Punnett square indicating possible combinations of alleles from heterozygous parents. Sperm and eggs produced from heterozygous parents contain either allele 1 or allele 2 of gene A. This results in four possible combinations of alleles in the offspring.

GENES, ALLELES, AND PHENOTYPE

Observable traits in an individual are referred to as the **phenotype**. The genetic makeup of an individual is the **genotype**. Phenotype is due to the expression of genes and therefore the genotype, including specific alleles, is a component of phenotype. However, genotype does not absolutely dictate the phenotype in that other factors regulate gene expression. For example, a butterfly and a caterpillar have the same genotype, but quite distinct phenotypes. This is due to different genes being expressed in the caterpillar stage of the life cycle compared with the butterfly stage of the life cycle. Similarly, the various cells and tissues making up the human body have the same genotype, even though the phenotypes of the cells are different. Again, this is due to differential gene expression in different cell types. Furthermore, environmental factors can also influence gene expression.

However, individuals have the same genes that are generally expressed equivalently according to the developmental stage or cell type. Therefore, differences in phenotypes between individuals are due largely to differences in their alleles. For example, one individual may have alleles for blue eyes, whereas another individual has alleles for brown eyes. The composition of a person's alleles is their genotype. The situation of heterozygous alleles that encode different versions of a trait raises questions about how alleles interact within an individual to produce traits. The three major ways in which alleles interact are referred to as: (1) dominant-recessive, (2) incomplete dominance, and (3) co-dominance (TABLE 5.1).

Table 5.1 **Relationships between Genotypes and Phenotypes as Determined by Alleles***

Genotype	Dominant-recessive	Incomplete dominance	Co-dominance
	Phenotypes		
A1A1	A1	A1	A1
A1A2	A1	A1.5	A1 + A2
A2A2	A2	A2	A2

*The genotype and phenotype match if the alleles are homozygous. If heterozygous, the phenotype will be the same as the dominant allele (A1), or the phenotype will be between the two alleles if dominance is incomplete, or both alleles will be expressed if co-dominant.

Some alleles are dominant in that they mask the expression of recessive alleles

For many genes it is common for one allele to be dominant and the other allele to be recessive. The dominant allele masks the recessive allele, so that the trait encoded by the recessive allele is not observed. In other words, the same trait is observed in individuals who are homozygous for the dominant allele and individuals who are heterozygous. Even though these heterozygous individuals have different genotypes, they have the same phenotype (see Table 5.1). Only individuals who are homozygous for the recessive alleles exhibit the phenotype of the recessive allele. Dominant-recessive traits are rather common in genetic diseases (Chapter 11). In these cases, the allele associated with disease is non-functional and being homozygous for the non-functional allele results in the disease. Having a single functional allele is often sufficient to preclude the disease. Another common example of a recessive trait is blue eyes. Many forms of blue eyes exhibit a recessive phenotype and are due to the lack of pigment production associated with brown eyes (BOX 5.2).

BOX 5.2

The Genetics of Blue Eyes

Eye color is largely determined by pigmentation and light scattering in the iris of the eye. Brown eyes are determined by the amount of pigmentation in the iris and can range from light to dark brown. Blue and other light-colored eyes are due to the lack of pigment and the blue appearance is due to a phenomenon associated with light scattering similar to what makes the sky appear blue. The presence of pigmentation masks the light scattering and thus brown eyes are dominant over blue eyes. Historically it was believed that eye color was determined by a single gene that exhibited a dominant-recessive phenotype. It is now known that several genes involved in pigmentation of the eyes contribute to eye color. For example, the *OCA2* gene is involved in the production of the pigment associated with brown eyes. A polymorphism in the regulatory region of this gene prevents its expression and therefore blocks pigment production. Therefore, pigment is lacking in those individuals who are homozygous for this polymorphism. Similarly, the *HERC2* gene regulates the expression of the *OCA2* gene and some *HERC2* polymorphisms repress the expression of *OCA2*. Polymorphisms in another gene involved in the synthesis of pigment have also been associated with differences between blue eyes and green eyes.

One consequence of dominant-recessive traits is that heterozygous parents can have offspring who exhibit the recessive phenotype, even though both parents exhibit the dominant phenotype. As half of the sperm and half of the eggs produced by heterozygous parents contain the recessive allele, there is a one in four chance that a sperm with the recessive allele combines with an egg with the recessive allele as illustrated by a Punnett square (see Figure 5.7). In this scenario, three-quarters of the offspring are expected to have the same dominant phenotype as the parents and one-quarter of the offspring are expected to have the recessive phenotype.

Incomplete dominance results in an intermediate phenotype

Sometimes neither allele is dominant and both alleles contribute to the phenotype in the heterozygote. An example from plants is the situation in which crossing a red-flowered plant with a white-flowered plant results in plants with pink flowers (FIGURE 5.8). In other words, the phenotype is between the phenotype of the red homozygote and the white homozygote. The red phenotype is due to enzymes that produce a red pigment. In the case of the heterozygote, half as much pigment is produced due to expression of a single copy of the gene responsible for the formation of the pigment. Offspring of two heterozygotes with pink flowers consist of one-quarter red flowers (homozygous for the red allele), one-quarter white flowers (homozygous for the white allele), and one-half pink flowers (heterozygous).

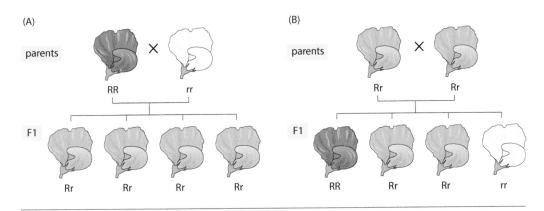

FIGURE 5.8 Incomplete dominance results in a blending of traits. (A) A cross between a homozygous red flower (RR) and a homozygous white flower (rr) results in an intermediate pink phenotype in the heterozygous offspring (Rr). (B) A cross between heterozygous pink flowers results in all three phenotypes in the ratio of 1(RR):2(Rr):1(rr). Dominant alleles are usually designated with a capital letter (e.g., R) and recessive alleles in lower case (e.g., r).

Both traits associated with co-dominant alleles are observed in heterozygotes

In co-dominance there is no apparent blending or intermediate state in the heterozygotes as observed in incomplete dominance. Instead, both alleles are fully detectable in the heterozygous individual. An example of co-dominance is the ABO blood types in humans that dictate blood transfusions between individuals (BOX 5.3). Human blood types are determined by a single gene located on chromosome 9 encoding an enzyme called glycosyltransferase, which transfers a carbohydrate to glycoproteins on the red blood cell surface. There are three major alleles, designated A, B, and O, of this glycosyltransferase gene in the human population. Alleles A and B are co-dominant and allele O is recessive to both alleles A and B.

Allele A transfers galactosamine to the glycoproteins on the surface of red blood cells and allele B transfers galactose. Allele O is non-functional and does not transfer any carbohydrate to the glycoproteins resulting in type O blood group. A person homozygous for the A allele or

BOX 5.3

ABO Blood Groups and Blood Transfusion

The four different human blood groups, designated as A, B, AB, and O, were first described in the early 1900s and are the basis for successful blood transfusions. Blood types are due to two different alleles (A and B) of a glycosyltransferase gene that result in two different glycoproteins on the surface of red blood cells, and a third allele (O) that is non-functional and recessive. These cell surface glycoproteins function as antigens in that the glycoproteins are recognized by antibodies in blood serum. During fetal development the immune system learns to tolerate self-antigens (Chapter 9) and does not produce antibodies that recognize antigens present on an individual's cells. Therefore, a person with type A blood does not have antibodies that recognize the type A glycoprotein. However, that person does have antibodies that recognize the type B glycoprotein. Similarly, a person with type B blood has antibodies to the type A glycoprotein, but not to the type B glycoprotein.

If blood from a person with type A blood is transfused into a person with type B blood, or vice versa, the antibodies present in the recipient's blood serum bind to the donated red blood cells and cause the blood cells to clump together. These clumps can block blood vessels and cause organ damage and even result in death. Therefore, it is critical that persons with type A or type B blood receive transfusions with a matching blood type. Persons with type AB blood do not produce antibodies to either type of glycoprotein, and therefore, can receive blood from donors of any blood type. Accordingly, persons with type AB blood are known as universal recipients. Similarly, the blood cells of persons with type O blood do not have either type A or type B glycoproteins on their cells and thus can donate blood to anyone since there are no antigens that are recognized by the antibodies. Thus, persons with type O blood are universal donors. However, persons with type O blood have antibodies against both type A and type B glycoproteins and therefore can only receive blood from type O donors.

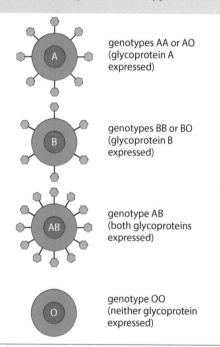

genotypes AA or AO
(glycoprotein A
expressed)

genotypes BB or BO
(glycoprotein B
expressed)

genotype AB
(both glycoproteins
expressed)

genotype OO
(neither glycoprotein
expressed)

FIGURE 5.9 Human blood types are determined by the alleles of the ABO glycosyltransferase gene. The ABO glycosyltransferase catalyzes the addition of either a galactosamine (A allele, blue) or galactose (B allele, green) to proteins found on the surface of red blood cells. The O allele is non-functional and does not transfer either galactosamine or galactose to the surface of the red blood cells. This lack of activity makes the O allele recessive to both the A and the B alleles. The six possible genotypes between these three alleles result in four phenotypes (A, B, AB, or O) due to the co-dominance of the A and B alleles and the recessive nature of the O allele.

heterozygous for the A allele and the O allele expresses galactosamine on the red blood cell surface and has type A blood (FIGURE 5.9). Similarly, a person homozygous for the B allele or heterozygous for the B and O alleles has type B blood due to the galactose on the surface of red blood cells. A person who is heterozygous for the co-dominant A and B alleles expresses both types of glycoproteins on the red blood cell surface and therefore has type AB blood. Since the O allele is recessive, type O blood is only observed in individuals who are homozygous for the O allele.

Many traits are influenced by several genes with numerous alleles

Traits determined primarily by a single gene are called **monogenic**. The inheritance patterns exhibited by monogenic traits, such as dominant-recessive, incomplete dominance, and co-dominance, are often referred to as Mendelian inheritance in honor of Gregor Mendel, who first described these inheritance patterns. However, many, if not most, traits are influenced by multiple genes, with each gene making a relatively small contribution to the overall trait. Such traits are referred to as **polygenic**. This means that the combined effects of alleles from several genes determine the observed trait. Polygenic traits are typically expressed along a continuum rather than as distinct traits and, thus, a continuous range of phenotypes is usually observed. For example, at least four genes influence height and, accordingly, height exhibits a continuous range in the human population. Quite often such polygenic traits have fewer individuals at the extremes of the trait and more individuals expressing an intermediate level of the trait in a population (FIGURE 5.10) and thereby resulting in the classic bell-shaped curve.

The number of phenotypes found in a population is further complicated by the fact that most genes have numerous possible alleles. Even

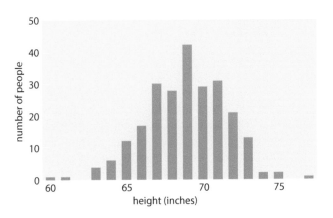

FIGURE 5.10 Example of continuous phenotypes in polygenic traits. Polygenic traits often exhibit a continuum of phenotypes in the population with higher frequencies of individuals with intermediate phenotypes compared with individuals with extreme phenotypes (i.e., the bell-shaped curve) as illustrated with this example of heights among male secondary-school seniors. [From Kimball, JW. Unit 8.6 Quantitative trait loci. In Kimball, JW (ed.), Biology. Available at https://bio.libretexts.org/@go/page/4857. Published under CC BY 3.0.]

though an individual can only have two alleles of any gene, there are many more alleles for any specific gene in a population. For example, the aforementioned ABO glycosyltransferase gene has three alleles. Many genes exhibit numerous alleles and are highly polymorphic within the human population. An example of highly polymorphic genes is the human leukocyte antigen (HLA) genes that have hundreds of alleles in the human population. The HLA proteins play a role in immunity (Chapter 9), and it is these HLA proteins that need to be matched between donor and recipient in organ transplantation (see Box 17.3). The matching of donor and recipient involves identifying individuals who have the same alleles of these HLA genes. Because there are several genes with numerous alleles, finding suitable matches can be difficult.

Some genes can affect multiple traits

Another complexity of inheritance involves genes that have multiple phenotypic effects, a phenomenon called pleiotropy. For example, genes for transcription factors or other regulatory proteins affect the expression of multiple genes and proteins. A polymorphism in such a regulatory gene in turn can affect the phenotype associated with several genes and thus affect several traits. Similarly, some metabolites are utilized by multiple metabolic pathways. Polymorphisms in enzymes that affect such metabolites may affect several potential end-products, and therefore influence multiple traits.

GENETIC LINKAGE

The recombination, or crossing over, that occurs during meiosis results in the mixing of the maternal and paternal genes on a single chromosome. Without crossing over, all the genes on a single chromosome would be derived from either the mother or the father and would always be inherited together. Crossing-over events occur randomly along the paired chromosomes and the chance that a recombination event occurs between any two genes is inversely correlated with the distance between those two genes on the chromosome. If the two genes are far apart on a chromosome, then it is quite likely that one or more crossing-over events will occur between those two genes during meiosis. However, if the genes are close together, then it is less likely that a crossing-over event will occur between them (FIGURE 5.11). Genetic linkage refers to the observation that alleles of genes that are close together on the same chromosome tend to be inherited together. In practical terms this means that the specific traits encoded by these linked alleles have a strong tendency to be inherited together.

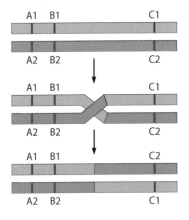

FIGURE 5.11 Genetic linkage. Specific alleles of genes tend to be inherited together if the genes are physically close to each other on a chromosome (e.g., genes A and B) since a crossover event between the two genes is less likely. Therefore, haplotypes (A1/B1 and A2/B2) occur more frequently in a population than expected. If the genes are far apart (e.g., gene C), recombination between the chromosomes is likely and gene C is not likely to form a haplotype with genes A and B since the alleles C1 and C2 are equally likely to be found with either the A1B1 haplotype or the A2B2 haplotype.

SIGNPOST

Genome-wide association studies are described in detail in Chapter 6, pages 124–130.

Linked genes define a haplotype

The alleles from linked genes that tend to be inherited together are called a **haplotype**. Because of the genetic linkage, the genes forming a haplotype are inherited from a single parent and can remain linked for many generations. Because of their stability across many generations, haplotypes can be useful in genealogy studies. Haplotypes can also be defined at the level of DNA sequence and the current technology allows a detailed analysis of genetic linkages at the nucleotide level. For example, single nucleotide polymorphisms (SNPs) in the DNA sequence between individuals can also form haplotypes and are used in genome-wide association studies to map traits and diseases to specific locations on chromosomes.

Linkage disequilibrium can be used to identify genes encoding a particular trait and map genes to specific locations on chromosomes

In addition to being useful in genealogy studies, linkage studies and haplotypes can be used to identify and map genes. Such gene identification is accomplished through **linkage disequilibrium**. Linkage disequilibrium is defined as a non-random association of alleles from different genes in a population. If genes are not linked, then the association of alleles between different genes is proportional to the frequencies of those alleles in the population. For example, in the simple case presented in Figure 5.11, if both alleles 1 and 2 for both genes A and B are distributed in the population at 50% and the two genes are not linked, then one would expect the distribution of haplotypes A1/B1, A1/B2, A2/B1, and A2/B2 to all be 25% in the population. A deviation from this expected distribution is defined as linkage disequilibrium and implies a physical association of the two genes on the same chromosome.

Linkage disequilibrium can also be used to identify genes that encode a particular trait. For example, if a particular trait in which the gene is not known exhibits a linkage disequilibrium with a trait in which the gene is known, then both genes are likely to be located close to each other on the same chromosome. The DNA sequence in this region of the chromosome can then be searched for the unknown gene. Linkage disequilibrium can also be used to assemble chromosome maps defining the locations of genes on the chromosomes.

ENVIRONMENTAL INFLUENCES AND EPIGENETICS

A person's genotype plays a large role in their phenotype, but not an absolute role. In addition to the exact characteristics associated with a particular allele, phenotype is also determined by the combinations of the genes that are expressed and the level of expression of those genes. For example, identical twins have identical genotypes, but can still exhibit different phenotypes. In fact, many traits are determined by a complex combination of heredity and environment. Even though one may inherit genes for tallness, poor nutrition can limit growth and an individual may not reach their true potential in height. The fact that many traits are polygenic combined with the external influences on gene expression sometimes makes it difficult to discern the exact roles that genes and environment play in determining some traits.

Chromosomal regions can adopt stable and heritable states that affect gene expression

Conventional wisdom is that inheritance is based on the information encoded within the nucleotide sequence of the genome. However, there are some heritable changes that are not encoded within the DNA sequence but are due to effects on gene expression. **Epigenetics** refers to inheritable changes that affect gene expression and phenotype but are not due to alterations in the nucleotide sequence. Such epigenetic changes are due to changes in chromosome structure that result from modifications of DNA and chromosomal proteins. For example, cytosine residues in DNA can be methylated by specific enzymes (FIGURE 5.12). Heavily methylated regions of DNA tend to be transcriptionally inactive. This DNA methylation pattern can be maintained after cell division and therefore the transcriptional propensity of a particular region of DNA can be passed onto daughter cells.

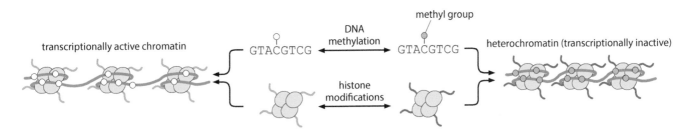

FIGURE 5.12 Epigenetic changes affect chromatin structure. Specific enzymes can add (white circles) or remove (pink circles) methyl (CH$_3$) groups to cytosine residues in DNA. Other enzymes can add or remove acetyl groups (blue) or methyl groups (purple) to histones (yellow). DNA and histone methylation lead to the formation of heterochromatin that is transcriptionally inactive.

Another potential epigenetic alteration is the post-translational modifications of the histones, which organize DNA into chromosomes (see Figure 5.1). Like DNA methylation, the post-translational modification of histones also affects regional chromosome structure and affects transcription in the modified regions. Some of these histone modifications result in the formation of heterochromatin, which is a condensed form of chromatin that is transcriptionally inactive. Histones exist in either an acetylated state or a methylated state and acetylation is generally associated with transcriptionally active chromosomes, whereas histone methylation is generally associated with heterochromatin (see Figure 5.12). Although the mechanism is not clear, these regions of heterochromatin can be inherited when cells replicate, meaning that the daughter cells have the same patterns of heterochromatin. The pattern of DNA methylation and histone modifications is sometimes called the 'histone code'. This epigenetic inheritance of heterochromatin plays a role in the differentiation of cells into specialized tissues and organs (Chapter 8). In addition, there is some evidence that such epigenetic inheritance can exhibit a transgenerational inheritance from parent to offspring in that in utero environmental exposures can have an effect on multiple generations.

GENETIC VARIATION IN POPULATIONS AND EVOLUTION

The mechanisms of inheritance discussed above generally focus on the flow of genetic information through the generations as genes pass from parents to offspring. Gene flow can also be followed in populations. This specialty, often called population genetics, studies the distribution and changes in the frequency of alleles within populations. The **gene pool** refers to all the alleles for all the genes in a population. Some alleles may be relatively common, whereas other alleles may be quite rare. The relative abundance of an allele is defined as the allele frequency, which is the prevalence of that allele in the population expressed as a fraction or percentage of the total alleles in that population. As part of evolution allele frequencies within biological populations change over time and, thus, characteristics of a population also change through the generations. Population genetics provides information about genetic diversity within a population, as well as evolutionary trends.

Forces of evolution include the introduction of new alleles and changes in allele frequencies

Polymorphisms in the DNA sequence are continuously introduced into genomes via mutations and errors in DNA replication. These changes in the DNA sequence result in new alleles. Polymorphisms that are acquired in somatic cells during mitosis do not enter the gene pool. However, such polymorphisms and mutations can play a role in the development of cancer (Chapter 16). For a DNA change to enter the gene pool, the change needs to occur during meiosis and the production of gametes.

The introduction of new alleles and the fate of specific alleles through the generations determine the evolution of a species (FIGURE 5.13). For example, some individuals leave more offspring and descendants than other individuals. Thus, the alleles of the more prolific individuals increase in frequency within the gene pool, and the alleles of the less prolific individuals decrease. The change in allele frequencies can be random or non-random. Changes in allele frequencies that are due to chance events or random mating are referred to as genetic drift. In contrast with this random genetic drift, certain alleles may increase or decrease in frequency because they are advantageous or disadvantageous to the individual. This non-random change in allele frequency is known as **natural selection**.

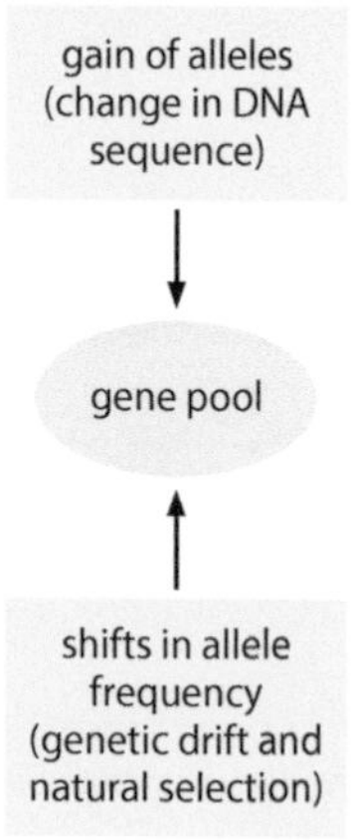

gain of alleles
(change in DNA
sequence)

↓

gene pool

↑

shifts in allele
frequency
(genetic drift and
natural selection)

FIGURE 5.13 Forces affecting the gene pool and evolution. New alleles can be introduced in the gene pool through changes in DNA sequence due to errors in replication. Genetic drift and natural selection influence allele frequencies.

Alleles may be subjected to natural selection

Allele frequency in a population depends on the survival and reproductive success of those individuals with a particular allele. For example, alleles that offer advantages to an individual in terms of survival or reproductive success result in that individual producing more descendants, and such an allele tends to increase in frequency in the population over time. Conversely, if an allele has an adverse effect on survival or reproductive success, then that allele tends to decrease in frequency.

However, whether an allele is advantageous or disadvantageous may be conditional as environment also influences evolution. An example of such natural selection due to a changing environment is the development of drug resistance in pathogens. Pathogens that have drug-resistant alleles will out-reproduce those individuals that are drug sensitive in the presence of drugs. This results in drug-resistant pathogens replacing drug-sensitive pathogens. However, in the absence of drugs, there is no selective advantage for the drug-resistant alleles and allele frequency is determined by genetic drift. As environments can rapidly change, genetic diversity within a population is a buffer against environmental changes. Without diversity in a population, a major environmental change could lead to the extinction of a population.

SIGNPOST

Mechanisms of drug resistance are described in detail in Chapter 19, pages 379–382.

Multiple genes and environmental pressures influence evolution

Natural selection and evolution are complex processes in that multiple alleles and multiple genes can be involved, as well as environmental factors. Interactions and linkages between individual genes further complicate the process. For example, neighboring genes on a chromosome are physically linked and tend to be inherited together as a haplotype. This means that it is possible for an allele to increase or decrease in frequency due to this linkage. This is often referred to as genetic hitchhiking and means that alleles under no selective pressure can change in frequency due to selective forces on neighboring genes. In addition, genes that are not physically linked on chromosomes may be functionally linked, requiring that certain alleles of the two genes be inherited together since the two alleles complement each other in some fashion.

Opposing environmental forces may also promote the selection of alleles that are favorable under some circumstances and unfavorable under other circumstances. This is called **balanced polymorphism**. One example of such a balanced polymorphism is the hemoglobin allele resulting in sickle cell anemia (see Box 11.1). Individuals who are homozygous for the sickle cell allele produce hemoglobin molecules that clump and weaken the erythrocytes, causing a severe disease that generally results in premature death. Normally there would be a strong negative selection against this allele and the allele would be of low frequency in the population, or even disappear completely. However, in some areas of tropical Africa the sickle cell allele is unexpectedly common. Interestingly, the areas in which the sickle cell allele is higher than expected coincide with areas that are endemic for malaria. This suggests that the sickle cell allele offers some protection against malaria (BOX 5.4).

Balance Polymorphism and Sickle Cell Disease

The frequency of sickle cell disease is higher in African populations and specifically in geographical areas that are endemic for malaria. This distribution of the sickle cell allele seems to be due to a protective effect the sickle cell allele has against malaria. Individuals who are heterozygous for normal hemoglobin and the sickle cell allele do not suffer from malaria as badly as those who are homozygous for normal hemoglobin. In addition, the heterozygotes only suffer a mild form of sickle cell anemia and therefore the sickle cell allele is considered a recessive trait. Thus, individuals who are heterozygous for the two alleles do not suffer from severe sickle cell disease and are partially protected from developing severe malaria. This means that, in the presence of malaria, the heterozygous individuals survive better and leave more offspring than those individuals who are homozygous for normal hemoglobin or homozygous for the sickle cell trait (**BOX 5.1 FIGURE 1**).

During evolution in areas endemic for malaria, the heterozygotes survived better than either of the two types of homozygotes and therefore left more descendants. As a result, both the normal hemoglobin allele and the sickle cell allele are preserved in the gene pool in the population exposed to endemic malaria. Without the pressure of malaria, the sickle cell allele would not offer any survival or reproductive advantages and therefore would decrease in the gene pool. This explains the high prevalence of sickle cell anemia in some African populations and other populations from areas endemic for malaria.

BOX 5.4

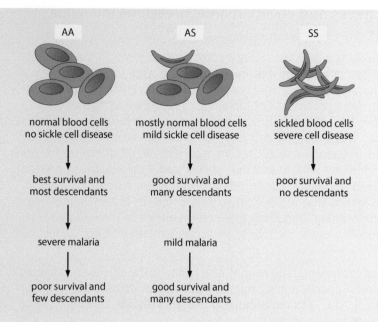

AA

normal blood cells
no sickle cell disease

↓

best survival and
most descendants

↓

severe malaria

↓

poor survival and
few descendants

AS

mostly normal blood cells
mild sickle cell disease

↓

good survival and
many descendants

↓

mild malaria

↓

good survival and
many descendants

SS

sickled blood cells
severe cell disease

↓

poor survival and
no descendants

BOX 5.4 FIGURE 1 Effect of malaria on maintaining the sickle cell allele. Maintenance of both the normal β-hemoglobin allele (A) and the sickle cell allele (S) is due to the influence of malaria and the protection from severe malaria in the heterozygous (AS).

PHYLOGENETICS

Phylogenetics is the study of the evolutionary history and relationships among individuals or groups of organisms, such as species or populations within a species. Such phylogenetic analyses can also be used to address questions about the origin of infectious diseases and to address the evolution of pathogens. In addition, molecular phylogenetics is useful in determining the roles of genes in disease, as well as elucidating our relationships with other organisms.

Generally, phylogenetic information is expressed in the form of phylogenetic trees. Phylogenetic trees graphically show evolutionary relationships via a branching structure (**FIGURE 5.14**). The ends of branches represent individuals, species, or other higher taxa. A taxon is a group of similar organisms that are generally related evolutionarily and quite often placed in hierarchical groups such as species, genera, families, orders, classes, phyla, and kingdoms. The point where the two branches meet, called the node, represents a hypothetical common ancestor shared by the two individuals or taxa. All the individuals derived from a common ancestor form a clade and share a node on a phylogenetic tree. Trees can be rooted, in which there is a clear bottom of the tree. In rooted trees the order of the branching depicts the evolutionary history.

FIGURE 5.14 Components of a phylogenetic tree. The ends of the branches represent an individual or species. The junction point, called the node, between branches represents a common ancestor shared by all individuals, subsequently branching from that point. All individuals joined by a common node form a clade. For example, all mammals would be included in the clade branching from the first hypothetical mammal. Similarly, the node at the root, or base of the tree, would include all vertebrates.

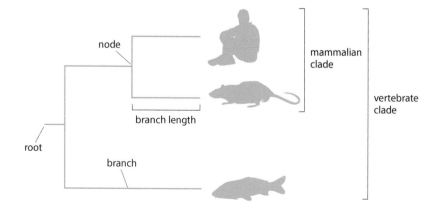

Unrooted trees demonstrate relationships between individuals or species without implications about the order of evolution. In some types of trees the branch lengths depict the evolutionary distance between species.

Evolutionary relationships can be determined by shared characteristics

Phylogenetic trees are developed by comparing characteristics between organisms. Characteristics can be any measurable or observable trait, and can include morphological, behavioral, biochemical, or molecular features. Each characteristic in the dataset exists in multiple states. For example, eye color could be a characteristic and the characteristic states would be blue and brown. Or the states could broaden to include lighter and darker shades of blue and brown and variations such as green. In the most simplistic terms, phylogenetic trees are generated by adding up the shared characteristic states between the various organisms in a pair-wise fashion and then constructing a tree to reflect these similarities and differences. Two organisms with the most shared characteristic states are presumed to be the most closely related and share a node. For example, humans and rats share the common features of mammals and the node between them represents the original mammal (see Figure 5.14). Similarly, fish share a common node with mammals represented by the first vertebrate.

Molecular phylogeny utilizes sequence data to determine evolutionary relationships

Molecular sequence data, such as the nucleotide sequence of DNA and RNA or the amino acid sequence of proteins, can also be used to generate phylogenetic trees. In this case, homologous sequences are compared and the similarities and differences in the sequences are used to determine how closely two organisms or individuals are related. Homologous sequences refer to genes and proteins that have a common ancestry and generally a common function. The differences in sequences are due to polymorphisms that are introduced and accumulate in the DNA sequence through the generations. Therefore, the sequence polymorphisms between homologous genes or proteins are an approximation of the number of generations that have passed since the organisms or individuals shared a common ancestor. The interpretation of analyses based on molecular sequence data is more straightforward and often less ambiguous than data based on other characteristics.

Homologous sequences can be aligned

The first step in comparing sequences is to align the sequences. Sequence alignment consists of arranging the sequences to be compared in linear order beginning with the first nucleotide at the 5'-end or the first amino acid at the N terminus. The sequences being compared are put into rows so that columns are formed by positions in the sequence that correspond to equivalent nucleotides or amino acids. For example, the first amino acid in the sequences in FIGURE 5.15 is leucine (L). The next column is the second amino acid in the proteins being compared and each subsequent column represents the next amino acid. As there can be insertions or deletions in the sequences, gaps may need to be inserted into some of the sequences to obtain the optimal alignment. The optimal alignment is defined as that alignment with the most positions (columns) that have identical or similar residues (amino acids or nucleotides).

```
              10          20          30          40          50
               |           |           |           |           |
(1) LTPEEKSAVTALWGKVN--VDEVGGEALGRLLVVYPWTQRFFESFGDLST
(2) LSGEEKAAVLALWDKVN--EEEVGGEALGRLLVVYPWTQRFFDSFGDLSN
(3) LSPADKTNVKAAWGKVGAHAGEYGAEALERMFLSYPTTKTFFPHF-DLS-
(4) LSAADKTNVKAAWSKVGGHAGEYGAEALERMFLGYPTTKTFFPHF-DLS-
    *:   :*. * *.* **    * *:*** *::: ** *..** * ***
```

FIGURE 5.15 Alignment of protein sequences. A graphic showing the alignment of four proteins using the single-letter amino acid code in which each letter represents an amino acid. Gaps (–) are inserted at positions where there is not a corresponding amino acid to align with the other sequences. The colors indicate groups of amino acids that are chemically similar (non-polar: yellow; polar: blue or purple; negatively charged: red; positively charged: gray). Below the alignment * indicates positions with identical amino acids, the colon (:) indicates positions with amino acids that are highly similar, and the period (.) indicates positions with amino acids that are somewhat similar.

Alignments reveal the degree of conservation at each position in the sequence. Positions along the alignment can range from being highly conserved, in which there is no or little variation at that position, to being non-conserved in which a wide range of nucleotides or amino acids is found at that position. In the case of proteins there can be partial conservation if an amino acid is replaced with an amino acid with similar properties. For example, the replacement of a hydrophobic amino acid with another hydrophobic amino acid, or the replacement of a charged amino acid with a similarly charged amino acid, may not have a major impact on protein structure and function. Long stretches in the sequence alignment that are highly conserved reflect domains of the protein that are important for the overall structure and function of the protein.

Aligned sequences are used to generate phylogenetic trees

Phylogenetic trees can be generated from aligned sequences using the same methodology based on characteristics and characteristic states. Positions (i.e., columns) in the alignment correspond to characteristics and the various nucleotides or amino acids in that position are the characteristic states. For example, for any position in a DNA or RNA sequence alignment there are five possible characteristic states corresponding to the four nucleotides (A, G, C, or T/U) or none if there is a gap at that position. Genetic distances between the aligned sequences are estimated from the number of mismatches in the aligned sequences. Protein sequence alignments are more complicated as mismatches can be differentially weighted based on chemical or physical similarities between the amino acids and some amino acid residues are weighted more heavily than others.

Computer programs are used to align the sequences and other computer programs generate the phylogenetic trees from the aligned sequences (FIGURE 5.16). There are many different programs available for the generation of phylogenetic trees from aligned sequences. These various programs may generate different trees depending on the exact algorithms used to generate the tree and how the different types of mismatches are weighted. Furthermore, due to the length and complexity of the aligned sequences, the same sequence alignment may produce multiple trees with different topologies. Therefore, the phylogenetic programs also include statistical analyses to determine which tree is the most likely for any given dataset, and to give scores that are a measure of the relative robustness of any generated tree.

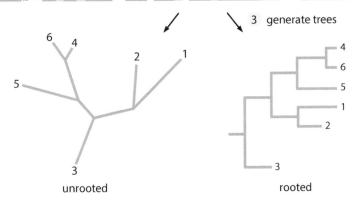

1 choose a gene set depending on question
being addressed or hypothesis being tested

2 aligned sequences

```
FNIRGGKASQLGIFISKVIPDSDAHRAG-LQEGDQVLAVNDVDFQDIEHSKAVEILKTAR-EISMRVR  1
ISITGGKEHGVPILISEIHPGQPADRCGGLHVGDAILAVNGVNLRDTKHKEAVTILSQQRGEIEFEVV  2
ISIKGGRENRMPILISKIFPGLAADQSRALRLGDAILSVNGTDLRQATHDQAVQALKRAGKEVLLEVK  3
FNVMGGKEQNSPIYISRIIPGGVAERHGGLKRGDQLLSVNGVSVEGEHHEKAVELLKAAKDSVKLVVR  4
FNIMGGKEQNSPIYISRVIPGGVADRHGGLKRGDQLLSVNGVSVEGEQHEKAVELLKAAQGSVKLVVR  5
FNIMGGKEQNSPIYISRIIPGGIADRHGGLKRGDQLLSVNGVSVEGEHHEKAVELLKAAQGKVKLVVR  6
```

3 generate trees

unrooted

rooted

FIGURE 5.16 Generation of phylogenetic trees. The first step when making a phylogenetic tree is to collect the sequences of genes or proteins that are going to be used to generate the tree. The choice of genes or proteins depends on the type of question being asked. A computer program is used to align the sequences. The aligned sequences are then used to generate phylogenetic trees using other computer programs. The trees can be presented in different topographies, such as rooted or unrooted, depending on the questions being addressed or additional knowledge about the sequences being compared.

Molecular phylogenetics can be used to determine evolutionary relationships between organisms or individuals in a population. Regarding human health, such analyses can be used to track the spread of a pathogen in the human population and possibly reveal the evolution of virulence or drug resistance in that pathogen. The phylogenetic approach can also be used to examine the evolution of a specific gene or gene family. Such information may provide insight into drug development.

HUMAN EVOLUTION

Human biology is the product of human evolution. Humans, like all other organisms, are subject to selective forces and these forces have also shaped social and cultural attributes associated with being human. Human physiology is overall similar to other mammals and especially other primates, and thus has not been extensively altered by evolution. Distinct human features include bipedalism and walking upright, a rather sophisticated symbolic language, advanced cognition, and manual dexterity. This evolution to modern humans has not been straight lines with branches, but somewhat of a meshwork involving the appearance and disappearance of several archaic human species. These various species have intermingled, and sometimes interbred, to produce the single human species that currently exists.

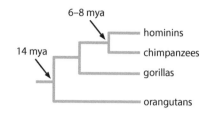

FIGURE 5.17 Phylogenetic relationships of hominids. The original ancestor of the great apes, or hominids, arose around 14 million years ago (mya). The branch leading to the evolution of humans (hominins) diverged from the chimpanzees between 6 mya and 8 mya.

Humans arose in Africa and are closely related to the great apes

The similarities of gorillas, chimpanzees, and humans have been long recognized and phylogenetic evidence suggests that chimpanzees and humans shared a common ancestor 6–8 million years ago (mya). Humans and chimpanzees have nearly 99% DNA sequence identity in regions that can be directly compared and overall exhibit approximately 96% sequence identity. On average, homologous proteins from humans and chimpanzees only have a single amino acid polymorphism. The great apes, which include humans, chimpanzees, gorillas, and orangutans, diverged from other primates approximately 14 mya (FIGURE 5.17). The next closest relative to humans is the gorilla, and the orangutan is the most distant relative.

The similarities of humans to the African apes and other data indicate that the human lineage arose in Africa. Collectively, members of this human lineage are called hominins and the great ape lineage are called hominids. Information about hominins is largely derived from fossils and thus is somewhat fragmentary. In addition, genetic analysis is generally not possible and, therefore, it is difficult, if not impossible, to determine the phylogenetics of the early hominins. One particularly prevalent early hominin was *Homo erectus* who lived from 1.8 mya to 70,000 years ago. *H. erectus* is also associated with the use of complex tools and possibly fire, as well as a substantial increase in brain capacity. *H. erectus* also expanded out of Africa into western Asia and ultimately into eastern Asia and Indonesia. It is believed that all hominins before *H. erectus* only lived in Africa.

Archaic modern humans include *Homo sapiens*, Neanderthals, and Denisovans

The earliest fossils that closely resemble modern humans date back about 300,000 years. Among the early modern humans, *Homo sapiens*, *H. neanderthalensis*, and the Denisovans are particularly notable. The Denisovans are the most recently discovered and have a close affinity with the Neanderthals. Therefore, they have not been given a species designation yet. Up until approximately 30,000 years ago these three distinct hominins with characteristics of modern humans coexisted contemporarily. Neanderthals lived in Europe 400,000–28,000 years ago and Denisovans lived in Asia 500,000–30,000 years ago. Fossil evidence also indicates that early *H. sapiens* were living outside of Africa 100,000–200,000 years ago. The phylogenetic relationships of these three early modern humans to *H. erectus* are not known.

Interbreeding and migration contributed to the evolution of modern humans

DNA sequence analysis indicates that extant non-Africans can trace their ancestry to a major migration out of Africa that occurred 50,000–60,000 years ago. This may have coincided with climatic events that lowered sea levels to facilitate passage from Africa to the Arabian Peninsula and the Middle East. These later *H. sapiens* eventually dispersed to all habitable places of the world and likely overwhelmed the prevailing *H. sapiens*, Neanderthals, and Denisovans of that time. However, this overwhelming was not solely an obliteration of the archaic hominins but included interbreeding possibly over tens of thousands of years as revealed by DNA sequence analysis. Extant non-African humans have 1.5–2.6% of their DNA derived from Neanderthal DNA, whereas extant Africans have significantly less Neanderthal DNA. Similarly, some Oceanian and southeast Asian populations have 4–6% of their DNA derived from Denisovans and Denisovan-derived ancestry is largely absent from current populations from Africa and western Eurasia.

H. sapiens, Neanderthals, and Denisovans likely share a common ancestral hominin (FIGURE 5.18). Migration of the Neanderthals to Europe and the migration of the Denisovans to Asia permitted the divergence of these three species, so that they exhibited distinct characteristics. However, the divergence was not so great as to prevent interbreeding. Subsequent periodic migrations would lead to spatial coexistence that permitted interbreeding. The genetic data suggest that most interbreeding occurred after the more recent mass migration out of Africa, which occurred approximately 50,000 years ago. However, gene flow between archaic hominins before the mass migration is also likely. Thus, genes flowed between these species and ultimately contributed to the evolution of humans.

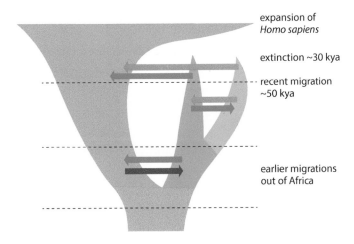

FIGURE 5.18 Gene flow and migrations in human evolution. Divergence of *Homo sapiens* (blue) from Neanderthals (green), from an ancestral hominin (gray) and later divergence, and from Denisovans (orange) was followed by periods of interbreeding resulting in gene flow (block arrows) between these hominins. Continued human migrations ultimately resulted in the expansion of *Homo sapiens* throughout the world. The widths of the shaded areas represent the population sizes and genetic diversity. Despite the extinction of Neanderthals and Denisovans approximately 30,000 years ago (30 kya) some of their DNA sequences are still found in extant humans.

The increased cognition and dexterity of hominins allow for migration to new geographical areas and adaptation to a wide range of environments. Unlike most other organisms, humans are not restricted to a defined biological niche but have expanded to nearly all parts of the world. The development of sophisticated language skills permitted enhanced cooperation that would lead to the development of agriculture approximately 10,000 years ago and urbanization approximately 5000 years ago. Agriculture and urbanization increased the reproductive capacity of humans. Human migration continues, and at a greatly accelerated rate due to technological advances in travel. Thus, interbreeding and gene flow now occur on a global scale.

SUMMARY OF KEY CONCEPTS

- Chromosomes are composed of a DNA molecule and proteins providing structural and functional support to the DNA

- DNA is the hereditary material and the sequence of the total DNA determines the genotype

- The phenotype is determined by the expression of specific genes encoded by the DNA

- Genes exist in multiple forms called alleles and the various alleles contribute to the phenotype

- Humans are diploid and have two copies of most genes, which can be either homozygous or heterozygous

- Sexual reproduction involves the formation of gametes through a specialized type of cell division called meiosis that results in single copies of all the genes in the gametes

- Independent chromosome assortment and intrachromosomal recombination are features of meiosis that generate genetic diversity between gametes

- Sexual reproduction results in distinct inheritance patterns of phenotypes such as dominant-recessive, incomplete dominance, and co-dominance for monogenic traits.

- Polygenic traits tend to exhibit a range of phenotypes with a bell-shaped curve distribution.

- Epigenetics is a type of inheritance that is not directly encoded by the DNA sequence but influenced by modifications in chromosome structure

- The introduction of new genes and alleles into a population, and changes in allele frequencies, are major components of evolution

- Genetic drift and natural selection are major factors promoting changes in allele frequency in a population during evolution

- Molecular phylogenetics is a tool to assess evolutionary and genetic relationships between genes, individuals in a population, or groups of organisms that can lead to a better understanding of biology and disease

- Humans share an ancient ancestry with the great apes of Africa and have evolved over hundreds of thousands of years

- Evolution of modern humans involved migration and interbreeding between distinct lineages of human ancestors

KEY TERMS

allele	One of several possible forms of a gene that is defined by the encoded trait or the DNA sequence.
balanced polymorphism	The maintenance of two alleles for a single gene in a population because the heterozygote is more fit than either of the homozygotes.
chromosome	A structure composed of a DNA molecule and associated proteins that carries the hereditary information of an organism in the form of genes.
epigenetics	Refers to heritable changes in gene function that occur without a change in the DNA sequence. Can include the modification of DNA or chromosome structure.
gene pool	All the alleles for all the genes in a specified population of interbreeding individuals.
genotype	The complete genetic composition of an organism including heterozygous alleles.

haplotype	A group of gene variants or DNA polymorphisms that tend to be inherited together due to genetic linkage.
heterozygous	Referring to a diploid cell or individual with two different alleles of a gene.
homozygous	Referring to a diploid cell or individual with the same alleles of a gene.
linkage disequilibrium	The non-random association of alleles at two or more loci due to the proximity of the genes on a chromosome and the infrequency of recombination (i.e., crossing over) between the genes.
meiosis	Special type of cell division involved in oogenesis and spermatogenesis. Comprises two successive rounds of nuclear division and cytokinesis with only one round of DNA replication.
monogenic	A trait influenced by a single gene. Also referred to as a Mendelian trait.
natural selection	The differential survival and reproduction of individuals that is a key process in evolution as species adapt to changing environments.
phenotype	The observable traits or characteristics of a cell or organism that are determined by gene polymorphisms and gene expression.
phylogenetics	Relating to or based on the evolutionary history among a group of organisms.
polygenic	A trait influenced by multiple genes.
polymorphism	The presence of genetic variation in a population that can be defined at the level of individuals, cells, proteins, or DNA sequence.

REVIEW QUESTIONS

1. Describe independent chromosome assortment and recombination that occurs during meiosis and the potential impact on inheritance.

2. Describe the various ways in which alleles can interact to determine the phenotype.

3. What is meant by heterochromatin and how is heterochromatin related to epigenetics?

4. What is the gene pool and what are the factors that influence the gene pool?

5. What are phylogenetic trees and how are they generated?

ADDITIONAL QUESTIONS

In addition to the Review Questions provided above, there is a range of free online questions designed to enable students to further test their understanding of the chapter material. In order to access these interactive questions, please visit the book's website: https://routledge.com/cw/wiser

Genomics and the 'Omic Revolution

6

The **genome** refers to the complete DNA sequence (genetic material) of an organism. The ability to accurately and rapidly sequence large amounts of DNA has revolutionized the study of genomes and led to the development of genomics as a field. Sequencing an entire genome can now be easily accomplished and this allows for all the genes of an organism to be analyzed and studied, thus expanding upon classical genetics in which one or a few genes are analyzed at a time. This concept of looking at all the genes has also expanded into the analysis of RNA, proteins, and metabolites and, accordingly, has led to the development of transcriptomics, proteomics, and metabolomics. The vast amount of data generated by these technologies is stored in databases that can be analyzed through computational biology.

Genomic research, including transcriptomics, proteomics, and metabolomics, will lead to a better understanding of health and disease. The information gained from these studies can be applied to the development of better prognostics, diagnostics, and treatments. For example, the identification of new drug targets may lead to the development of better drugs. These technologies are also ushering in an era of personalized medicine in which disease prevention and treatment are tailored to the genetic makeup of an individual.

DNA SEQUENCE ANALYSIS

Nucleic acids are long polymers of nucleotides, and the order of those nucleotides defines the sequence (Chapter 1). Major advancements in the ability to determine the sequence of DNA have greatly accelerated the field of genomics and led to the Human Genome Project (BOX 6.1). Knowing the DNA sequence has revealed the approximate number of human genes and other basic features of the human genome (TABLE 6.1). Much of the work on interpretation and analysis of genomic data, especially regarding gene function and the role of genes in disease, is still to come. Numerous other genome projects on model organisms, pathogens, and agriculturally important plants and animals have also been carried out. This knowledge of the human genome and the genomes of other organisms impacts fields ranging from molecular medicine to human evolution.

BOX 6.1

Human Genome Project

The Human Genome Project started in 1990 with the objective of determining the entire nucleotide sequence of the human genome. The project was an international consortium involving scientists and laboratories from around the world concentrating on the various human chromosomes. It was anticipated that the project would take 15 years to complete. However, in 2000 the first working draft of the human genome was announced and the data were published in 2001. In 2003 an essentially complete genome was released, thus completing the project 2 years ahead of schedule. Advances in the field of sequence analysis and computer technology, as well as a high level of international collaboration, contributed to the success of the Human Genome Project. The sequence of the DNA is stored in databases freely available via the internet. Sequences corresponding to genes are annotated to include ancillary information about the gene. Computer programs can be used to search for specific genes or gene features and to analyze the data.

Table 6.1 Human Genome Features	
Size (haploid)	3.1 billion base-pairs
Protein-encoding genes	Approximately 20,000
Non-protein-encoding genes	Approximately 12,000
Gene complexity	35–65% of genes may undergo alternate splicing
Gene function	24% of genes unclassified by function
Pseudogenes	Approximately 13,000
Protein-coding regions	Approximately 1.5% of the genome encodes proteins
Regulatory regions	Up to 8% of the genome involved in regulation of gene expression
Transposon-related sequences	36% of the genome is related to repetitive sequences of viral origin called transposons
Individual variation	A single nucleotide polymorphism (SNP) occurs approximately every 1000–2000 base-pairs

Computational biology and databases are key elements of genomics

The vast amount of DNA sequences in a genome – 3.1 billion base-pairs in humans – and the four-letter alphabet necessitates computers to analyze sequence data. Searching databases or carrying out molecular analyses would be exceedingly tedious and time-consuming without the use of computers, and it would be impossible to interpret genomic sequence data. Various computer programs are available to carry out a wide range of analyses of genomic data. For example, there are algorithms that can identify genes, translate a DNA sequence into a protein sequence, make predictions about protein structure, or generate sequence alignments. Sequence alignments are especially useful in searching for homologous genes or in determining phylogenetic relationships.

The vast amount of sequence and related molecular information generated by genome projects are stored in databases. Most of these databases are freely accessible and there are numerous databases that store various types of biological data. Some of the databases have primary sequence data that can be searched for specific DNA or protein sequences. Other databases take the primary data and add value to it by associating it with additional information or confirmatory data (FIGURE 6.1). These various secondary databases can also be searched with computer programs to extract useful information and carry out analyses. For example, if a specific gene is suspected to be involved in a particular disease, all the

SIGNPOST

For more on molecular phylogenetics see Chapter 5, pages 104–107.

databases can be searched for information related to that gene. This information can be used to guide future research on that gene.

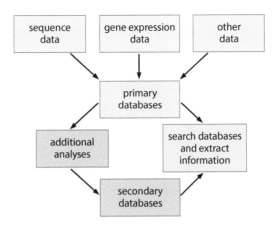

FIGURE 6.1 Flow of information in biological databases. Primary data are submitted by investigators to databases. Refinements of these data made by other investigators are stored in secondary databases. Databases can be searched with computer programs to identify information and publications about sequences, genes, transcripts, proteins, and metabolites of interest.

Metagenomics allows for the study of environmental samples and microbiota

It is also possible to sequence DNA samples containing a mixture of species. For example, samples from the environment can be sequenced to determine the microorganisms that are present. Previously it was necessary to isolate and culture the collection of microorganisms in a sample and then determine the sequence of the individual microorganisms. The individual microorganisms making up a microbial community are called the **microbiota**. Prior to the advances in DNA-sequencing technologies, it was not possible to completely define a microbiota since many of the individual microorganisms cannot be cultured. It is now possible to simply collect a sample and determine the DNA sequences that are present as a means of identifying the organisms. This analysis of DNA sequences from environmental samples has been referred to as environmental genomics, or more commonly **metagenomics**. The same approach can be applied to determine the human microbiota (BOX 6.2).

Human Microbiota and Metagenomics

A microbiota is an ecological community of microorganisms. The wide range of bacteria and other microorganisms that colonize the various tissues and organs of the human body can also be viewed as a microbiota. Most of these microorganisms are not harmful and are viewed as commensals. Many commensals may be mutualistic in that they provide a metabolic benefit such as assisting in digestion or synthesizing vitamins. However, some members of the microbiota are pathogenic and cause disease. In addition, non-pathogenic species may be harmful if they are overly abundant. It is increasingly being recognized that the human microbiota plays a large role in human health.

Historically it has not been possible to learn much about the role that the microbiota plays in health and disease since much of the human microbiota has not been successfully cultured, identified, or otherwise characterized. Metagenomics is a tool to characterize the human microbiota and the Human Microbiome Project was started in 2008 to better characterize the human microbiota. The microbiome refers to the collective genomes of the microbiota. One of the goals of the project is to determine a normal range of the microbiome among humans, with a particular focus on the microbiota that normally inhabit the skin, respiratory tract, digestive tract, and urogenital tract. This information can then be used to explore the relationship between disease and changes in the human microbiota.

BOX 6.2

There are two basic approaches to metagenomics (FIGURE 6.2). One approach involves the targeted sequencing of a specific gene such as the 16S ribosomal (r) RNA gene. All organisms have the gene for the 16S rRNA and its evolution is well characterized. The other approach is to randomly sequence DNA fragments that are present in the sample. In both approaches the acquired sequences are compared with sequences in the databases. Exact sequence matches between a DNA sequence in the sample and a known sequence in the database identify the presence of a particular organism. If the DNA of a particular organism is present in the sample, then the organism is likely to be present. Since many species present in the microbiota have not yet been identified, many of the sequences in the sample may not exactly match a known sequence in the databases. In that case, molecular phylogenetic analysis (Chapter 5) identifies related species so that the unknown organism can be placed in a relevant taxonomic group.

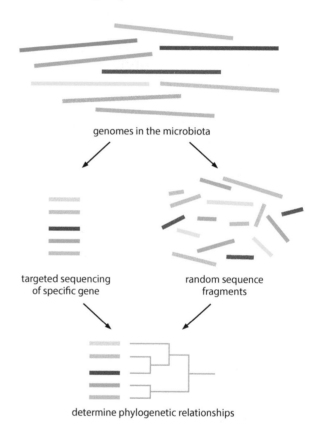

genomes in the microbiota

targeted sequencing
of specific gene

random sequence
fragments

determine phylogenetic relationships

FIGURE 6.2 **Metagenomic analysis of the microbiota.** DNA containing a mixture of organisms is analyzed by either sequencing a targeted gene such as 16S small subunit rRNA or randomly sequencing DNA present in the sample. The sequences are then compared with known sequences in databases to identify the organisms that are present or to determine their relationship to known organisms.

OTHER 'OMICS

High-throughput DNA-sequencing technology and computational biology have led to the development of genomics. A primary focus of genomics is accruing DNA sequence data and annotating the gene sequences stored in the databases. To fully appreciate the power of genomics, information about RNA, proteins, and metabolism is also needed. As genomics has allowed genetics to look at all the genes instead of just a few genes, the advances in biotechnology and information technology are allowing an 'omic approach to be applied to RNA, proteins, metabolites, and chromatin structure. For example, the collection of total RNA transcripts from an organism, tissue, or cell type are the **transcriptome** (TABLE 6.2). Similarly, the collection of total proteins from an organism, tissue, or cell type are the **proteome**. Transcriptomics and proteomics are sometimes referred to as functional genomics since information about gene function is revealed.

Table 6.2 Genomics and Its Derivatives

'Omics	Descriptions	Technologies
Genome	Complete DNA sequence	High-throughput DNA sequencing
Transcriptome	RNA expression profile	Microarrays (gene chips)
Proteome	Protein profile	Chromatography and mass spectrometry
Metabolome	Catalog of metabolites	Chromatography, NMR, mass spectrometry
Epigenome	Collection of chromatin modifications	Immune precipitation, microarrays, restriction mapping

The advances in chromatography, mass spectrometry, and nuclear magnetic resonance (NMR) are also being applied to the analysis of metabolites. Instead of analyzing one or a few metabolites at a time, the newer technologies allow for the characterization of a large proportion of the total metabolites in a sample. This metabolic profile, or **metabolome**, gives a complete picture of the metabolism of an organ, tissue, or organism. The metabolome can be further subdivided into specific classes of metabolites, such as lipids, or the lipidome. Epigenetic modifications are reversible modifications to a cell's DNA or histones that affect gene expression without altering the DNA sequence. Epigenomics is the study and characterization of all the epigenetic modifications, called the **epigenome**, in a cell.

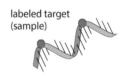
Epigenetics is discussed in more detail in Chapter 5, pages 100–101.

SIGNPOST

TRANSCRIPTOMICS

Transcriptomics is based on detecting the RNA transcripts expressed by a specific cell, tissue, organ, or organism. Traditional methods to detect a specific RNA molecule rely upon the hybridization of a complementary DNA probe to an immobilized RNA sample (see Box 4.1). This approach only allows for the characterization of a few RNA transcripts at a time and cannot accommodate the thousands of transcripts that might be present. Immobilizing the probes instead of the samples allows for a high-throughput method to analyze transcripts. The probes are single-stranded DNA molecules complementary to all the potential transcripts in a sample and are called reporters (FIGURE 6.3). These reporters could correspond to all the genes of an organism, or a particular subset of genes. The various reporters are placed in rows and columns, forming an array of spots in which each spot serves as the reporter for a specific gene. This **DNA microarray** is also known as a gene chip. Tens of thousands of reporters can be placed on a single gene chip.

To detect the RNA molecules in a sample it is first necessary to convert the RNA into DNA. This is done with an enzyme called reverse transcriptase (BOX 6.3). The resulting DNA is known as copy or complementary DNA (cDNA). This cDNA is made in the presence of nucleotides that are

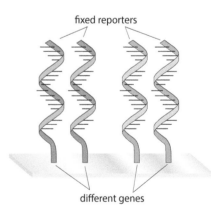

labeled target (sample)

fixed reporters

different genes

FIGURE 6.3 Microarray technology. Single-stranded DNA molecules called reporters are immobilized on a solid surface such as a thin sheet of glass in spots arranged in rows. This glass slide containing the reporters is called either a microarray or a gene chip. Each spot corresponds to a different reporter that is complementary to a specific gene. A sample containing a mixture of RNA transcripts is labeled with a fluorescent dye (purple circles) and incubated with the gene chip. The RNA transcripts hybridize to their complementary reporters and analysis for fluorescence determines which specific RNA molecules are present in the sample.

labeled with a fluorescent dye that can readily be detected and measured by illuminating with ultraviolet light. The sample containing a mixture of fluorescently labeled cDNA molecules is incubated with the gene chip and the various cDNA molecules bind to the spots on the chip that correspond to their complementary reporter sequences. The gene chip is then illuminated with ultraviolet light and spots that fluoresce correspond to transcripts that are present in the sample. No fluorescence means that there is no transcript corresponding to that reporter. Furthermore, the amount of fluorescence can be quantified and that corresponds to the relative abundances of the various transcripts in the sample.

BOX 6.3

Reverse Transcriptase and the Analysis of RNA

DNA and RNA are quite similar chemically and many of the methodologies work on either DNA or RNA with only minor differences. However, in general, DNA is usually easier to work with and some of the techniques such as sequencing can only utilize DNA. It is possible, though, to convert RNA into DNA and carry out sequencing and other analyses with the DNA molecule. This conversion is accomplished with an enzyme found in retroviruses called reverse transcriptase. Retroviruses, such as human immunodeficiency virus (HIV), have an RNA genome and as part of their life cycle convert the RNA genome into DNA (Chapter 22). Reverse transcriptase is an RNA-dependent DNA polymerase that utilizes an RNA template to synthesize a complementary double-stranded DNA molecule.

Reverse transcriptase can also be used in the test tube to convert RNA into DNA. This is accomplished by incubating isolated RNA with reverse transcriptase and nucleotides. The product is called complementary or copy DNA (cDNA). The resulting cDNA can then be sequenced, utilized in recombinant DNA technology (Chapter 7), or analyzed by any other DNA methodology. For example, another approach to analyzing the transcriptome is to convert the total RNA of a cell or tissue into cDNA and carry out high-throughput sequencing as done in metagenomics. In addition, the detection of viruses with RNA genomes is highly dependent on converting the RNA genomes into cDNA and carrying out subsequent analyses.

Conditional changes in transcription profiles are easily determined

The gene expression profile (transcriptome) may change due to some pathological condition or disease state. For example, cancerous tissue can be compared with normal tissue to determine differences in gene expression associated with the development of cancer. Similarly, drug treatment may alter the transcription profile. This means that the transcription profile might be diagnostic or prognostic of a particular disease or indicate whether a particular treatment is working.

Differences in transcriptional profiles can be easily determined using control and experimental samples (FIGURE 6.4). The control and experimental samples are labeled with different fluorescent dyes that emit different wavelengths of light (i.e., colors) which can be easily distinguished. For example, the control sample could be labeled with a dye that fluoresces red and the experimental sample could be labeled with a dye that fluoresces green. The two samples are then co-incubated with the same DNA microarray. The microarray is analyzed with an instrument that quantifies the amount of fluorescence at both wavelengths. This analysis will indicate whether the expression of all the individual genes on the gene chip has increased, decreased, or remained the same in the experimental or diseased state.

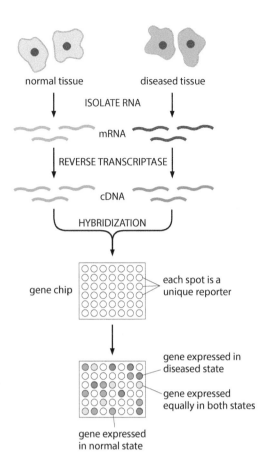

PROTEOMICS

The proteome is the complete set of proteins in a cell, tissue, or organism. High-throughput techniques, such as those used to sequence DNA or analyze RNA transcripts (gene chips), are not as readily available for the analysis of proteins. Traditionally protein analysis and identification relied heavily upon separating proteins by gel electrophoresis (see Box 1.5) and sequencing the N terminus of the individual proteins. Gel electrophoresis is quite labor intensive and not amenable to automation, and thus is not capable of high-throughput analysis of entire proteomes. Liquid chromatography is another technology to separate proteins (BOX 6.4) that is more amenable to automation. Furthermore, chromatography can be combined with mass spectrometry to determine partial amino acid sequences of proteins.

Sequential chromatography and mass spectrometry are used to determine proteomes

The basic procedure is to treat a protein sample with enzymes called site-specific proteases. Site-specific proteases break the peptide bonds between specific amino acid residues within a protein and generate peptides of defined size and chemical properties. These peptides are then separated by liquid **chromatography** (FIGURE 6.5). As the peptides exit the chromatography column, they are automatically sent to an instrument called a mass spectrometer (BOX 6.5) which determines the amino acid sequence of the peptide. These peptide sequences are then automatically used to search databases for matching sequences. Assuming that an identical or homologous sequence has been submitted to a database, the combination of chromatography with **mass spectrometry** will lead to

BOX 6.4

Liquid Chromatography of Peptides and Proteins.

Each individual protein has a unique set of chemical and physical properties that is determined by its amino acid composition and sequence. These properties include hydrophobicity, polarity, charge, size, and ligand-binding activity. Individual proteins can be separated from each other by liquid chromatography, which exploits the differences in these chemical properties. In liquid chromatography a solution with a mixture of proteins is passed through a porous resin in the shape of a column (**BOX 6.4 FIGURE 1**). As the solution flows through the column the proteins interact with the resin based on their chemical or physical properties. Those proteins that interact more strongly with the resin pass through the column more slowly and are separated from other proteins. For example, if the resin contains hydrophobic moieties, then proteins that are more hydrophobic interact with the column more strongly than less hydrophobic proteins. Therefore, hydrophobic proteins are retained longer as they pass through a hydrophobic column. Similarly, ion-exchange chromatography can be used to separate proteins based on their overall charge, and gel filtration can be used to separate proteins based on their size.

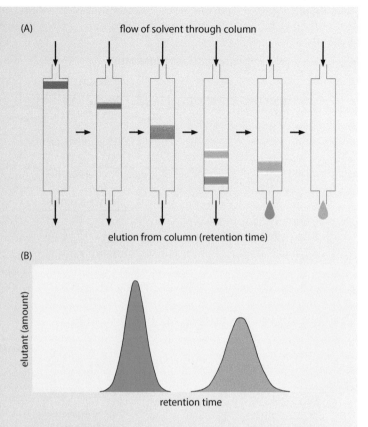

(A) flow of solvent through column

elution from column (retention time)

(B) elutant (amount) / retention time

BOX 6.4 FIGURE 1 Principles of liquid chromatography. (A) A solution containing a mixture of solutes is passed over a column. Those solutes interacting with the column material more strongly (blue) are retained on the column longer and elute later and can be separated from solutes that interact weakly with the column material (red). (B) The eluting substances (eluant) can also be measured as they exit the column, and a chromatograph can be generated (lower panel).

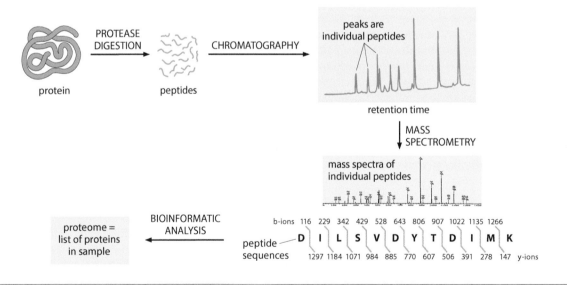

peaks are individual peptides

PROTEASE DIGESTION

CHROMATOGRAPHY

retention time

protein

peptides

MASS SPECTROMETRY

mass spectra of individual peptides

BIOINFORMATIC ANALYSIS

proteome = list of proteins in sample

peptide sequences

b-ions 116 229 342 429 528 643 806 907 1022 1135 1266

D I L S V D Y T D I M K

1297 1184 1071 984 885 770 607 506 391 278 147 y-ions

FIGURE 6.5 Determination of a proteome. A sample containing proteins is digested with a protease that cleaves peptide bonds between specific amino acids to generate peptides. The sample is then analyzed by an instrument that sequentially carries out chromatography (see Box 6.4), mass spectrometry (see Box 6.5), and bioinformatic analysis. Chromatography separates the peptides according to their chemical properties and the individual peptides are subjected to mass spectrometry. Analysis of the mass spectrum for each peptide reveals its sequence. The identities of the proteins in the original sample can then be determined by searching sequence databases for sequences either identical or homologous to the peptides.

the identification of the individual proteins that are found in a sample containing a mixture of proteins. This protein profile of the sample is defined as the proteome.

Mass Spectrometry and Peptide Sequencing

The mass spectrometer is an instrument that can precisely determine the masses of the molecules in a sample containing a mixture of molecules. Knowing the exact mass of a molecule can potentially lead to its identification. Furthermore, a molecule can be fragmented into smaller molecules by bombarding it with atoms of an inert gas. Since some covalent bonds are more likely to be broken than other covalent bonds, the fragmentation generates a defined mixture of smaller molecules that is unique to the original parental molecule.

These smaller molecules generate a spectrum that can serve as a fingerprint of the original molecule. Thus, an unknown molecule can be identified by comparing its unique fingerprint to known molecules. This technology can be adapted to the analysis of peptides since fragmentation is most likely to occur at the peptide bonds between amino acids and each amino acid has a unique mass. The unique spectrum generated by a peptide of unknown sequence can be analyzed with a computer program to determine its sequence.

BOX 6.5

POTENTIAL APPLICATIONS OF GENOMICS AND ITS DERIVATIVES

The 'omics approach to studying disease offers a paradigm shift in the identification of the determinants of disease. Traditional approaches are generally driven by a hypothesis regarding a specific gene, protein, or metabolic pathway that may be involved in a disease process. Experiments to test the hypothesis are then carried out. Genomics and its derivatives allow for an unbiased approach to compare all the genes, transcripts, proteins, or metabolites between the healthy state and the diseased state. Any differences that are found can then be further investigated as determinants of disease. Furthermore, differences between the normal and disease states may provide better biomarkers for the diagnosis and prognosis of disease, as well as provide targets for the development of therapeutics.

Molecular biomarkers can better define disease risks and provide earlier diagnosis

Chronic diseases such as cardiovascular disease and diabetes have a complex etiology involving a genetic predisposition combined with environmental and behavioral factors. Much of this genetic predisposition is likely to be associated with specific alleles of a defined set of genes. It may be possible through genomics to identify and characterize such alleles and these alleles could then be used as biomarkers to define baseline risks for developing certain diseases (FIGURE 6.6). Baseline risks are not deterministic, and the development of disease often involves initiation events associated with behavior or environmental exposures. These initiation events can involve changes in gene expression. Transcriptomics, proteomics, and metabolomics may be able to identify these changes in gene expression. Furthermore, any such changes are potential biomarkers for preclinical disease development.

Having biomarkers that define baseline risks or provide earlier diagnosis can greatly reduce disease burden in terms of both human suffering and medical costs associated with treating disease. For example, blood glucose is a biomarker for diabetes (Chapter 13) and blood pressure is

FIGURE 6.6 Potential contribution of 'omics to the development of diagnostic biomarkers. Clinical diagnosis and therapeutic decisions are generally made relatively late in the progression of chronic diseases. Through genomics it may be possible to identify DNA sequences that define baseline risks of developing chronic disease. Since the initiation of disease is often associated with environmental factors and gene expression, transcriptomics, proteomics, and metabolomics may identify markers for the preclinical stages of disease. [Modified from Snyderman, R (2009) The role of genomics in enabling prospective health care. In Willard, H and Ginsburg, GS (eds.), *Genomic and personalized medicine*, Elsevier, Durham, NC, pp. 378–385. With permission from Elsevier.]

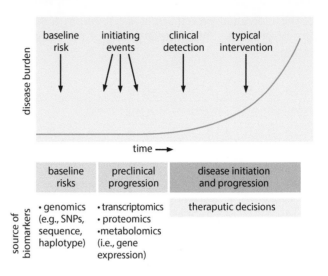

SIGNPOST

Drug action is discussed in detail at the end of Chapter 3, pages 66–69.

a biomarker for hypertension (Chapter 14). Through the use of these markers, conditions known as pre-diabetes and pre-hypertension can be defined before clinical symptoms are evident. Preventive and therapeutic measures can be initiated to slow the progression and to minimize the pathology associated with these diseases. Molecular biomarkers could allow for an even earlier diagnosis and a further lessening of the disease burden.

Genomics and its derivatives play a key role in drug development

Many drugs function by binding to specific proteins and altering the function of those proteins. Effective drugs are more toxic for the diseased state than the normal state and for many diseases the drug target is some molecule or process that is different in the diseased state. The unbiased approach of genomics and its derivatives, such as transcriptomics, proteomics, and metabolomics, potentially identifies disease determinants and other features that are unique to the disease process.

Molecular differences between the normal and disease states can include sequence polymorphisms or differences in transcripts, proteins, or metabolites (FIGURE 6.7). These differences can be further validated experimentally in vitro and in animal models. Not only can these differences serve as potential biomarkers for diagnosis, but also such differences may be potential drug targets. For example, it may be possible to design inhibitors of activities that are overexpressed or stimulators of activities that are underexpressed. Thus, one of the goals of genomic research and its derivatives is to identify potential drug targets and to develop more effective and less toxic drugs.

Genomics is leading the way to a future era of personalized medicine

All human disease arises from interactions between genetic components and environmental determinants. With the availability of the entire genomic sequence, it may be possible to better define the genetic components of disease. The new methods of DNA sequencing have increased the efficiency of DNA sequencing by orders of magnitude and dramatically lowered the costs. For example, sequencing the entire human genome in 2001 cost US$100 million. With the advancement in sequencing technology the cost lowered to US$10,000 in 2011 and in 2016 the cost was approximately US$1500. At this price it is feasible

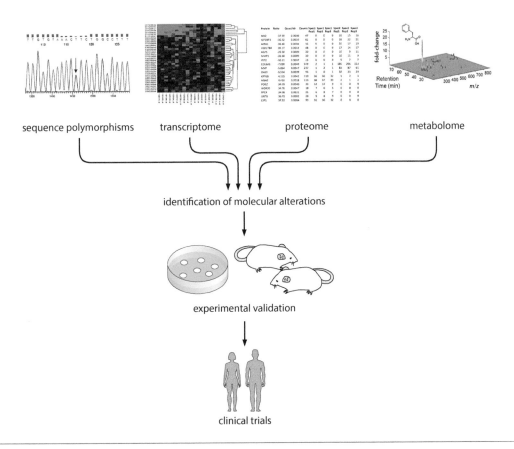

FIGURE 6.7 Identification of molecular alterations associated with disease through genomic research. Genomics and its derivatives provide an approach to identify molecular changes that are associated with disease. Sequence analysis can identify sequence polymorphisms and copy number variants found in the disease state. Similarly, transcriptomics, proteomics, and metabolomics can identify changes in the expression of genes, alterations in protein levels and function, and metabolic perturbations that correlate with disease. These molecular alterations can serve as biomarkers for the diagnosis of disease or as potential drug targets for the treatment of the disease. Potential biomarkers and drug targets can then be validated experimentally in vitro or in animal models and further validated in clinical trials.

for individuals to determine their genome sequence. Furthermore, gene chip technology can also be used to determine single nucleotide polymorphisms (SNPs) that may be associated with disease. Knowing the genetic polymorphisms of an individual could theoretically define all the genetic components associated with disease. However, it will be quite some time before all the polymorphisms that are associated with disease are known.

Many diseases are associated with altered levels of gene transcription rather than sequence polymorphisms. Therefore, gene chips can be used to screen for variations in the transcription of key genes as a means to diagnose disease or monitor the progression of disease. Microarray technology will likely be adapted to clinical applications in the future. This ability to determine a person's genotype (SNPs) and phenotype (expressed genes) via gene chip technology may usher in an era of personalized medicine. Preventive and therapeutic measures can be prescribed based on an individual's genetic makeup, including the transcriptome or proteome.

Personalized medicine can improve therapeutic decision-making

Probably the area that will be most impacted by personalized medicine is drug treatment. Currently drug treatments are prescribed based on clinical trials that show which drugs are the most efficacious for the majority of people. However, there are always some individuals who do not respond to the drug and a few individuals who have adverse reactions to the drug (FIGURE 6.8). People who do not respond to a drug, or respond adversely, are then prescribed another treatment until the treatment that works best is determined. This variation in how a person responds to drugs probably has a genetic component.

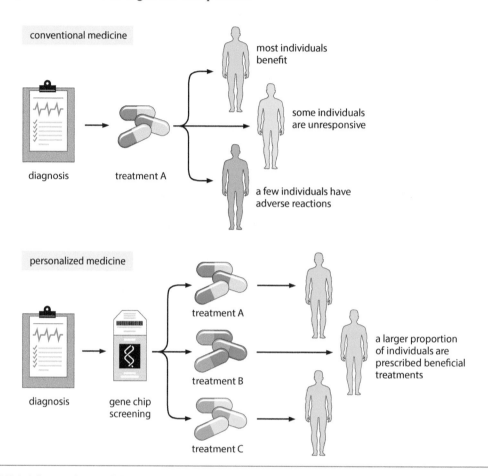

FIGURE 6.8 Personalized medicine. In conventional medicine treatments are prescribed based on what works for the majority of people. Personalized medicine opens the possibility of prescribing treatments that are best for each individual person.

Pharmacogenomics, previously called pharmacogenetics, seeks to identify genetic variations that determine how a person responds to a particular drug. Through genomic studies it may be possible to identify genetic polymorphisms that can serve as biomarkers for whether a drug will be efficacious for an individual or result in adverse reactions. Once these biomarkers are identified it will be possible to screen individuals before prescribing a treatment and then prescribe a treatment that works best for that individual (see Figure 6.8). The hope of personalized medicine is to deliver the right treatment to the right person at the right time.

GENOME-WIDE ASSOCIATION STUDIES

Genes responsible for diseases have been traditionally identified through family-based linkage studies or through population-based studies by looking at specific genetic markers. In a linkage study, members of a family exhibiting a particular genetic disease or trait are analyzed for associations between the disease phenotype and other genetic markers. Such genetic linkage studies may lead to the identification of the chromosomal location of the gene responsible for a disease. The chromosomal location provides a means to identify a candidate gene that contributes to the disease state. Alternatively, candidate genes that contribute to a disease can be hypothesized based on the phenotypic characteristics of a disease. A population-based study is then carried out to determine whether there is a correlation between specific alleles of the candidate gene(s) and the disease. The two approaches can also be used together. These approaches work well for diseases caused by a rare allele of a single gene with high **penetrance**. Penetrance refers to the percentage of individuals with a particular allele that express the trait associated with that allele.

For details on genetic linkage and linkage disequilibrium see Chapter 5, pages 99–100.

SIGNPOST

Discerning the genetic contributions of diseases with complex etiologies that involve a combination of environmental and genetic influences is quite difficult using these traditional approaches. Complex diseases are often polygenic in that multiple genes contribute to the disease and no single gene is predominant. Furthermore, disease-specific alleles generally exhibit a low penetrance and are common in the population. The limitations of family-based linkage studies and single allele population-based studies can be overcome through a technology called **genome-wide association** (GWA). GWA studies utilize **single nucleotide polymorphisms** (SNPs) to map genes throughout the human genome. This provides a much higher resolution in mapping genes than traditional genetic linkage studies. In addition, since the GWA studies are carried out on large populations the background noise of low penetrance and environmental influences can be filtered out through statistical analyses.

SNPs can serve as genetic markers for chromosomal locations

An SNP is a variant in the nucleotide sequence between individuals at a specific location in the genome. Most SNPs exist in two allelic forms corresponding to the two different nucleotides found at that position of the DNA sequence (FIGURE 6.9). It is possible to have three or four alleles at any specific location in the DNA sequence, however, this rarely happens. Some SNPs are extremely rare in the human population, whereas other SNPs are common to and may even approach an equal distribution in the human population. The SNP that is found most frequently in the

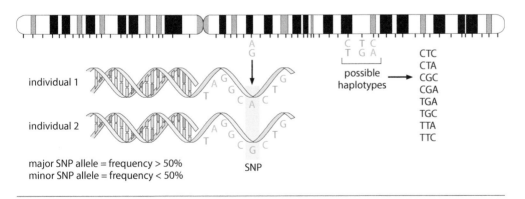

major SNP allele = frequency > 50%
minor SNP allele = frequency < 50%

SNP

possible haplotypes

CTC
CTA
CGC
CGA
TGA
TGC
TTA
TTC

individual 1

individual 2

FIGURE 6.9 SNPS and haplotypes. SNPs used in GWA studies are distributed throughout the chromosomes (denoted by hash marks below the chromosome diagram) and correspond to SNPs between individuals. The more frequent SNP found in a population is called the major allele and the minor allele is the less frequent variant. Groups of SNPs can also define haplotypes.

population is called the major allele and the other SNP is the minor allele. Since SNPs occur approximately every 1000–2000 base pairs in the human genome it is possible to use SNPs as chromosomal markers for gene-mapping studies.

Gene chips can determine an individual's SNP haplotype

The gene chip technology used in transcriptomics can be adapted to detect SNPs. For each of the SNPs being analyzed, reporter molecules that are complementary to the major and minor alleles are present in the microarray. The single base difference in the DNA sequence of the major and minor allele reporters impacts the hybridization between the reporter and sample DNA (FIGURE 6.10). The reporter that is not an exact complement to the sample DNA gives a weaker signal, allowing the major and minor alleles of an SNP to be distinguished. Furthermore, since reporters for both alleles are present on the gene chip, it can be determined whether an individual is homozygous for either the major or the minor allele or is heterozygous for the two alleles.

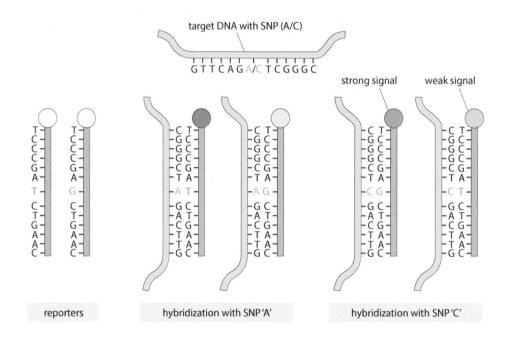

FIGURE 6.10 Detection of SNPs by hybridization with reporters. Shown is a hypothetical region of DNA surrounding an SNP that exists as either A (red) or C (green). Reporters complementary to the region encompassing the two SNPs are synthesized. Genomic DNA from an individual that is homologous for the A allele of the SNP binds more efficiently to the reporter with the T than the reporter with the G and thus gives a stronger signal (bright red vs. dull green). Conversely, an individual with the C allele binds more efficiently to the G-reporter than the A-reporter (bright green vs. dull red). Both signals are strong in a heterozygous individual (not shown).

Genomic DNA is isolated from an individual, processed, and incubated with the gene chip. As in transcriptomics, an instrument is then used to assess the amount of fluorescence associated with each reporter. The gene chip reader is interfaced with a computer and for each SNP it is recorded whether the individual is homozygous for the major allele, homozygous for the minor allele, or heterozygous. The number of SNPs needed to give complete coverage of the human genome is approximately one million. The end result generated by the gene chip analysis is an entire genome SNP haplotype for that individual.

Well-characterized SNPs are needed for GWASs

In addition to needing sufficient SNPs to cover the entire human genome, the SNPs also need to be well characterized and exhibit a frequency of at least 5% in the population. The frequency of the major and minor alleles in the human population also needs to be known, as well as knowledge about the distribution of the alleles in subpopulations such as people of African, Asian, or European descent. Many SNPs are specific to certain human subpopulations and can be used to determine ancestry (BOX 6.6). As part of the characterization of SNPs to be used in GWASs, the International HapMap Project was started in 2002. HapMap is short for haplotype mapping. One goal of the HapMap project is to characterize human SNPs in regard to their frequency, distribution in different subpopulations, and their utility in GWASs. The overall goal of this project is to identify all human SNPs and, like other genome projects, the data are stored in databases and are freely accessible.

SNPs and Ancestry

Hidden in a person's DNA sequence is a roadmap of their ancestry. Human evolution has been characterized by sporadic changes in the DNA sequence and migrations throughout the world (Chapter 5). Since single nucleotide polymorphisms (SNPs) are relatively rare and make up only 0.1% of a person's genome, SNPs are maintained through numerous generations and can serve as biomarkers for a person's biogeographical ancestry. This means that a group of ancestry-informative SNPs can be used to determine a person's ancestry. For example, some ancestry-informative SNPs can distinguish across African, European, and Asian ancestries. Other ancestry-informative SNPs can distinguish subgroups within these major groups. There are several commercial services that determine a person's ancestry and help to identify relatives. As part of that service, the proportion of ancestry that is derived from these various biogeographical ancestral groups is estimated.

BOX 6.6

Linkage disequilibrium between traits and SNPs can identify the chromosomal location of genes

The technical ability to screen the entire genome for SNPs with gene chips can be used to determine the location of genes associated with a particular trait. This is based on linkage disequilibrium between a particular SNP and the trait in question. Linkage disequilibrium refers to unexpected variations in haplotype frequencies between genetic markers. For example, three pairs of SNPs can define eight haplotypes (see Figure 6.9). If each of the alleles for all three SNP pairs has a frequency of 50%, then each of the eight haplotypes is expected to exhibit a frequency of 12.5% in the population. A deviation from this expected frequency distribution is a linkage disequilibrium, which may indicate that those genetic markers are located close to each other on the chromosome.

The primary question being addressed in a GWAS is whether the major or minor allele of a defined SNP is found more often than expected in individuals with the phenotype of interest. A linkage disequilibrium between a particular SNP allele and the disease trait implies that the gene influencing the disease is located close to the SNP (FIGURE 6.11). Thus, a genome-wide screen of SNPs located across the entire genome can reveal the approximate chromosomal location(s) of gene(s) associated with a particular disease. Furthermore, in the case of polygenic traits, chromosomal locations for multiple genes associated with the trait can potentially be identified. Once the chromosomal locations of the genes are known, the genes in those regions of the chromosomes can be analyzed for their potential role in the trait.

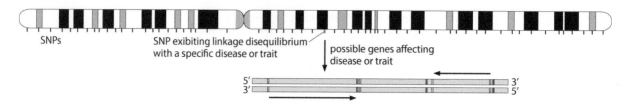

FIGURE 6.11 SNP linkage disequilibrium can reveal chromosomal locations of potential genes of interest. SNPs located throughout the chromosomes can be used as chromosomal markers to identify potential genes associated with a particular phenotype or disease. If a specific SNP is found more frequently in people with a particular disease (red hash mark) than people without the disease, then there is a linkage disequilibrium between that SNP and the disease. This implies that the gene contributing to the disease is located near that SNP. The candidate genes can then be further investigated regarding possible roles in the disease.

Case–control studies are most often used in GWA studies

In case–control studies, individuals exhibiting the trait or disease (cases) being investigated are put into the case group. The control group consists of similar individuals who do not exhibit the trait or disease. DNA from all the study participants is then analyzed with an SNP microarray to determine their genotypes for each of the SNPs on the chip. In other words, for each SNP it is determined whether an individual is homozygous for the major allele, heterozygous, or homozygous for the minor allele (FIGURE 6.12). The frequencies of the major and minor alleles for each SNP are then determined for the cases and for the controls. A difference in the frequencies of the major and minor alleles between the cases and controls is indicative of linkage disequilibrium.

FIGURE 6.12 Case–control GWA study. (A) The study population is divided into a case group of individuals with the disease or trait and a control group of individuals without the trait. The genotype of each individual is then determined with an SNP microarray. (B) For each SNP the frequencies of the major and minor alleles (e.g., C vs. T) are determined. A difference in allele frequency between the case and control groups is indicative of a linkage disequilibrium.

(A)   genotype study subjects with SNP microarray

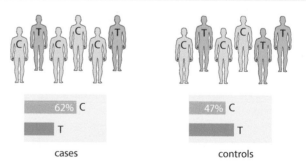

In addition to calculating the differences in frequencies, *P* values for each SNP are also calculated. The *P* value is a statistical measure assessing the reliability of any difference observed between cases and controls. A low *P* value means that the difference observed between cases and controls is more than likely true and not due to random chance. Typically, the data are plotted as the negative log of the P value for each SNP on the y axis and the position of the SNP in the genome on the x axis (FIGURE 6.13). The x axis essentially represents the length of the human genome. Since P values are less than 1, the negative log values on the x axis increase with an increasing statistical significance of the difference between controls and cases. Therefore, a value above a predefined cutoff value is likely to represent a linkage disequilibrium associated with that specific SNP. This plot has been nicknamed a Manhattan plot in that it resembles the skyline of a city with tall skyscrapers.

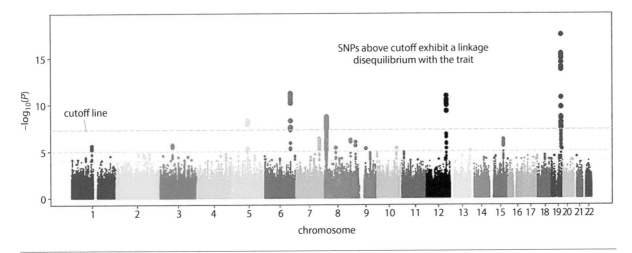

FIGURE 6.13 Manhattan plot of GWA study. A scatter plot of each SNP based on its chromosome location (x axis) and its negative log of the P value associated with the difference in frequency between cases and controls. Those SNPs with values above the cutoff value are considered significant and subject to further analyses. [Modified from Ikram MK, Xueling S, Jensen RA et al. (2010). Four novel Loci (19q13, 6q24, 12q24, and 5q14) influence the microcirculation in vivo. *PLoS Genetics* 6, e1001184. Published under the Creative Commons Public Domain declaration.]

Further studies are needed to define the genetics and biology of disease

A linkage disequilibrium between a particular SNP and a disease or trait implies that a gene involved in that trait is located near the SNP. Therefore, the DNA sequence surrounding that SNP needs to be further analyzed. A typical first step is to examine the function(s) of the gene(s) near the SNP and ask questions about whether the biological role(s) of those gene(s) can explain the manifestations of the disease. For example, high cholesterol is associated with an increased risk of developing cardiovascular disease (Chapter 14). In GWA studies of people with high cholesterol, the SNPs exhibiting linkage disequilibria are near genes encoding various proteins involved in the metabolism of cholesterol. Therefore, polymorphisms in these genes could contribute to developing high cholesterol.

It is unlikely that the SNPs exhibiting the linkage disequilibria are the actual polymorphisms associated with the disease. Therefore, the DNA sequence of the suspected gene is examined for other SNPs or sequence polymorphisms such as insertions or deletions. These other polymorphisms can then be compared in the cases and controls. Those polymorphisms that are associated with cases can then be further analyzed by molecular and biochemical methods to determine the biological effects of the polymorphism. Ultimately these studies may lead to the identification of the polymorphism that is directly correlated with the disease.

Since the first GWA study in 2005 on age-related macular degeneration (BOX 6.7) more than a 1000 GWA studies have been published on over 200 diseases and thousands of associations have been identified. Obviously GWA studies will be extremely useful in identifying common alleles of genes with low penetrance that are involved in complex and common diseases. Subsequent research on these genes and alleles can address their exact role in disease etiology and progression. This knowledge can then be applied to the development of better diagnostics and therapeutics.

BOX 6.7

Complement Factor H and Age-related Macular Degeneration

Age-related macular degeneration (AMD) is a common cause for loss of vision in people over 50 years old. It is characterized by progressive destruction of the retina's central region, called the macula, and the accumulation of extracellular deposits called drusen. In addition to age, the other two major risk factors are smoking and family history. However, no consistent genetic markers had been identified in previous studies. Klein et al. (2005) carried out a genome-wide screening of 96 AMD cases and 50 controls.[1] An SNP located within the gene for complement factor H (CFH) located on chromosome 1 was highly correlated with AMD. Resequencing and further studies revealed a tyrosine-to-histidine mutation at amino acid 402 (Y402H) that was strongly associated with the development of AMD.

CFH is a key regulatory protein of the complement system and inhibits the activation of complement. The complement system is a component of innate immunity and inflammation (Chapter 9).

The identification of complement as a potential etiological agent of AMD is consistent with other characteristics of AMD. Most importantly, complement is a component of drusen. Thus, AMD may result from an aberrant inflammatory process that includes inappropriate complement activation. In addition, smoking is associated with higher levels of circulating CFH. Similarly, dietary zinc supplementation slows the progression of AMD and zinc affects the function of CFH. It is not known how the Y402H mutation affects the function of CFH and leads to AMD. This amino acid is in a domain of CFH that binds to C-reactive protein (CRP), a marker for inflammation, as well as heparin, an anticoagulant. Elevated levels of CRP have also been observed in AMD. Thus, the identification of CFH by GWA is consistent with the known manifestations and features of the disease.

1. Klein RJ, Zeiss C, Chew EY, et al. (2005) Complement factor H polymorphism in age-related macular degeneration. *Science* 308: 385–389. doi: 10.1126/science.1109557.

SUMMARY OF KEY CONCEPTS

- High-throughput DNA-sequencing technologies and bioinformatics have led to the realization of genomics as a field of genetics

- The complete sequence of the human genome, as well as the genomes from many other organisms, is known

- Metagenomics is the sequencing of complex communities of organisms and is being used to analyze the human microbiota

- Other high-throughput technologies complementing genomics are transcriptomics, proteomics, and metabolomics

- Knowledge gained from genomics and its derivatives may lead to a better understanding of the etiology and pathogenesis of disease and the development of better diagnostics and therapeutics

- A future era of personalized medicine may be possible through gene chip and DNA sequencing technologies

- Genome-wide association studies can identify genetic loci associated with traits or diseases

KEY TERMS

chromatography	The process of separating components in a mixture, such as solutes dissolve in a liquid, by passing the mixture over a stationary phase that differentially interacts with the various components.
DNA microarray	A piece of glass or plastic with an array of different single-stranded DNA reporters that is used to detect specific nucleic acids in a sample. Also called a gene chip.
epigenome	The complete set of chemical changes to the DNA and histone proteins in the genome of an organism that are inherited during cell division and influence the formation of heterochromatin.
genome	The complete DNA sequence (genetic material) of an organism.
genome-wide association (GWA)	Association between SNPs across an entire genome and a trait such as a disease using gene chip technology.
mass spectrophotometry	An analytical technique used to determine the chemical composition of a sample by accurately measuring the molecular masses of the components
metabolome	The total number of metabolites present within an organism, organ, tissue, or cell.
metagenomics	The study of genetic material recovered from environmental or clinical samples. Also called environmental genomics.
microbiota	The entire collection of the microbes isolated from a specific habitat. The habitat can also include symbiotic organisms isolated from a host organism.
penetrance	The proportion of individuals in a population who carry a specific gene polymorphism and who also express the related trait.
proteome	The entire set of proteins present within an organism, organ, tissue, or cell.
single nucleotide polymorphism (SNP)	A nucleotide substitution at a specific location in a genome found in some individuals.
transcriptome	The entire set of RNA transcripts present within an organism, organ, tissue, or cell.

REVIEW QUESTIONS

1. Identify and briefly describe the derivatives of genomics.

2. Describe how GWA studies are carried out.

3. What are gene chips and how does this technology differ from the standard hybridization methods used to detect specific genes or transcripts?

4. How are chromatography and mass spectrometry used to determine a proteome?

5. Discuss how genomics, transcriptomics, proteomics, and metabolomics may lead to the development of new drugs to treat diseases.

ADDITIONAL QUESTIONS

In addition to the Review Questions provided above, there is a range of free online questions designed to enable students to further test their understanding of the chapter material. In order to access these interactive questions, please visit the book's website: https://routledge.com/cw/wiser

Genetic Engineering

7

The advancements in DNA technology since DNA was first revealed as hereditary material nearly seven decades ago have been astounding. Among these advancements is the ability to manipulate and modify DNA in a test tube and to introduce this modified DNA back into organisms. In general, this technology is referred to as **recombinant DNA**, and recombinant DNA has advanced our understanding of genes and molecular biology. Recombinant DNA technology provides a means to isolate a specific fragment of DNA or RNA, such as a gene, and molecularly clone that specific sequence of nucleic acid. This allows for the study and characterization of a gene in isolation. Furthermore, proteins encoded by cloned genes can be expressed and the recombined proteins can be studied in situations in which analyzing the natural protein is not feasible.

In addition to serving as an important research tool, recombinant DNA technology can be applied to many commercial endeavors and has opened the field of genetic engineering. For example, recombinant DNA technology can be used to produce large quantities of proteins that may have industrial applications or can function as biopharmaceuticals. Indeed, many drugs and vaccines are now produced via recombinant DNA technology. Similarly, recombinant DNA technology is being applied to agriculture. Genes can be introduced into plants and animals to produce transgenic organisms that can increase and improve the production of food. In addition, recombinant DNA technology can increase the nutritional value of food and lead to improvements in health. And genetic material can also be directly introduced to humans in the treatment and prevention of disease in a process called gene therapy.

RECOMBINANT DNA TECHNOLOGY

PRODUCTION OF BIOPHARMACEUTICALS

AGRICULTURAL APPLICATIONS

GENE THERAPY

RECOMBINANT DNA TECHNOLOGY

The basic premise of recombinant DNA is to combine the nucleic acid of interest, typically a gene, with another DNA molecule called the **vector**. The vector functions as a carrier of the DNA of interest which is also called the target DNA. The process of combining a target gene with the vector is often called gene cloning. The recombinant vector containing the cloned gene is introduced into another organism called the host. If the cloned gene is from a different species, then the host carrying the foreign gene is called a **transgenic organism**.

Some of the impetus for carrying out recombinant DNA is to gain knowledge about the functions of specific genes. For example, the introduction of foreign or modified genes into a cell may affect a cell's phenotype in such a way as to provide clues about the function of that gene. It is also possible to express the cloned genes and produce recombinant proteins in the host organism. This provides a means to produce large quantities of a protein that can be used in studies on the structure and function of that protein. In addition to increasing our basic knowledge about biology, recombinant DNA technology is also being applied to pharmaceutics, agriculture, and medicine (FIGURE 7.1).

FIGURE 7.1 Applications of recombinant DNA technology. The ability to manipulate DNA and create transgenic organisms is widely used in the basic sciences, agriculture, the pharmaceutical industry, and medicine.

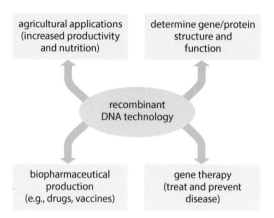

The two basic types of recombinant DNA vectors are viruses and plasmids

Viruses are pathogens that consist of either a DNA or an RNA genome and associated proteins (Chapter 19). As part of their life cycle, viruses introduce their genomes into host cells and the viral genomes take over the machinery of the host cell to promote their replication. It is possible to replace some of the viral genome with a foreign gene and, therefore, the foreign gene is carried into the host cell and replicated with the viral genome. The other major type of recombinant DNA vector is a **plasmid**. Plasmids are found primarily in bacteria and are typically small circular DNA molecules that replicate independently from the bacterial chromosomal DNA. In nature, plasmids function to exchange genetic material between bacteria. Thus, this natural ability of plasmids to function in the transfer of DNA between organisms is exploited in recombinant DNA technology. As in the case of viral vectors, target DNA can be combined with the plasmid DNA and introduced into a host cell.

Vectors also have features to facilitate the gene-cloning process (FIGURE 7.2). For example, vectors contain a specific site for the introduction of foreign DNA called the cloning site. If the intent is to express the foreign gene, then the vector also needs an appropriate promoter adjacent to the cloning site. Some promoters allow for the overexpression of the protein and large quantities of the protein can be produced. An antibiotic-resistance gene is often included in vectors as a selectable marker. This allows for the selection of host cells containing the vector by growing the cells in the presence of an antibiotic. Only cells containing the vector survive and cells without the vector die. Reporters can also be used to easily determine which cells contain the vector. Common reporters are genes for fluorescent proteins or bioluminescence enzymes. In both cases, cells with the vector emit light and can be easily distinguished from cells without the vector.

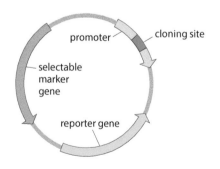

FIGURE 7.2 Features of a typical plasmid vector. Vectors designed for molecular cloning contain a cloning site with sequences of several restriction enzymes. If the vector is to be used for transcription an appropriate promoter is placed on the 5′ side of the cloning site. Most vectors also contain a gene for a selectable marker such as antibiotic resistance. Some vectors contain a reporter gene to facilitate the detection of cells containing the vector.

Enzymes are used to construct recombinant DNA molecules in the test tube

The general process of recombinant DNA involves preparing the vector, preparing the target DNA, and mixing the two DNA molecules so that covalent bonds form between the two DNA molecules. This process utilizes enzymes that carry out chemical reactions on DNA, such as restriction endonucleases and DNA ligase. Restriction endonucleases, more commonly known as restriction enzymes, can break DNA molecules at specific sequences called restriction sites (BOX 7.1). Treating large genomic DNA molecules with restriction enzymes breaks the DNA molecule into smaller fragments that are more amenable to molecular cloning. For example, it may be possible to obtain DNA fragments that correspond to a single gene. Furthermore, many restriction enzymes leave single-stranded overhangs that facilitate the recombining of vector and target DNA.

Restriction Enzymes and Molecular Cloning

Restriction enzymes are nucleases found in bacteria that protect bacteria from infection with viruses. Nucleases are proteins that break the phosphate bonds between nucleotides in DNA and RNA molecules. Restriction enzymes are distinguished from other nucleases by only breaking the DNA molecule at specific sequences called restriction sites. Typically, the recognition site is 4–8 nucleotides in length and is palindromic in that the sequence is the same on both strands when read from 5' to 3' (**BOX 7.1 FIGURE 1**). The natural function of restriction enzymes is to destroy the DNA of viruses by breaking them into small fragments and rendering them non-functional. The host bacterium protects its DNA by modifying the restriction site with another enzyme such as a methylase. The methylated restriction sites are not cleaved by the restriction enzyme.

More than 3000 different restriction enzymes from various bacterial species that recognize a wide range of restriction sites have been characterized. This provides a rather large toolkit for the fragmentation of DNA molecules, and restriction enzymes have proved to be extremely powerful tools in molecular biology and played a key role in the development of recombinant DNA technology. In addition, the high specificity of restriction enzymes for their restriction

site has been exploited to detect single nucleotide polymorphisms (SNPs). An SNP within a restriction site can result in the gain or loss of a restriction site that creates a restriction fragment length polymorphism (RFLP). RFLPs were previously used to carry out DNA fingerprinting studies.

BOX 7.1

BOX 7.1 FIGURE 1 Restriction enzyme recognition site and cleavage. A typical restriction site consists of six nucleotides in which the sequence is the same on both strands when read from the 5' to the 3' direction. For example, GAATTC is the recognition site of the restriction enzyme called EcoR1. Restriction enzymes also cleave between the same two nucleotides on both strands and in the case of EcoR1 the cleavage occurs between the G and A residues (arrows). Cleavage of the DNA at positions off-center of the restriction site results in single-stranded overhangs as indicated by the 5'-ends with the sequence of AATT. Another enzyme that methylates the restriction site blocks the action of the restriction enzyme.

Vector DNA and target DNA are prepared so that both have compatible restriction sites (FIGURE 7.3). The two DNA molecules are mixed in the presence of DNA ligase. DNA ligase catalyzes the formation of a covalent phosphate bond between the terminal nucleotides of the vector and target DNA molecules to form a single hybrid DNA molecule. This hybrid DNA molecule has the same chemical properties as any other DNA molecule even though the target DNA and vector DNA originate from different sources.

FIGURE 7.3 Preparation of recombinant DNA. The vector DNA is isolated and treated with a restriction enzyme that breaks the DNA within the cloning site (yellow). A compatible restriction site is added to the target DNA. The vector and target DNA are mixed in the same test tube in the presence of DNA ligase. The single-stranded overhang (AATT) of the restriction site from the vector can pair with the complementary overhang from the target DNA. DNA ligase forms a covalent phosphate bond between the adjacent nucleotides of the vector and target. This results in the production of a recombinant DNA molecule.

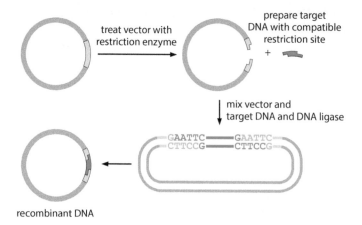

recombinant DNA

Sources of target DNA for molecular cloning include fragmented genomic DNA, complementary DNA, or PCR products

One convenient source of target DNA is genomic DNA fragmented with restriction enzymes (TABLE 7.1). This is dependent on restriction enzyme sites near the gene of interest that generate fragments of appropriate length. Treatment of DNA with restriction enzymes produces well-defined fragments of DNA with ends corresponding to the restriction sites. Furthermore, the use of genomic DNA is not dependent on the active expression of the gene by a particular cell type or assessable tissue. One major limitation of this approach is that the target gene represents a very small percentage of total genomic DNA and, therefore, the target DNA needs to be isolated. This isolation is usually accomplished by a molecular cloning process which can be arduous and time-consuming. Furthermore, genes from genomic DNA may contain introns that may affect the ability to express the target gene.

Table 7.1 Sources of Target DNA for Molecular Cloning

Source	Advantages	Disadvantages
fragmented gDNA	Target gene present in all cells and tissues	Target gene may contain introns Target gene is a very small proportion of total gDNA
RNA/cDNA	Contains only the coding region of the gene Abundant copies of mRNA in cells expressing the target gene	Target gene may not be expressed in available cells or accessible tissues
PCR products	Does not require isolation of target	Difficult to generate long DNA fragments

cDNA, copy or complementary DNA; gDNA, genomic DNA; PCR, polymerase chain reaction.

SIGNPOST

Introns are described in Chapter 4, pages 81–82.

Another potential source of target DNA is the messenger RNA (mRNA) that encodes the gene of interest. The mRNA is converted to complementary DNA (cDNA) with reverse transcriptase (see Box 6.2). An advantage of using RNA is that the mature mRNA has the introns removed and corresponds to the protein-coding sequence. One limitation of using RNA as the nucleic acid source is that the gene of interest needs to be expressed in obtainable cells or assessable tissues. This approach works well for genes that are transcribed at high levels in the cells or tissues of interest. In addition, like using genomic DNA fragments, the gene of interest also needs to be separated from other genes in the sample.

An approach that obviates the need to isolate the gene of interest is the **polymerase chain reaction** (PCR). PCR specifically amplifies the gene of interest (BOX 7.2) and the PCR product can be directly ligated into the vector. This approach has largely been replaced using genomic DNA fragments or cDNA. One minor limitation of PCR is that it is generally difficult to generate long DNA fragments. This is often not a problem since the production of DNA fragments in the size range of a single gene usually works well.

PCR

BOX 7.2

The description of the polymerase chain reaction (PCR) in 1985 by Kary Mullis revolutionized recombinant DNA technology and the analysis of nucleic acids as molecular biomarkers. In recognition of the profound impact PCR has on the advancement of molecular biology, Kerry Mullis was awarded a Nobel Prize in 1993. Regarding biomarkers, the detection of specific nucleic acid molecules relies on the use of a labeled probe complementary to the target molecule (see Box 4.1). Although this method is highly specific, it lacks sensitivity as a detectable amount of target DNA or RNA must be present in the sample. PCR amplifies the target molecule so that a minute amount – perhaps even a single molecule – can be detected.

The process involves using two primers that are complementary to the target sequence. One primer is complementary to the sense strand and the other primer is complementary to the antisense strand. Separation of the two DNA strands and incubation with the primers results in each primer binding to one of the target DNA strands (**BOX 7.2 FIGURE 1**). DNA polymerase and nucleotides are added and the complement of each strand is synthesized, thus creating two double-stranded DNA molecules containing the target sequence. Repeating this process results in the newly synthesized strands of

DNA also serving as templates for the synthesis of complementary strands and the reaction produces four molecules of the target DNA. Each repetition of this process results in a doubling of the target DNA. After many cycles this geometric amplification produces a large amount of a specific DNA molecule defined by the sequence between the primers that is easily detected and analyzed. The method can also be used to analyze RNA by first converting the RNA to cDNA using reverse transcriptase (see Box 6.2). It is also possible to quantify the amount of DNA or RNA in a sample by a modification of the procedure called real-time PCR.

In addition, to being a highly sensitive method for the detection of nucleic acids, PCR is routinely used in recombinant DNA technology. For example, the amplified DNA fragment, often called an amplicon, can be used as the source of the target DNA. Since the amplicon is not substantially contaminated with other nucleic acids that are not of interest, this approach does not require the isolation of target DNA and accelerates the production of recombinant DNA molecules. In addition, restriction enzyme site sequences can be incorporated into the PCR primers and added to the ends of the amplicon to facilitate the gene cloning process.

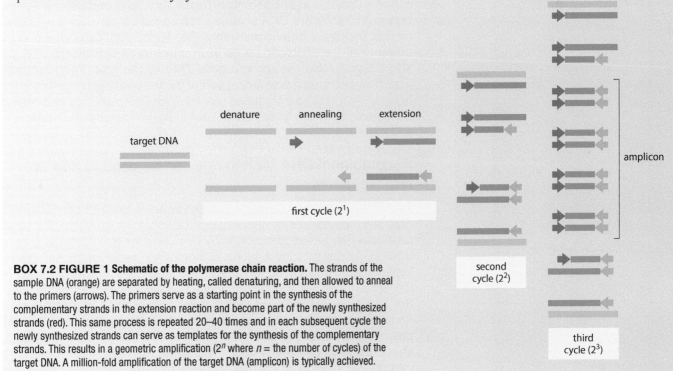

BOX 7.2 FIGURE 1 Schematic of the polymerase chain reaction. The strands of the sample DNA (orange) are separated by heating, called denaturing, and then allowed to anneal to the primers (arrows). The primers serve as a starting point in the synthesis of the complementary strands in the extension reaction and become part of the newly synthesized strands (red). This same process is repeated 20–40 times and in each subsequent cycle the newly synthesized strands can serve as templates for the synthesis of the complementary strands. This results in a geometric amplification (2^n where n = the number of cycles) of the target DNA. A million-fold amplification of the target DNA (amplicon) is typically achieved.

Recombinant DNA can replicate as episomes or integrate into chromosomal DNA

The recombinant vector containing the target gene is introduced into a host cell. Both viral vectors and plasmids exhibit an autonomous replication that is independent of host cell replication. This means that the vector can generate many copies of itself and the target gene is amplified (FIGURE 7.4). At the same time, host cells can also replicate and increase their numbers, leading to further amplification of the gene.

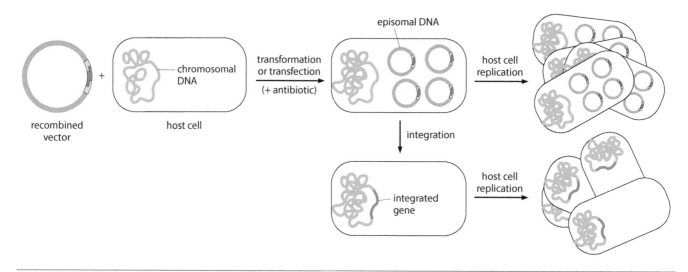

FIGURE 7.4 Incorporation of recombinant DNA into host cells. The recombinant DNA is mixed with appropriate host cells under conditions that allow the DNA to be taken up by the host cells. This process is called transformation if plasmid vectors are used or transfection in the case of viral vectors. Typically, the host cells are grown in the presence of an antibiotic corresponding to a selectable marker on the vector. The vector can replicate as episomal DNA independent of chromosomal replication, or the introduced gene can integrate into the chromosomal DNA and become an integral part of the host cell genome.

Vectors replicating in host cells independent of the chromosomal DNA are called episomes. Episomes can be lost from the host cell unless the episome contains a gene necessary for the survival of the host cell. For this reason, recombinant DNA vectors often contain a gene for a selectable marker such as antibiotic resistance (see Figure 7.2). Cells grown in the presence of the antibiotic retain their episomes. Recombinant DNA can also integrate into the chromosomal DNA of the host. The introduced gene now becomes a permanent part of the host cell's genome and is less likely to be lost. The introduced recombinant DNA, whether episomal or integrated, functions as any other DNA in the cell and can be transcribed and translated.

Recombinant DNA technology can be used to add or remove genes

Recombinant DNA technology can also be used to affect specific genes. This is accomplished by introducing recombinant DNA into the target organism that has sequence homology to regions that flank the gene of interest. These flanking regions of homology allow for recombination between the introduced DNA and the target gene which results in the replacement of the target gene (FIGURE 7.5). For example, often the target gene is replaced with a selectable marker or reporter gene. Removing a gene is often referred to as knocking out a gene and this process may result in phenotypic changes that can provide information about the function of that gene. This same technology can be used to introduce a new gene that is not present in the genome of the host organism. In this case there may be a gain of function associated with the newly 'knocked-in' gene. As with knockouts, knock-ins may provide information regarding the function of that gene. In addition, there are numerous practical applications for transgenic organisms that have had genes removed or genes added via genetic engineering.

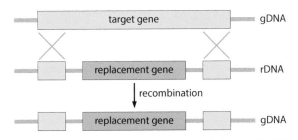

FIGURE 7.5 Targeting genes for replacement. Recombinant DNA (rDNA) is prepared having regions of homology with the target gene that flank the replacement gene (yellow blocks). Recombination by a double crossover between the rDNA and genomic DNA (gDNA) results in the replacement of the target gene with a different gene – often a selectable marker or reporter gene. The target gene is knocked out and the replacement gene is knocked in.

Gene editing via CRISPR increases the efficiency of modifying genes

Another powerful genetic engineering tool is **CRISPR**, which is an acronym for 'clustered regularly interspaced short palindromic repeats.' CRISPR functions as an antiviral immunity mechanism in bacteria. After infection with a virus, bacteria retain portions of the viral genome in their genome and use these viral sequences to generate guide RNAs that are incorporated into a nuclease called CRISPR-associated protein 9 (Cas9). In subsequent infections with the same virus, the Cas9 specifically recognizes the virus DNA and causes double-stranded breaks in the viral DNA that inactivate the virus. This ability to target a nuclease to a specific sequence has been exploited as a mechanism to edit genes. CRISPR exhibits much greater specificity and efficiency in modifying genes than other methods and the use of CRISPR technology is accelerating progress in genetic engineering and gene therapy. The 2020 Nobel Prize in Chemistry was awarded to Jennifer Doudna and Emmanuelle Charpentier for developing this methodology only 8 years after their landmark publication.

Gene editing via CRISPR is accomplished by transforming target cells with the *Cas9* gene and a guide sequence. The guide sequence encodes for an RNA molecule consisting of the scaffold sequence that associates with Cas9 and a sequence complementary to the target gene (FIGURE 7.6). The portion of the guide RNA complementary to the target gene directs Cas9 to catalyze a double-stranded DNA break in the target gene. The cell then repairs the double-stranded break in a process called non-homologous end joining. As part of this non-homologous repair process, a random number of nucleotides is inserted or deleted at the break site. This generally results in a frameshift in the coding sequence that renders the gene non-functional. Thus, the gene is knocked out.

It is also possible to knock in a functional gene by including a third component called the repair template. The repair template contains a functional gene that is flanked by sequences that are complementary to the sequences flanking the guide sequence. The repair template is incorporated into the target gene during the repair of the double-stranded break by homologous recombination (see Figure 7.6). Inclusion of the repair template increases the fidelity of the repair and allows precise gene editing.

Frameshift mutations are described in detail in Chapter 11, pages 224–225.

SIGNPOST

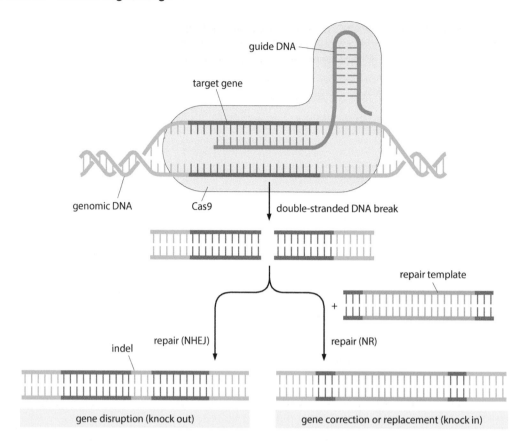

FIGURE 7.6 Gene editing with CRISPR. The guide RNA that is complementary to the target gene directs Cas9 to catalyze a double-stranded break in the DNA of the target gene. Repair of this break by non-homologous end joining (NHEJ) results in the insertion or deletion (indel) of a random number of nucleotides that leads to a disruption of the target gene (knock out). The repair can proceed by homologous recombination (HR) if a repair template is also introduced to the host cell with *Cas9* and the guide sequence. Homologous recombination can correct an error in the gene or replace the gene with another gene (knock in).

PRODUCTION OF BIOPHARMACEUTICALS

Many proteins such as hormones, blood clotting factors, enzymes, and vaccines have direct medical applications in the prevention and treatment of disease. However, producing pharmaceutical proteins via chemical synthesis is essentially impossible except for rather short peptides. Furthermore, there are several issues associated with isolating proteins from blood or tissues (TABLE 7.2). In particular, human blood or tissues are of limited supply and proteins isolated from animals may not optimally function in humans. There are also safety issues with pharmaceutical proteins prepared from human blood and tissues. For example, many hemophiliacs were inadvertently infected with HIV (Chapter 22) due to treatment with a clotting factor that was prepared from contaminated human blood. Similarly, there are safety issues associated with the manufacture of vaccines from pathogenic organisms. All these issues make the production of protein-based pharmaceuticals by conventional biochemistry rather expensive.

Recombinant DNA technology increases efficiency in biopharmaceutical production

Since the introduction of Humulin®, a human insulin produced via recombinant DNA technology, numerous recombinant DNA biopharmaceuticals have been introduced (TABLE 7.3). A major advantage

Table 7.2 Limitations in the Production of Protein-based Pharmaceuticals

Limitation	Recombinant DNA solution
Limited supply of source material	Overexpression of target in the convenient host
Suboptimal performance of non-human source	Clone and use human gene
Contamination of product with human pathogens	Human material not used in production
Safety issues with growing large amounts of a pathogen	Target genes are expressed in non-pathogenic host and a living pathogen is not used during production

provided by recombinant DNA technology is that transgenic organisms essentially provide an unlimited supply of the biopharmaceutical that is easily obtained from a convenient source. This results in reduced production costs compared with conventional methods. Recombinant DNA also obviates the need to work with intact and living pathogens as part of vaccine production. Instead, the relevant genes are expressed as recombinant proteins in a host that does not require biosafety containment. Not only does recombinant DNA technology lower production costs, but in some cases the technology allows the production of a biopharmaceutical that would have been impossible by other methods.

Table 7.3 Examples of Pharmaceuticals Produced via Genetic Engineering

Product (trade name(s))	Therapeutic use	Produced by
Hormones/Cytokines		
Insulin (Humulin®, Insuman®, Novolin®)	Manage diabetes	Bacteria, yeasts
Erythropoietin (Epogen®)	Treat anemia (stimulate production of red blood cells)	Mammalian cells
Epidermal growth factor (Heberprot-P®)	Promote wound healing in chronic diabetes	Bacteria
γ-Interferon (Actimmune®)	Treat infections or cancer	Bacteria
Interleukin-2 (Proleukin®)	Treat cancer	Bacteria
Proteins/Enzymes		
Tissue plasminogen activator (Retavase®, Activase®)	Treat heart attack or stroke (clot-busting drug)	Bacteria, mammalian cells
Factor VIII (Recombinate®)	Manage hemophilia (blood-clotting factor)	Mammalian cells
DNAse-1 (Pulmozyme®)	Reduce mucus during cystic fibrosis	Mammalian cells
Anti-thrombin III (Atryn®)	Prevent blood clotting (e.g., during surgery)	Goats
Vaccines		
Hepatitis B vaccine (Engerix-B®)	Prevent infection/disease	Yeast
Influenza vaccine (FluBlok®)	Prevent infection/disease	Insect cells

Numerous vectors and hosts are available for biopharmaceutical production

The basic premise of biopharmaceutical production via recombinant DNA is to clone the gene of interest into an appropriate vector and to express the protein product in an appropriate host (FIGURE 7.7). The vectors used in biopharmaceutical production are designed to express high levels of the target protein, sometimes up to 10% of the total protein in the host. Bacteria, and especially the well-characterized laboratory strains of *Escherichia coli*, provide a convenient host. These bacteria grow well in culture and are easily adapted to large vessels called bioreactors. The recombinant protein is purified from the transgenic bacteria and manufactured into a drug.

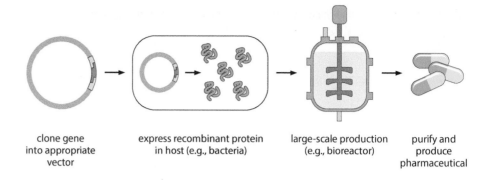

clone gene
into appropriate
vector

express recombinant protein
in host (e.g., bacteria)

large-scale production
(e.g., bioreactor)

purify and
produce
pharmaceutical

FIGURE 7.7 Production of recombinant therapeutic proteins. The gene of interest is cloned into an appropriate vector so that it can be expressed in a host cell such as bacteria. The transgenic host is cultured in large bioreactors and the protein is purified from the cultures.

The expression of human proteins is not always optimal in bacteria. Even though prokaryotes and eukaryotes have the same genetic code and eukaryotic genes can be transcribed in prokaryotes, it is not always possible to obtain functional eukaryotic proteins in prokaryotic systems. One reason is that sometimes eukaryotic proteins do not fold correctly in the bacterial cell. Probably more important, though, eukaryotic proteins are often subjected to post-translational modifications, such as glycosylation, which are not carried out by bacteria. Therefore, it is often necessary to utilize eukaryotic systems for the expression of functional recombinant human proteins (TABLE 7.4).

SIGNPOST

Post-translational modifications are described in Chapter 3, page 56.

Table 7.4   Eukaryotic Hosts for Expressing Recombinant Proteins

Hosts	Notes
Saccharomyces cerevisiae	Well-characterized laboratory model
Pichia pastoris	High level of protein expression
Baculovirus/insect cells	High level of protein expression
Cultured mammalian cells	Similar to human cells
Transgenic plants	Large-scale production in crop plants
Transgenic animals	Large-scale production of milk

Included among possible eukaryotic hosts are yeasts such as *Saccharomyces cerevisiae* and *Pichia pastoris*. Yeasts are single-cell eukaryotes that are easily grown and manipulated in culture and can be scaled up to bioreactors for the production stage. Baculovirus is a recombinant DNA vector specifically engineered for high levels of protein expression in insect cells. Mammalian cell lines have also been used as host cells. It is also possible to produce transgenic plants and animals that can be used for the production of biopharmaceuticals. In terms of large-scale production, fields of transgenic plants expressing the pharmaceutical provide a large source of material. In theory, transgenic plants could also be engineered to deliver an edible vaccine. Instead of purifying the recombinant protein from plants, edible portions of the plant could be eaten as a means of vaccination.

Pharming is the production of biopharmaceuticals in transgenic animals

It is also possible to produce transgenic animals. Recombinant DNA is injected into the nucleus of a fertilized egg (FIGURE 7.8) and the transgenic egg develops into an embryo in vitro. This embryo is implanted into a surrogate mother and develops into a transgenic offspring. These transgenic animals can be exploited in the production of human therapeutic proteins. To accomplish this, the target genes are recombined with promoters and other transcription regulatory elements so that the engineered protein is expressed in mammary gland cells and secreted into the milk. The recombinant biopharmaceutical is isolated from milk. This approach allows for large-scale production of a therapeutic protein produced in a mammalian system. In a play on the words farming and pharmaceutical, the production of biopharmaceuticals in transgenic animals is called **pharming**.

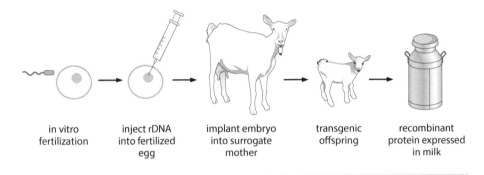

| in vitro fertilization | inject rDNA into fertilized egg | implant embryo into surrogate mother | transgenic offspring | recombinant protein expressed in milk |

FIGURE 7.8 **Production of transgenic animals.** Recombinant DNA (rDNA) is injected into the nucleus of a fertilized egg and the resulting embryo is implanted into a surrogate mother. The resulting transgenic offspring expresses the trait encoded by the introduced DNA. Pharmaceuticals can be produced and harvested from the milk.

Genetic engineering can improve the quality of biopharmaceuticals

In addition to lowering production costs and increasing safety, recombinant DNA technology can produce more efficacious biopharmaceuticals. For example, historically insulin was extracted from the pancreas of cows or pigs obtained from slaughterhouses. Such insulin functions satisfactorily for most people since there is only a single amino acid sequence difference between porcine and human insulins and only three amino acid differences between bovine and human insulins (FIGURE 7.9). Nonetheless, animal insulin does not perform as well as human insulin and a small percentage of patients with diabetes have an allergic reaction to animal insulin. In addition, antibodies might form against the foreign insulin and decrease its efficacy. Thus, producing recombinant human insulin results in a more efficacious drug.

It is also possible to increase the efficaciousness or stability of many biopharmaceuticals by altering the sequence of the cloned DNA. For example, several derivatives of recombinant insulin have been created by making changes in the amino acid sequence of insulin that results in insulin that is either faster acting or longer lasting (see Figure 7.9). Faster-acting insulin better controls blood sugar levels after a meal and longer-lasting insulin provides a steady basal level of insulin during the day and particularly overnight. Three faster-acting analogs have been created by changing a few amino acids in the insulin sequence that

FIGURE 7.9 Insulin analogs. Recombinant DNA technology has been used to modify insulin to make it either faster acting (Aspart, Lispro, and Glulisine) or longer lasting (Glargine, Detemir, and Degludec). Shown are the various amino acids that have been modified by genetic engineering to produce insulin derivatives. The three amino acids that are different in bovine insulin and the single amino acid difference in porcine insulin.

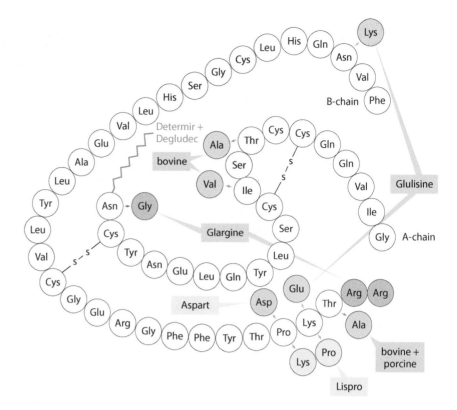

decreases insulin dimers and hexamers. The decreased interaction of insulin with itself increases the bioavailability of the active insulin monomers. Longer-lasting insulin was produced by making the insulin less soluble through the addition of fatty acids so that the bioavailable form is slowly released.

AGRICULTURAL APPLICATIONS

The ability to produce transgenic plants and animals is having a major impact on agriculture. This is especially true for plants, and many of the major farm crops throughout the world are now genetically modified. Most of the traits engineered into transgenic plants are aimed at improving yields and lowering the cost of food production (TABLE 7.5). In addition to increasing yield and lowering production costs, genetic engineering in plants can potentially be used to improve the quality of food. For example, through genetic engineering it may be possible to improve flavor or reduce toxin levels. Some efforts have also been directed toward improving the nutritional value of food and, in particular, rice (BOX 7.3).

Table 7.5   Examples of Traits Engineered into Crop Plants

Trait	Examples	Advantages
Insect resistance	Bt-corn, Bt-cotton, and several others	Prevention of crop loss and less chemical insecticide use
Pathogen resistance	Virus-resistant papaya, squash, potatoes	Prevention of crop loss and less chemical insecticide use
Herbicide resistance	Numerous crop plants, especially corn and soybeans	Lower production costs and increased yields
Stress resistance	DroughtGard corn	Better tolerance of extremes in temperature, moisture, salinity
Delayed ripening	FlavrSavr tomato	Allows vine ripening without compromising shelf-life
Increased nutrition	Golden Rice	Reduction of malnutrition in many parts of the world

Golden Rice

Rice is a major food source in much of the world and vitamin A deficiency is relatively common in regions that rely heavily on rice as a food source. Vitamin A deficiency lowers immunity and can lead to vision problems (Chapter 12). The edible portion of rice, called endosperm, does not produce vitamin A. Genes for the synthesis of β-carotene, the precursor to vitamin A, have been introduced into rice so that β-carotene is synthesized in the endosperm. This gives the rice grains a yellow color (**BOX 7.3 FIGURE 1**) and hence the moniker golden rice. Golden rice has proven to be safe for human consumption but has not yet been produced commercially. Commercial farmers are hesitant due to lower yields compared with other rice variants and questions about the market demand for such a product. Nonetheless, the development of golden rice illustrates that it is possible to improve the nutritional quality of food through genetic engineering.

BOX 7.3 FIGURE 1 Golden rice. Expression of β-carotene in transgenic rice gives it a yellow color compared with normal rice. [From International Rice Research Institute (IRRI), published under CC BY 2.0.]

BOX 7.3

Another factor in the application of genetic engineering to crop plants is the relative ease at which transgenic plants can be created. To create transgenic plants, cells from the actively growing root tips are disaggregated and grown in culture as single cells. Therefore, the methodology of introducing recombinant DNA into cells is similar to those used for bacteria and eukaryotic microbes. More importantly, the transformed plant cells can readily be induced to develop back into plants that are capable of flowering and producing seeds. Plants obtained from seeds of transgenic plants continue to express the traits encoded by the introduced gene(s). Although creating transgenic plants is relatively easy, there are some safety and environmental concerns that must be addressed (TABLE 7.6).

Table 7.6	Potential Concerns about Genetically Modified Foods
Food safety	Possible toxicity or allergies associated with introduced genes
Non-target organisms	Non-pest insects and other organisms may be adversely affected
Promotion of resistance	Pest and pathogens may develop resistance to introduced genes and weeds may develop herbicide resistance
Genetic pollution	The introduced genes may transfer to other organisms via horizontal gene transfer
Domination of 'Big Ag'	Small and low-income farmers may not be able to capitalize on transgenic technologies and be at an economic disadvantage

Tolerance to pests, pathogens, and herbicides greatly improves crop yields

Common types of genetic modifications introduced into crop plants are insect pest resistance, pathogen resistance, and herbicide tolerance. Insect resistance is engineered by introducing a gene called Bt toxin from the bacterium *Bacillus thuringiensis* into plants (FIGURE 7.10). Bt toxin has been used as a natural insecticide for decades due to its high specificity for insects and low toxicity for animals. The toxin binds to specific receptors on the intestinal epithelium of insects and disrupts the epithelial layer. Insects die a few days after ingesting Bt toxin. The use of Bt crops has reduced losses due to direct insect damage, as well as indirect loss due to fungal infections resulting from the insect damage. Furthermore, the use of Bt crops greatly reduces the use of chemical pesticides and the associated environmental effects of chemical pesticides. It is estimated that the application of 450,000 tons of insecticide has been avoided due to the use of Bt toxin genes in crops.

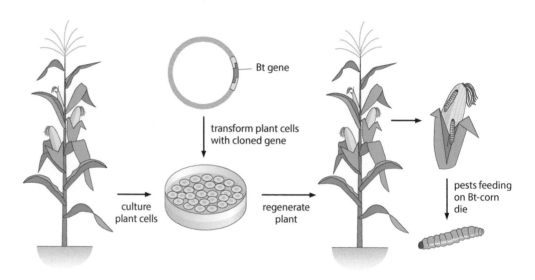

FIGURE 7.10 Genetic engineering in plants. Plant cells grown in culture are transformed with a cloned gene. Plants are regenerated from the transformed cells and those plants express the trait of the introduced gene. For example, transgenic plants with the Bt toxin gene are toxic to insect pests resulting in greater crop yields without the application of chemical pesticides.

Similarly, many plants have been engineered to be less susceptible to pathogens such as viruses. The loss due to pathogen infection can be 20–80% of some crops. Unlike animals, there are no drugs or vaccines to control infectious diseases in plants. Currently the control of plant viruses is through insecticides since many plant viruses are transmitted by insects. Thus, plants engineered to be pathogen resistant not only result in higher production due to less loss, but also result in fewer chemicals in the environment.

The first genetically modified plants to be widely used were herbicide tolerant. Weed control is a major challenge in crop production in that poorly controlled weeds drastically reduce crop yield and quality. Numerous crop plants have now been genetically engineered to be tolerant to herbicides. Therefore, such crops can be treated with the appropriate herbicide to control weeds without affecting transgenic plants. In general, yields are increased by approximately 10% through the use of herbicide-resistant crops and production costs are lowered due to less effort at controlling weeds. Another benefit of herbicide-tolerant crops is that they can be directly planted into weedy fields, and this reduces the need for plowing, which means less soil erosion. However, the use of herbicide-resistant crops is counter-productive toward reducing the use of chemicals in agriculture.

Transgenic animals potentially increase the efficiency of meat production

Research is also being done on producing transgenic animals as a food source. The rationale is the same as in using transgenic plants in agriculture, namely increasing productivity and quality. For example, livestock that grows larger or faster can lower the cost of producing meat. Thus far, the only transgenic animal approved by the United States Food and Drug Administration for human consumption is a salmon called AquAdvantage®. Through the replacement of a growth hormone gene, this farm-raised salmon can reach harvest size in 16–18 months instead of the typical 3 years.

As in plants, it is theoretically possible to genetically engineer animals to be resistant to infectious diseases and minimize loss due to disease. Similarly, it may be possible to genetically engineer cows to produce more milk or improve the nutritional quality of milk. This, in turn, can lead to higher growth rates and earlier weaning in young animals, and thereby increases in productivity. Attempts are also under way to genetically engineer cows to produce milk that is more like human milk. Such milk may be an improvement over existing formulas for mothers unable to breastfeed.

GENE THERAPY

Genetic material can also be introduced into humans to treat or prevent disease, and this is referred to as **gene therapy**. Currently, gene therapy is primarily experimental and there are very few approved gene therapies (TABLE 7.7). Nonetheless, gene therapy is an active area of research and more gene-based therapies will be developed in the future. An obvious application of gene therapy is the treatment of genetic diseases. Many diseases are due to defective genes (Chapter 11) and it is at least theoretically possible to replace defective genes with functional genes. On a technical level, this has proven to be rather difficult. Another active area of gene therapy research is cancer treatment via immunotherapy. The basic approach is to genetically modify cells of the immune system so that these immune effector cells become more efficient at detecting and destroying cancer cells (see Box 16.4). Several therapies against specific blood cancers based on using a receptor called the chimeric antigen receptor have been approved.

Immunotherapy against cancer is discussed in Chapter 16, pages 315–316.

SIGNPOST

Table 7.7	Examples of Approved Gene Therapies	
Therapy (year approved)	**Disease**	**Description**
Luxturna (2017)	Hereditary retinal dystrophy due to defective *RPE65* gene (converts light to an electrical signal), causing vision loss or blindness	Functional *RPE65* gene in adeno-associated virus vehicle is injected into the retina, leading to restored vision
Zolgensma (2019)	Spinal muscular atrophy (a neuromuscular degenerative disease) due to defective *SMN1* gene	Functional *SMN1* gene in adeno-associated virus vehicle is infused into a vein
CAR-T-cell therapies (several, 2017–2021)	Various leukemias and lymphomas (cancers) of B-cell origin	Patients' T cells are modified ex vivo with a retroviral vectors containing chimeric antigen receptor (CAR) which directs T cells to kill cancer cells
COVID-19 vaccines (several, 2021)	A coronavirus that emerged in 2019 and caused a pandemic in 2020 (Chapter 20)	Adenovirus vector recombined with modified mRNA of the spike protein gene delivered with lipid nanoparticles

Viral vectors and lipid nanoparticles are common DNA delivery vehicles

A critical element in the success of gene therapy depends upon the ability to deliver the recombined DNA to the appropriate target cell. Viruses are efficient at delivering nucleic acids to cells and several vectors for the delivery of genes to humans have been developed. These viral vectors are modified so that they do not cause disease and that they can carry foreign DNA. The three most widely used viral vectors in gene therapy are adenoviruses, retroviruses, and adeno-associated viruses (TABLE 7.8). The gene of interest is recombined with the vector genome and the recombinant virus is used to infect humans. The virus carries this recombined gene into the host cell and the gene is transcribed and translated into a protein. Adenoviruses replicate as episomes in the host cell, whereas retroviruses integrate into the host cell genome. Adeno-associated viruses can either replicate as episomes or integrate into the host genome depending on how the vector is modified. Episomal DNA may be lost over time and thus may only be transiently expressed.

Table 7.8	Viral Vectors for Gene Therapy			
Virus	**Structure**	**Replication**	**Advantages**	**Disadvantages**
Adenovirus	dsDNA, non-enveloped	Episomal	Large capacity to carry foreign DNA	Transient expression and potential immune response against the vector
Retrovirus	dsRNA, enveloped	Integrated	Integration into the host genome	Potential for insertional mutagenesis
Adeno-associated virus	ssDNA, non-enveloped	Episomal, integrated	No disease potential and limited immune response	Limited capacity to carry foreign DNA

A major concern of viral vectors in gene therapy is the potential for inflammatory responses against the vector. This is especially true for adenoviruses, which are common and to which most people have been exposed several times. This means that a sensitized immune system may exhibit a strong immune response and cause adverse side effects. Conversely, prior immunity to the vector could result in the neutralization of the vector and render the therapy ineffective. A concern about retroviruses is the potential for insertional mutagenesis due to the random integration of the viral genome into the host genome. Integration into or near an important gene could adversely affect the function of that gene and retroviral vectors have been associated with higher risks of developing cancer. In contrast with the random integration of retroviral genomes, the integration of adeno-associated viruses is primarily site specific in chromosome 19.

Another delivery vehicle for nucleic acids is **lipid nanoparticles**. Lipid nanoparticles circumvent many of the problems associated with viral vectors since they cannot cause disease and are not recognized by the immune system. The gene of interest, in the form of either DNA or RNA, is encapsulated into a spherical mass of lipids (FIGURE 7.11). Encapsulation of the nucleic acid into lipids protects the nucleic acids from degradation and the lipid nanoparticles can fuse with the plasma membrane of cells to deliver the genes inside the cell. It is also possible to incorporate ligands on the surface of the nanoparticle that binds to receptors on target cells. Such ligands can minimize the delivery of the recombinant gene to non-target cells.

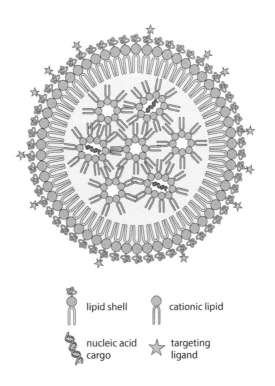

FIGURE 7.11 Lipid nanoparticles as nucleic acid delivery vehicles. Nucleic acids such as DNA and RNA can be incorporated into lipid nanoparticles. The lipid shell protects the nucleic acid cargo and cationic (positively charged) lipids help neutralize the negatively charged nucleic acids. Ligands that target the nanoparticles to cells with specific receptors can also be integrated into the outer shell.

lipid shell　　cationic lipid

nucleic acid cargo　　targeting ligand

Gene therapy often involves an ex vivo approach

An obstacle of both viral vectors and liposomes as DNA delivery vehicles is getting the recombined gene to the correct organs or cells. The simplest approach is to simply inject the recombinant DNA vehicle into the patient. This approach works for COVID-19 RNA vaccines since there is no specific target cell. However, if there is a specific target cell, the recombined gene may never be delivered to the cell where it is most needed. In addition, delivery of the target gene to the wrong cell could have adverse effects. Sometimes it is possible to inject the vehicle directly into the affected organ or tissue as is the case with Luxturna which is injected into the retina (see Table 7.7). Directly injecting the recombinant DNA vehicle into the patient is called in vivo gene therapy (FIGURE 7.12). Another approach is ex vivo gene therapy.

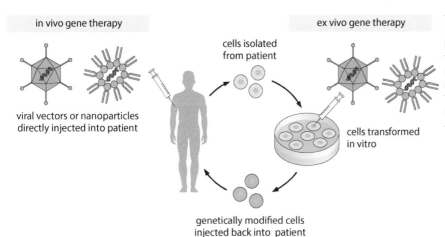

in vivo gene therapy

ex vivo gene therapy

cells isolated from patient

viral vectors or nanoparticles directly injected into patient

cells transformed in vitro

genetically modified cells injected back into patient

FIGURE 7.12 Introduction of genes into humans. Viral vectors or nanoparticles containing the therapeutic gene(s) can be directly injected into either the patient or the target organ. In the ex vivo approach cells are isolated from the patient and the therapeutic gene(s) are introduced in vitro. The modified cells are then returned to the patient.

The ex vivo approach has shown moderate success in immunotherapy against cancer (see Table 7.7) and in situations in which white blood cells are the target. For example, the first clinical trial of gene therapy utilized the ex vivo approach (BOX 7.4). White blood cells are immune effector cells that circulate throughout the body and help defend against infections and cancer (Chapter 9). Due to their location in the blood, white blood cells are readily accessible and are easily removed from an individual. It is also possible to isolate a specific type of immune effector cell from the blood. The isolated cells are cultured in vitro and, while in culture, the target cells are transformed with the recombinant DNA. This is a more controlled environment to deliver genes to specific cells and monitor the success of the delivery. These genetically modified cells are then injected back into the same individual. This approach works well for white blood cells, but is not widely applicable to cell types that are not easily removed from the body and cultured.

BOX 7.4

The First Gene Therapy Clinical Trial

The first clinical trial testing the feasibility of gene therapy was carried out in 1990. The targeted disease was severe combined immunodeficiency disease (SCID) due to adenosine deaminase (ADA) deficiency. ADA catalyzes the conversion of the nucleotide adenosine to the nucleotide inosine. Toxic levels of adenosine accumulate in individuals who have defective ADA genes. Cells of the immune system, and in particular lymphocytes (Chapter 9), are sensitive to high levels of adenosine and cease to function. This leads to a severe immunodeficiency in which the affected individual has essentially no immunity and is susceptible to severe infections from even the most innocuous microbes. This is sometimes called the 'bubble boy' disease in that children with this condition are housed in a germ-free and sealed environment as much as possible. The disease is treatable by the injection of a modified ADA protein. This does not completely restore immunity and the modified protein must be injected every 2–3 days. Furthermore, the production of

the modified ADA protein is expensive, and this treatment can cost more than US$100,000 per year.

Two patients with ADA-SCID were treated as part of the first clinical trial. The ex vivo approach was used in which lymphocytes were obtained from the patient's blood and then infected with a retrovirus containing the ADA gene. The transfected lymphocytes were cultured to expand their numbers and then injected back into the patients. One of the patients showed modest improvement, whereas the other patient exhibited a dramatic improvement that persisted and allowed a normal life. Although subsequent clinical trials did not always show such dramatic success, the trials did demonstrate that the concept of gene therapy was feasible. This success paved the way for continued research into gene therapy and numerous clinical trials involving many different genes have been subsequently carried out.

Recombinant DNA technology has advanced vaccine development

Vaccination is a process to stimulate the immune system by inoculating with a non-virulent infectious agent or a component of an infectious agent (Chapter 9). Immunogens contained within the vaccine activate the immune system and a vaccinated individual acquires immunity against that infectious agent without experiencing the disease. Recombinant DNA technology is utilized at many levels in vaccine production and administration. For example, some proteins of the pathogen are good immunogens and can induce protective immunity. The genes for such immunogens can be cloned and expressed in a transgenic organism and produced as a recombinant vaccine.

Recombinant DNA technology can also be used to create chimeric viruses that function as vaccines. In this case, the immunogen from a pathogenic virus is recombined with an avirulent viral vector such as adenovirus (FIGURE 7.13). Viral vaccine vectors can either be replicating or non-replicating. Non-replicating viruses simply deliver the immunogen gene to the host cell. The immunogen is expressed and secreted by the host cell, and the immune system recognizes this foreign protein (FIGURE 7.14). This is essentially gene therapy, since it is the introduction of a gene to treat or prevent disease. Some viral vectors are capable of limited replication. However, the replication does not reach levels that can cause disease. In the case of replicating viral vectors the immune system recognizes the chimeric virus and induces immunity against the pathogen.

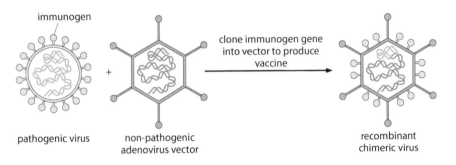

FIGURE 7.13 Chimeric viruses as vaccines. The gene for an immunogen from a pathogen is cloned into a vector such as an adenovirus. The vector remains non-pathogenic, but now expresses the immunogen, which can activate a protective immune response against the pathogen.

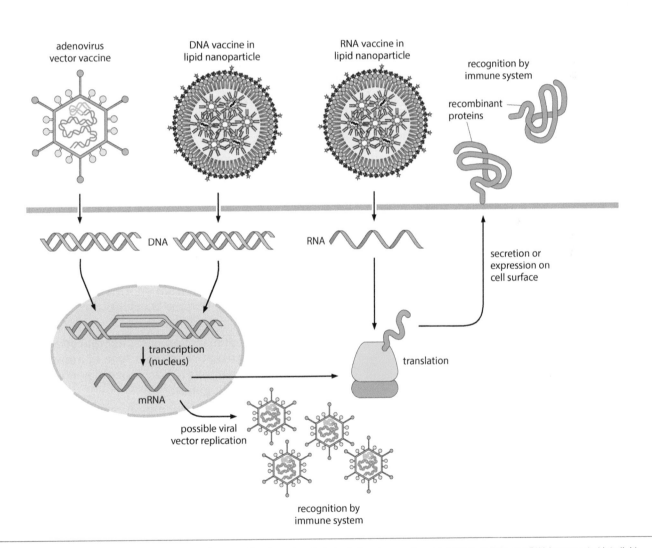

FIGURE 7.14 Recombinant vaccines. Vehicles for recombinant vaccines include viral vectors, such as adenovirus, or DNA or mRNA incorporated into lipid nanoparticles. Both the adenovirus vector and DNA vaccine deliver the cloned gene of the immunogen to the inside of the cell and the DNA enters the nucleus and is transcribed. The resulting mRNA is translated by the host cell. RNA vaccines contain mRNA that can be directly translated. The recombinant protein is secreted or expressed on the cell surface where it is recognized by the immune system. This recognition activates a protective immune response. Some viral vectors are capable of replication and produce chimeric viruses that are recognized by the immune system.

It is also possible to vaccinate with the cloned gene or mRNA of the immunogen delivered with lipid nanoparticles. The cells of the individual then produce the immunogen that activates the immune system (see Figure 7.14). These DNA- or RNA-based vaccines obviate the need to purify proteins or work with pathogens and can greatly reduce production costs. For a DNA vaccine the strategy is to clone the gene of interest and place it adjacent to a promoter that allows for a high level of expression in human cells. After entry into a cell the DNA enters the nucleus and is transcribed into mRNA and translated into protein. The host cell secretes or exports the foreign protein to its surface and the immune system recognizes the protein as foreign, resulting in the activation of the immune system. It is also possible to vaccinate with the mRNA of the immunogen that is directly translated in the cytoplasm of the host cell. Several COVID-19 vaccines utilizing either adenovirus vectors or mRNA have been approved (see Table 7.7).

SUMMARY OF KEY CONCEPTS

- DNA molecules can be recombined and modified in the test tube and this recombinant DNA can be introduced into organisms

- Transgenic organisms contain genetic material from another species and are widely used in research to elucidate the function of genes and proteins

- Transgenic organisms are also widely used to produce biopharmaceuticals

- Genetic engineering is improving the efficiency of production in agriculture and can potentially increase the nutrition of food

- Gene therapy has the potential to treat or prevent many diseases through the introduction of genetic material into humans

- A gene therapy approach can also be used to vaccinate and induce protective immunity via viral vectors or RNA vaccines

KEY TERMS

complementary DNA (cDNA)	Refers to DNA synthesized from an RNA template, such as messenger RNA, using reverse transcriptase. Also called copy DNA.
CRISPR	An acronym for clustered regularly interspaced short palindromic repeats that refers to a type of antiviral immunity found in bacteria. This technology provides a precise method to edit genes.
gene therapy	Treatment involving the introduction of genetic material into an individual for the purpose of curing, managing, or preventing disease.
lipid nanoparticle	A vehicle for the delivery of genetic material to cells during gene therapy or vaccination with DNA or RNA.
polymerase chain reaction (PCR)	A technique by which a specific nucleic acid is amplified through the use of specific DNA primers and repeated rounds of synthesis.
pharming	The process of producing pharmaceuticals in transgenic plants or animals.
plasmid	A small autonomously replicating DNA found primarily in bacteria. Plasmids are widely used as vectors in recombinant DNA.

recombinant DNA	Refers to the joining together of DNA molecules from different sources to produce new genetic combinations that are used in science, medicine, agriculture, and industry.
transgenic organism	An organism in which DNA from another organism has been introduced.
vector	An autonomously replicating DNA, such as a virus or plasmid, which is combined with the DNA of interest during recombinant DNA procedures.

REVIEW QUESTIONS

1. Describe the in vivo and ex vivo approaches of gene therapy.

2. Describe the steps involved in the production of pharmaceutical proteins in animals via recombinant DNA technology.

3. Describe how the polymerase chain reaction can be used to identify specific nucleic acid molecules.

4. What are some of the potential advantages of using transgenic plants in food production and agriculture?

ADDITIONAL QUESTIONS

In addition to the Review Questions provided above, there is a range of free online questions designed to enable students to further test their understanding of the chapter material. In order to access these interactive questions, please visit the book's website: https://routledge.com/cw/wiser

Cell Differentiation and Stem Cells

8

The human body is composed of a wide variety of cells performing distinct roles in anatomy and physiology. Every individual started out as a fertilized egg and this single cell, through the process of mitosis and differentiation, developed into a complex organism containing an estimated 37 trillion cells. Although all the cells of an individual have essentially the same genotype, there are many distinct cellular phenotypes. These different phenotypes are determined by differential gene expression in that a specific cell type expresses specific genes that are associated with the specific functions of that particular cell. This process by which cells differentiate into specific cell types is complex and not completely understood.

Most cells in the human body are no longer capable of cell division and are referred to as terminally differentiated cells. However, there is a need to generate replacement cells and repair damaged tissues. Regeneration and repair processes are carried out by stem cells. Stem cells retain their ability to divide and are responsible for the replacement of senescent or damaged cells. In addition, it may be possible to use stem cells in cell-based therapies to treat some diseases.

ORGANIZATION OF CELLS

The organization of cells can be considered on two levels: intracellular and intercellular. The intracellular level represents how the molecules and subcellular compartments making up the cell are organized (Chapter 2). Intercellular organization is concerned with how the cells are organized to form the organism. For example, the various types of cells making up the human body are organized tissues. **Tissues** represent the intercellular organization between cells (FIGURE 8.1). A tissue is a collection of similar cells that carries out a specific function. Different types of tissues come together and form organs. The tissues function in a coordinated fashion to carry out the specific functions of the organ. Likewise, many organs function together to form an organ system. This combination of cells, tissues, organs, and organ systems makes up the complete organism.

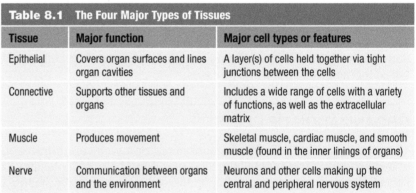

FIGURE 8.1 Organizational relationships starting from atoms to organism. Diagram of hierarchical organization of the human body.

There are four basic types of tissues in humans

The various tissues found in humans and other animals can be grouped into four major types: epithelial, connective, muscular, and nervous (TABLE 8.1). Epithelial tissues generally form layers of cells that cover organ surfaces, such as the skin and the lining of the digestive, respiratory, and reproductive tracts. Connective tissue is the most abundant and widely distributed in the human body and generally functions in support. It is typically found between other tissues and often includes extracellular material. Muscle tissue functions to produce force and cause movement. Nervous tissue consists of nerve cells that make up the brain and other parts of the nervous system. The nervous system functions to send signals throughout the body and coordinate activities in the various organs. Within each of these four basic types of tissues there are also numerous tissue subtypes.

squamous

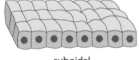

cuboidal

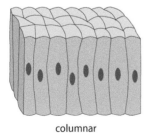

columnar

FIGURE 8.2 Examples of simple epithelium. Simple epithelium consists of a single layer of cells and the three basic types of epithelial cells are defined by relative shape. Squamous epithelium consists of flattened cells, cuboidal epithelial cells are somewhat squarish, and columnar epithelium is composed of tall cells that form a thick epithelial layer.

Table 8.1	The Four Major Types of Tissues	
Tissue	**Major function**	**Major cell types or features**
Epithelial	Covers organ surfaces and lines organ cavities	A layer(s) of cells held together via tight junctions between the cells
Connective	Supports other tissues and organs	Includes a wide range of cells with a variety of functions, as well as the extracellular matrix
Muscle	Produces movement	Skeletal muscle, cardiac muscle, and smooth muscle (found in the inner linings of organs)
Nerve	Communication between organs and the environment	Neurons and other cells making up the central and peripheral nervous system

Epithelial tissue primarily forms barriers

Epithelial cells are tightly packed together and form layers. A single layer of cells is called simple **epithelium** and epithelium consisting of multiple layers is called stratified. There are three different subtypes of epithelial cells based on the morphology of the cells (FIGURE 8.2). Squamous epithelium consists of flattened cells that permit diffusion and movement of substances across the layer. Examples include the cells that line the blood vessels and the epithelial cells that line the alveoli of the lungs. Skin is composed of multiple layers of stratified squamous cells. As the name implies, cells of cuboidal epithelium are squarish in shape. Cuboidal epithelium typically forms ducts and tubules. Columnar epithelium, which primarily lines the digestive tract, is composed of cells that are relatively tall and make a thick layer of single cells.

The cells forming epithelial layers are firmly connected to each other via junctions known as tight junctions. This tight connection contributes to the barrier function of most epithelial layers. For example, the outer layer of the skin and the inner linings of the digestive and respiratory tracts are strong physical barriers against pathogen invasion. In addition, some epithelial tissues, such as the epithelial cells lining the digestive tract, may also exhibit specialized functions in excretion and absorption. Similarly, sweat and oil glands are found in the skin and these glands, as well as most glands, are composed of epithelial cells. Blood vessels are also found throughout organs and tissues. The inner walls of blood vessels are composed of endothelial cells. Endothelial cells are a type of epithelial tissue that lines the blood vessels and the inner chambers of the heart. In addition, endothelial cells provide a barrier between the organ and the blood, as well as controlling substances that can pass between the blood and organs.

Connective tissue is diverse in structure and function

Connective tissues support other tissues. For example, directly below the outer layer of skin and other epithelial layers is a layer of connective tissue. This layer of connective tissue separates the epithelium from deeper tissue layers. In addition, the connective tissue layer is filled with capillaries that nourish the epithelium and therefore provides nutritional support as well as physical support. Connective tissue also consists of extracellular material that holds the cells together and gives the tissue structure. Cells within connective tissue are often embedded in fibrous protein material known as the **extracellular matrix**. Collagen is the primary protein found in the extracellular matrix.

There are many forms of specialized connective tissues, such as tendons, ligaments, cartilage, and bone, that play major structural roles in the body (TABLE 8.2). Generally, such tissues are composed of cells surrounded by large amounts of extracellular material responsible for the strength and unique structures of these tissues. Adipose, more commonly known as fat, is a specialized connective tissue that does not have a large fibrous extracellular component but has a loose network of extracellular matrix. Some consider blood and lymph as connective tissue, whereas others do not. For those that consider blood and lymph as connective tissue, the extracellular component of blood and lymph is fluid.

Table 8.2 Specialized Connective Tissues

Connective tissue	Primary function	Principal cellular component	Principal extracellular component
Tendons	Connect muscle to bone	Fibroblasts	Collagen
Ligaments	Connect bone to bone	Fibroblasts	Collagen
Cartilage	Structural and protects bones	Chondrocytes	Collagen and proteoglycans
Bone	Skeletal system	Osteoblasts and osteocytes	Collagen and calcium phosphate
Adipose	Energy storage, insulation, cushion organs	Adipocytes	Minimal loose extracellular matrix
Blood*	Transport nutrients and oxygen	Red and white blood cells	Fluid (plasma)
Lymph*	Transport fats and immune effector cells	White blood cells	Fluid (lymph)

*Blood and lymph are not considered by everyone to be connective tissue.

Muscle tissue functions in movement and force generation

Many organs need to generate force or movement as part of their function, and this is accomplished with muscle tissue. Muscle cells are long and thin and generate force by shortening upon stimulation. Force is generated by the cytoskeletal protein actin and an associated motor protein called myosin. The three major types of muscle tissue are skeletal, cardiac, and smooth. Skeletal muscle and the heart are composed primarily of muscle tissue and their primary function is to generate force. The walls of other organs and large blood vessels have smooth muscle. For example, two layers of smooth muscle are found in the walls of the intestine to assist in moving food through the digestive system.

Nerve cells mediate communication throughout the body

Nerve tissue makes up the brain, spinal cord, and peripheral nerves that connect to all parts of the body. The primary function of nerve tissue is to receive and send signals across the various organs, tissues, and cells throughout the body. The two primary types of cells in nerve tissue are neurons and glial cells. A typical neuron consists of a cell body, dendrites, and an axon (FIGURE 8.3). The cell body is responsible for the normal maintenance of the cell. The dendrites receive signals from other neurons or the environment, and then transmit that signal along the axon to the target cell. The signal transmitted along the axon is often an electrical impulse. The glial cells surround, protect, and support the neurons. For example, some glial cells form a myelin sheath around the axon to insulate and protect the axon.

SIGNPOST

Actin, myosin, and microfilaments are described in Chapter 2, pages 40–41.

FIGURE 8.3 A typical neuron. Neurons consist of a cell body, dendrites, and an axon. Neurons receive signals from other neurons at the dendrites. This signal is then propagated to the target cell down the axon. Axons are surrounded by a myelin sheath formed from glial cells.

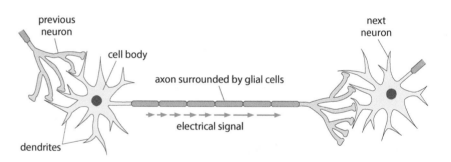

Organs consist of two or more tissue types

Organs carry out specific functions through the interaction and cooperation of the various cells that make up a particular organ. The cells making up the organ are organized into tissues and the various tissues have distinct roles in the overall function of an organ. This phenomenon can be illustrated by the alimentary canal of the digestive system. The alimentary canal essentially consists of a long hollow tube running from the mouth to the anus. Several distinct regions of the alimentary canal with distinct specific functions include the esophagus, stomach, and the small and large intestines.

Distinct layers make up the wall of the alimentary canal (FIGURE 8.4). The inner layer is called the **mucosa** and is composed of epithelial cells. The mucosa forms a barrier as is the generalized role of epithelial layers. In addition, the mucosa has secretory and absorption roles, and the specific secretory and absorption roles are distinct in the different regions of the alimentary canal. Directly beneath the mucosa is a layer of loose connective tissue called the submucosa. This is followed by two muscle layers in which the muscle fibers either run along the length of the alimentary canal or circle the alimentary canal. The muscle layer provides the force to move the food through the alimentary canal as it is being digested. The outermost layer is the serosa, which consists of epithelial cells and a layer of connective tissue. Blood and lymph vessels, as well as nerves, run throughout these various layers. The nerves function to regulate muscle contraction and the secretion of substances from various glands associated with the alimentary canal. Thus, all four tissue types make up the alimentary canal and the various cells of these tissues have distinct roles in the overall function of the alimentary canal.

Physiological functions of the body are carried out by organ systems

The various organs in the human body are grouped into organ systems. Organ systems consist of several distinct organs that each have a specific role in an overarching physiological function (TABLE 8.3). The organ systems are not completely independent but depend on the other organ systems for their function. For example, all organ systems depend upon the circulatory system to deliver oxygen and nutrients and to remove waste products. Similarly, the acquisition of oxygen and nutrients depends on the respiratory and digestive systems. Thus, these various organ systems function in a coordinated manner. This coordination is mediated, to a large extent, by hormones secreted by the endocrine system and the nervous system.

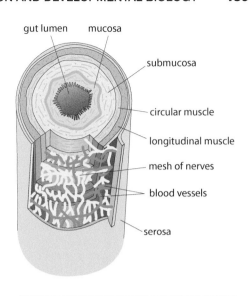

FIGURE 8.4 Cross-section of the small intestine as a representative organ. The outermost serosa layer and inner mucosa layer consist of epithelial cells. Between those epithelial layers and the muscle layers are layers of connective tissue. Interspersed between all the layers are blood vessels, lymph vessels (not shown), various glands and ducts (not shown), and nerves.

Table 8.3	Major Organ Systems	
System	**Main components**	**Primary physiological roles**
Circulatory	Heart, blood vessels, blood	Transports nutrients, gases, hormones, metabolic wastes
Digestive	Alimentary canal (esophagus, stomach, intestines), liver, pancreas	Food processing (ingestion, digestion, absorption, waste elimination)
Respiratory	Lungs, respiratory tract	Gas exchange (oxygen uptake, carbon dioxide elimination)
Nervous	Brain, spinal cord, nerves, sensory organs (e.g., eyes, ears)	Senses environment, directs behavior, coordination of body activities
Endocrine	Hormone-secreting glands (hypothalamus, pituitary, adrenal, pancreas)	Coordination of physiological processes
Excretory	Kidneys, bladder, ureters, urethra	Removal of waste, regulation of osmotic balance
Immune and lymphatic systems	Bone marrow, thymus, spleen, lymph nodes and vessels, white blood cells	Defense against infections and cancer, transporting fat, regulation of blood volume
Reproductive	Testis or ovaries and associated male or female organs	Reproduction (sperm and egg production)
Musculoskeletal	Muscles, bones, cartilage, tendons, ligaments	Support and movement
Integumentary	Skin and derivatives (hair, nails, glands)	Protection against injury, infection, and dehydration

CELLULAR DIFFERENTIATION AND DEVELOPMENTAL BIOLOGY

The human body is composed of a wide variety of distinct cell types with distinct functions. Within an individual all these distinct cells have essentially the same genotype, since all the cells originated from a single fertilized egg. Through the process of mitosis, this single fertilized egg produces all the cells that make up an individual. Therefore, except for errors made during DNA synthesis, all cells within an individual have the same DNA sequence. The differences in phenotypes between these various cell types are due to differences in the specific genes that are expressed in the individual cells. As cells differentiate into specialized cells, the genes necessary to carry out those specialized functions are transcribed and translated. Cells with other specialized functions express a different set of genes. For example, muscle cells express high levels of the proteins associated with contraction, whereas non-muscle cells do not express high levels of these proteins. The distinct combination of genes that are expressed in a particular cell determines its overall structure and function (i.e., phenotype).

The early steps of development are repeated rounds of cell division with minimal cell differentiation

After fertilization, the **zygote** undergoes several rounds of mitosis to produce a solid mass of 16–32 cells called the morula (FIGURE 8.5). The formation of the morula occurs in the oviduct during the 4 days following fertilization. The cells of the morula continue dividing and move around it to form a hollow ball called the **blastocyst**. The cells forming the outer wall of the blastocyst are called trophoblasts and are ultimately involved in the formation of the placenta. This hollow ball is somewhat lopsided, because on one side is a mass of cells called the **inner cell mass**. These inner mass cells subsequently form the embryo. On day 6 or 7 the blastocyst implants into the wall of the uterus.

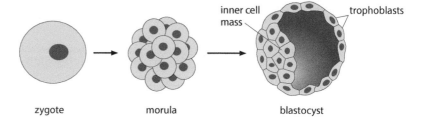

FIGURE 8.5 Early steps in embryogenesis. The fertilized egg, called a zygote, undergoes several rounds of mitosis to form a solid mass of cells called the morula. As the cells in the morula continue to divide, they take the shape of a hollow ball with trophoblasts forming the outer wall. On one side is a mass of cells called the inner cell mass that ultimately develops into the embryo.

Specialization of cells begins in the early embryo

After implantation, the inner cell mass continues growing through the cell division process and splits to form two fluid-filled sacs called the amnion and the yolk sac (FIGURE 8.6). The amnion eventually grows around the embryo and provides a watery environment for the developing embryo and later the fetus. Between the amnion and yolk sac is the embryonic disk. The embryonic disk develops three layers called the **endoderm**, **mesoderm**, and **ectoderm**. The endoderm is derived from the yolk sac and the ectoderm is derived from the amnion. As part of development, the ectoderm and endoderm layers of the embryonic disk separate and cells from the ectoderm migrate into this space between the layers to form the mesoderm. These three layers are called the primary germ layers and ultimately develop into the organs and tissues found in the body.

FIGURE 8.6 Formation of embryonic disk and primary germ layers. A yolk sac and amnion form after implantation of the blastocyst into the uterine wall. The regions in contact between the yolk sac and amnion form the embryonic disk. Within the embryonic disk cells of the amnion develop into ectoderm and cells of the yolk sac develop into endoderm. Some cells from the ectoderm move into the space between the ectoderm and endoderm to form the mesoderm. Trophoblast cells interact with cells of the uterus to form the placenta.

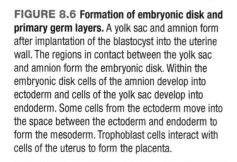

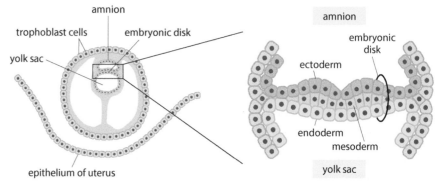

The embryonic disk cells continue to divide, and the cells of the endoderm, mesoderm, and ectoderm begin to rearrange themselves into the organs. These primary layers give rise to specific tissues that make up the organs (TABLE 8.4). For example, the ectoderm forms the outer layer of skin, called the epidermis, and nerve tissue. The endoderm forms the lining of the digestive and respiratory tracts and some glands, and the mesoderm forms muscles, bones, the circulatory system and other organs. This process of organogenesis involves continued cellular replication, as well as the movement of cells or the folding of entire cell layers. Through this morphogenesis, the organs begin to take shape and the tissues and cells of the organs start to differentiate into their final states.

Table 8.4 Developmental Fate of Primary Germ Layers		
Endoderm	**Mesoderm**	**Ectoderm**
Lining of the digestive system	Muscle	Nervous system and brain
Lining of the respiratory system	Outer covering of organs	Epidermis and related structures
Liver	Excretory system	
Pancreas	Gonads	
Other glands	Circulatory system	
	Bones and cartilage	
	Dermis	

Cell determination and differentiation are two processes of the developmental pathway

As development progresses from zygote to organism, the cells become more specialized. In addition, as cells become more specialized, they become locked into a specific cell fate. For example, cells produced during the first few cleavages are often described as being totipotent because they have the capacity to develop into the entire organism. This is not true beyond this stage as the blastocyst has cells that have become more specialized. For example, trophoblast cells are fated to become part of the placenta and the cells of the inner cell mass develop into the embryo. At the blastocyst stage, the trophoblast cells cannot become embryonic cells and the inner mass cells cannot become part of the placenta. Similarly, as development proceeds, cells of the three primary germ layers are fated to become specific tissues and organs (see Table 8.4).

At the end of the developmental process, cells differentiate into highly specialized cells that carry out specific functions. This process is called **terminal differentiation** as these cells have essentially reached their final destination. Terminally differentiated cells express the genes necessary to carry out their specific functions, but do not express genes that are not associated with their specific cell function. At this point the cells cannot differentiate into other types of cells under normal circumstances. For example, muscle cells cannot convert into nerve cells and vice versa. Terminally differentiated cells are also no longer capable of undergoing cell division and are often described as being in the G0 state of the cell cycle (see Box 2.5).

Cell determination and differentiation are complex and highly regulated processes

The pathway going from the zygote to an adult organism is fascinating and mysterious. Many factors play a role in this process and, for the most part, the details of the process are not known. It is also fraught with danger, in that perturbation of the process can have severe consequences. Environmental factors, such as infections or drugs, can

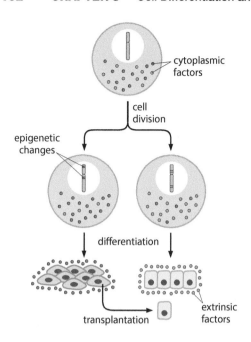

FIGURE 8.7 Factors influencing the developmental fate of cells. Numerous factors potentially influence the differentiation of cells. Cytoplasmic factors unequally distributed between daughter cells may affect the specific differentiation of those cells. Changes in the heterochromatin patterns that are inherited in an epigenetic fashion can restrict some developmental pathways. However, it is also known that developmental fate can sometimes be changed through the transplantation of cells, implying that extrinsic factors such as hormones also influence differentiation.

SIGNPOST

Epigenetics and heterochromatin are described in more detail in Chapter 5, pages 100–101.

negatively impact the developmental process and result in the death of the embryo or fetus or cause severe deformities. It is well known that, during pregnancy, women need to be especially cautious about exposure to drugs, chemicals, or pathogens. For example, the consumption of alcohol during pregnancy can lead to fetal alcohol syndrome (see Box 18.2).

Both intrinsic and extrinsic factors regulate cellular determination and differentiation. Intrinsic factors refer to elements within the cells that regulate their differentiation. Intrinsic elements could include cytoplasmic elements or epigenetic changes in chromosome structure. These intrinsic elements could potentially be distributed differently in the two daughter cells following cell division. Although, the daughter cells contain identical DNA, the cytoplasmic elements and heterochromatin patterns may not be equally distributed between the two daughter cells. For example, if some regulatory substance is unequally distributed in the cytoplasm, then the two resulting daughter cells contain different levels of this factor (FIGURE 8.7). The daughter cell with more of this factor may be subjected to one fate, whereas the daughter cell with less of the factor undergoes a different fate.

However, determination is not completely controlled by intrinsic factors, as extrinsic cues also play a large role in cellular determination and differentiation. It is also known, through experiments in animals involving the transplantation of cells from one region of an early embryo to another region, that the fate of the transplanted cells is changed. These cells now adapt to the fate determined by the neighboring cells, rather than retain their original fate (see Figure 8.7). In other words, neighboring cells also influence the developmental fate of cells. This mechanism could involve cell–cell contacts or signaling molecules secreted by cells. Chemical gradients have also been shown to play a role in organogenesis with the exact fate of any specific cell determined by the distance between that cell and the source of the gradient.

Epigenetics also impacts determination and differentiation

Modifications to chromatin also play a role in the developmental process. Through DNA methylation and modification of histones, sometimes called the 'histone code,' chromatin is converted into heterochromatin. Regions of chromosomes that have been converted to heterochromatin are no longer capable of being transcribed. As cells proceed through developmental pathways, defined areas of the chromosomes are converted into heterochromatin and fewer genes are available for transcription. The silencing of the transcription of certain genes could explain the inability of terminally differentiated cells to develop into other cell types. This silencing of genes associated with specific cell types restricts the developmental repertoire of a cell. Furthermore, the patterns of heterochromatin are inherited in that the methylated regions of DNA are reproduced in the daughter cells as well as the regions of the chromosomes containing modified histones (FIGURE 8.8). This type of inheritance is called **epigenetics**. Thus, a complex combination of factors and signals controls the developmental fate of a cell.

STEM CELLS

Stem cells are cells that are capable of undergoing mitosis and are not terminally differentiated. The two primary types of stem cells are embryonic and adult (TABLE 8.5). Embryonic stem cells refer specifically to the cells of the inner cell mass in the blastocyst and do not include all embryonic cells. Inner mass cells can become any type of cell and are therefore also referred to as **pluripotent** stem cells. This means that embryonic stem cells have the capacity to develop into any type of cell found in the human body. Totipotent refers to the ability to generate the complete organism. Pluripotent stem cells of the inner cell mass cannot generate the entire organism because they have lost the ability to form the trophoblast cells of the blastocyst stage. Totipotent cells include the zygote as well as the individual cells from the first few divisions of the zygote.

Table 8.5	Types of Stem Cells and Potency	
Type	**Source**	**Potency**
Embryonic (early)	Zygote, early morula (≤ 8 cells)	Totipotent
	Inner cell mass (blastocyst)	Pluripotent
Adult (mature)	Various tissues (especially hematopoietic, skin, intestinal and respiratory epithelium)	Multipotent
		Oligopotent
		Unipotent

As cells progress through the differentiation process, they lose potency with regard to the type of cells they can develop into. For example, cells of the three primary germ layers have less potency than the inner cell mass cells since each of the three primary germ layers is restricted to the formation of specific organs (see Table 8.4). However, each of these three primary germ layers can still develop into the wide variety of cell types that are associated with those tissues and organs. Similarly, stem cells are also found in the adult organism. Adult stems cells have a restricted repertoire of cell types they can develop into and are classified as **multipotent**.

Stem cells exhibit self-renewal and differentiation capacities

The replication of an adult stem cell can result in both daughter cells remaining as stem cells with the same potency. Alternatively, one of the daughter cells can remain, as a stem cell with the same potency as the parental cell and the other progeny can undergo differentiation (FIGURE 8.9). The daughter cell undergoing differentiation may be able to choose between multiple developmental pathways. Cells progressing down the different pathways may still be multipotent. However, they are less potent than the original stem cell as they can no longer develop into cells of the alternative pathway. This decrease in potency is often called oligopotency. Stem cells that can only develop into a single type of differentiated cell are called unipotent. Progeny cells of a unipotent stem cell either replace that particular stem cell or terminally differentiate.

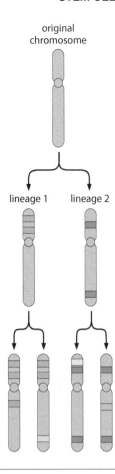

FIGURE 8.8 Heterochromatin and epigenetics. Regions of heterochromatin are depicted by the various colors and are formed during cell differentiation. The chromosomal regions corresponding to heterochromatin are maintained in the daughter cells by epigenetics. As cells continue to divide and differentiate additional regions of the chromosome are converted to heterochromatin. Genes associated with heterochromatin are transcriptionally silent and, as cells differentiate, their repertoire of expressible genes decreases. This limits the developmental capacity of cells as they become more specialized.

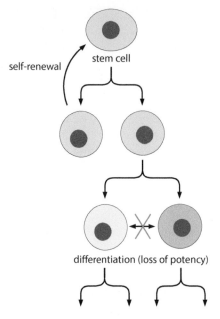

FIGURE 8.9 Stem cells and differentiation.
Replication of stem cells maintains the population of stem cells and provides a source of stem cells for differentiation into different cell types. As stem cells differentiate, they lose potency and have less capacity to develop into other cell types.

The various types of blood cells are derived from a single lineage of stem cells

The process of generating blood cells, called hematopoiesis, illustrates the concept of stem cells and potency (FIGURE 8.10). All blood cells, including red blood cells, white blood cells, and platelets, arise from progenitor cells found in the bone marrow called multipotent **hematopoietic stem**

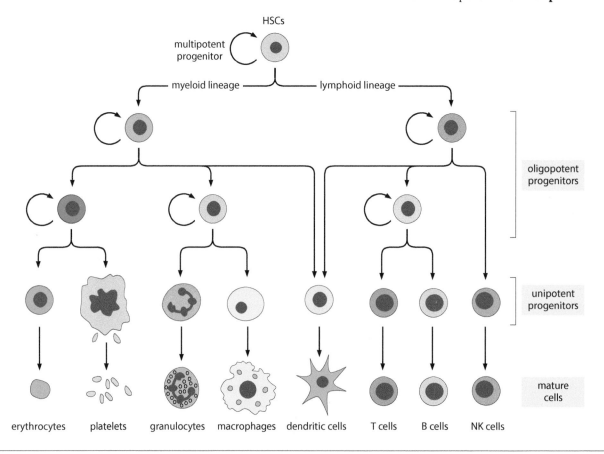

FIGURE 8.10 Hematopoietic stem cell differentiation. Schematic representation of the various differentiation pathways of multipotent hematopoietic stem cells (HSCs) into the various types of cells found in blood and lymphoid tissues. Most of these cells are white blood cells that function in immunity (Chapter 9). NK, natural killer.

cells. Progeny resulting from the replication of hematopoietic stem cells can remain as hematopoietic stem cells to generate more stem cells or can undergo differentiation. Cells undergoing differentiation can develop along either the myeloid lineage or the lymphoid lineage. Myeloid cells include various phagocytic cells, red blood cells, and platelets, whereas various types of lymphocytes develop from the lymphoid lineage. Cells in both the myeloid and lymphoid lineages are still stem cells, but they have less potency than the hematopoietic stem cells since cells in the lymphoid lineage cannot develop into cells of the myeloid lineage, and vice versa. This process of replication and differentiation continues until reaching lineage-specific unipotent stem cells that can only produce one type of differentiated cell. These unipotent progenitor cells renew themselves and produce terminally differentiated cells.

Hematopoietic stem cells differentiate into a wide variety of cells to meet the cellular needs of the individual. This process is highly regulated. For example, if an individual becomes anemic more cells are diverted down the myeloid lineage and, similarly, at all subsequent branch points down the lineage leading to the erythrocyte progenitor cell. As the level of erythrocytes becomes normalized the flow of cells down the erythrocyte lineage is decreased. Similarly, when the immune system (Chapter 9) needs more lymphocytes or other immune effector cells to fight infectious disease, the numbers of cells in those lineages are increased. The regulation of these various differentiation pathways is primarily controlled by hormones.

Adult stem cells function in cell replacement and tissue repair

During development from zygote to adult, the size of the individual is increasing and this increase in size is due to an increase in cell number. Upon reaching adult size new cells are not needed as part of the growth process. However, a continuous supply of new cells is still needed to replace cells and regenerate damaged tissues. For example, circulating blood cells have a limited life span and, therefore, need to be continuously replaced. Similarly, epithelia, such as skin and the lining of the digestive and respiratory tracks, are also continuously replaced. Therefore, epithelial tissues contain a comparatively large number of stem cells to serve as a source of replacement cells. Such tissues are called labile (TABLE 8.6). Other tissues contain fewer stem cells and are stable in that they do not exhibit a continuous replacement. These stable tissues primarily need stem cells to replace damaged cells. Some tissues contain very few stem cells and are considered permanent.

Table 8.6	Tissues and Stem Cells		
Tissue	**Examples**	**Number of stem cells**	**Characteristics**
Labile	HSC (bone marrow), skin and other epithelia	Numerous	Continuous, rapid, normal regeneration
Stable	Liver, kidney	Moderate	Regenerative activity associated with injury
Permanent	Brain and nerve, muscle	Few	Limited regeneration, primarily scarring

HSC, hematopoietic stem cell.

Stem cells are also needed to regenerate tissue damaged by disease or injury. For example, loss of blood results in the production of more blood cells by the hematopoietic stem cells. In labile tissues, complete regeneration is possible if the damage is not too extensive (FIGURE 8.11). Stable tissues, such as liver and kidney, have some regenerative capacity and can recover from relatively minor injuries. Nerve and muscle tissues, however, contain virtually no regenerative capacity. This small regenerative capacity is why there are often permanent or long-term sequelae associated with heart attacks and strokes (Chapter 14) .

FIGURE 8.11 Repair of tissue damage after injury or disease. Tissues with sufficient stem cells are capable of complete regeneration in cases of limited injury. Repair of extensive injuries or in tissues with limited stem cells is through scarring.

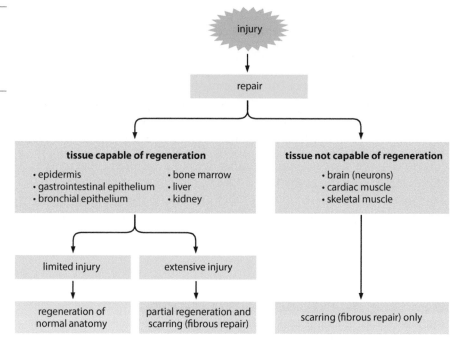

Scarring is a repair process involving specialized stem cells

Regeneration is a process by which the normal anatomy and function of a tissue are restored after injury. If regeneration is not possible the tissue can be repaired through a healing process known as scarring. Scarring does not restore the original anatomy or functionality of the tissue but does serve to hold the tissue together. In this type of repair, cells known as fibrocytes migrate to the wound and proliferate. The fibrocytes secrete collagen, which is an extracellular matrix protein, and the collagen forms fibers that tie the disrupted tissue back together. The collagen fibers also provide a structure to support additional cell growth for continued healing. Over time more collagen is deposited and cells disappear, leaving a strong extracellular matrix at the wound site. This deposited extracellular matrix is often visible as a scar also called fibrosis.

POTENTIAL OF STEM CELLS IN CELL-BASED THERAPIES

The ability of stem cells to regenerate damaged tissues opens the possibility of **cell-based therapies** in which stem cells are used to treat diseases. One such example of stem cell-based therapy is hematopoietic stem cell transplantation. Historically, the major source of hematopoietic stem cells was the bone marrow and, hence, the procedure was previously known as bone marrow transplantation. However, now peripheral blood is the more common source of hematopoietic stem cells. Other sources of hematopoietic stem cells are amniotic fluid and umbilical cord blood. Healthy multipotent hematopoietic stem cells from a matched donor are injected and these stem cells migrate to the bone marrow and reconstitute hematopoiesis.

Cancers such as leukemia, some anemias, and immunodeficiency diseases, can be treated by hematopoietic stem cell transplantation. In the case of leukemia, the cancerous hematopoietic cells are first destroyed by chemotherapy or radiation. The hematopoietic stem cells repopulate the ablated bone marrow. After several weeks of cell replication, the hematopoietic stem cells and their progeny are sufficient to normalize blood cell counts and reconstitute the immune system.

The limited potency of hematopoietic stem cells restricts their use in treating blood disorders. However, in principle, stem cells with the proper potency could be used to treat diseases or damage associated with any tissue. For example, it is possible to grow skin stem cells in vitro to generate sheets of skin that can be transplanted onto patients with severe burns. This newly generated skin lacks hair follicles and sweat and oil glands, and therefore is far from ideal. However, the procedure can be used as a life-saving measure. Currently there is a lot of interest in mesenchymal stem cells for cell-based therapy (BOX 8.1).

Pluripotent embryonic stem cells are a potential source of cells for the regeneration of stable and permanent tissues

Permanent tissues, such as muscle and nerves, contain very few stem cells, and therefore the prospect of treating damage or disease to these tissues is limited. Nonetheless, there is much interest in potentially using stem cells to regenerate permanently damaged tissue. The limited potency of readily available adult multipotent stem cells, such as hematopoietic stem cells or skin cells, precludes their use in the regeneration of muscle and nervous tissues. Embryonic stem cells, conversely, are pluripotent and capable of developing into any type of cell. Therefore,

Mesenchymal Stem Cells

Mesenchymal stem cells (MSCs) are another type of multipotent adult stem cell. MSCs are of mesodermal origin and can differentiate into other mesodermal cells such as cartilage cells, bone cells, or fat cells. Clinical trials are ongoing investigating the possibility of using MSCs to repair damaged bone and cartilage. Cartilage does not repair itself well and quite often surgery is required to replace the damaged joint. In addition, MSCs could prevent or delay the development of osteoarthritis. Furthermore, transplanted MSCs may release substances that stimulate the repair of the damage by the existing cells. MSCs also have immunomodulatory and anti-inflammatory effects and may be useful in the treatment of autoimmune diseases (Chapter 17).

Despite their mesodermal origin, MSCs have displayed the capacity of transdifferentiation into cells of endodermal and ectodermal lineages (**BOX 8.1 FIGURE 1**). This transdifferentiation is primarily a phenomenon in the Petri dish and there are some questions about whether MSCs are truly pluripotent stem cells. If indeed MSCs can be converted into cells with more potency, then MSCs may have the potential to treat diseases beyond damage to the skeletal system. Another favorable feature of MSCs for cell-based therapy is that they are found in numerous tissues including bone marrow, adipose tissue, amniotic fluid, umbilical cord, peripheral blood, and skin. In general, MSCs are accessible and easy to isolate and grow in culture. Some studies suggest that MSCs isolated from the umbilical cord may have more potency than those isolated from differentiated tissues.

BOX 8.1

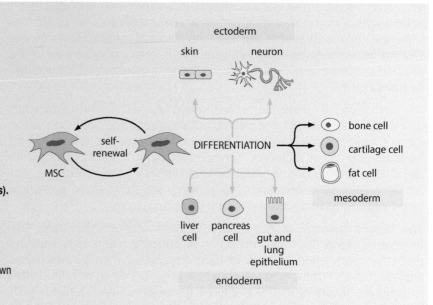

BOX 8.1 FIGURE 1 Mesenchymal stem cells (MSCs). MSCs replicate to produce more MSCs and have the capacity to differentiate into many cell types. Normally MSCs differentiate into mesodermal tissues such as bone cells, cartilage cells, and fat cells (black arrows). Some studies suggest that MSCs may also be able to trans-differentiate into ectodermal and endodermal tissues (brown arrows).

these pluripotent stem cells could be used to regenerate essentially any tissue, including permanent tissues. The general approach would be to isolate the embryonic stem cells from the inner cell mass of the blastocyst and expand the cells in culture (FIGURE 8.12). These cells could then be injected into the affected tissue. The environment of that tissue would direct the stem cells to differentiate into the required cell type, leading to the regeneration of the damaged tissue.

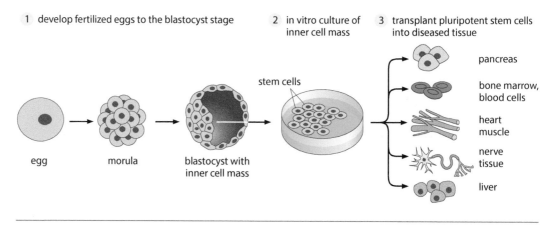

FIGURE 8.12 Potential strategy for cell-based therapy using embryonic stem cells. Eggs are fertilized in vitro and allowed to develop into blastocysts. The cells of the inner cell mass are propagated by in vitro culture. These pluripotent stem cells are then introduced into the affected tissue where they potentially develop into the correct cell type, leading to a regeneration of the damaged tissue.

Although in theory the use of embryonic stem cells to treat degenerative diseases appears straightforward, there are several issues that may preclude their use. Some desired features of stem cells for therapeutic use include: (1) proliferative capacity to make sufficient tissue, (2) differentiation into desired cell type, (3) survival and integration into tissues after transplantation, (4) functioning correctly for the duration of the recipient's life, and (5) causing no harm to the recipient. In addition to these safety and technological obstacles, there are also supply issues and potential ethical issues involved in the use of embryonic stem cells.

Potential sources of embryonic stem cells are unused embryos from fertility clinics. In vitro fertilization involves the production of several embryos that are frozen to be used in the event a first implantation and subsequent implantations are not successful. Once a successful pregnancy and birth are achieved, the remaining stored embryos could be used as a source of stem cells. However, the supply of these cells is rather limited. In addition, the use of embryos as a source of stem cells presents ethical problems for some people.

Multipotent stem cells can be reprogrammed into induced pluripotent stem cells

Recent advances in stem cell technology have possibly provided a solution to the limitations of embryonic stem cells and multipotent adult stem cells. Through the addition of genes via recombinant DNA technology (Chapter 7), multipotent stem cells can be converted to pluripotent stem cells (FIGURE 8.13). The introduced genes are transcription factors that play a role in maintaining the pluripotent state. During cellular differentiation, the loss of expression of these transcription factors results in a loss of potency due to changes in chromatin (see Figure 8.8). The introduction of these transcription factors and their continuous expression reverse the changes in chromatin structure associated with differentiation. This introduction results in a reprogramming of the cell back to the pluripotent state. These reprogrammed cells are called **induced pluripotent stem cells** and could possibly be used in stem cell therapy.

The strategy of stem cell therapy using induced pluripotent stem cells involves isolating stem cells from the individual and converting those cells to induced pluripotent stem cells in vitro (FIGURE 8.14). For example, stem cells from the skin can easily be obtained from a biopsy punch. The reprogramming genes would then be introduced to the cells and these induced pluripotent stem cells can then be introduced back into the patient to replace the damaged tissue. However, due in part to our rudimentary understanding of the biology of stem cell differentiation and replication, the routine clinical practice of such stem cell therapy is still a long way off. In addition, all the desired features discussed for embryonic stem cells are also needed for induced pluripotent stem cells. Nonetheless, there is hope that such an approach will someday provide a means to treat diseases that are currently considered terminal, without any prospects of recovery.

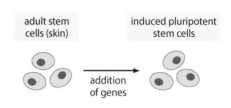

FIGURE 8.13 Induced pluripotent stem cells. Addition of genes for specific transcription factors to multipotent adult stem cells can reprogram the epigenetic changes and allow those cells to develop into pluripotent stem cells.

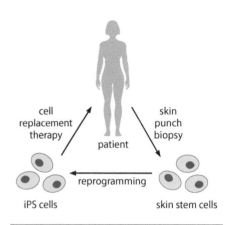

FIGURE 8.14 The potential use of induced pluripotent stem cells in cell replacement therapy. Multipotent stem cells derived from the skin or other tissues are reprogrammed to become induced pluripotent stem (iPS) cells. These iPS cells can then be introduced into the patient to repair damaged tissues.

SUMMARY OF KEY CONCEPTS

- The human body is composed of a wide variety of cell types that interact and carry out all functions associated with life

- Similar cells organize into tissues that interact to form organs and organ systems

- During development, cells expand in number through mitosis and terminally differentiate into the specific cells making up the tissues and organs

- Terminal differentiation is associated with epigenetic modifications of chromosomes

- Stem cells are capable of undergoing replication and serve as a source for cells during growth and development, as well as for replacement, repair, and regeneration

- Stem cells can potentially be used in cell-based therapies to repair and replace damaged or diseased tissues

KEY TERMS

blastocyst	A stage in the early embryonic development of mammals that contains the inner cell mass which will develop into the embryo
cell-based therapy	The treatment of disease by introducing cells, such as stem cells, from a donor or modifying an individual's cells and reintroducing those cells.
connective tissue	Tissue that connects, supports, binds, or separates other tissues or organs that typically have relatively few cells embedded in non-cellular matrix.
ectoderm	The primary embryonic tissue layer derived from the amnion that develops into the epidermis and nervous system.
endoderm	The primary embryonic tissue layer derived from the yolk sac that develops into the inner lining of the digestive and respiratory systems and other organs and glands.
epigenetic	Refers to heritable changes in gene function that occur without a change in the DNA sequence and that is mediated by the modification chromosome structure.
epithelium	A sheet of cells that covers or lines a body surface, e.g., skin or the layer of cells lining the gastrointestinal tract or the respiratory tract.
extracellular matrix	A network of proteins and other molecules that surround, support, and give structure to cells and tissues in the body.
hematopoietic stem cell	A multipotent stem cell that can develop into all types of blood cells, including white blood cells, red blood cells, and platelets.
induced pluripotent stem cells	Derived adult stem cells that have been reprogrammed via genetic engineering back into an embryonic-like pluripotent state for possible therapeutic purposes.
inner cell mass	A group of cells on one side of the blastocyst that forms the embryo.
mesoderm	The middle embryonic tissue, lying between the ectoderm and endoderm, that develops into various specialized tissues including muscle and skeletal.
mucosa	The mucus-secreting epithelium lining the gastrointestinal, respiratory, and urogenital tracts.
multipotent	Stem cells capable of forming many different types of cells in a particular cell lineage, but incapable of forming cells from other cell lineages. Also called adult stem cells.
pluripotent	Stem cells capable of forming any other type of cell but cannot form the entire organism. Embryonic cells from the inner cell mass of the blastocyst are the primary source.
stem cell	Any cell that has not terminally differentiated and is capable of undergoing cell division.
terminal differentiation	A process by which undifferentiated cells become highly specialized and no longer proliferate nor have the potential to further differentiate into other types of cells.
tissue	A group of similar cells and intercellular material that are a component of organs. The four types of tissues in animals are epithelial, connective, muscle, and nervous.
zygote	A fertilized egg resulting from the fusion of a female gamete (ovum) with a male gamete (sperm).

REVIEW QUESTIONS

1. What are the four major types of tissues of animals and their general features and roles?

2. What are the three primary germ layers of the embryonic disk and which organs and tissues are formed from each of these primary tissues?

3. Describe and discuss multipotent and pluripotent stem cells.

4. What are two potential sources of pluripotent potent stem cells for cell-based therapy? Discuss the strengths and limitations of these pluripotent stem cells and multipotent stem cells.

ADDITIONAL QUESTIONS

In addition to the Review Questions provided above, there is a range of free online questions designed to enable students to further test their understanding of the chapter material. In order to access these interactive questions, please visit the book's website:
https://routledge.com/cw/wiser

The Immune System

9

The immune system and inflammation play a central role in disease. Most notable is the role of the immune system in protecting us against infectious diseases. The immune system recognizes pathogens and eliminates those pathogens, thereby minimizing the diseases caused by infectious organisms. In addition, the immune system participates in the healing process following damage caused by pathogens or other injuries. Stimulation of the immune system by pathogens or tissue damage results in an inflammatory response which, when functioning correctly, leads to healing. However, an inappropriate inflammatory response can be the cause of disease or contribute to pathogenesis.

Defense against pathogens is often described as being three lines of defense. The first line of defense includes various barriers that prevent entry and proliferation of the pathogen. Infectious agents that break through these protective barriers then encounter the second line of defense, which is called innate immunity. Elements of the innate immune response recognize broad classes of pathogens and eliminate those pathogens. If the innate immune response is unable to eliminate the intruders in a relatively short period of time, the adaptive immune response is activated. This third line of defense recognizes specific pathogens and brings more sophisticated weapons to the fight against pathogens. In addition, the adaptive response has immunological memory and can lead to naturally acquired protective immunity, which protects an individual against subsequent encounters with the pathogen. Through the process of vaccination, protective immunity can also be artificially induced.

COMPONENTS OF THE IMMUNE SYSTEM

The immune system does not exhibit the normal hierarchy of cells, tissues, and organs that is normally associated with organ systems (see Figure 8.1). Instead, a large component of the immune system consists of soluble proteins and cells that are found throughout the body. In this way, the immune system is positioned to detect and fight pathogens wherever they may occur. Furthermore, these proteins and cells can move around the body via the circulatory and **lymphatic** systems. In particular, the white blood cells are part of the immune system and many proteins found in the fluid portion of blood, called plasma, play a role in immunity. These proteins and cells also circulate through the lymphatic system. The lymphatic system intersects with the circulatory system and plays a role in maintaining blood volume as well as transporting lipids and fat-soluble vitamins from the digestive system. In addition to these roles, the lymphatic system plays a major role in immunity and, thus, the lymphatic system is often considered a part of the immune system.

Specialized organs of the lymphatic system play a central role in the immune system

The lymphatic system consists of a branching network of lymph vessels and lymphoid organs, as well as the fluid within the lymph vessels, called lymph. At the ends of this network are small capillary-like vessels that are closed and in proximity to the capillaries of the circulatory system. The lymphatic capillaries allow the exchange of fluids and cells with the blood capillaries and tissues. Movement of lymph through the lymph vessels is passive, as there is no organ such as the heart to pump lymph through the vessels. The lymphatic capillaries merge to form larger lymph vessels and the larger vessels merge to form even larger vessels (FIGURE 9.1). Periodically, at the junctions between lymph vessels, there are small lymphoid organs called lymph nodes. The lymph vessels also connect to other lymphoid organs and tissues.

FIGURE 9.1 Lymphatic system and lymphoid organs. Lymphatic vessels form a branching network of vessels found throughout the body and connect to the lymphoid organs and tissues. Primary lymphoid organs include the thymus and bone marrow. Secondary lymphoid organs are the spleen, lymph nodes, and mucosa-associated lymphoid tissues (MALT) such as tonsils and adenoids.

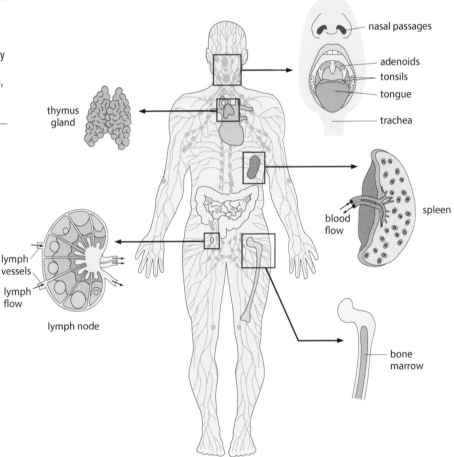

The lymphoid organs are classified as either primary or secondary based on their overall function in immunity (TABLE 9.1). Primary lymphoid organs are the sites where immune effector cells mature. The two primary lymphoid organs are bone marrow and the thymus. Secondary lymphoid organs are involved in the detection and elimination of pathogens. In other words, the secondary lymphoid organs carry out surveillance. For example, the spleen is the largest of the lymphoid organs and functions to cleanse the blood by removing old and damaged blood cells, as well as detecting and destroying pathogens. In a similar fashion, the lymph nodes cleanse the lymph by removing microorganisms, damaged cells, and cellular debris.

Table 9.1	Lymphoid Organs and Tissues	
Class	**Organ/Tissue**	**Function/Examples**
Primary	Bone marrow	Maturation of B cells
	Thymus	Maturation of T cells
Secondary	Spleen	Filter blood
	Lymph nodes	Filter lymph
	Mucosa-associated lymphoid tissue (MALT)	Examples include tonsils, adenoids, Peyer's patches, appendix

Secondary lymphoid tissue is also found associated with various mucosal layers throughout the body. In many of these cases, the lymphoid tissue is not a highly organized organ, but rather a loose collection of lymphoid tissue. For example, throughout the digestive tract are small regions of lymphoid tissue known as Peyer's patches which function in immune surveillance. Similarly, the tonsils and adenoids of the throat are lymphoid tissue. Collectively, these tissues are known as mucosa-associated lymphoid tissue (MALT). Some cells within MALT continuously sample their immediate surroundings and, when these cells encounter a pathogen, they activate other cells of the immune system.

Various types of immune effector cells function in immunity

Surveillance and destruction of pathogens are mediated by white blood cells, also called **leukocytes**. There are several distinct types of leukocytes that have distinct roles, as well as shared roles, in immunity (TABLE 9.2). The major types of leukocytes are lymphocytes, macrophages, granulocytes, and dendritic cells. All these various leukocytes are derived from hematopoietic stem cells (see Figure 8.10) and thus have a common origin. However, they all have distinct cellular phenotypes and carry out specific functions in immunity. For example, some of these immune effector cells have cytotoxic activity and directly kill pathogens. Other effector cells engulf and kill pathogens by phagocytosis.

Table 9.2	Types of Immune Effector Cells	
Major type	**Subtype**	**Major role**
Lymphocytes	B cells	Antibody production
	T cells	Cytotoxicity
	Natural killer (NK) cells	Cytotoxicity
Macrophages		Phagocytosis
Granulocytes	Neutrophils	Phagocytosis
	Mast cells	Anti-helminth immunity
Dendritic cells		Pathogen surveillance

Cytokines are hormone-like molecules that mediate immunity

The various immune effector cells work together in their quest to detect and eliminate pathogens. This means that these various cells must be able to communicate with each other, as well as communicate with other cells of the body. This communication is accomplished primarily through soluble mediators secreted by a specific type of cell that binds to receptors on other cells. Instead of calling these molecules hormones, the communication molecules involved in immunity are called **cytokines**. Numerous cytokines have been described and a wide range of functions is attributed to cytokines (TABLE 9.3). A specific cytokine tends to promote or inhibit a particular type of immune response. For example, some cytokines may promote an inflammatory response, and therefore are called proinflammatory cytokines. In contrast, other cytokines are anti-inflammatory. In general, cytokines activate or suppress specific immune effector cells to evoke an immune response directed at a specific pathogen. Chemokines are a subclass of cytokines that attract immune effector cells to the site of infection or injury.

Table 9.3 Examples of Cytokines

Cytokine	Sources	Target cells	Actions
Interleukin-1 (IL-1)	Macrophages Dendritic cells	Lymphocytes Endothelial cells	Proinflammatory (e.g., fever)
Tumor necrosis factor-alpha (TNF-α)	Macrophages Dendritic cells Helper T cells	Endothelial cells Neutrophils	Proinflammatory, fever, increases vascular permeability
Chemokines	Many cell types	Many cell types	Attract immune effector cells (chemotaxis)
IL-10	Macrophages T-helper cells	Macrophages Dendritic cells	Anti-inflammatory
Interferon-gamma (INF-γ)	Lymphocytes Macrophages	T and B cells Macrophages	Lymphocyte development Macrophage activation
IL-2	T cells	Lymphocytes	Induce proliferation and activate
INF-α/β (type 1)	Virus-infected cells	Most cells NK cells	Induce antiviral activity

PROTECTIVE BARRIERS

The first line of defense against infectious organisms is non-immunological and involves physical, chemical, or biological barriers (TABLE 9.4). Among the most important of these barriers is skin. The outer layer of skin, composed of keratin and dead epithelial cells, is a formidable physical barrier for microbes and few pathogens can penetrate intact skin. Similarly, the epithelial layers that line the digestive, respiratory, and urogenital tracts also form physical barriers. These internal epithelial layers also secrete mucus, and this thick gel-like material impedes microbes from reaching the epithelial cells. Furthermore, the epithelial cells of the respiratory tract have small hair-like projections called cilia that sweep the mucus with the trapped particles and microbes up to the throat and out of the respiratory tract. Similarly, earwax traps microbes and impedes their invasion deeper into the ear.

Saliva and tears contain an enzyme called lysozyme. This enzyme breaks down the cell wall of bacteria, thereby weakening them and rendering them more susceptible to death, thus slowing their proliferation. The acidic pH of the stomach and vagina is another example of a chemical barrier that is either toxic to microbes or at least provides an unfavorable environment for many microbes. Commensal bacteria also protect us from more pathogenic organisms by providing an unfavorable environment for the proliferation of pathogens. For example, the endogenous microflora found associated with our various epithelial layers compete for resources with harmful pathogens and limit their growth. In fact, the introduction of beneficial microbes is sometimes used to treat diseases or to improve health. Such beneficial microbes are called probiotics.

Table 9.4 Protective Barriers	
Barrier	**Protective mechanism**
Skin and other epithelial layers	Forms a physical barrier that prevents pathogens from gaining access to internal tissues
Mucus and earwax	Traps microorganisms
Cilia	Remove mucus with trapped microorganisms from the respiratory tract
Lysozyme in tears and saliva	Kills bacteria by disrupting cell wall
Acidic pH	Provides less hospitable conditions for microorganisms in the stomach, vagina, urine, and skin
Endogenous microflora (probiotics)	Commensal organisms compete with pathogens for resources and create a less hospitable environment

INFLAMMATION AND MEDIATORS OF INNATE IMMUNITY

Despite the various protective barriers, some pathogens still gain entry into the body and cause damage. In fact, pathogens are defined, in part, by their ability to break through barriers and cause damage. After breaking through a protective barrier, these pathogens encounter elements of the innate immune response, also called the second line of defense. **Innate immunity** is also referred to as non-specific immunity, in that it does not target specific pathogens, but rather targets broad classes of pathogens and injury. A key element of innate immunity is **inflammation**. Any type of injury, whether it is due to microbes, physical trauma, burns, or chemical irritants, triggers an inflammatory response (FIGURE 9.2). The overt signs of inflammation are redness, warmth, swelling, pain, and often loss of function of the affected tissue. These conditions provide a less hospitable environment for pathogens, as well as attract immune effector cells and promote healing.

Acute inflammation is associated with the healing process

Damaged cells or the presence of pathogens serves as signals that stimulate immune effector cells to secrete cytokines and other chemicals. These various cytokines and chemicals cause the physical manifestations of inflammation. Pyrogens cause an increase in body temperature, either locally at the site of injury or throughout the entire body. Sensory nerves are stimulated to produce pain. Some of these chemicals are vasoactive and cause blood vessels in the immediate area to dilate. In addition, these vasoactive compounds affect the endothelial cells that line the blood vessels, so that the tight connections between them are loosened. This results in blood vessels becoming more permeable and plasma and cells now leak from the interior of blood vessels into the surrounding tissue. As a by-product of this leakage and blood-vessel dilation, the tissue swells and becomes red, which are hallmark signs of inflammation. The accumulation of blood results in increased warmth and the swelling causes pain. Although counterintuitive, the damage caused by inflammation is an important part of the healing process.

Major mediators of innate immunity are phagocytic cells, complement, interferon, and natural killer (NK) cells (TABLE 9.5). Some of these

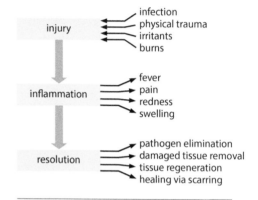

FIGURE 9.2 Inflammatory responses and resolution. The immune system initiates an inflammatory response characterized by clinical manifestations (fever, pain, redness, swelling) after an injury. The elements of the immune response resolve the cause of the inflammation and promote healing.

mediators of inflammation also help to remove damaged cells and tissues and stimulate the healing process. Generally, inflammation is a transient phenomenon. As pathogens are eliminated and damaged tissue is repaired, the inflammation decreases and completely resolves with healing (see Figure 9.2). Sometimes, this acute inflammation does not resolve and develops into chronic inflammation. Chronic inflammation is associated with some immunological disorders (Chapter 17), as well as many other chronic diseases.

Table 9.5	Major Effector Mechanisms Involved in Innate Immunity
Effector	**Activity**
Phagocytosis	Pathogens and damaged cells are engulfed and killed by phagocytic cells such as neutrophils and macrophages
Complement system	A set of proteins that assist in immunity by directly killing microorganisms or enhancing phagocytosis
Interferons	Cytokines that activate an antiviral state
Natural killer (NK) cells	A class of lymphocyte that recognize and directly kill virus-infected cells and tumor cells

Specialized cells carry out phagocytosis of microbes

Neutrophils and macrophages are the two primary types of phagocytic cells. Neutrophils are relatively abundant in the blood and are activated by infection. Inflammatory mediators, such as chemokines and cytokines, attract neutrophils from the blood to the tissues. These neutrophils are short-lived and dedicated killers that migrate to the tissues. **Macrophages**, conversely, are longer lived, reside in essentially all tissues, and carry out surveillance. Both neutrophils and macrophages engulf pathogens and other particles, such as dead cells and cellular debris, by a type of endocytosis called phagocytosis. As part of the endocytic pathway, lysosomes fuse with the vacuole containing the pathogen and the contents of the lysosome kill the pathogen (FIGURE 9.3).

SIGNPOST See Chapter 2, pages 35–37 for more on endocytosis.

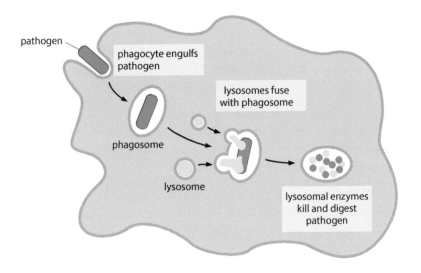

FIGURE 9.3 Phagocytosis. Phagocytes such as macrophages and neutrophils engulf microbes. Lysosomes fuse with the phagosome and destruction of the microbes occurs within the phagolysosome.

The contents of the lysosome include antimicrobial peptides, enzymes called oxidases that generate reactive oxygen species (ROS), and degradative hydrolytic enzymes. These enzymes and toxic substances – combined with the acidic environment of the lysosome – kill the pathogen and break it down into molecules. For example, antimicrobial peptides called defensins disrupt the membrane of microbes, leading to their death. Similarly, ROS damage lipids, proteins, and nucleic acids (see Box 1.2), and thereby contribute to microbe killing. This production of ROS is called a respiratory burst. Hydrolytic enzymes present in the lysosome catabolize the macromolecules of the microbe.

Macrophages also play a role in inflammation and activation of other immune effector cells

Macrophages reside in tissues and carry out surveillance for pathogens. Upon encountering a pathogen, the macrophage engulfs and destroys the pathogen and, at the same time, releases a battery of cytokines and chemokines resulting in local inflammation (FIGURE 9.4). Some cytokines affect the vascular endothelium, causing dilation of the blood vessels and an increase in permeability across the endothelial cells. Immature macrophages in the blood, called monocytes, are activated and attracted to the infection site. In addition, chemokines attract neutrophils and other immune effector cells to the location of the infection. Neutrophils leave the capillaries due to increased permeability and enter the tissue to seek out and kill the invading microbes. After pathogen elimination, the production of inflammatory cytokines ceases and cytokines associated with healing are secreted.

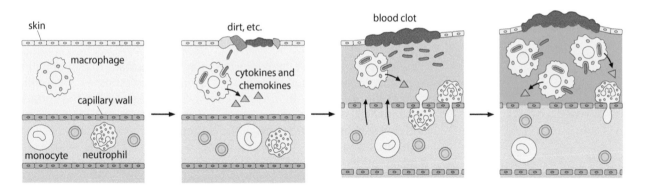

FIGURE 9.4 Macrophages and inflammation. Upon encountering a pathogenic microbe, macrophages secrete cytokines and chemokines that results in localized inflammation and recruitment of neutrophils and other immune effector cells to assist in killing the pathogen.

Complement proteins promote inflammation, enhance phagocytosis, and directly kill pathogens

The **complement** system consists of several proteins that are found in blood, lymph, and extracellular fluid and that participate in both innate and adaptive immune responses. These proteins circulate as inactive proteins that are activated by the presence of a microbe. The various complement proteins have several roles in immunity. These roles include: (1) stimulating immune effector cells to secrete inflammatory cytokines and chemokines, (2) targeting microbes for phagocytosis, (3) directly killing microbes by poking holes in their membranes, and (4) interacting with antibody molecules as part of the adaptive immune response. The name complement refers to the role of these proteins in enhancing, or complementing, other elements of the immune system.

The direct killing of microbes is facilitated by a group of proteins that assemble into a pore called the membrane attack complex. This membrane attack complex is assembled on the membrane of microbes in a stepwise fashion (FIGURE 9.5). Recognition of the microbe by the immune system results in the binding of a complement protein called C3b to the surface of the microbe. This C3b protein attracts another complement protein called C5. C5 is converted to C5a and C5b by a protease called C5 convertase. C5a functions to attract neutrophils to the site and the C3b/C5b complex serves as an initiation site to assemble a pore from other complement proteins. These large pores disrupt the integrity of the membrane of the microbe and this loss in membrane integrity causes the death of the microbe.

FIGURE 9.5 Formation of membrane attack complex. Complement protein C5 binds to the activated complement protein C3b on the surface of the bacterium or eukaryotic microbe. An enzyme called C5 convertase cleaves C5 into C5a and C5b. C5a functions to attract neutrophils to the site. The activated C3b/C5b complex binds to other complement proteins that facilitate the assembly of the pore called the membrane attack complex. The pores disrupt the membrane and kill the pathogen.

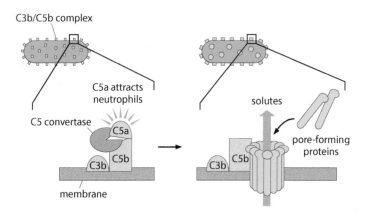

Innate immunity against viruses is mediated by interferons and natural killer cells

Type 1 **interferons** are cytokines that play an important role in antiviral immunity. Cells infected with a virus secrete interferons as a warning system for other cells. Interferon binding to the receptor on the surface of nearby cells induces an antiviral state in those cells (FIGURE 9.6). This antiviral state is induced through the expression of proteins that interfere with virus replication, hence the name. Virus replication is completely dependent upon the host cell to translate viral mRNA into viral proteins. The induction of the antiviral state involves the expression of proteins that specifically degrade viral RNA. In addition, protein synthesis is blocked by interferon. Therefore, the virus is unable to replicate in interferon-activated cells. Interferon also activates **natural killer (NK) cells**, as well as other immune effector cells, to increase surveillance and killing activities.

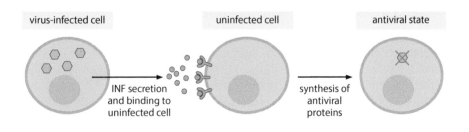

FIGURE 9.6 Interferon mechanism of action. Virus-infected cells secrete alpha (α) and beta (β) interferons (INF). Binding of interferon to receptors on neighboring cells induces the expression of antiviral proteins that specifically degrade viral RNA and block translation, leading to an antiviral state.

NK cells are a type of lymphocyte that recognize alterations in the plasma membrane of virus-infected cells and tumor cells. Upon binding to the target cell, the NK cell releases toxic proteins from cytoplasmic granules (FIGURE 9.7). Included among these proteins is a pore-forming protein called perforin. Perforins function similarly to the membrane attack complex and form pores on the target cell membrane. This results in cell lysis and death and thereby eliminates the source of a new virus. Interferons induce the proliferation of NK cells and increase the killing activity associated with NK cells. NK cells also secrete inflammatory cytokines after detecting virus-infected cells to alert other immune effector cells about the infection.

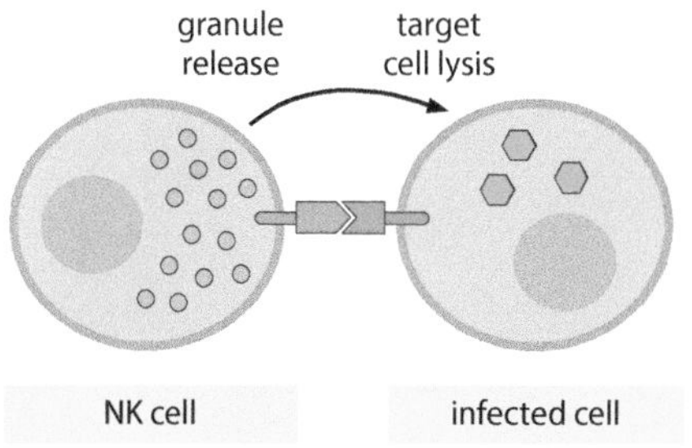

FIGURE 9.7 Cytotoxicity of natural killer (NK) cells. Binding of NK cells to altered membranes of virus-infected cells or tumor cells triggers the release of the contents of cytoplasmic granules. The toxic proteins in these granules lyse and kill the target cell.

LIGANDS AND RECEPTORS OF INNATE IMMUNITY

Some components of innate immunity are receptors that recognize pathogens and cellular damage. The ligands recognized by these receptors of innate immunity are called **pathogen-associated molecular patterns (PAMPs)** or damage-associated molecular patterns (DAMPs). DAMPs are molecules released from damaged or dying cells that function as danger signals and initiate an inflammatory response. PAMPs are molecules that are unique to pathogens. Molecules that function as PAMPs are generally shared by a wide range of pathogens, relatively abundant, invariant, and often contain a repeating molecular structure. These features result in a rapid response by the host to low levels of pathogens.

PAMPs are often common structural elements of pathogens

Most PAMPs are major structural features of pathogens (TABLE 9.6). For example, bacteria are characterized by a cell wall composed of peptidoglycan. Mammalian cells do not have a cell wall and do not exhibit any molecular structures like peptidoglycan. Furthermore, the structure of the bacterial cell wall is similar among most bacteria without extensive variation. Therefore, elements of the immune system that recognize bacterial cell wall components recognize bacteria in general. The abundance of the cell wall material and its repeating molecular structure also serves to amplify this recognition by the immune system. Other common bacterial PAMPs are lipopolysaccharide (LPS), flagellin, and mannose-6-phosphate-conjugated lipids and proteins. LPS, also called endotoxin, is well known for inducing strong inflammatory reactions.

Endotoxin is described in detail in Chapter 19, pages 369–370.

SIGNPOST

Table 9.6	Examples of Common PAMPs
PAMP	**Description**
Peptidoglycan cell wall	All bacteria have a similar cell wall structure that is not found in eukaryotes
Lipopolysaccharide (LPS)	Forms a glycolipid capsule around some bacteria that is not found in mammals
Flagellin	The protein component of bacterial flagella that is distinct from eukaryotic flagella
Mannose-6-phosphate	A carbohydrate conjugate of surface proteins and lipids found on many microbes, but not mammalian cells
Double-stranded RNA	Many viruses have dsRNA genomes or intermediates, and dsRNA is rare and rather transient in cells
Unmethylated CpG	DNA oligonucleotides of cytosine and guanosine that are methylated in vertebrates but unmethylated in bacteria and DNA viruses

PAMPs of intracellular pathogens are often nucleic acids. For example, double-stranded RNA (dsRNA) can serve as a PAMP. Some viruses have a genome consisting of dsRNA or they exhibit a dsRNA intermediate during their replication. However, dsRNA is not abundant in normal cells. Similarly, some viruses have uniquely modified DNA or RNA that functions as a PAMP. Another common PAMP for intracellular pathogens is unmethylated CpG. CpG oligodeoxynucleotides are short single-stranded synthetic DNA molecules consisting of guanine and cytosine. Unmethylated CpG is rare in vertebrate genomes, but common in microbial genomes and DNA viruses.

Some receptors that recognize microbial PAMPs activate complement and promote phagocytosis

The various PAMPs are recognized by host proteins that function as receptors. There are distinct classes of PAMP receptors, which include acute-phase proteins, receptors on phagocytic cells, and toll-like receptors (TABLE 9.7). Acute-phase proteins are synthesized by the liver and bind to pathogens. This binding activates complement or promotes phagocytosis. For example, mannose-binding lectin is a protein found in serum and other extracellular fluids that recognizes a wide range of bacteria. Binding of this protein to the surface of the pathogen activates the membrane attack complex of complement (see Figure 9.5). Similarly, C-reactive protein binds to LPS found on the surface of some bacteria and fungi and activates complement.

Table 9.7   PAMP Receptors

Class	Examples	Ligand
Acute-phase proteins	Mannose-binding lectin	Mannose
	C-reactive protein	LPS
Receptors on phagocytes	Mannose-binding protein	Mannose
	LPS receptor	LPS
	Complement receptor	Complement
	Fc receptor	Antibody
Toll-like receptors	TLR1–TLR10	Various

Proteins similar to mannose-binding lectin and C-reactive protein are also found on the surface of phagocytes and these surface proteins function as receptors. Thus, phagocytes directly recognize and bind to microbes via receptors that recognize PAMPs. After binding, the microbe is then phagocytosed and killed (FIGURE 9.8). In addition, macrophages and neutrophils also have receptors that recognize complement and antibody molecules. Antibodies are proteins produced as part of the adaptive immune response and some antibodies bind to microbes. These host proteins bound on the surface of microbes allow the indirect recognition of the pathogen. Thus, proteins such as complement can have multiple roles in the immune response. For example, complement can initiate pathogen killing by activating the membrane attack complex or by enhancing phagocytosis of the pathogen. Similarly, phagocytes have receptors called Fc receptors that bind to antibody molecules. Substances that enhance phagocytosis are called **opsonins**.

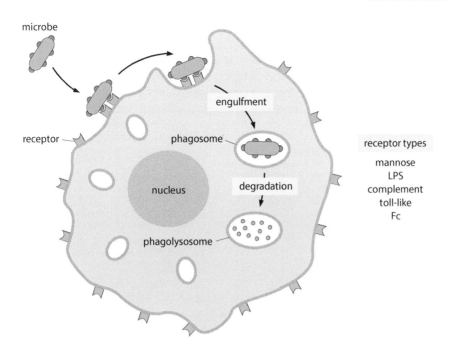

receptor types

mannose
LPS
complement
toll-like
Fc

FIGURE 9.8 Phagocyte receptors bind microbes. There are several different types of receptors found on the surface of macrophages and neutrophils that participate in phagocytosis. Some receptors, such as mannose receptors, LPS receptors, or toll-like receptors, bind directly to components on the microbe surface. Other receptors bind to host molecules, such as complement or antibody (Fc receptor), that are bound to the microbe. After binding, the microbe is then engulfed and degraded in the phagolysome.

Binding of PAMPs to toll-like receptors activates gene transcription of cytokines

Toll-like receptors (TLRs) are part of a gene family of proteins. At least 10 TLRs have been identified in humans and they are numbered TLR1–TLR10. The various TLRs recognize distinct PAMPs that tend to be associated with broad classes of pathogens (TABLE 9.8). TLRs either are found on the cell surface or are intracellular. Cell surface receptors recognize microbial PAMPs of extracellular pathogens, whereas intracellular receptors generally recognize viruses, or other intracellular pathogens. Some TLRs, particularly TLR2 and TLR4, also recognize DAMPs.

Table 9.8	Toll-like Receptors		
Receptor (location)	**Host cells with receptor**	**Major ligands (PAMPs)**	**Pathogens recognized**
TLR1/TLR2 (cell surface)	Neutrophils, macrophages, dendritic cells	Bacterial cell wall components	Numerous bacteria
TLR3 (intracellular)	Dendritic cells, lymphocytes	dsRNA	Viruses
TLR4 (cell surface)	Neutrophils, macrophages, dendritic cells, lymphocytes, intestinal epithelium	LPS	Gram-negative bacteria
TLR5 (cell surface)	Macrophages, intestinal epithelium	Bacterial flagellin	Motile bacteria
TLR6 (cell surface)	Macrophages	Lipopeptides	Mycobacteria
TLR7/TLR8 (intracellular)	Macrophages, dendritic cells, lymphocytes	Viral ssRNA	Viruses
TLR9 (intracellular)	Macrophages, dendritic cells, lymphocytes	Unmethylated CpG	Bacteria and DNA viruses

Binding of the ligand to the TLRs activates a signal transduction pathway involving protein kinases. The result of this signal transduction is the activation of a transcription factor that induces the expression of

Signal transduction and protein kinases are described in detail in Chapter 3, pages 63–66.

SIGNPOST

cytokine genes (FIGURE 9.9). Extracellular TLRs turn on the expression of proinflammatory cytokines that stimulate an increase in phagocytic activity. This helps to eliminate bacteria or other microbes. Intracellular TLRs activate the expression of type 1 interferons which help to suppress viral replication. Thus, the immune response activated by TLRs tends to be directed at broad classes of pathogens such as bacteria in general or viruses in general.

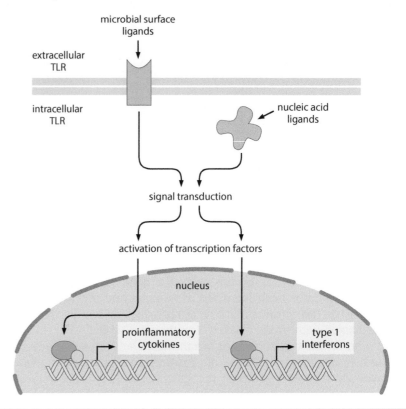

FIGURE 9.9 Signaling pathways of toll-like receptors (TLRs). Binding of a PAMP to a TLR initiates a signal transduction pathway, which activates transcription factors and leads to the expression of specific cytokine genes. PAMP binding to extracellular TLR activates proinflammatory cytokines that increase phagocytic activity. PAMP binding to intracellular TLR activates the expression of type 1 interferons and suppresses viral replication.

ADAPTIVE IMMUNITY

Another element of immunity is the adaptive immune response. The adaptive response follows the innate immune response and typically occurs days to weeks after the infection (TABLE 9.9). Therefore, the adaptive response is often called the third line of defense. Another big difference between innate immunity and **adaptive immunity** is specificity. In contrast to innate immunity, which recognizes broad classes of pathogens, adaptive immunity is highly specific in that it is directed against a specific pathogen. In addition, the adaptive immune response improves during the course of the infection and exhibits an immunological memory. Immunological memory is a phenomenon in which each subsequent exposure to a pathogen results in a stronger immune response and in many cases can lead to **protective immunity**. This improving immunity is the basis for the name adaptive. The adaptive immune response adds to the ongoing innate immune response and many of the mechanisms for pathogen elimination are shared between innate and adaptive immunity.

Table 9.9   Comparison of Innate and Adaptive Immunity

Innate immunity	Adaptive immunity
Rapid response (hours)	Slow response (days to weeks)
Receptors have broad specificity and recognize PAMPs or DAMPs	Receptors have narrow specificity and recognize antigens
Ligands are often polysaccharides or polynucleotides	Ligands are primarily proteins (highly selective specificity)
No memory	Memory of prior exposure
Constant response	Improving response
Local response	Entire body
Mediated by phagocytes and NK cells	Mediated by B and T lymphocytes
Overlapping effector mechanisms for pathogen elimination	

The primary cellular elements of adaptive immunity are B cells and T cells

Lymphocytes are derived from the lymphoid lineage of hematopoietic progenitor cells found in the bone marrow (see Figure 8.10) and include NK cells, B cells, and T cells. The progenitor B cells complete their development in the bone marrow, and hence the name B cells. Progenitor T cells, conversely, migrate to the thymus to complete their maturation. Both mature B cells and T cells migrate to the various secondary lymphoid tissues and carry out surveillance for pathogens. B cells and T cells also differ in the nature of their response to pathogens. B cells synthesize proteins called antibodies that recognize pathogens and primarily mediate the **humoral** immune response. In contrast, T cells are primarily involved in a **cell-mediated** immune response. However, B cells and T cells work together along with the elements of the innate immune response to eliminate pathogens.

Both B cells and T cells have receptors that recognize antigens associated with pathogens. Generally, the antigens are protein molecules, but they can also be complex carbohydrates or lipids. Proteins, due to their complex structure and the uniqueness of the amino acid sequence, exhibit a high level of specificity in the antigen–receptor interaction. The B-cell and T-cell receptors do not recognize the entire macromolecule but recognize a small portion of the molecule called the epitope. The epitope is also called the antigenic determinant and, in the case of protein antigens, the epitope generally consists of 5–17 acids. Therefore, the adaptive immune response is directed at specific pathogens rather than a broad class of pathogens, as is the case in innate immunity.

HUMORAL IMMUNITY

Humoral immunity is mediated by **antibodies**. Antibodies are proteins synthesized and secreted by mature **B cells**, called plasma cells. Antibodies are found in blood, various bodily secretions, and the spaces between cells in tissues. Binding of the antibodies to pathogens or pathogen products assists in eliminating the pathogen or preventing the development of disease.

Antigen binding stimulates B cells to develop into plasma cells and secrete antibodies

B cells express an antibody molecule on their cell surface that functions as a receptor for an antigen. There are potentially billions of different specific antibodies, and each specific antibody recognizes a single epitope

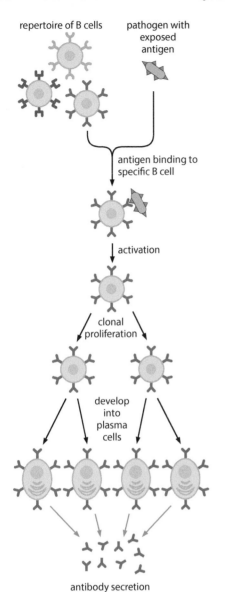

FIGURE 9.10 Activation and clonal proliferation of B cells. Binding of antigen to B-cell receptors induces a proliferation of those B-cells that recognize that antigen. This expanded population of B-cells develops into plasma cells that secrete soluble antibody molecules with the same antigen specificity as the IgD receptor on the cell surface.

on an antigen. Thus, an individual B cell expresses specific antibody molecules on its surface that recognize a single antigen. This means that an individual B cell only recognizes one antigen. However, there are billions of distinct B cells based on the epitopes recognized. This large repertoire of B cells carries out surveillance for pathogens containing antigens that are recognized by the various B-cell antibody receptors.

Binding of an antigen to the antibody on the B-cell surface activates the B cell (FIGURE 9.10). Only those B cells that recognize the antigen are activated. Other B cells that do not recognize the antigen continue with their surveillance activity. The activated B cells start to replicate and undergo clonal proliferation. This clonal proliferation expands the population of specific B cells that recognize the target antigen. Following clonal proliferation, these B cells differentiate into plasma cells that secrete antibodies. The soluble antibodies secreted by plasma cells have the same antigen specificity as the antibody which function as the IgD receptor on the surface of the B cell. These secreted antibodies then mediate immunity by binding to the antigen.

Various antibody isotypes have distinct biological properties and roles in humoral immunity

A typical antibody molecule consists of four polypeptides: two heavy chains and two light chains. The names heavy and light are reflective of the larger size of the heavy chain polypeptide. There are five distinct classes of heavy chains leading to five distinct antibody isotypes called IgA, IgD, IgE, IgG, and IgM. The Ig refers to immunoglobulin and these proteins are all part of a large superfamily of proteins. The five antibody isotypes differ in their biological properties, functional locations, and role in humoral immunity (TABLE 9.10). For example, IgD is the antigen receptor found on the surface of B cells and the other isotypes are secreted antibodies. In addition, IgA and IgM form complexes consisting of two antibody monomers and five antibody monomers, respectively. There are also two subtypes of IgA and four subtypes of IgG.

Table 9.10	Antibody Isotypes		
Isotype	**Subtypes**	**Complex**	**Distinctive feature**
IgA	2	Dimer	Found in the mucosa of digestive, respiratory, and urogenital tracts, as well as tears, saliva, and breast milk
IgD	1	Monomer	Functions as an antigen receptor on B cells
IgE	1	Monomer	Protects against parasitic worms and is involved in some allergic reactions
IgG	4	Monomer	Found in blood and tissues and provides the majority of humoral immunity
IgM	1	Pentamer	Expressed early in humoral immunity preceding IgG

The polypeptides making up antibody molecules are held together by covalent disulfide bonds. Both the heavy and the light chains consist of constant regions and variable regions. The amino acid sequences of the constant regions are conserved within an isotype, whereas the variable regions exhibit sequence variation between the various antibody molecules. The variable regions of the heavy and light chains come together to form a pocket that binds the antigen (FIGURE 9.11). The shape of this pocket is highly specific for a particular epitope and accounts for the high specificity of antibody–antigen interactions. This interaction is often described as being like a lock-and-key fit, since the shape of the epitope and the shape of the antigen-binding site are complementary. In addition, the interaction between the antibody and antigen is of high affinity. Thus, the antibody binds to the antigen tightly and stable antigen–antibody complexes are formed.

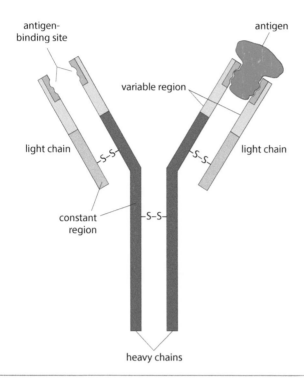

FIGURE 9.11 Antibody structure. The antibody molecule consists of two heavy chains and two light chains held together by disulfide bonds. Each polypeptide consists of constant regions with conserved amino acid sequences and variable regions. The variable regions form the antigen-binding site.

It is estimated that the immune system can generate up to 10 billion different antibody molecules that recognize distinct epitopes. This variability is generated through DNA recombination events, which occur during B-cell maturation in the bone marrow. The genes for the light and heavy chain proteins are in segments that can rearrange and recombine into different sequences. During B-cell maturation the progenitor B cells replicate and each of the individual progeny of these replications has a different arrangement of the antibody gene segments. Therefore, each individual B cell has its own unique sequence of both the heavy and the light chain genes that makes up the antigen-binding region of the antibody molecule.

Antibodies improve during the course of an infection

The first exposure of naive B cells to an antigen generally results in the production of IgM. If the infection persists, the continued exposure of B cells to an antigen results in class switching (FIGURE 9.12). In class switching the genes for the antibody molecule rearrange, so that the variable regions are now associated with a different constant region. It is the constant region that determines the antibody isotype and subtype. The plasma cells generated after class switching now produce a different subtype of antibody with the same antigen specificity. The specific antibody subtype, called an isotype, is determined by specific cytokines in conjunction with the continued presence of antigen.

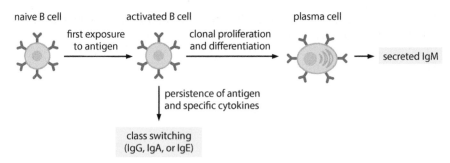

FIGURE 9.12 Class switching. Generally activated B cells on their first exposure to antigen produce plasma cells that secrete IgM. Persistence of the antigen and the presence of specific cytokines result in class switching and the production of a different subtype of antibody molecule with the same antigen specificity.

In addition, after class switching the DNA sequences making up the variable regions are subjected to hypermutation, resulting in more variations in the sequence of the variable region. Those B cells with mutations producing antibodies that bind more strongly to the antigen are selected and expand via clonal proliferation and replace those producing antibody of lower affinity. This process of increasing antibody affinity during an infection is referred to as **affinity maturation** (FIGURE 9.13). Thus, the antibody response improves as the infection proceeds, hence the name adaptive immunity.

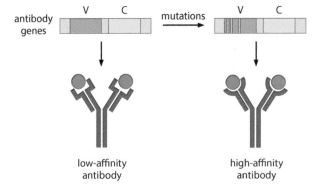

FIGURE 9.13 Affinity maturation. With the continued presence of the antigen, the variable (V) regions of antibody genes are subjected to a hypermutation and antibodies with higher affinity for the antigen are selected. C, constant region.

low-affinity antibody

high-affinity antibody

Antibodies mediate immunity through several mechanisms

Secreted antibodies bind to antigens and quite often this binding has an inhibitory effect on the pathogen or disease progression. Binding of an antibody to a pathogen may inhibit pathogen replication or block the ability of a pathogen to adhere to host cells or tissues. For example, antibodies binding to a virus might block its entry into a host cell and thereby stop virus replication. Similarly, antibodies can bind to toxins produced by some pathogens and block the action of those toxins. The process by which antibodies directly inhibit a pathogen in some fashion or block pathogenesis is referred to as **neutralization**.

Antibodies can also interact with elements of the innate immune response to enhance pathogen elimination (TABLE 9.11). For example, some proteins of complement recognize the constant region (Fc) of the antibody and can initiate a cascade reaction, leading to the formation of the membrane attack complex and the subsequent lysis of the microbe (FIGURE 9.14). Similarly, many effector cells of the immune system have receptors that recognize the Fc region of antibody molecules. For example, some macrophages have Fc receptors and phagocytose microbes with antibodies bound to their surface. This enhancement of phagocytosis is called opsonization. Similarly, NK cells also have Fc receptors and recognize antibodies bound to the surface of infected cells or tumor cells and secrete cytotoxic substances in a process called antibody-dependent cell-mediated cytotoxicity (ADCC). The four different subtypes of IgG differentially enhance complement fixation, phagocytosis, and ADCC. Thus, these antibody subtypes have specialized roles with regard to pathogen elimination.

Table 9.11	Mechanisms of Antibody Action
Mechanism	**Action (antibody isotypes/subtypes primarily involved)**
Neutralization	Antibody binds to antigen and directly inhibits pathogen replication, spread, or pathology (IgG1–IgG4, IgA1, IgA2)
Complement fixation	Membrane attack complex is initiated on microbes covered with antibodies (IgM, IgG1, IgG3)
Opsonization	Receptors on phagocytic cells recognize microbes covered with antibodies (IgG1, IgG3)
ADCC*	Receptors on NK cells target virus-infected cells or tumor cells for elimination (IgG1, IgG3)

*Antibody-dependent cell-mediated cytotoxicity.

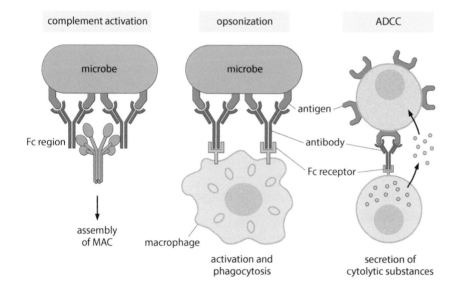

FIGURE 9.14 Enhancement of innate immunity by antibodies. Antibodies bound to antigens on pathogens or infected cells can activate elements of innate immunity. Some complement proteins bind to the Fc region of the antibody molecule and activate the assembly of the membrane attack complex (MAC) to kill microbes. Many immune effector cells have Fc receptors that bind to microbes or infected cells covered with antibody. In the case of macrophages, the antibody serves as an opsonin and promotes phagocytosis of the microbe. Natural killer (NK) cells also have Fc receptors and release substances that kill the cell with the bound antibody in a process called antibody-dependent cell-mediated cytotoxicity (ADCC).

IMMUNOLOGICAL ASSAYS AND SEROLOGY

Antibodies bind to antigens with high affinity and high specificity. These high-affinity and highly specific interactions can be exploited to measure antibodies found in either blood or other bodily fluids, or to measure an antigen recognized by a specific antibody (BOX 9.1). In general, the diagnostic identification of antibodies in serum or other bodily fluids is called **serology**. Detection of specific antibodies in a clinical sample indicates a possible exposure to a pathogen or other immunogen. Therefore, serology can be used to diagnose some infectious diseases and historically has been used to diagnose viral infections since direct observation of viruses via microscopy is not possible. One limitation of serology for the diagnosis of infectious diseases is that a positive serology cannot distinguish a current infection from a past infection due to the long-lived nature of some antibodies. Serology is also useful in the diagnosis of autoimmune disorders such as rheumatoid arthritis and in determining exposure to immunogenic substances.

BOX 9.1

Analysis of Antibodies and Antigens

Immunoassays can detect and measure either antibody or antigen. Most assays for the analysis of antigens or antibodies involve immobilizing the antigen on a solid surface. Antibodies can also be immobilized on a surface and in this case the assay is called antigen capture. Sera, or other fluids containing antibodies, are added to the antigen-containing substratum. If antibodies are being measured or characterized, then a defined antigen is used (**BOX 9.1 FIGURE 1**). In this case only the antibody that recognizes the specific antigen binds and the other antibodies are washed away. A defined antibody is used to evaluate and characterize antigens. The antibody only binds to the specific antigen that it recognizes. In both approaches the antigen–antibody complex is measured and detection of the complex provides information about whether either the antibody or the antigen is present. In many cases the amount of bound antibody can be quantified and provide a measure of the relative abundance of either antibody or antigen.

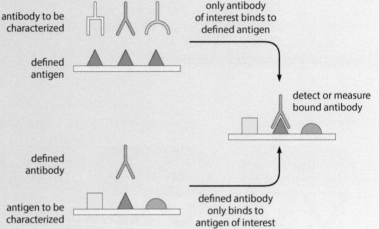

BOX 9.1 FIGURE 1 Basic principles of immunoassays. Sera contain a mixture of antibodies with a wide range of specificities. It is possible to analyze a specific antibody using a defined antigen. Alternatively, a defined antibody can be used to detect or measure a specific antigen.

Antibodies can be used as reagents to evaluate proteins and other antigens

Many of the assays used to measure antibodies can be easily converted into assays to detect and measure antigens. In this case, a well-characterized antibody is used to either isolate or detect the antigen (see Box 9.1). Detection of antigens with antibodies is both highly sensitive and highly specific. The antibody can detect small amounts of antigen in a sample that has numerous other proteins and macromolecules. Many rapid diagnostic tests that are based on antigen detection can be easily and quickly performed without the need for a laboratory. Home pregnancy tests are examples of a rapid diagnostic test based on antigen detection. Monoclonal antibodies are a type of antibody that is particularly well characterized and used in many applications based on antigen detection (BOX 9.2).

Antigen detection methods can also be combined with microscopy to determine the location of a specific protein within a cell or tissue or combined with protein gel electrophoresis (see Box 1.5) to identify specific proteins. This latter method is sometimes referred to as western blotting. Scientists routinely use antibodies to learn more about protein function.

Serotyping is used to distinguish variants within a pathogen species or blood to tissue types

The high specificity of antibodies means that relatively minor differences between antigens can be distinguished. This situation, combined with

Monoclonal Antibodies

BOX 9.2

A monoclonal antibody is produced by a B-cell clone that synthesizes a specific antibody that recognizes a single epitope. Immunogens typically contain multiple epitopes and the immune response activates several different B cells, each producing a distinct antibody that recognizes the immunogen. Such antibodies are called polyclonal and contain a mixture of antibodies with different specificities. It is possible to select and clone the individual B cells to produce a monoclonal antibody in vitro. Most often the production of monoclonal antibodies involves the immunization of mice and the subsequent cloning of B cells, producing the monoclonal antibody of the desired specificity. This technology allows an unlimited supply of a highly specific reagent that binds to an antigen. Thus, monoclonal antibodies are important tools in biochemistry, molecular biology, and medicine.

Due to the high specificity and high affinity of antibody binding to antigens, monoclonal antibodies provide a highly sensitive means to analyze and characterize proteins and other antigens. Monoclonal antibodies are also widely used in diagnostic tests and offer advantages over polyclonal antibodies produced by immunizing animals. For example, polyclonal antibodies often differ between individual animals. In addition, monoclonal antibodies can be used in therapies, such as passive immunization. Therapeutic monoclonal antibodies need to be derived from human sources. It is also possible to 'humanize' murine monoclonal antibodies through recombinant DNA technology.

the relative ease of carrying out most immunoassays, means that variants within a pathogen species can be easily distinguished. Different pathogen variants distinguished by their reactivity to antibodies are called **serotypes**. The determination of blood types and tissue types for transplantation is also based on serotyping. Serotypes are determined using a set of reference antibodies. Each reference antibody recognizes a specific serotype (FIGURE 9.15). The sample is tested for reactivity against a panel of reference antibodies to determine the serotype of the sample.

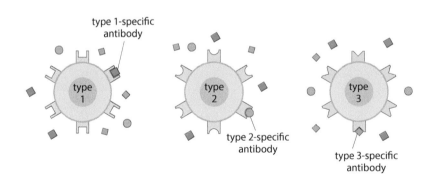

type 1-specific antibody

type 1

type 2-specific antibody

type 2

type 3-specific antibody

type 3

FIGURE 9.15 Serotyping is based on reference antibodies. A biological or clinical sample is tested against a panel of reference antibodies to determine its serotype. Each reference antibody recognizes a specific variant of the antigen being used to define the serotype (e.g., type 1, type 2, or type 3).

The ease of carrying out serological assays makes serotyping a convenient biomarker for the diagnosis of disease and other health-related issues. For example, the various serotypes of a particular pathogen may exhibit different characteristics in terms of their transmissibility or virulence. In the case of influenza (Chapter 20), many distinct serotypes of the influenza virus have been identified in birds. However, only a few of these serotypes are capable of infecting humans. Similarly, certain serotypes of the normally harmless bacterium *Escherichia coli* are virulent and can cause severe symptoms such as bloody diarrhea. A notable example is *E. coli* O157:H7 (see Box 21.2). O157:H7 is the designation of the serotype based on bacterial antigens called O and H with the number designating the specific serotype.

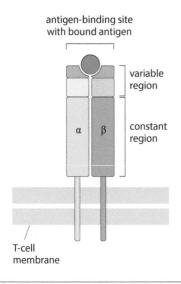

antigen-binding site
with bound antigen

variable
region

constant
region

α   β

T-cell
membrane

FIGURE 9.16 T-cell receptors. The T-cell receptor consists of an α-chain (yellow) and β-chain (blue). The variable regions of both chains come together to form the antigen-binding site (pink). Parts of the constant regions span the plasma membrane of T cells.

CELL-MEDIATED IMMUNITY

Cell-mediated immunity refers to immune responses involving cellular elements of the immune system – especially **T cells**. One type of T cell, called a **cytotoxic T cell**, directly kills pathogen-infected cells or tumor cells. Cytotoxic T cells are characterized by a surface protein called CD8 and therefore are also called CD8+ cells. The other type of T cell expresses the CD4 protein on its surface and these CD4+ cells affect immune responses by secreting cytokines that affect other cells of the immune system. CD4+ cells are called **helper T cells** due to their role in activating or regulating other cells.

T cells have an antigen receptor on their surface called the T-cell receptor. The T-cell receptor is composed of an α-chain and a β-chain (FIGURE 9.16). T-cell receptors are part of the immunoglobulin family and have a structure similar to that of antibody molecules which includes constant and variable regions. The variable regions of the two subunits fold into a pocket that recognizes the epitope. Like B cells, an individual has a large repertoire of T cells that all have unique epitope-specific T-cell receptors on their surface. This large repertoire is generated through recombination events that occur during T-cell maturation in the thymus. Following maturation, the T cells migrate to the peripheral lymphoid tissues to carry out surveillance. Mature T cells that have not yet been exposed to antigens are called naive T cells.

T cells recognize processed antigens associated with antigen-presenting cells

Although T-cell receptors and antibodies have a similar structure, there is a fundamental difference in how T cells recognize antigens. Unlike B cells, which recognize the epitope on an intact or native antigen, T-cell antigens first need to be unfolded and subjected to proteolysis to break the protein down into small peptides (FIGURE 9.17). This proteolysis is called antigen processing and occurs within cells. The peptide making up the epitope binds to a protein that is a member of the **major histocompatibility complex** (MHC) gene family and the MHC protein

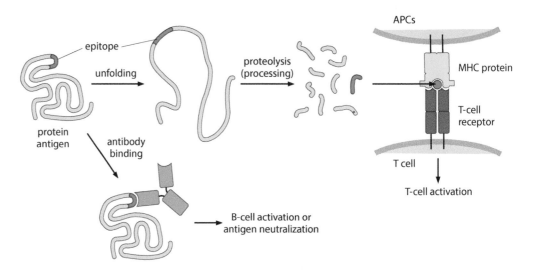

FIGURE 9.17 Antigen processing and presentation. B-cells recognize an epitope (red) on an intact antigen in its native state. T-cell antigens are unfolded and proteolytically processed into peptides. Peptides containing the epitopes bind to major histocompatibility complex (MHC) proteins that are present on the surface of cells called antigen-presenting cells (APCs). The T-cell receptor binds to the epitope that is bound to the MHC protein.

with the bound epitope is displayed on the cell surface. The display of the epitope on the cell surface is called antigen presentation. T-cell receptors recognize the displayed antigen and only recognize the antigen that is bound to MHC proteins. Binding of the T-cell receptor to the antigen–MHC protein complex creates an interaction between the antigen-presenting cell and the T cell which activates the T cell.

Internal and external antigens are processed differently

The MHC is a gene family divided into two major classes (see Box 17.3). MHC class I proteins are expressed by most types of cells and are primarily involved in immunity against intracellular pathogens such as viruses and immunity against tumor cells (TABLE 9.12). Some viral proteins synthesized by the host cell pass through the endoplasmic reticulum (ER) and the secretory pathway. Within the ER, these viral proteins may be recognized as being foreign and are degraded by proteases (FIGURE 9.18). Some of the peptides produced during this proteolysis bind to MHC proteins and these MHC proteins are subsequently translocated to the surface of the cell. This exposes the peptide on the cell surface and the epitope serves as a warning flag, indicating that the cell is infected with a virus. Cytotoxic T cells express a T-cell receptor that recognizes a particular epitope bound to the MHC–epitope complex. After binding to the MHC-epitope complex, T cells then secrete toxic substances that kill the infected cell in a manner similar to natural killer (NK) cells (see Figure 9.7). Antigens associated with tumors can be similarly processed and utilized by the immune system to kill cancerous cells.

The secretory pathway is described in greater detail in Chapter 2, pages 36–37.

SIGNPOST

Table 9.12   Major Histocompatibility Complex (MHC) and Antigen Presentation

Type	Antigen source	Cells expressing	Activates
MHC class I	Internal antigens such as viruses, other intracellular pathogens, and tumors	Most cells	Cytotoxic T cells
MHC class II	External antigens taken up by exocytosis	Professional antigen-presenting cells such as dendritic cells or macrophages	Helper T cells

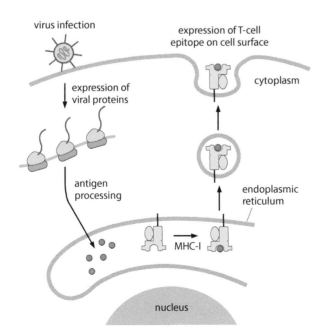

FIGURE 9.18 Presentation of internal antigens by MHC class I proteins. Internal antigens, such as viral proteins, are broken down by proteases in the endoplasmic reticulum and peptides bind to an MHC class I protein. The MHC protein with the bound peptide is transported to the cell surface where it can be recognized by T-cell receptors.

MHC class II proteins are expressed primarily on the surface of **dendritic cells**, macrophages, and B cells. These cells take up external antigens via endocytosis and present the processed epitopes on their surface. Thus, they are often called professional antigen-presenting cells. The primary function of dendritic cells is antigen presentation, and dendritic cells are especially abundant in tissues that are in contact with the external environment, such as the skin and the submucosal layers of the respiratory tract, the digestive tract, and the urogenital tract. Dendritic cells continuously sample their local environment for pathogens and other foreign materials. Antigens are taken up by endocytosis and processed as part of the endocytic pathway (FIGURE 9.19). The endocytic pathway intersects the secretory pathway to return membranes back to the cell surface. As part of this process, vesicles with MHC class II proteins fuse with the endosome containing processed proteins and the peptide epitopes bind to an MHC protein and are translocated to the cell surface.

FIGURE 9.19 MHC class II antigen presentation. External antigens are taken up by endocytosis and processed into peptides in the endosome. MHC class II proteins are transported to the endosomes from the endoplasmic reticulum. Peptides corresponding to epitopes bind to MHC class II proteins and are translocated to the cell surface where they can be recognized by T-cell receptors.

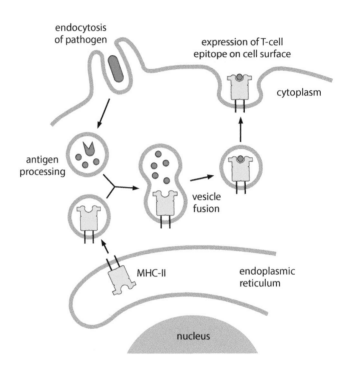

Activation of T cells leads to clonal proliferation

Dendritic cells with exposed MHC–epitope complexes migrate to the secondary lymphoid tissues such as lymph nodes, mucosa-associated lymphoid tissue, and spleen. These secondary lymphoid organs contain numerous naive T cells that have not yet been activated. A naive T cell that recognizes the presented antigen is activated and those T cells that do not recognize the presented antigen remain in a naive state (FIGURE 9.20). Like B-cell activation, activated T cells undergo clonal proliferation. This produces a large cadre of T cells that recognizes a specific epitope. These activated T cells then leave the secondary lymphoid tissues and migrate throughout the body looking for cells expressing the epitope that they recognize. When encountering the epitope, these activated T cells mediate the elimination of the pathogen. For example, cytotoxic T cells directly kill cells expressing the epitope, such as virus-infected cells or tumor cells.

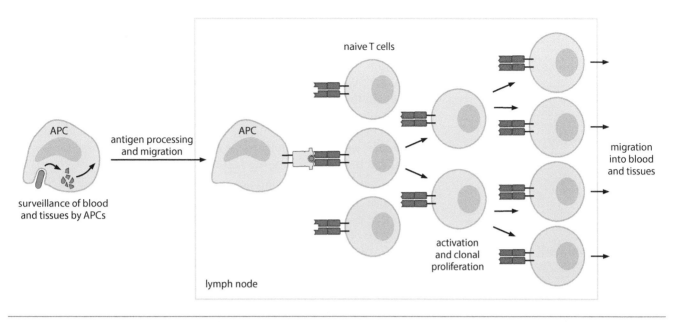

FIGURE 9.20 T-cell activation. Professional antigen-presenting cells (APCs) carry out surveillance in the blood and tissues. After antigen processing the APC migrates to secondary lymphoid organs such as the lymph nodes and presents the antigen to naive T cells. T cells recognizing the epitope are activated and undergo clonal proliferation. These activated T cells leave the secondary lymphoid tissues and migrate throughout the body searching for cells presenting the same epitope.

Activation of helper T cells stimulates the secretion of cytokines that stimulate other immune effector cells

Helper T cells do not directly participate in the elimination of pathogens. Instead, these cells stimulate other immune effector cells by secreting cytokines. There are different subtypes of helper T cells that are designated as Th1, Th2, Th17, and Treg (TABLE 9.13). These subtypes of helper T cells are activated by exposure to the presented antigen in combination with specific cytokines. Once activated, these helper T cells secrete cytokines upon subsequent encounter with the specific epitope that they recognize. The different subtypes of helper T cells secrete specific cytokines according to their subtype, and these secreted cytokines activate other elements of the immune response. For example, Th1 cells secrete interferon-γ, which stimulates macrophages and cytotoxic T cells (FIGURE 9.21). Similarly, Th2 cells secrete interleukin-4 that stimulates B cells and antibody production. Th17 cells are characterized by the secretion of interleukin-17 and play a role in immunity at mucosal surfaces by activating neutrophils. Generally, a particular pathogen will only activate one of these three subtypes of helper T cells. This means that the helper T cells tend to activate either a cell-mediated immune response or a humoral response, depending on the pathogen.

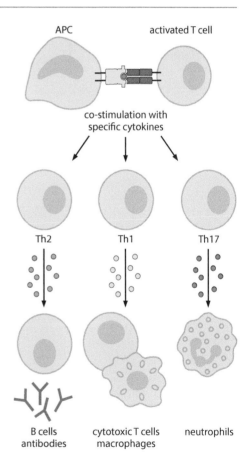

FIGURE 9.21 Helper T cell function. Co-stimulation of activated T cells with antigen via antigen-presenting cells (APCs) and subtype-specific cytokines leads to the secretion of cytokines (colored circles). The cytokines secreted by the different subtypes of helper T cells determine the type of immune response that is enhanced.

Table 9.13	Subtypes of CD4⁺ Cells		
Subtype	**Stimulating cytokine(s)**	**Secreted cytokine(s)**	**Target cells**
Th1	IL-12, IL-2	INF-γ, TNF-β	Macrophages, cytotoxic T cells
Th2	IL-4	IL-4, IL-5, IL-13	B cells, mast cells
Th17	TGF-β plus IL-6	IL-17	Neutrophils
Treg	TGF-β minus IL-6	IL-10, TGF-β	T cells

The primary role of helper T cells is to enhance an ongoing immune response against pathogens. For example, macrophages and neutrophils are activated during the innate immune response. If the innate response is unable to clear the infection, then helper T cells are subsequently activated. The cytokines secreted by these activated helper T cells further stimulate the phagocytic activity and attract more immune effector cells to the sites of infection. Similarly, specific cytokines secreted by helper T cells can stimulate the further clonal proliferation of cytotoxic T cells or B cells, resulting in an increase in the number of immune effector cells available to fight the invading pathogen.

It should be noted, that there is a destructive element to the immune response and the activation of T cells can lead to an intense immune response that may start to do more damage than good. Therefore, there is a subtype of T cells that can suppress the immune response in an antigen-specific manner called a regulatory T cell (Treg cell). These regulatory cells are also important in controlling adverse immune reactions that can lead to diseases such as autoimmunity (Chapter 17).

IMMUNOLOGICAL MEMORY AND PROTECTIVE IMMUNITY

Activation of the adaptive immune response results in the production and activation of large numbers of lymphocytes that are directed at a specific pathogen. These B cells and T cells evoke a wide variety of immune mechanisms to efficiently eliminate pathogens. However, there is a delay in producing these activated lymphocytes that allows the pathogen to initially proliferate (FIGURE 9.22). Once the adaptive response is activated the pathogens are generally eliminated. The activated lymphocytes associated with the adaptive immune response require the presence of antigen and, once the pathogen has been cleared, the activated lymphocytes undergo senescence and disappear. However, some of the activated lymphocytes remain and develop into long-lived **memory cells** (FIGURE 9.23). Some of these memory cells may even be retained throughout life.

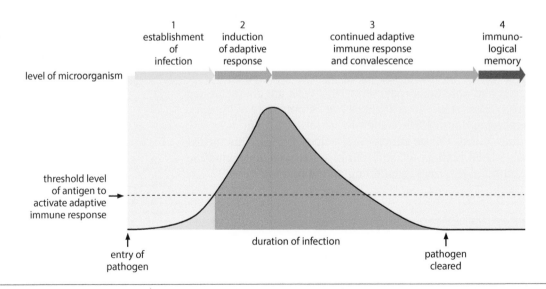

FIGURE 9.22 Progression of the adaptive immune response. Activation of the adaptive immune response requires a threshold level of antigen. The clonal proliferation of lymphocytes associated with activation takes time and there is a delay in the development of adaptive immunity. This delay allows for the proliferation of the pathogen to levels that cause disease. The fully activated adaptive response clears the pathogen and the manifestations of the disease diminish (convalescence). After pathogen clearance, some of the activated lymphocytes become memory cells that lead to protective immunity.

A subsequent encounter with the same pathogen often results in milder disease, or even no disease at all. This is due to the **protective immunity** established by the long-lived memory cells produced during the first infection. These memory cells have the same antigen specificity and, in the case of B cells, any affinity maturation that occurred is retained. In addition, these antigen-specific memory lymphocytes are generally 10–100 times more abundant than antigen-specific naive lymphocytes. The memory cells are also primed to respond more strongly against the pathogen. Therefore, subsequent encounters with the same antigen cause a quicker and stronger adaptive immune response due to these memory cells. The second encounter with the same antigen is called the secondary response. The higher levels of antigen-specific lymphocytes result in a shorter time for clonal proliferation to produce sufficient effector lymphocytes. Furthermore, antibodies produced by memory B cells are generally of higher affinity than those produced by naive B cells. Thus, the secondary immune response is more rapid and more intense than the primary immune response (FIGURE 9.24). This stronger and more rapid immune response during subsequent exposures to a pathogen often eliminates the pathogen before it can proliferate to levels that cause disease. This phenomenon is referred to as protective immunity.

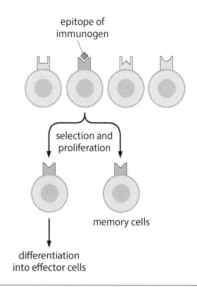

FIGURE 9.23 Development of memory cells. Exposure of lymphocytes to antigens results in clonal proliferation of antigen-specific cells. Most of these cells develop into effector cells that participate in pathogen elimination. Some of the activated lymphocytes develop into long-lived memory cells that provide protective immunity.

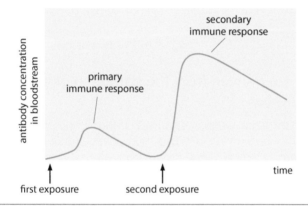

FIGURE 9.24 The basis of protective immunity. The primary immune response associated with the first exposure to a pathogen is delayed and decreases after the pathogen is eliminated. Long-term memory cells result in a more rapid and stronger secondary immune response during the second exposure to the same pathogen.

VACCINATION

Protective immunity can also be artificially induced through **vaccines**. Vaccines mimic the pathogen in some fashion and activate specific lymphocytes to undergo clonal proliferation. Some of these activated lymphocytes develop into long-lived memory cells, which may provide partial or complete protective immunity. This means that an individual can acquire at least some immunity without experiencing the disease. The process of artificially inducing protective immunity with a vaccine is called **active immunization**. Since the time of the first immunization described by Jenner (BOX 9.3), vaccination has saved many lives and protected many more people from the morbidity associated with infectious diseases. Active immunization is generally to prevent disease (prophylaxis). However, stimulating the immune response could be used to treat diseases. Currently, only a few therapeutic vaccines are available, but they are the subject of research particularly in cancer treatment.

BOX 9.3

Smallpox and the First Vaccine

The first disease to be routinely prevented by vaccination was smallpox. Smallpox is a viral disease caused by variola major and variola minor. Variola major causes severe disease and has a mortality rate of 20–30%. It had been noted that milkmaids and other animal handlers suffered from smallpox less often than other people and typically developed a milder disease. In particular, milkmaids developed a milder disease called cowpox which was probably acquired from cows. In 1796 Edward Jenner took a scrapping from a cowpox lesion of a milkmaid and inoculated the scrapping into an 8-year-old boy named James Phipps. After recovery from the mild infection, Jenner then inoculated (challenged) James Phipps with actual smallpox and the boy showed no signs of infection or adverse effects. Jenner repeated the experiment several times and published the results. Other people had probably used cowpox to immunize against smallpox before Jenner. Jenner, however, is credited with the discovery due to his documentation of the process and for conducting an actual challenge.

Cowpox was called variola vaccinea, which means smallpox of the cow. Thus, the word vaccine is derived from the Latin word for cow, and Jenner coined the term vaccination to describe the process. The success of Jenner's vaccinations spread throughout Europe and immunization against smallpox with cowpox replaced a process called variolation. Variolation was a process of immunizing with actual smallpox, which had a high risk of producing severe disease and a case-fatality rate of 0.5–2%. Smallpox vaccination became compulsory in many countries and, in 1967, the World Health Organization initiated a smallpox elimination campaign. The last smallpox epidemic occurred in Somalia in 1977 and in 1979 it was concluded that smallpox had been eradicated.

Vaccines are derived from the pathogen or components of the pathogen

There are three basic types of vaccines (TABLE 9.14). One type of vaccine is the live-attenuated vaccine. This type of vaccine consists of a related pathogen that is less virulent, or a pathogen that has been mutated in some fashion to be less virulent. With this type of vaccine, the pathogen can establish an infection and replicate. However, pathogen replication is limited and vaccination only results in mild disease, or no disease at all. The use of the cowpox virus to immunize against smallpox is an example of such a vaccine (see Box 9.3). Related or modified pathogens share many antigens with the pathogen and the vaccine activates memory lymphocytes that recognize the actual pathogen. Exposure to the pathogen after vaccination results in a secondary immune response and the pathogen is eliminated before causing serious disease.

Table 9.14	**Types of Vaccines**		
Type	**Description**	**Advantages**	**Limitations**
Live attenuated	A related or modified pathogen that only produces mild disease	Mimics disease process and induces strong protective immunity	Vaccine can mutate to become more virulent or can cause serious disease in immunocompromised people
Intact killed	An intact pathogen inactivated by heat, chemicals, or radiation	Numerous immunogens and vaccine cannot mutate into more virulent forms	Adverse side effects common due to complex mix of antigens
Subunit	Isolated components such as proteins or capsular polysaccharides from a pathogen	Less likely to be associated with adverse side effects	Often poorly immunogenic and requires boosters, adjuvants, or conjugates

Another approach to producing a vaccine is to kill or inactivate the pathogen by treating it with chemicals, radiation, or heat. Even though the killed pathogen is no longer capable of replicating, many of the antigens associated with the pathogen can activate lymphocytes and induce protective immunity. Specific antigens can also be purified from the pathogen and used as a vaccine. The use of pathogen components is called a subunit vaccine. Examples of subunit vaccines are purified pathogen proteins or complex polysaccharides that make up the outer capsules of some bacteria. In the case of protein-based vaccines, recombinant DNA technology (Chapter 7) is often utilized to produce the vaccine (TABLE 9.15). This approach includes producing the vaccine in transgenic organisms, chimeric viruses, generating virus-like particles, and gene therapy approaches.

Table 9.15	Applications of Recombinant DNA Technology to Vaccines
Subunit vaccine production	Recombinant protein is expressed in transgenic organisms for the large-scale production of subunit vaccines without working with large quantities of a pathogen
Chimeric viruses	Gene for immunogen of pathogen virus is incorporated into the genome of avirulent viral vector to create an attenuated vaccine
Virus-like particles	Viral structural proteins are assembled without the viral genome and therefore are non-infectious. Elicit strong B-cell and T-cell responses
Gene therapy approaches	Gene for immunogen is delivered into host cells via a viral vector or lipid nanoparticle. Host cell expresses the gene to elicit an immune response

The applications of gene therapy to vaccination and chimeric viruses are described in detail in Chapter 7, pages 150–152.

SIGNPOST

Immunogenicity of vaccines can be increased by conjugates and adjuvants

Quite often subunit vaccines are not effective at activating lymphocytes. In particular, polysaccharides are not very immunogenic. One approach to increase the immunogenicity is to couple the polysaccharides with a toxoid. These conjugated vaccines stimulate helper T cells that further activate B cells to produce a stronger antibody response. The antibodies produced by the vaccine bind to the surface of the bacterial capsule and lead to clearance of the bacteria through complement fixation or opsonization. Another element of a good immune response is inflammation and many subunit vaccines do not induce an inflammatory response. Therefore, substances called adjuvants are often added to vaccines to stimulate an inflammatory response and increase the immunogenicity of the vaccine.

Antibodies can be used in the treatment of disease by passive immunization

The administration of preformed antibodies can also be used to treat disease. This is referred to as **passive immunization** (TABLE 9.16). For example, serum can be taken from individuals who have recovered from an infectious disease and that serum is injected into a person with the same infection. Antibodies in the donor serum bind to the pathogen and

Table 9.16	Examples of Passive Immunization	
Natural	Mother-to-child transfer of IgG across the placenta	
	IgA in breast milk protects against intestinal infections	
Artificial	Donor serum from people surviving an infectious disease	
	Monoclonal antibodies targeting pathogen or disease mediator	

assist in the elimination of the pathogen from the recipient. Monoclonal antibodies (see Box 9.2) are also available for the treatment of some diseases. Passive immunization can also be used to protect an individual who was exposed to a pathogen. However, this is only short-term protection, and the recipient does not benefit from memory B cells and does not develop protective immunity.

There is also a natural transfer of passive immunity that occurs from mother to child. IgG crosses the placenta in utero and enters the circulatory system of the child. This protects the child from infections during the first 6 months of life until these maternal antibodies disappear. In addition, IgA is secreted in breast milk, providing further protection to the infant. This allows the infant to survive infectious diseases until developing immunity due to natural exposures.

Developing a safe and effective vaccine has many limitations

In principle, one could imagine vaccines being able to prevent and resolve all infectious diseases. However, vaccine development can be costly and complicated. It is difficult to identify antigens that are sufficiently immunogenic to induce protective immunity and, at the same time, not cause adverse events. In general, vaccines that mimic the pathogen and disease process are more likely to induce protective immunity. For example, live-attenuated vaccines establish an infection that is controlled by the immune system. However, attenuated vaccines can mutate or evolve into more virulent pathogens that cause serious disease (see Table 9.14). In addition, a person with a compromised immune system may not be able to control the infection and therefore experience a serious disease.

Killing the pathogen eliminates the ability to mutate and evolve. However, without the ability to replicate, larger, and perhaps multiple, doses of the vaccine are required. This becomes a problem since some components of this complex mixture of antigens may have toxic effects or stimulate inappropriate immune reactions, leading to adverse reactions. These side effects can be mild such as minor pain, fever, or a rash. In rare cases the adverse reaction can be quite severe, leading to permanent tissue damage or even death. Subunit vaccines minimize such side effects as the toxic elements can be removed from the vaccine. However, subunit vaccines using purified antigens tend to be less immunogenic and may not induce a protective response.

Pathogens also evolve or undergo antigenic variation resulting in a vaccine that is no longer effective. For pathogens to survive they need mechanisms to avoid the sophisticated adaptive immune response of the host. This is often accomplished by varying the antigens that are associated with protective immunity. Variation of the antigens might mean that the memory lymphocytes no longer recognize the pathogen. This is true whether immunity is naturally induced or artificially induced through vaccines. Thus, over time, variants of a particular pathogen may emerge that are no longer susceptible to the protective immunity induced by the vaccine. Therefore, some vaccines need to be periodically upgraded to account for the evolution of the pathogen. For example, the flu vaccine that needs to be adjusted every flu season (Chapter 20).

SUMMARY OF KEY CONCEPTS

- The immune system consists of specialized cells and proteins that are found throughout the body as integral parts of the circulatory system and lymphatic system

- The immune system is regulated by hormone-like molecules called cytokines

- Various physical, chemical, and biological barriers block the entry and proliferation of pathogens as a non-immunological defense mechanism

- Innate immunity recognizes and eliminates broad classes of pathogens via phagocytosis, complement, and natural killer cells

- Phagocytic cells engulf and kill bacteria and other microbes

- Complement proteins bind to microbes and initiate a cascade reaction, leading to the lysis of the pathogen or the phagocytosis of the pathogen

- Innate immunity against viruses is mediated by type 1 interferons and natural killer cells

- The innate immune system recognizes molecular patterns associated with pathogens or cellular damage

- Adaptive immunity is mediated by B cells and T cells that recognize specific epitopes on antigens and targets specific pathogens

- The humoral immune response consists of B cells that produce antibody molecules

- Antibodies bind to antigens associated with pathogens and directly neutralize pathogens, activate complement, enhance phagocytosis, or induce cell killing by natural killer cells

- Antibodies are widely used in serological assays in disease diagnosis, as well as in research

- T cells recognize epitopes associated with antigen-presenting cells and play a fundamental role in cell-mediated immunity

- Cytotoxic T-cells target cells infected with specific viruses, other intracellular pathogens, or tumor cells

- Activated helper T cells secrete cytokines to stimulate various elements of the immune response

- After pathogen clearance some of the activated B cells and T cells develop into long-lived memory cells that lead to a protective immunity

- Vaccines mimic the pathogen or disease process in some fashion and can artificially induce protective immunity

KEY TERMS

active immunization	The induction of immunity following exposure to or inoculation with antigen(s).
adaptive immunity	Immunity based on responses of antigen-specific B cells or T cells and includes the development of immunological memory.
affinity maturation	The process by which antibodies increase their anti-pathogen activity through mutations in the variable regions of antibody genes in B-cells.
antibody	A protein synthesized and secreted by B cells that recognizes a specific antigen and mediates the elimination of pathogens expressing that antigen.
antigen	A molecule recognized by components of the adaptive immune system such as antibodies or T-cell receptors.

antigen-presenting cell (APC)	A cell that expresses MHC proteins complexed with peptides derived from antigens (i.e., epitopes) on their surface to activate T cells.
B cells	A type of lymphocyte that develops in the bone marrow and makes antibodies in response to an antigen.
cell-mediated immunity	An adaptive immune response involving cytokines and the activation of phagocytic cells or lymphocytes.
complement	A set of extracellular proteins that function to kill pathogens or enhance other elements of the immune response.
cytokine	Hormone-like molecules made by some cells that affect the behavior of immune effector cells and mediate the immune response.
cytotoxic T cells	T cells that directly kill virus-infected cells, tumor cells, or other abnormal cells as part of cell-mediated immunity.
dendritic cells	A phagocytic cell found in tissues that presents antigens to T cells resulting in activation of those T cells that recognize the antigen.
helper T cells	A subset of T cells that functions to stimulate or activate other immune effector cells by secreting cytokines.
humoral immunity	An adaptive immune response involving B cells and antibodies directed against specific antigens.
innate immunity	Immunological defense mechanisms that are not directed against a specific pathogen, but rather at a class of pathogens. Also called non-specific immunity.
inflammation	An immune response resulting from injury or infection characterized by swelling, redness, pain, and warmth due to the accumulation of fluid and leukocytes.
interferon (type 1)	Cytokines that trigger an antiviral state in cells and potentiates innate and adaptive immune responses against infected cells and tumor cells.
leukocytes	A general term that includes all white blood cells and cells that mediate immunity.
lymphatic system	A major component of the immune system that is composed of fluid called lymph, vessels that transport lymph and leukocytes, and lymphoid tissues and organs.
lymphocytes	Cells derived from the lymphoid lineage of hematopoietic progenitor cells found in the bone marrow and include natural killer cells, B cells, and T cells.
macrophages	Phagocytic cells found in most tissues that kill pathogens, remove contaminants, clean up cellular debris, and activate other immune effector cells.
major histocompatibility complex (MHC)	A set of linked genes encoding proteins that present antigens to T cells.
memory cells	Long-lived lymphocytes that recognized antigens they were previously exposed to and that are a major component of the secondary immune response following subsequent exposures to the same pathogen.
natural killer cells	Cytotoxic lymphocytes that circulate in the blood and target virus-infected cells and tumor cells as part of innate immunity, as well as mediating antibody-dependent cellular cytotoxicity in adaptive immunity.

neutralization	The ability of antibodies, without the assistance of other components of the immune system, to block the ability of pathogens or their products to cause disease.
opsonin	An antibody or other substance which binds to a pathogen and makes it more susceptible to phagocytosis.
passive immunization	The administration of antibodies to treat or prevent an infection or disease.
pathogen-associated molecular pattern (PAMP)	Molecules associated with pathogens that serve as ligands to activate innate immunity.
protective immunity	The acquisition of adaptive immunity against a specific pathogen or vaccine that results in little or no disease upon a subsequent encounter with that pathogen due to long-lived memory lymphocytes.
serology	Evaluation of blood serum, especially with regard to the antibody response to pathogens or immunogens.
serotype	A group of related microbes or viruses distinguished by a specific set of antigens as determined by reference antibodies.
T cell	A subtype of lymphocyte that develops in the thymus and participates in cell-mediated immunity.
toll-like receptor	A class of proteins that bind PAMPs and mediate the inflammatory response associated with innate immunity.
vaccine	A biological agent that can evoke protective immunity and protect an individual against a disease.

REVIEW QUESTIONS

1. What are toll-like receptors and how do they mediate innate immunity?

2. Describe class switching and affinity maturation with regard to the antibody response in the continued presence of a pathogen.

3. How do antibodies mediate the elimination of a pathogen?

4. Describe the two fundamental types of antigen presentation and the types of pathogens affected.

5. What is immunological memory and how is it related to protective immunity?

6. What are the four types of helper T cells and their distinct roles in immunity?

7. What are the different types of vaccines and how can immunogenicity be enhanced?

ADDITIONAL QUESTIONS

In addition to the Review Questions provided above, there is a range of free online questions designed to enable students to further test their understanding of the chapter material. In order to access these interactive questions, please visit the book's website: https://routledge.com/cw/wiser

Non-infectious Disease

Disease is generally defined as something abnormal that affects a person's health and well-being. Section 2 gives an in-depth description of the diseases responsible for the majority of human morbidity and premature mortality. The first chapter in this section defines terminology often used in the description of diseases and describes aging and degenerative diseases. Also included in this chapter is a discussion of biomarkers and their use in disease diagnosis. Each of the subsequent chapters in this section is a comprehensive description of a particular type of disease. For example, there are chapters on diabetes, cardiovascular disease, chronic kidney disease, cancer, and immunological disorders such as allergy and autoimmune disease. The focus of each chapter is on the etiology and pathogenesis of these diseases including discussions about possible molecular and cellular mechanisms involved in disease progression. Also included for each of these diseases are the clinical manifestations, treatment options, and disease management.

The remaining chapters in this section are genetic diseases, nutritional disorders, and addictive disorders. Genetic diseases refer to disorders in which defects in single genes are both necessary and sufficient to cause the disease. A few specific diseases are described in detail with a focus on how a defect at the genetic level results in pathology at molecular and cellular levels and how this pathology causes disease manifestations. The chapter on nutritional disorders focuses on overnutrition and obesity since obesity is an underlying cause of many diseases such as diabetes and cardiovascular disease. Addictive disorders are also a major underlying cause of disease. An overview of what is known about addiction is given and diseases associated with tobacco use and alcohol use are described in detail.

Disease Overview and Molecular Biomarkers

10

Disease is generally defined as an impairment of health or an abnormal condition associated with human anatomy or physiology. Damage to cells or tissues is often an inherent feature of the disease state. As is often the case with definitions, the definition of disease implies a black-and-white state of being either healthy or sick. However, disease is not a black-or-white condition and there are a lot of gray areas between diseased and healthy states. Disease and health can also be viewed as a continuum between ideal health and severe illness (FIGURE 10.1). On one end of the spectrum is physical and mental well-being and on the other extreme is severe illness characterized as being disabling or life threatening. Even though not extensively disabling, diseases lying between this spectrum nonetheless affect a person's ability to carry out normal activities.

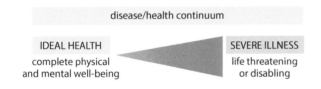

disease/health continuum

IDEAL HEALTH
complete physical and mental well-being

SEVERE ILLNESS
life threatening or disabling

FIGURE 10.1 **Continuum between ideal health and disease.** Ideal health is defined as complete physical and mental well-being. Disease is a deviation from this well-being and can range from mild to severe.

Pathology is the study of structural and functional changes that are associated with disease. Pathology also refers to the actual damage associated with disease and pathophysiology reflects functional alterations observed with disease. Other important elements of disease are **etiology** and **pathogenesis**. Etiology refers to the origin and cause of disease and pathogenesis refers to the development and progression of disease. Disease etiology and pathogenesis are generally complex processes involving numerous factors. These factors can include environmental or behavioral elements, as well as genetic and immunological elements. Disease etiology and pathogenesis can often be described on a molecular or cellular level. As our understanding of disease on these levels advances, the use of molecular biomarkers is becoming increasingly important in the diagnosis and study of disease. Biomarkers result in a more accurate diagnosis of disease and the use of biomarkers is increasingly being used to better define disease determinants.

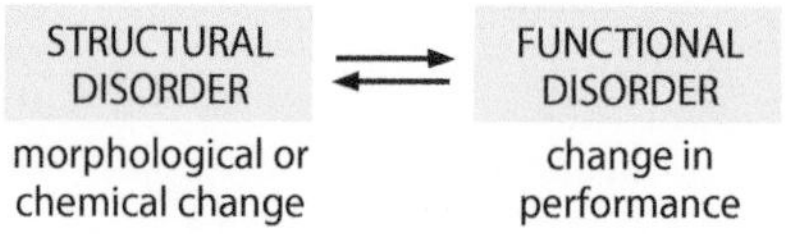

FIGURE 10.2 Interplay between structural and functional disorders. Damage to the structure of a cell, tissue, or organ can alter physiological functions. Similarly, function disorders can lead to structural damage.

DISEASE MANIFESTATIONS

Diseases present themselves by causing observable and measurable changes. Changes can be either structural or functional in nature. Structural disorders are changes in the normal morphology or chemistry of a cell, tissue, or organ. Morphological or chemical alterations associated with disease are called **lesions**. Functional disorders are changes in the performance of a cell, tissue, or organ. High blood pressure or mental disorders are examples of functional disease. However, structural disease and functional disease are often intertwined in that structural defects can lead to functional abnormalities and vice versa (FIGURE 10.2). For example, structural damage to a heart valve can eventually lead to congestive heart failure, a functional disorder. Similarly, persistent high blood pressure can cause enlargement of the heart, a structural disorder.

Disease manifestations are usually described in terms of signs and symptoms

Symptoms are complaints reported by the patient such as pain or nausea. Signs are observations made by an examiner such as a doctor, nurse, or physician assistant. Signs can be visible and observable, such as redness or swelling, upon a physical examination of a patient. Signs and symptoms along with other data, such as the medical history of the patient, family history, and test results, are used to diagnose a particular disease. If there are no apparent signs and symptoms the disease is referred to as subclinical. In other words, subclinical diseases are asymptomatic, at least in the early stages of the disease, but can develop into symptomatic disease. For example, there are often no symptoms associated with the early stages of cancer and cancer is often discovered as a result of other complaints by the patient.

Signs and symptoms are not always visible to the naked eye and other procedures are needed to diagnose the disease. If the lesion or abnormality causing the disease symptoms is at the cellular or tissue level, a histological examination is needed in which a sample of the affected tissue is removed and examined with a microscope (TABLE 10.1). Removal of a tissue specimen is called a biopsy. In other cases, the lesion may be associated with an internal organ and thus not visible to the naked eye. There are various imaging techniques, such as X-rays, computed tomography (CT) scans, and magnetic resonance imaging (MRI), that can be used to assess the well-being and function of internal organs and tissues. In addition, lesions can be chemical or molecular in nature and laboratory tests are needed to diagnose the disease. Typically, blood or urine is utilized in such laboratory tests.

Table 10.1	Diagnostic Procedures
Gross examination	Manifestations that are visible with the naked eye such as swelling or bleeding
Histology	Examination of biopsied material with the microscope for abnormalities at the cellular level
Imaging techniques	X-rays, MRI, CT scans, ultrasound, and other techniques to examine internal organs and tissues
Laboratory tests	Testing of bodily fluids, especially blood and urine, for metabolites or other biomarkers of diseases

Diagnosis of disease is accompanied by prognosis and treatment options

The observations and data collected by the medical practitioner may lead to a diagnosis of a specific disease (BOX 10.1). A collection of specific signs, symptoms, and other data is called a syndrome. If a specific disease or syndrome is diagnosed, the practitioner can offer a prognosis and prescribe appropriate treatment. A prognosis is an opinion on how the disease may progress. Treatment takes two general forms: specific treatment and symptomatic treatment. Specific treatment targets the cause of the disease and attempts to cure the disease. For example, antibiotics can be used to eliminate bacterial infections. Symptomatic treatment is aimed at making the patient more comfortable and may not address the underlying cause of the disease manifestations. Pain relievers, cough suppressants, and anti-inflammatories are examples of symptomatic treatments. For some diseases there are no cures, and the only recourse is symptomatic treatment.

Principles of Diagnosis

Disease diagnosis by a physician or other health practitioner has two major components: the history and the physical examination. The clinical history brings out details about the severity, time of onset, and other characteristics of the disease (**BOX 10.1 FIGURE 1**). Included in the history are questions about past symptoms and seemingly unrelated symptoms to provide information about past illnesses and overall general health. Family history of disease is also important since many diseases tend to run in families and have an inherited component. Also important is the patient's social history which includes occupation and habits such as smoking and drinking. The physician then carries out an examination placing particular emphasis on the parts of the body affected by the illness and any detected abnormalities are correlated with the clinical history. At this point the physician may be able to make a provisional diagnosis. Sometimes, more than one possible diagnosis needs to be considered and this is called a differential diagnosis. In a differential diagnosis the physician considers multiple diseases that could coincide with the patient's symptoms and history. Based on the provisional diagnosis, various diagnostic tests

can then be ordered to confirm a diagnosis or to distinguish between possible diagnoses. In difficult cases the physician often confers with a specialist or refers the patient to a specialist. Based on the final diagnosis, appropriate treatment can be initiated.

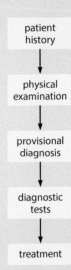

BOX 10.1 FIGURE 1 Flow chart of disease diagnosis. The typical protocol used in disease diagnosis and treatment.

BOX 10.1

DISEASE ETIOLOGY AND PROGRESSION

Progression from a state of ideal health to one of disease implies that something causes disease and influences its development. Generally, the origin and progression of disease are complex and involve many factors. These factors can generally be categorized as endogenous factors, exogenous factors, and behavior (FIGURE 10.3). Endogenous factors are an integral part of the individual such as their physiology and genes. Exogenous factors are external to the individual and part

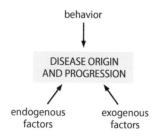

FIGURE 10.3 Factors influencing disease origin and progression. Many factors influence the development of disease. These factors can be associated with a person's genetics (endogenous), environmental factors (exogenous), or behavior.

of the physical and social environment of an individual. Behavior also plays a role in disease origin and progression. In general, endogenous factors, exogenous factors, and behavior all simultaneously contribute to disease origin and progression. However, for any specific disease their contributions are not equally weighted. For example, some diseases may be primarily genetic in origin, whereas other diseases may be primarily due to behavior or environmental exposures.

Etiological agents directly cause disease

In many cases the cause of a disease can be explained. For example, many diseases are caused by pathogens, such as bacteria and viruses. These pathogens represent an exogenous factor that directly causes the disease and are called **etiological agents**. Etiological agents are not necessarily exogenous factors but can also be endogenous factors. For example, defective genes are known to cause many specific diseases and, in these cases, the defective gene is the etiological agent of that disease. In general, an etiological agent is a substance or phenomenon that is both necessary and sufficient to cause a particular disease. However, disease origin can also involve factors that may not directly cause the disease but increase the risk of developing disease or make a person more susceptible to disease. Such factors are called **risk factors** or **predisposing factors**.

Risk factors and predisposing factors can increase the likelihood of disease development and progression

Some diseases are associated with certain lifestyle conditions. For example, there is a strong correlation between smoking cigarettes and developing heart or lung disease. In this case, smoking is a risk factor for developing heart disease or lung cancer. Smoking is not an etiological agent since it is neither necessary nor sufficient to cause disease. There are plenty of examples of people who smoke but do not develop cardiovascular or lung disease. Similarly, certain behaviors are risk factors for acquiring certain infectious diseases. For example, promiscuous sex increases the risk of acquiring a sexually transmitted infection. In general, risk factors are exogenous conditions or behaviors.

Endogenous conditions such as age, sex, and genetics can also influence disease origin and progression. These conditions increase susceptibility to certain diseases and are called predisposing factors. For example, infants and children are more susceptible to infectious diseases because their immune system is not yet fully developed. Some individuals may have a genetic predisposition to develop certain diseases. But this genetic predisposition does not mean that a person will develop the disease and therefore predisposing factors do not directly cause disease. The predisposing factor creates a situation in which the disease can develop or progress. The distinction between risk factors and predisposing factors can be subtle and quite often the two terms are used interchangeably. Disease etiology is a complex combination of etiological agents, risk factors, and predisposing factors (FIGURE 10.4).

Diseases can be defined by their progression and duration

Pathogenesis refers to disease development and includes describing how etiological agents, risk factors, and predisposing factors cause the clinical and pathological manifestations associated with a particular disease. Quite often there is a delay between exposure to an etiological agent

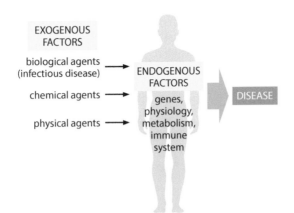

and the development of disease manifestations. This delay is called the incubation period. In some diseases this incubation period can be rather long – lasting up to decades – and this long delay is referred to as a latent period. Diseases can progress rapidly or slowly or last a short time or a long time. **Acute** diseases are diseases in which the symptoms rapidly appear and are generally of short duration (TABLE 10.2). In contrast, **chronic** diseases develop slowly over a long period of time and can last for the lifetime of the patient. These patterns of disease progression are a part of the natural history of a disease.

Table 10.2 Features of Acute and Chronic Diseases	
Acute	**Chronic**
• Arises rapidly • Usually distinct symptoms • Lasts a short time (days to weeks) • Sometimes acute conditions become chronic	• Usually begins slowly • Often vague signs and symptoms • Persists for a long time (months to years) • Generally difficult to prevent and treat

Diseases are also associated with **morbidity** and **mortality**. Morbidity is the effect that the disease has on the patient and mortality refers to the possibility of causing death. Morbidity refers to having the disease or symptoms and can range from being annoying to causing severe disabilities. Extremely morbid diseases can be mortal. In acute diseases morbidity, and possibly mortality, occurs soon after disease initiation. Sometimes acute diseases self-resolve and cure themselves (FIGURE 10.5). Acute conditions can also persist and develop into chronic conditions if not treated. Chronic diseases are characterized by a slow increase in morbidity and often do not exhibit a high degree of mortality. In many cases the slow onset of the symptoms can go unnoticed and quite often the symptoms are somewhat vague. In addition, chronic diseases rarely cure on their own and typically are difficult to treat. Over time this can lead to a high degree of morbidity, even though mortality may be low.

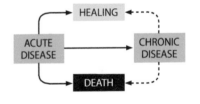

FIGURE 10.5 Acute and chronic diseases. Acute diseases usually self-cure or respond to treatment. Treatment failure can result in death, which can happen over a short period of time. Acute diseases that persist can develop into chronic diseases. Chronic diseases persist for a long time and generally do not self-cure and are difficult to treat (dashed arrow). Progression to death typically takes a long time (dashed arrows).

CLASSIFICATION OF DISEASE

Diseases share common features that allow them to be separated into different types of diseases. For example, diseases can be classified based on the organ afflicted, such as heart disease or liver disease. This is convenient in that the organ system affected is identified and the symptoms and manifestations can be related to the affected organ. However, an organ may be afflicted with a wide variety of diseases in terms of pathogenesis and etiology. Another way to classify diseases is based on common etiologies or pathophysiology (TABLE 10.3). In this case seemingly disparate diseases may be grouped together due to similar lesions and disease progression. No system of disease classification is perfect and any specific disease may be classified in multiple ways. Quite often disease etiology is not known and such diseases are called idiopathic diseases.

Table 10.3 Classification of Diseases Based on Etiology or Pathogenesis	
Class	**Description**
Infectious	Caused by pathogens such as viruses, bacteria, and parasites. Are often transmitted from person to person (Chapters 19–24)
Genetic	Due to defects in genes that are passed from parents to offspring. Primarily refers to diseases caused by a defect in a single gene that is inherited (Chapter 11)
Congenital	Diseases that are present at birth in the neonate. This includes many genetic diseases. However, not all congenital diseases are inherited, since some may be acquired in utero due to infections, toxicity, or other causes
Neoplastic	Characterized by uncontrolled and abnormal cellular replication such as tumors and other cancers (Chapter 16)
Immunological	Due to malfunction of the immune system such as allergies or autoimmune disease (Chapter 17)
Homeostatic	Generally due to disturbances of the endocrine system or important metabolic processes that disrupt normal physiology such as fluid and electrolyte balance
Degenerative	The progressive loss of tissues and impairment of their function. Most often associated with aging but can have genetic or environmental components
Nutritional	Results from inadequate intake of nutrients. Includes protein-energy malnutrition as well as lack of a specific nutrient such as a vitamin (Chapter 12)
Traumatic	Results from a physical injury and includes extremes of heat or cold, exposure to radiation, and electrical shock
Toxic	Caused by ingestion or absorption of poisons encountered in the environment
Psychogenic	Refers to mental diseases with a strong psychological or emotional component (Chapter 18)
Iatrogenic	Results from the treatment of another disease. For example, drugs or surgeries used to treat one disease may cause another disease
Idiopathic	Diseases of unknown cause

AGING AND DEGENERATIVE DISEASE

It could be argued that aging is not a disease since it is a normal process. Furthermore, aging has no apparent or obvious cause that is associated with a specific event. However, there is morbidity associated with aging and the pathology is due to the deterioration of organs and physiological functions. This progressive deterioration over time is called **senescence**, or biological aging. This is distinct from chronological aging, which simply reflects a person's age. Biological aging proceeds at different rates in different individuals and is influenced by many factors. An example of biological aging is the loss of muscle mass as a person gets older. This loss is due to both a decrease in the number of muscle cells and individual muscle cells becoming smaller. A decrease in cell or organ size is called atrophy. As expected, atrophy associated with aging is an element in most human disease, since the deterioration of organs ultimately contributes to disease progression.

Senescence is also observed at the cellular level

A possible contributing factor to cellular senescence is the accumulation of mutations in DNA. For example, oxygen radicals are produced as a toxic by-product of metabolism and can damage DNA (see Box 1.2). Chemicals and radiation also damage DNA. The balance between the accumulation of DNA mutations and the repair of these mutations can influence the longevity of cells. As mutations accumulate over time cells may die or decrease in their functionality.

It is also known that human cells grown under laboratory conditions can only undergo 50–90 successive rounds of cell division and then the cells lose their ability to replicate. This cessation of cell division is due in part to a phenomenon called telomere attrition. Telomeres are the ends of chromosomes and with each round of cell division a small piece of the telomere is lost (BOX 10.2). Cells quit replicating when telomeres become too short. However, it is not clear whether telomere attrition plays a role in human aging. The cessation of cell division does not completely explain aging, since the loss of cellular functions in non-dividing cells also contributes to deterioration and aging.

Telomeres and Cell Division

The ends of linear chromosomes are capped with repetitive DNA called telomeres. Human telomeres consist of thousands of copies of the sequence TTAGGG added to the ends of chromosomes in tandem. These TTAGGG polynucleotide fragments are added to the ends of DNA molecules by an enzyme called telomerase. A co-factor of telomerase is an RNA molecule that is complementary to the telomere sequence (i.e., CCCUAA). This RNA component of telomerase serves as a template for the synthesis of the telomere repeats. The activity involved in synthesizing DNA from an RNA template is called reverse transcriptase.

One function of the telomere is to protect the chromosome from shortening during each round of DNA replication and cell division. This shortening is due to the inability to completely copy the lagging strand during DNA replication (Chapter 4). Through the addition of the TTAGGG repeats the telomeres are replenished and continue to protect the ends of the chromosomes. Lack of telomerase expression may play a role in cellular senescence, because after a defined number of cell divisions the telomeres have shortened to such an extent that chromosome stability is affected. Accordingly, telomerase activity is usually higher in cells undergoing rapid cell division such as stem cells (Chapter 8) or cancer cells (Chapter 16). Abnormal telomere maintenance may contribute to the increased replication potential of malignant cancerous cells.

BOX 10.2

Genes play a major role in premature degenerative diseases

Many degenerative diseases are related to aging (TABLE 10.4). However, in some cases, these degenerative changes occur prematurely. Life experiences and genetics also play a big role in degenerative diseases. For example, osteoarthritis may be initiated by an injury and obesity contributes to its progression. Many degenerative diseases have a strong genetic component and exhibit Mendelian inheritance patterns. Huntington's disease (see Box 11.3) is one such example. Degeneration of nerve and muscle tissue is particularly problematic due to the paucity of stem cells in these tissues and their limited ability to regenerate.

Stems cells and regeneration are discussed in detail in Chapter 8, pages 163–166.

SIGNPOST

Table 10.4   Examples of Degenerative Diseases

Disease	Manifestations	Etiology
Osteoarthritis	Breakdown of joint cartilage and underlying bone	Mechanical stress and insufficient repair
Huntington's disease	Neurodegeneration affecting mental abilities and coordination	Genetic disease due to defective *huntingtin* gene
Alzheimer's disease	Neurodegeneration leading to dementia	Genetic predisposition involving many genes
Amyotrophic lateral sclerosis	Progressive loss of motor neurons leading to loss of control of voluntary muscles	Unknown etiology involving both genes and environment
Muscular dystrophies	Progressive weakening and breakdown of muscle tissue	Several different types involving different genes
Age-related macular degeneration	Blurred vision and blindness due to chronic inflammation of the retina	Strong association with polymorphisms in complement factor H gene

POPULATION MEASURES OF DISEASE

Epidemiology is the study of the distribution and determinants of disease at the population level. The amount of disease in a population can be expressed in terms of prevalence and incidence (TABLE 10.5). **Prevalence** is defined as the total number of cases in a specified population at a specific time. The specified population could range from a small community to the entire world. It is usually expressed as a percentage calculated by dividing the number of cases by the population at risk. **Incidence** is the rate of new cases of a disease in a specified population over a defined time period. Typically, incidence is expressed as the number of new disease cases in the specified population per year.

Table 10.5   Terms Used to Describe Disease in Populations

Term	Definition
Prevalence	The total number of disease cases in a specified population at a specified time
Incidence	The rate of new disease cases in a specified population over a defined time period (e.g., per year)
Morbidity	The number of persons affected by a particular disease in a specified population (i.e., prevalence or incidence)
Mortality	The death rate due to a particular disease in a specified population over a defined time period
YLL	Years of potential life lost due to a fatal disease
DALY	Disability-adjusted life-years calculated from years of life lost added to years living with a disability or disease
Endemic	Disorder caused by health conditions constantly present in the population at a stable incidence
Epidemic	A sudden rise in disease incidence clearly above normal incidence levels
Pandemic	An epidemic spreading across international borders

Mortality is a measure of the impact and importance of a specific disease

Morbidity and mortality are also used to describe disease in populations and have different definitions for the epidemiologist and the pathologist. For the pathologist, morbidity refers to a diseased state or the degree of pathology associated with a disease. In epidemiological terms morbidity can refer to either the prevalence or the incidence of a disease in a specified population. In other words, it reflects the proportion of individuals in a specified population that is affected by a particular disease. Likewise, mortality for the epidemiologist refers to the death rate caused by a particular disease in a specified area or population. It is often expressed as a percentage or ratio of deaths compared with the total population being measured during a specified time period. Mortality is quite often used to evaluate the importance of a disease since death is a relatively easy outcome to measure, and death is obviously an indicator of the severity of a specific disease.

In that all humans are mortal, the true impact of disease mortality is death occurring sooner than expected. An epidemiological measure of premature death in a population is years of potential life lost (YLL). The YLL reflects the number of additional years that probably would have occurred in the absence of disease. It is calculated by subtracting the age of death from a reference age reflecting the average age of death for that population. The YLL emphasizes diseases that tend to be mortal for children, such as infectious diseases (FIGURE 10.6). For example, the YLL is approximately the same for heart attacks and respiratory infections even though heart attacks have much higher mortality. This is because heart attacks occur later in life and respiratory infections exhibit a high death rate in young children.

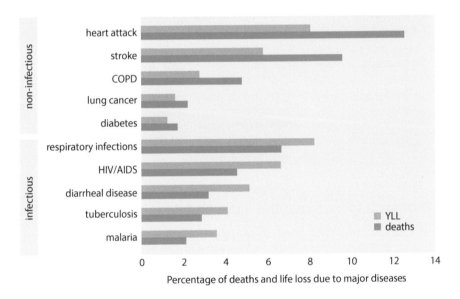

Percentage of deaths and life loss due to major diseases

FIGURE 10.6 **Comparison of deaths and years of life lost due to major diseases.** The deaths due to major non-infectious diseases tend to be late in life and therefore do not extensively shorten life. Children are more susceptible to and likely to die from infectious diseases than adults. Thus, infectious diseases have a greater impact on years of life lost (YLL). COPD, chronic obstructive pulmonary disease.

Many diseases exhibit a high rate of morbidity, but are not necessarily fatal. The impact of such diseases on populations is described using disability-adjusted life-years (DALYs). DALYs give an overall measure of disease burden and are the years of healthy life lost due to ill-health, disability, or early death. The DALY is calculated by adding the YLL to the years lived with a disability.

Disease prevalence can be relatively stable or fluctuate

Endemic and epidemic are most often used to describe infectious diseases in populations, but these terms can be applied to other types of disease. For example, one often hears about the epidemic of obesity. Endemic refers to diseases that occur at a constant and expected rate and are continuously present in the population (FIGURE 10.7). In other words, the disease incidence remains approximately the same from year to year. This stable incidence can range from low endemicity to high endemicity. The term 'epidemic' refers to a sudden and substantial rise in the level of endemicity above normal levels. For example, a sporadic outbreak of a particular disease is an epidemic. Epidemics that occur across international borders are called pandemics.

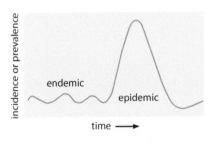

FIGURE 10.7 **Comparison of endemic and epidemic.** Endemic refers to diseases that are constantly present at approximately the same prevalence or incidence. An epidemic is a sudden increase in disease prevalence or incidence.

BIOMARKERS OF DISEASE

The diagnosis of disease relies heavily on overt signs and symptoms. Various biomarkers can also be used in disease diagnosis and to improve our understanding of disease and health. A **biomarker** is a biological substance or process that can predict disease occurrence or outcome. A wide range of molecules can be used as biomarkers including immune responses, metabolites, proteins, and nucleic acids (TABLE 10.6). The presence of a biomarker can be diagnostic of a specific disease. In addition, some biomarkers can be prognostic, since they may indicate exposure to disease agents, susceptibility to a disease, or responses to therapy. Biomarkers can also help with determining the complex etiology of disease by measuring the various environmental, genetic, physiological, and immunological elements of a disease. Evaluating these various elements allows the assessment of the individual contributions of these various elements in disease pathogenesis.

Table 10.6 Molecular Biomarkers		
Class	**Diagnostic applications**	**Subtyping**
Serological and immunoassays	Immune response indicates exposure Antigen detection	Serotypes
Metabolites, enzymes, and proteins	Abnormal metabolites or proteins define disease Define bacterial species	Enzyme variants
Nucleic acids	Genetic disease markers Detect pathogens (viruses)	Genetic fingerprinting

Biomarkers increase the ease and sensitivity of diagnosis

Biomarkers are often used to diagnose disease or confirm a diagnosis based on signs and symptoms. Very often these biomarkers are substances that are easily measured in the blood or urine and provide information about the disease state. For example, creatine levels in the blood and urine are used to assess kidney function (see Box 15.1). Similarly, glycated hemoglobin provides a measure of glucose levels in the blood over the previous 1–2 months (Chapter 13). In general, metabolites or proteins found in the blood can be used to measure physiological functions. Quite often these biomarkers are easily measured in a less invasive fashion than other diagnostic procedures. For example, testing for prostate-specific antigen (PSA) in the blood as a screen for prostate cancer is far less invasive than performing a prostate biopsy (see Box 16.3).

In many cases, biomarkers exhibit a high sensitivity of detection and allow the earlier diagnosis of a disease. This situation is especially true of biomarkers based on antibodies or the detection of nucleic acids. Antibodies bind to antigens with a high affinity and this high specificity and sensitivity of the antibody–antigen interaction can be exploited to develop sensitive assays for antibodies. Furthermore, the high specificity and affinity of antibodies can also be exploited to detect specific antigens. Similarly, the polymerase chain reaction (PCR) can detect a few molecules of DNA or RNA and provides a highly sensitive method to detect the presence of a specific nucleic acid (see Box 7.2).

SIGNPOST

Serology and the use of antibodies in assays are described in detail in Chapter 9, pages 187–189.

Biomarkers can be used to detect variants and polymorphisms

The increased specificity of biomarkers can also be used to identify variants. For example, many pathogens have different subtypes that exhibit different degrees of virulence. Identifying pathogen variants can provide information about virulence, drug resistance, and other characteristics of that pathogen. For example, there are variants of the common bacterium *Escherichia coli* that can be distinguished by serotyping, and some of these variants are rather pathogenic. Subtypes identified by differences in nucleic acid sequence are called biotypes. Like serotypes, biotypes within a species can exhibit differences in transmissibility, virulence, and drug resistance. Biomarkers for virulence factor genes or drug-resistance genes provide more information about the pathogen that can guide the treatment options. This increased knowledge about pathogenicity and drug sensitivities can lead to more efficacious and improved patient outcomes.

In summary, the use of biomarkers in the diagnosis of infectious diseases generally increases the sensitivity and specificity of the diagnosis and possibly provides more information (FIGURE 10.8). Quite often it is difficult – and sometimes impossible – to detect a pathogen by direct observation via microscopy or in vitro culture. The increased sensitivity of serology, antigen detection, or nucleic acid detection overcomes these limitations. Identifying specific serotypes or biotypes provides more information about the pathogen.

Nucleic acid-based methods are widely used in the diagnosis of viral infections

Over the past two decades advances in biotechnology have led to a dramatic increase in the use of biomarkers based on nucleic acids. Specific nucleic acid fragments are detected by hybridization with complementary probes (see Box 4.1) or PCR (see Box 7.2). Such technologies allow for the specific detection of nucleic acids associated with pathogens in a high background of contaminating human DNA. Detection of a pathogen-specific nucleic acid indicates that the pathogen is present and can serve as a sensitive and specific diagnosis. For example, many virus infections that were previously diagnosed via serology or antigen detection are now diagnosed by the detection of virus DNA or RNA via PCR. The increased sensitivity of PCR allows an earlier and more accurate diagnosis of the infection (FIGURE 10.9). Furthermore, diagnosis with serology cannot always distinguish between a past infection and a current infection, since antibodies remain in circulation after the pathogen has cleared.

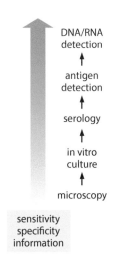

FIGURE 10.8 Relative sensitivities and specificities of various methods for diagnosis of infectious disease. Typical methods for the detection of pathogens in clinical specimens going from the least amount of sensitivity and specificity to the most sensitivity and specificity include direct observation by microscopy, in vitro culture of the pathogen, detection of antibodies specific for the pathogen (serology), using antibodies to detect pathogen antigens, or detection of pathogen nucleic acids using PCR. Methods based on biomarkers also may provide more information than methods based on microscopy or culture.

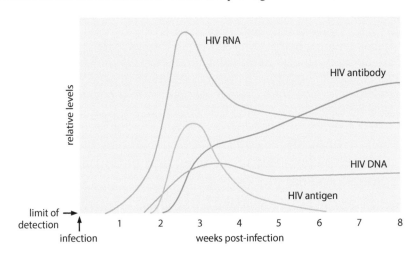

FIGURE 10.9 Comparison of methods to diagnose HIV infections. The weeks after infectious virus is first detectable in the blood are shown on the *x* axis and the *y* axis depicting the relative sensitivity of the various methods. The methods include the detection of viral RNA molecules in the blood using PCR, detection of HIV-specific antibodies in the blood, detection of HIV DNA incorporated into helper T cells by PCR, and detection of viral antigens in the blood.

SIGNPOST

Gene chips and transcriptomics are described in Chapter 6, pages 117–119.

Molecular biomarkers are especially useful for the diagnosis and analysis of genetic disorders

The most fundamental aspect of a genetic disorder is at the level of DNA sequence. Molecular assays can be designed to detect single nucleotide polymorphisms (SNPs) and other polymorphisms in the DNA sequence, thus allowing the identification of alleles. This subtyping of DNA-sequence variations can distinguish individuals and is often called genetic fingerprinting. In much the same way that a fingerprint left at a crime scene can lead to the identification of a suspect, DNA can identify a specific person. Gene chip technology can quickly and easily determine a genetic profile or a profile of genes that are transcribed. Detection of specific nucleic acid sequences and RNA transcripts can diagnose a specific disease or predict the likelihood of developing a specific disease.

Development of biomarkers involves technological issues and validation

Molecular biomarkers are clearly powerful tools in the characterization and diagnosis of disease. However, the development of biomarkers is not always straightforward and there are several issues that may confound their use (TABLE 10.7). Some of the issues are technological, such as the collection and handling of biological specimens. In addition, laboratory analyses must be amenable to handling large numbers of samples. Another issue with newly developed biomarkers is how they compare with the previous biomarkers. For example, biotypes based on nucleic acids may not correspond exactly to serotypes based on antibodies. Thus, the interpretation of data using different classes of biomarkers may be problematic. Biomarkers also need to be validated. Validity refers to the degree to which a biomarker truly reflects a particular disease determinant.

Table 10.7   Issues in Biomarker Development

- Collecting, handling, and storing biological specimens
- Transforming bench methods to handle large numbers of samples effectively
- Newer generations of biomarkers may not correlate exactly with previous biomarkers
- Validity of a biomarker

MOLECULAR EPIDEMIOLOGY

At first glance the phrase **molecular epidemiology** may seem like an oxymoron since one is focused on populations and the other on molecules. However, these two disparate approaches come together to better understand human disease. For example, epidemiological studies on large populations can be used to validate knowledge gained in laboratory studies or to validate biomarkers. Validated biomarkers can be used in both descriptive and analytical epidemiology. Descriptive epidemiology describes disease distribution in a population. Analytical epidemiology refers to studies that strive to identify disease determinants.

Molecular epidemiology utilizes molecular biomarkers, such as nucleic acids, proteins, and antibodies, in epidemiological studies

In actuality, molecular epidemiology is simply epidemiology. The use of biomarkers provides increased sensitivity and specificity in the diagnosis of disease. For example, the sensitivity of pathogen detection in a clinical specimen generally increases as the detection method moves from microscopy to in vitro culture to molecular and biochemical methods (see Figure 10.8). Similarly, molecular biomarkers can provide more sensitivity and specificity in the diagnosis of non-infectious disease. Furthermore, more detailed information about the disease and underlying factors may be provided by biomarkers. Thus, the application of biomarkers in epidemiological studies increases the accuracy of describing disease distribution in a population.

Biomarkers can be used to identify disease determinants

Disease etiology is rather complex and multiple factors determine disease (see Figure 10.4). The increased discriminatory power of molecular markers can be used to look at these various factors in isolation and determine their relative contributions to disease. For example, the role of specific genes or alleles in a particular disease can be defined using allele-specific biomarkers. The large sample sizes in epidemiology studies can correlate a particular allele with a particular disease. For example, genome-wide association studies (GWASs) can potentially identify alleles that correlate with disease. Similarly, biomarkers can more precisely define environmental exposures or factors that may be disease determinants. In summary, the increased accuracy and precision of biomarkers can identify and better define disease determinants and the greatest contribution of molecular epidemiology will likely be in analytical epidemiology.

GWAS is described in Chapter 6, pages 124–130.

SIGNPOST

SUMMARY OF KEY CONCEPTS

- Disease is an impairment of human health or abnormal condition that is often associated with damage to cells and tissues and manifests as signs and symptoms

- Disease origin and progression are complex and involve endogenous factors, exogenous factors, and behavior

- Etiological agents are necessary and sufficient to cause disease and can be considered direct causes of disease

- Risk factors and predisposing factors do not directly cause disease but increase the likelihood of developing a specific disease

- Diseases are often classified according to their etiology and pathogenesis

- Aging and degenerative diseases are associated with a progressive deterioration of organs and tissues

- Biomarkers are widely used in the diagnosis and investigation of disease

- Biomarkers based on nucleic acids provide information about the genetic basis of disease at the DNA-sequence level

- Molecular epidemiology is the use of molecular biomarkers in population-based studies to learn more about disease determinants

KEY TERMS

acute (disease)	A disease characterized by a rapid onset and short duration due to either a rapid cure or death.
biomarker	Any structure, substance, or process that can be measured that may influence or predict the occurrence or outcome of diseases.
chronic (disease)	A disease characterized by a slow progression and long duration due to lack of curing and low mortality.

disease	Any disturbance or anomaly in the normal functioning of the body that usually has a specific cause and identifiable symptoms.
etiological agent	Substance or phenomenon that is necessary and sufficient to cause disease.
etiology	Refers to factors that cause disease and mechanisms of disease transmission.
incidence	The number of disease cases that occurred during a defined period of time (e.g., cases per year) in a defined population.
lesion	A structural or chemical abnormality caused by disease or injury.
molecular epidemiology	The use of molecular markers, such as DNA, RNA, proteins, metabolites, or antibodies, to determine disease distribution or identify determinants of disease.
morbidity	The state of being diseased or having symptoms. In epidemiology morbidity is expressed as a rate (incidence) or percentage (prevalence) of the disease in a population.
mortality	The state of being mortal or the death rate in a population.
pathogenesis	Describes the development of the disease or, more specifically, how the etiological agent(s) acts to produce the clinical and pathological changes associated with that disease.
predisposing factor	Something that makes an individual more susceptible to the disease, such as a genetic predisposition.
prevalence	The total number of cases of a disease in a population or the proportion of a population with the disease at a specified time.
risk factor	Condition, such as a lifestyle or diet, that is associated with acquiring or developing the disease but does not directly cause the disease (i.e., an underlying cause of the disease).
senescence	Biological aging characterized by a gradual deterioration of function in living organisms or cells.

REVIEW QUESTIONS

1. What is the difference between a sign and a symptom and how are they used to diagnose diseases?

2. Define etiological agents, risk factors, and predisposing factors and describe their roles in disease origin and progression.

3. What are the various types of disease based on etiology or pathogenesis? Briefly describe each.

4. Describe the role of telomeres in cellular senescence and what role cellular senescence may play in biological aging.

5. What is a biomarker and what advantages do biomarkers offer in the diagnosis of disease?

ADDITIONAL QUESTIONS

In addition to the Review Questions provided above, there is a range of free online questions designed to enable students to further test their understanding of the chapter material. In order to access these interactive questions, please visit the book's website:
https://routledge.com/cw/wiser

Genetic Diseases

11

All diseases have a genetic component and are influenced by a person's genetic makeup. The etiologies of most diseases is rather complex and involves both environmental and genetic components. The contribution of the genetic component can range from the minor to the major etiological determinant of disease (FIGURE 11.1). Thousands of diseases that are due to a single defective gene have been described and, in such cases, the genetic defect is both necessary and sufficient to cause the disease. Most of these individual genetic diseases are quite rare. However, collectively, genetic diseases cause a substantial proportion of total diseases.

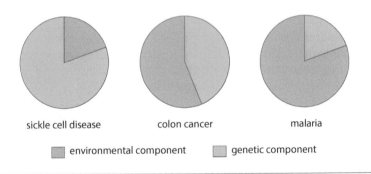

sickle cell disease   colon cancer   malaria

☐ environmental component   ☐ genetic component

FIGURE 11.1 Range of genetic contribution to disease. Disease results from a combination of genetic and environmental determinants and the genetic component can be major (sickle cell disease), minor (malaria), or somewhere in between (colon cancer).

Genetic diseases that are caused by mutations in single genes are inherited in predictable patterns and are often referred to as Mendelian traits in honor of Gregor Mendel who first described these inheritance patterns. For example, many monogenic traits can be described as dominant or recessive. Through studies of families with a genetic disorder it is possible to map and identify the gene responsible for the disease. In general, genetic disorders are due to alleles of genes that result in a non-functional or malfunctioning protein. This lack or malfunctioning of the protein manifests as a disease. In some cases, the defective protein can interfere with normal cellular processes and therefore has a toxic effect on cells.

SIGNPOST

Monogenic and polygenic traits are described in more detail in Chapter 5, pages 95–99.

TYPES OF GENETIC DISEASES

The focus of this chapter is on diseases caused by the inheritance of a single gene, or monogenic diseases. However, some diseases may be influenced by multiple genes with each of the genes contributing in some fashion to the disease state. Such diseases are called polygenic, and a range of phenotypes is expected. Polygenic diseases tend to run in families and this is often referred to as a genetic predisposition (TABLE 11.1). Genetic predisposition can also be viewed as having an increased risk of developing a particular disease. Furthermore, such diseases often have environmental determinants. Cancer, type 2 diabetes, and cardiovascular disease are examples of diseases in which heredity and environment contribute approximately equally to disease development and progression.

Table 11.1	Types of Genetic Disease	
Type	**Defect**	**Inheritance**
Monogenic	Mutation in a single gene	Exhibits distinct inheritance patterns in which the gene is sufficient and necessary to cause the disease
Polygenic	Mutations in multiple genes	Exhibits a familial genetic predisposition that can be influenced by environmental determinants
Cytogenetic	Chromosomal abnormalities	Is not inherited, but due to errors that occur during meiosis or fertilization

Diseases due to chromosomal abnormalities are rare and not inherited

Other diseases that are often included in genetic diseases are diseases due to chromosomal abnormalities such as extra or missing chromosomes (see Table 11.1). Sections of chromosomes can also be deleted or duplicated. Chromosomal abnormalities also include major translocations between two chromosomes in which a segment of one chromosome becomes joined with a different chromosome. Because many genes are affected by chromosomal abnormalities the consequences tend to be rather severe and most result in spontaneous abortion. There are some examples of chromosomal abnormalities that are found with some frequency (TABLE 11.2). Down's syndrome, or trisomy 21, in which individuals have three copies of chromosome 21 instead of the normal two copies, is one of the more common and better known.

Table 11.2	Examples of Chromosomal Abnormalities		
Syndrome	**Abnormality**	**Incidence**	**Main features**
Down's	Trisomy 21	1 in 1000 births	Physical and intellectual disabilities
Edward's	Trisomy 18	1 in 5000 births	Heart defects and survival beyond 1 year <10%
Turner's	Monosomy X	1 in 5000 births	Short-statured female, delayed or absent puberty, infertile
Klinefelter's	XXY	1 in 1000 births	Small testicles, reduced fertility, weaker muscles, increased breast tissue

Cytogenetic diseases are not passed through the generations and do not exhibit discernible inheritance patterns since the defects are not in specific genes. Instead, the defects are due to errors in oogenesis, spermatogenesis, or the fertilization process. Many of these cytogenetic diseases occur more frequently in mothers over the age of 45. Since diseases due to chromosomal abnormalities are diagnosed through karyotyping (see Box 5.1) it is possible to screen for chromosomal abnormalities in utero. Detection of a chromosomal abnormality in the fetus may lead to a decision to end the pregnancy.

INHERITANCE PATTERNS OF MONOGENIC DISEASES

Most monogenic diseases are classified as autosomal recessive, autosomal dominant, or sex linked. Sex linked refers to diseases due to genes found on either the X or the Y chromosomes. The X and Y chromosomes play a role in sex determination and individuals with two X chromosomes are female and individuals with one X chromosome and one Y chromosome are male. The other 22 human chromosomes are called autosomes. The human genome is diploid and normal individuals have 46 total chromosomes and two copies of most genes. An exception is in males who have only one copy of the genes found on the X and Y chromosomes.

Genes exist in different forms called alleles. This means that, for any gene, both copies can be the same, called homozygous, or the two copies can be different, called heterozygous (see Figure 5.6). In the case of monogenic disorders, a particular allele is responsible for the disease. In general, alleles that cause disease are referred to as defective genes and are due to mutations in the nucleotide sequence of that gene. This mutated gene can be passed down through the generations. Whether the mutation is dominant or recessive or associated with autosomes or sex chromosomes determines the inheritance pattern of the disease within families.

Autosomal recessive disorders appear sporadically in family genealogies

Recessive alleles are masked by dominant alleles and therefore heterozygous individuals have the phenotype of the dominant allele. Many genetic diseases are due to recessive alleles, which means only individuals who are homozygous for the recessive alleles exhibit the disease. Heterozygous individuals do not manifest the disease or do not express severe manifestations of the disease. Such heterozygous individuals are often referred to as carriers. Even though the carriers do not manifest the disease, it is possible for their children to have the disease and it would be expected that approximately 25% of their children would have the disease in cases in which both parents are heterozygous (FIGURE 11.2).

In addition to the possibility of unaffected carriers having affected children, an affected parent may not always have affected children. For example, if one parent is affected (homozygous recessive) and the other parent is homozygous dominant, then none of the children will be affected and all the children will be carriers (TABLE 11.3). If one parent is affected and the other parent is a carrier only half the children will be affected. In addition, both parents must have at least one allele (i.e., affected or a carrier) to have affected children. Therefore, recessive disorders often are not manifested in every generation in family genealogy and can skip generations (FIGURE 11.3). The occurrence of recessive disorders depends on the frequency of the recessive allele in the population and the likelihood of two unaffected carriers having children. A shared common ancestry increases the likelihood of two carriers procreating. Therefore, it is not uncommon for a particular recessive disorder to be found more frequently in a particular ethnic or racial group.

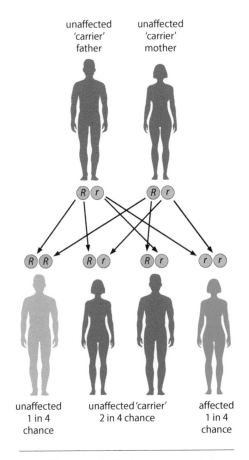

FIGURE 11.2 Inheritance of recessive traits. Parents who are heterozygous for a dominant-recessive disease are unaffected by the recessive allele (*r*). However, they can be carriers of the disease and one out of four children is expected to be homozygous for the recessive trait.

Table 11.3 Inheritance of Recessive Disorders

Parents	Children
Carrier + carrier	50% carriers, 25% unaffected, 25% affected
Unaffected + affected	100% carriers
Unaffected + carrier	50% unaffected, 50% carriers
Affected + carrier	50% affected, 50% carriers

Affected individuals exhibit the disorder and are homozygous for the recessive defective allele. Unaffected individuals are homozygous for the dominant normal allele. Carriers are heterozygous and do not exhibit the manifestations of the disorder.

FIGURE 11.3 Family tree demonstrating inheritance of recessive traits. Recessive traits do not appear in every generation since affected individuals need to be homozygous for the recessive allele. The frequency of recessive traits depends on the frequency of heterozygous carriers in the population.

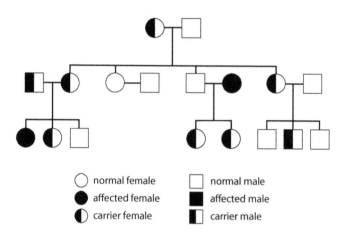

A single defective allele is sufficient in disorders due to dominant alleles

In contrast to recessive disorders, disorders due to dominant alleles do not skip generations and usually manifest in every generation (FIGURE 11.4). In the case in which an affected individual is homozygous for the defective allele, all the children are affected. Half of the children are affected in the case in which the affected individual is heterozygous. Many dominant disorders manifest later in life and individuals are not aware of the disorder during reproductive years. If this were not the case, dominant disorders that affected reproduction would be extremely rare and due primarily to spontaneous mutations.

FIGURE 11.4 Typical family tree of an autosomal dominant trait. Individuals who are heterozygous for a dominant trait are affected (filled circles and squares). Disorders due to dominant alleles tend to be seen in every generation.

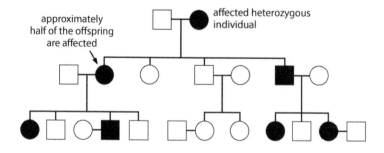

In addition, dominant disorders often have a reduced penetrance. Penetrance is the percentage of individuals with a particular allele of a gene that exhibits the phenotype associated with that allele. A reduced penetrance means that not all individuals with the defective allele have the disorder. For example, if an allele has 50% penetrance, then only half of the individuals with that allele exhibit the trait. Environmental factors, other genes, or epigenetic regulation can affect penetrance.

Defective alleles on the X and Y chromosomes exhibit sex-specific inheritance patterns

Traits due to genes on the X and Y chromosomes show some differences in inheritance between the sexes (TABLE 11.4). The most obvious is the Y chromosome, which is only passed from fathers to sons. In addition, dominant and recessive are somewhat subject to debate as there is normally only one copy of the Y chromosome. Therefore, a defective allele generally manifests as a disorder. Similarly, males only have one copy of the X chromosome. Therefore, defective alleles on the X chromosome in males result in the disorder, regardless of whether it is dominant or recessive. Females have two copies of the X chromosome and therefore can be heterozygous carriers of recessive traits. For this reason, recessive traits on the X chromosome manifest more frequently in males than females.

Table 11.4	Sex-linked Inheritance
X-linked recessive	• Males are affected more often than females • Heterozygous carriers only possible in females • Affected sons can be born to unaffected mothers (i.e., carriers) • Never passed from father to son • Can skip generations
X-linked dominant	• Both males and females are affected approximately equally • All daughters of affected fathers are affected • Affected mothers have both affected sons and daughters • Does not skip generations
Y-linked	• Females are never affected • All sons of affected fathers are affected • Does not skip generations

MOLECULAR BASIS OF INHERITED GENETIC DISORDERS

Genetic disorders are due to alterations in the DNA sequence of the affected individual. These alterations are copied during meiosis and become a part of the DNA of the gametes. Fertilization involving gametes with defective DNA results in the defective genes becoming part of the genome of the progeny. Therefore, all the cells of an affected individual contain defective genes. However, all the cells are not necessarily affected by the defective genes. Only those cells and tissues that express the defective genes exhibit manifestations of the disorder. Since the defective genes are also found in the germline of the affected individual, and the germline serves as the progenitor cells of the gametes, the defective DNA sequence can be passed down to subsequent generations.

Inherited diseases originate from mutations in the germline DNA

Errors can be introduced during the synthesis of DNA that result in a change in the DNA sequence. In addition, some chemicals, called mutagens, or radiation can damage DNA and increase the likelihood of an error occurring. These errors can occur during either mitosis or meiosis. However, only errors that occur during meiosis are inheritable (TABLE 11.5). For inherited diseases the original mutation that represents the origin of the disease occurs in the germline of an individual and this results in sperm or eggs carrying a defective gene. Thereafter, the defective genes can be passed to all subsequent generations. Mutations can also occur in somatic cells, and such mutations are important in the etiology of cancer (Chapter 16). However, somatic mutations are not passed down to subsequent generations and therefore are not inherited disorders.

Table 11.5 Inherited vs. Somatic Genetic Disease	
Inherited genetic disease	Caused by abnormal gene(s) passed down from one generation to the next due to mutations in the germline
Somatic genetic disease	Caused by the appearance of an abnormal form of a gene in one part of the body (e.g., cancer)

The type of DNA-sequence alteration and its location determine the potential biological effects

Variations in the DNA sequence are classified as single nucleotide polymorphisms (SNPs), insertion and deletion events, and copy number variants (FIGURE 11.5). A SNP (pronounced 'snip') is a substitution of one nucleotide for another. In cases in which the change in the nucleotide has a detrimental effect, the polymorphism is often called a **point mutation**. Another type of sequence alteration is the gain or loss of nucleotides, which are called indels (short for insertion or deletion). A specific type of indel is a copy number variant associated with tandem repeats in the nucleotide sequence. Blocks of tandem repeats are found throughout the genome and consist of several copies of a specific sequence (e.g., CAG) repeated multiple times in a row. The number of tandem copies making up the repeated DNA block often varies between individuals.

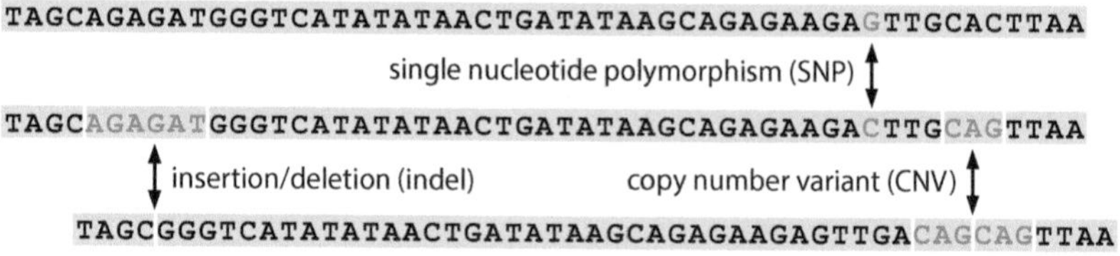

FIGURE 11.5 Types of variations in DNA sequence. A common variation in the DNA sequence is the substitution of one nucleotide for another (e.g., a SNP of either C or G). Nucleotides can be inserted or deleted and collectively these events are referred to as indels (e.g., the insertion or deletion of AGAGAT). Differences in the number of tandem repeats of a sequence are called copy number variants (e.g., variation in the number of copies of CAG).

TAGCAGAGATGGGTCATATATAACTGATATAAGCAGAGAAGAGTTGCACTTAA

single nucleotide polymorphism (SNP)

TAGCAGAGATGGGTCATATATAACTGATATAAGCAGAGAAGACTTGCAGTTAA

insertion/deletion (indel)   copy number variant (CNV)

TAGCGGGTCATATATAACTGATATAAGCAGAGAAGAGTTGACAGCAGTTAA

The effect of a DNA-sequence variation depends largely on the location of the variation. Variations in the protein-coding regions of DNA are more likely to have a detrimental effect than variations in non-coding regions of DNA. An exception would be variations affecting promoter regions or other regulatory regions of DNA that control gene transcription. An increase or decrease in the transcription of specific genes may cause disease, depending on the function and importance of the genes affected. For example, the overexpression of genes that promote cell division may contribute to the development of cancer (Chapter 16). Similarly, underexpression of specific genes could result in a loss of cellular functions.

Frameshift mutations or the introduction of stop codons generally has major impacts on protein structure and function

Inherited disorders result when DNA changes cause a complete loss of a protein or a protein loses significant functionality. Indels in non-multiples of three can be quite detrimental to protein structure and function. Such mutations are called **frameshift mutations**, since a shift in the reading

SIGNPOST

Regulation of transcription is described in Chapter 4, pages 78–81.

frame of the gene occurs. This is due to the triplet genetic code (see Figure 4.15) in which three sequential nucleotides encode for a single amino acid. Therefore, starting at the point of the indel, the amino acid sequence encoded by the defective gene is substantially different from the amino acid sequence encoded by the normal gene (FIGURE 11.6). This almost always results in a non-functional protein unless the frameshift occurred near the C terminus of the protein and therefore may not have a major impact on the overall protein structure. Indels can also introduce a premature stop codon resulting in a truncated protein. Unless this premature stop codon occurs near the C terminus of the protein, a large portion of the protein sequence will be missing and almost certainly will have a major impact on the structure and function of that protein.

AUG	GAA	GAA	UUG	AAU	UUU	UUU	AGA	GGG	GAU	ACA	AUU	UUA	UAA
M	E	E	L	N	F	F	R	G	D	T	I	L	*

AUG	GAA	GAA	UUG	AAU	UUU	UUU	UAG	AGG	GGA	UAC	AAU	UUU	AUA
M	E	E	L	N	F	F	*						

AUG	GAA	GAA	UUG	AAU	UUU	UUA	GAG	GGG	AUA	CAA	UUU	UAU	AA
M	E	E	L	N	F	L	E	G	I	Q	F	Y	

FIGURE 11.6 **Frameshift mutations.** Shown are hypothetical mRNA sequences and the corresponding protein sequence translations. Insertion of a uridine (U) residue into the sequence as represented by the underlined UU in the second row results in +1 frameshift that introduces a stop codon resulting in a severely truncated protein. Removal of the uridine underlined (U in the top row) results in a –1 frameshift (third row) and a protein sequence that is quite different from the original protein sequence (red).

SNPs that impact protein structure and function can cause disease

Generally, SNPs do not have a major effect on a protein's structure or function. In fact, due to the redundancy of the genetic code, some SNPs do not even change the encoded amino acid. In addition, changing a single amino acid in the protein sequence usually does not affect protein structure or function. Such a change in the amino acid sequence of a protein simply represents a polymorphism that is of no or little consequence. Even if the SNP results in a slight difference in protein function, this slight difference in function may not result in clinical manifestations.

However, there are examples of situations in which changing a single amino acid in a protein sequence results in a major loss of protein function. For example, changing an amino acid that is part of the active site of an enzyme can lead to that enzyme becoming non-functional. Similarly, changing an amino acid that plays a critical role in the folding or three-dimensional structure of the protein may result in a misfolded protein that is non-functional. The effect of changing a single amino acid in a protein sequence depends on the overall importance of that specific amino acid to the function and structure of that protein. A change in the protein sequence that results in a substantial loss of that protein's function or a major change in that protein's structure generally results in clinical manifestations.

A single mutation in hemoglobin has severe pathophysiological consequences

Hemoglobin is composed of two α-globin and two β-globin polypeptides (see Figure 3.8). Sickle cell disease is due to a single-point mutation in the β-globin gene that changes a glutamate residue at position six in the amino acid sequence to a valine residue. This change is designated as E6V in which the first letter is the single-letter code for the normal amino acid, the number is the position of the amino acid from the N-terminal amino acid, and the second letter is the single-letter code of the new amino acid. This E6V mutation does not affect the ability of hemoglobin to function in the exchange of oxygen and carbon dioxide, but rather, results in a change in the interactions of the polypeptides making up hemoglobin. The more hydrophobic valine in the β-globin polypeptide causes the hemoglobin molecules to clump together in long strands. This polymerization of the hemoglobin into long strands causes the erythrocytes to deform and take a more elongated shape. This change in erythrocyte shape results in the clinical manifestations associated with sickle cell disease (BOX 11.1).

BOX 11.1

Sickle Cell Disease

Sickle cell disease, also called sickle cell anemia, is a disorder of the blood due to a mutation in hemoglobin. Hemoglobin is the protein that carries oxygen from the lungs to the tissues and carbon dioxide from the tissues to the lungs. The defective hemoglobin forms long strands instead of the normal globular shape and these strands cause the erythrocyte to deform and take on a sickle shape, hence the name of the disease. These deformed erythrocytes cause circulatory problems and are removed by the spleen resulting in anemia (**BOX 11.1 FIGURE 1**). People with sickle cell disease are also subject to acute manifestations of the disease called a sickle cell crisis. The sickle cell crisis is characterized by extreme pain and can be triggered by stress, dehydration, or high altitude. The disease is particularly prevalent among people with African ancestry, which may be due to a balanced polymorphism associated with protection from malaria (see Box 5.4).

Sickle cell disease is considered recessive since a person needs two copies of the defective hemoglobin to have severe clinical manifestations of the disease. People with sickle cell disease generally have a shorter life span. Survival can be increased through proper management of the disease. People who are heterozygous with one defective copy and one normal copy of the gene are carriers and are referred to as having sickle cell trait. Since at least half of the hemoglobin produced by carriers is normal, the polymerization into the long

strands occurs less frequently and the disease is milder. Severe symptoms in carriers are infrequent and are typically only associated with situations such as oxygen deprivation at high altitudes, severe dehydration, or extreme physical exertion.

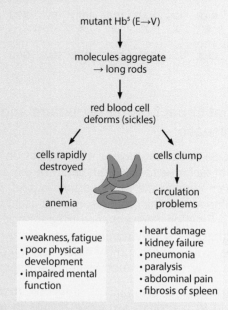

BOX 11.1 FIGURE 1 Sickle cell disease pathophysiology. The glutamate (E) to valine (V) conversion in hemoglobin causes the polypeptides to aggregate into long rods. This change, in turn, causes the cells to deform and they lose their normal biconcave shape. The deformed cells are removed by the spleen, resulting in anemia and the clinical manifestation of anemia (box on the left). In addition, the deformed cells tend to clump and obstruct the capillaries, resulting in manifestations of circulatory problems (box on the right).

DEFECTS IN ENZYMES OR TRANSPORTERS

Enzymes are proteins that catalyze chemical reactions without being consumed by the reaction. A defect in a gene sequence that renders a specific enzyme non-functional can result in the accumulation of the substrate for that enzyme. This accumulation could have toxic effects on cells. In addition, the product of the reaction is not available for subsequent reactions in metabolic pathways and this could affect important physiological functions. Disruptions in metabolism can cause a disease that ranges from mild to severe and life threatening. Clinical manifestations depend on the specific enzyme that is affected and how that enzyme deficiency affects cellular physiology in the cells, tissues, or organs (TABLE 11.6).

SIGNPOST

Enzymes are discussed in detail in Chapter 3, pages 57–58.

Table 11.6 Examples of Diseases due to Specific Enzyme Deficiencies

Disease	Deficient enzyme	Biochemical abnormality	Pathophysiology
SCID-ADA	Adenosine deaminase	Accumulation of deoxyadenosine inhibits DNA replication	Abnormal lymphocyte development leading to severe immune deficiencies
Tay–Sachs	β-Hexosaminidase A	Accumulation of GM2-ganglioside in lysosomes of neurons	Degeneration of neurons leading to mental and physical disabilities
PKU	Phenylalanine hydroxylase	High levels of phenylalanine in the blood and low levels of tyrosine	Blocks uptake of essential amino acids and hinders normal brain development

PKU, phenylketonuria; SCID-ADA, severe combined immunodeficiency disorder due to adenosine deaminase deficiency.

Enzymes are not consumed during the reaction and continue to convert the substrate to a product as long as substrate is present. Therefore, the amount of a specific enzyme within a cell may not be particularly crucial to keep up with the metabolic needs of the cell. The same is true for membrane transporters. For example, cystic fibrosis is due to a defect in an ion transporter (BOX 11.2). Since the number of enzyme or transporter proteins in a cell may not affect the overall cellular physiology, genetic disorders due to defects in enzymes or transporters are usually recessive. This is because one good copy of the gene produces sufficient enzyme to carry out the cellular function of that gene, even though only half as much of the enzyme or transporter may be present in the cells. It is only when such a gene is completely lacking or completely non-functional – as is the case in people who are homozygous for the defective alleles – that the disorder manifests itself. This phenomenon is called **haploid sufficiency**.

Cystic Fibrosis

Cystic fibrosis is a relatively common genetic disease that occurs in approximately one in 3000 births. It is most frequent in people of northern European ancestry. The disease is autosomal recessive and heterozygotes are completely normal. There is no cure for cystic fibrosis, but it can be managed. Historically, children born with cystic fibrosis rarely survived to adulthood. However, now with proper management it is common for people with cystic fibrosis to survive into adulthood.

The gene was initially identified through genetic mapping studies that localized the gene to a region on chromosome 7. Sequencing the gene indicated that it encodes for an ATP-regulated channel that was named the cystic fibrosis transmembrane conductance regulator (CFTR). The gene is expressed in epithelial cells and controls the movement of water and anions, especially chloride, in and out of cells. Many different mutations have been described in the *CFTR* gene. However, approximately two-thirds of the cystic fibrosis cases are due to a missing phenylalanine at position 508 in the protein sequence (designated as ΔF508). CFTR proteins with this defect do not fold correctly and are degraded in the endoplasmic reticulum, and thus never reach the plasma membrane. Therefore, the epithelial cells of people with two defective copies of the *CFTR* gene have no CFTR protein on their surface. However, people with one good copy of the *CFTR* gene have sufficient protein on the plasma membrane to regulate chloride and water transport, and therefore do not exhibit any clinical manifestations.

The lungs are particularly affected by the lack of CFTR protein. The inability of water and anions to be secreted by the epithelial cells results in thick mucus. The thick mucus can clog the airways resulting in breathing problems. The inability to clear the thick mucus also results in increased levels of lung infections and inflammation. Management of people with cystic fibrosis includes airway clearance and antibiotics to control infections. Other secretory tissues are also affected by thicker than normal secretions. In the gastrointestinal system digestion is affected. Sweat glands are also affected and early on children with cystic fibrosis were described as tasting salty. The high level of salt in sweat is the basis of a diagnostic test for cystic fibrosis.

BOX 11.2

Excess levels of a metabolite may inhibit normal cellular processes

Accumulation of a substrate due to a defective enzyme may affect other metabolic pathways that involve that substrate. For example, adenosine deaminase converts the nucleotide deoxyadenosine to inosine and a defective adenosine deaminase leads to the accumulation of deoxyadenosine. Excess levels of deoxyadenosine can inhibit another enzyme called ribonucleotide reductase since deoxyadenosine is a product of ribonucleotide reductase. Ribonucleotide reductase converts ribonucleotides into deoxyribonucleotides and the inhibition of ribonucleotide reductase results in lower levels of the other deoxyribonucleotides, which are needed for DNA synthesis and cellular replication (FIGURE 11.7).

FIGURE 11.7 A defective adenosine deaminase gene impacts DNA synthesis. A mutation in the adenosine deaminase (ADA) gene leads to an accumulation of deoxyadenosine. Deoxyadenosine is one of the products of ribonucleotide reductase, and therefore excess deoxyadenosine inhibits ribonucleotide reductase, leading to a deficiency in deoxyribonucleotides for the synthesis of DNA.

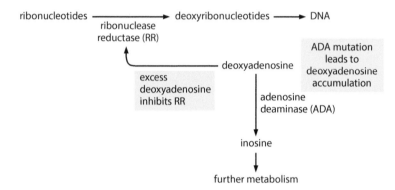

Lymphocytes are cells of the immune system that undergo clonal proliferation (Chapter 9) and therefore exhibit a high level of cellular proliferation, and thus are particularly sensitive to a deficiency in adenosine deaminase. Adenosine deaminase deficiency has an adverse effect on lymphocyte development and leads to a disease called severe combined immunodeficiency (see Table 11.5). People with this disease have an extremely compromised immune system and can die from infections with microbes that are generally quite innocuous (see Box 7.4).

Accumulation of metabolites can have toxic effects on cells and lead to disease

Some mutations that render enzymes non-functional can have a toxic effect on cells. For example, Tay–Sachs disease is due to a defect in β-hexosaminidase A (see Table 11.5). This enzyme is important for the metabolism of a class of lipids called gangliosides. A deficiency in the β-hexosaminidase leads to an accumulation of GM2-ganglioside in the lysosomes of neurons. Tay–Sachs disease is one of several lysosomal storage diseases (see Box 2.3). The accumulation of indigestible material in the neurons interferes with the normal biological processes of endocytosis and exocytosis as the lysosomes accumulate in the body of the cell. This accumulation of lysosomes eventually leads to neuron degeneration and death. Infants with Tay–Sachs disease appear to develop normally for the first 6 months after birth. However, as neurons start to degenerate, a relentless deterioration of mental and physical abilities ensues.

Diet can alleviate the manifestations of some enzyme deficiencies

Since a deficiency in a specific enzyme often leads to the accumulation of a substrate it may be possible to alleviate disease manifestations through diet. One such example is the disease phenylketonuria (PKU) which is due to a mutated phenylalanine hydroxylase gene (see Table 11.5). Phenylalanine hydroxylase catalyzes the chemical reaction that converts the amino acid phenylalanine to the amino acid tyrosine (FIGURE 11.8). Deficiency in phenylalanine hydroxylase leads to an accumulation of phenylalanine and a deficiency in tyrosine. The excess phenylalanine is toxic to the brain and results in intellectual disabilities and other neurological problems. Specifically, the excess phenylalanine hinders brain development by blocking the uptake of other amino acids that are needed by the brain. In addition, tyrosine is an important amino acid for the synthesis of some neurotransmitters and hormones such as dopamine, serotonin, and epinephrine. Deficiencies in neurotransmitters can also contribute to brain dysfunction. Tyrosine is also used in the production of melanin and hypopigmentation is a non-neurological manifestation of PKU.

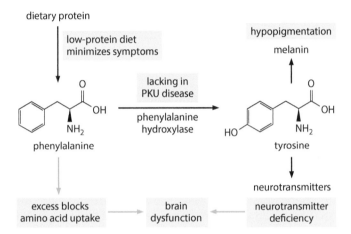

FIGURE 11.8 **Phenylketonuria.** People with phenylketonuria (PKU) lack a functional phenylalanine hydroxylase gene. This enzyme catalyzes the conversion of phenylalanine to tyrosine. The excess phenylalanine due to a defective phenylalanine hydroxylase affects amino acid uptake and protein synthesis, leading to brain dysfunction. Tyrosine is utilized in many biochemical processes including the synthesis of melanin and neurotransmitters. The deficiency in tyrosine results in hypopigmentation and deficiencies in neurotransmitters may affect brain function. PKU can be managed with a low-protein diet since phenylalanine is acquired through dietary protein and protein catabolism.

A diet low in phenylalanine reduces the clinical manifestations of PKU. Furthermore, supplementing the diet with the other required amino acids, in particular tyrosine, also reduces the clinical manifestations. The diet should begin as soon as possible after birth and is necessary throughout the person's life. Individuals who are diagnosed early and maintain this strict diet can have normal health and a normal life span.

DEFECTS IN STRUCTURAL PROTEINS

In contrast with defects in enzymes and transporters, which usually result in recessive diseases, defects in structural proteins often result in dominant diseases (TABLE 11.7). Some structural proteins assemble into complexes that provide mechanical support to a cell or tissue. For example, the internal cytoskeleton of cells is composed of a meshwork of fibrous structures called microfilaments, microtubules, and intermediate filaments. Strong interactions between the proteins that make up these structures provide mechanical stability to the cell and determine its shape. Similarly, some secreted proteins assemble into the extracellular matrix that helps hold cells together in tissues. The amount of protein making up these structural elements is critical to the integrity of the overall structure and, therefore, it is not unexpected that one copy of a defective gene can result in a disorder. This would be analogous to building a brick wall with half as many bricks or half of the bricks are deformed. Obviously, such a wall would have less stability and structural integrity than a wall built with a complete set of non-defective bricks.

The cytoskeleton is described in Chapter 2, pages 37–41.

SIGNPOST

Table 11.7 Examples of Autosomal Dominant Diseases

Disease	Manifestation	Gene(s)	Pathology
Marfan's syndrome	Long limbs and digits, flexible joints	Fibrillin-1 (collagen component)	Defects in elastic tissue (ligaments)
Hereditary spherocytosis	Misshaped and fragile red blood cells	Submembrane cytoskeleton proteins	Weak erythrocyte membrane
Huntington's disease	Neurodegeneration (onset at 35–45 years)	*Huntingtin* gene	Toxic effect of the mutant protein (nerve cells)

Mutations in structural proteins affected morphology and structural integrity

Marfan's syndrome is a disease due to a defect in a collagen protein. Collagen is part of the extracellular matrix and plays a key role in the formation of ligaments and tendons, as well as other connective tissue (see Table 8.2). The fibrillin-1 protein is particularly important for elastic fibers such as ligaments and people with Marfan's syndrome have hyperflexible joints. In addition, patients with Marfan's syndrome are generally tall and thin with long bones, especially the arms, legs, and fingers. The heart and aorta are also affected and defects in the heart valves and aneurysms are seen at higher frequencies in people with Marfan's syndrome. Similarly, a displaced lens due to stretching of the ligaments in the eye is also observed.

Hereditary spherocytosis is a genetic disease characterized by misshaped erythrocytes in which the erythrocytes are smaller and more spherical in shape than the normal biconcave erythrocyte. Mutations in genes that make up the submembrane cytoskeleton or the integral membrane proteins to which the cytoskeleton attaches results in hereditary spherocytosis (FIGURE 11.9). Spectrin, which is the major component of the submembrane cytoskeleton, is the most often affected gene. However, other genes encoding proteins that make up the submembrane cytoskeleton or proteins that are involved in the attachment of the cytoskeleton to the membrane can also result in this disorder. Defects in the submembrane cytoskeleton also make the erythrocytes less flexible and more prone to lyse and, as is the case in sickle cell disease, the spleen removes the misshaped erythrocytes. Therefore, anemia is a common manifestation of hereditary spherocytosis.

FIGURE 11.9 Hereditary spherocytosis. Defects in the proteins of the submembrane cytoskeleton (blue dashes), such as spectrin and ankyrin, or defects in integral membrane proteins that attach to the submembrane cytoskeleton (red), such as band 3 or band 4.2, lead to a loss of structural integrity of the affected erythrocyte. Such erythrocytes lose their normal biconcave shape and become more spherical resulting in a weakening of the membrane. Spherocytes are removed by the spleen and more prone to hemolysis causing anemia.

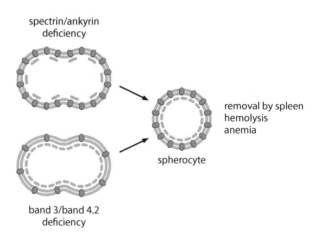

spectrin/ankyrin deficiency

band 3/band 4.2 deficiency

spherocyte

removal by spleen hemolysis anemia

TRINUCLEOTIDE REPEAT EXPANSION DISORDERS

Diseases associated with copy number variants have also been described. One notable disease due to copy number variants is Huntington's disease (BOX 11.3). In the case of Huntington's disease, the mutation is an expansion of repeats of the trinucleotide CAG within the *huntingtin* gene. This results in tracts of sequential glutamine residues of various lengths in the huntingtin protein. These diseases are also called polyQ diseases in reference to Q being the single-letter code for glutamine. The lengths of these polyQ tracts determine the disease state. If the number of sequential glutamine residues is within the normal range, then the phenotype is normal and there is no disease. Expansion of the repeats beyond this normal range results in a mutant protein that is toxic and causes Huntington's disease (FIGURE 11.10). Since the discovery of this phenomenon in Huntington's disease several other neurological disorders due to the expansion of polyQ tracts in proteins have been described and these are often referred to as trinucleotide repeat expansion disorders or polyQ diseases. The pathology in these diseases is likely due to the tendency of the polyQ tracts to aggregate in the neurons and adversely affect neuron function.

Huntington's Disease

Huntington's disease, also called Huntington's chorea, is an inherited neurological disorder characterized by the degeneration of brain cells. The onset of the symptoms usually occurs between the ages of 30 and 50. The earliest symptoms are usually subtle changes in mood, personality, mental abilities, and physical skills. The physical symptoms are generally the first to be manifested and both the physical and mental abilities worsen as the disease progresses. The most characteristic physical symptoms are jerky, random, and uncontrollable movements called chorea. Cognitive abilities are also impaired and progressively worsen as the disease progresses, usually resulting in dementia.

Huntington's disease exhibits an autosomal dominant type of inheritance since one mutant copy of the gene is sufficient to cause the disease. Genetic mapping studies identified the responsible gene on chromosome 4 and the gene was named the *huntingtin* gene. The huntingtin protein may facilitate vesicular transport, synaptic transmission, and neuronal gene transcription, but its exact function is not clear. The mutation causing the disease is an expansion of repeats of the trinucleotide CAG, which encodes glutamine. This results in the huntingtin protein containing tracts of poly-glutamine (polyQ) of various lengths. Generally, no disease manifestations are observed if the polyQ tract is less than 36 and the disease will not be passed to the offspring. As the tandem glutamines increase, the risk of exhibiting disease manifestations and passing on the disease to offspring increases.

The disease is not due to inadequate production or a loss of function of the huntingtin protein, but rather due to the toxicity of the mutant form of the gene. As the polyQ tract expands, the protein is more prone to break down into smaller fragments only containing the polyQ tracts. These polyQ fragments tend to aggregate and, over time, large aggregates called inclusion bodies are formed. These inclusion bodies are similar to the β-amyloid seen in other protein deposition diseases such as Parkinson's disease (see Box 2.3). These protein deposits within the cells interfere with the normal function of the neurons and therefore are toxic.

BOX 11.3

FIGURE 11.10 Pathology of Huntington's disease. The healthy version of the *huntingtin* gene contains a limited number of CAG repeats encoding the poly-glutamine (polyQ) tract (red circles). A replication error results in an expansion of the polyQ tract and produces a toxic version of the huntingtin protein that is responsible for the neurodegeneration associated with Huntington's disease. Over time the polyQ regions of the mutated huntingtin protein form aggregates in the cytoplasm of neurons which leads to neuron dysfunction and death.

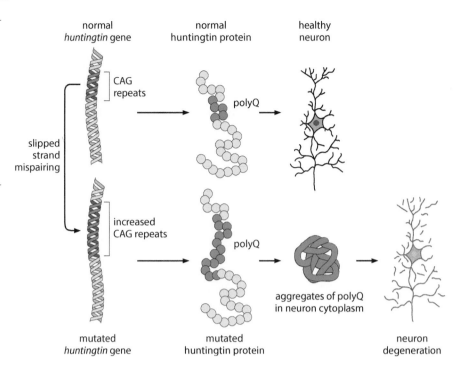

The number of CAG repeats impacts the clinical manifestations of trinucleotide repeat expansion disorders

In the case of Huntington's disease individuals with 36 or fewer CAG repeats in their *huntingtin* genes are not affected and do not exhibit symptoms associated with Huntington's disease (TABLE 11.8). However, individuals with 40 or more repeats in one of the *huntingtin* gene alleles develop Huntington's disease. In other words, the mutant allele is completely penetrant since all individuals with a mutant gene containing 40 or more repeats exhibit the disease. Individuals with a *huntingtin* gene containing an intermediate number of repeats (36–39) exhibit an incomplete penetrance with some individuals developing the disease, whereas others do not. In general, the number of repeats also influences the severity of the manifestations and age of onset. The number of CAG repeats is inversely correlated with age of onset and accounts for about 60% of the variation in age of onset. Similarly, the severity of the disease and risk of passing the trait to offspring are correlated with the number of CAG repeats.

Table 11.8 CAG Repeats and Huntington's Disease Manifestation and Inheritance		
Repeat number	**Disease status**	**Risk to offspring**
≤26	None	None
27–35	None	<50%
36–39	Incomplete penetrance	50%
≥ 40	Complete penetrance	100%

Errors in DNA replication result in an expansion or contraction of tandem repeats

During DNA replication or repair there is a possibility that tandem repeats of DNA sequence expand or contract. Changes in tandem repeat numbers are due to a phenomenon called slipped-strand mispairing. If the template strand and newly synthesized nascent strand temporarily separate during DNA replication it is possible that the two strands misalign during the reformation of the double-stranded molecule (FIGURE 11.11). This creates a bulge in the DNA molecule and, depending on whether the bulge is in the template strand or the nascent strand, there can be a gain or loss of repeats depending on how the error is resolved. The expansion and contraction of repetitive DNA sequences are well known and occur with some frequency. For the trait to be passed down to offspring this slipped-strand mispairing would need to occur during spermatogenesis or oogenesis.

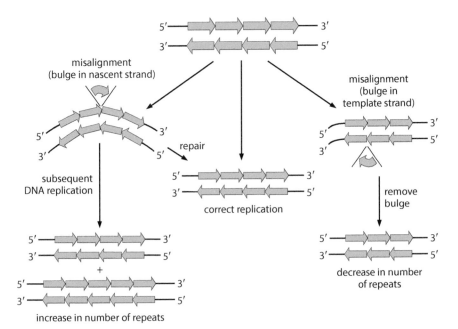

FIGURE 11.11 Slipped-strand mispairing.
During DNA replication the template and the nascent DNA strands may temporarily separate. Due to the repeating structure (block arrows), it is possible that the two DNA strands come back together with a misalignment of the repeats that creates a bulge in either the nascent strand or the template strand. Removal of the bulge in the nascent strand repairs the DNA error. If the bulge is not removed and the DNA is subsequently replicated again, half of the progeny will have the correct number of repeats and half will have an increase in the number of repeats. Removal of the bulge in the template strand decreases the number of repeats. Failure to repair the bulge in the template strands results in half of the progeny having the correct number of repeats and half the progeny having a decrease in the number of repeats (not shown).

Increasing the number of repeats leads to genetic anticipation

Although both expansion and contraction of trinucleotide repeats are possible, there is a greater tendency for the repeats to expand rather than contract. Furthermore, as the repeat region lengthens there is increased instability and a greater risk of further repeat expansion. Interestingly, for reasons that are not clear, there is a tendency for the number of CAG repeats to increase more during spermatogenesis than during oogenesis. The tendency for the repeats to expand, means that a mutated gene can be passed down from parents – particularly fathers – who have a functionally normal gene. For example, individuals with 27–35 CAG repeats in the *huntingtin* gene are phenotypically normal and do not have Huntington's disease. However, such individuals have an increased risk of passing a mutant *huntingtin* gene to their offspring due to the increased chance of repeat expansion. However, this risk is well below the normal 50% risk of passing a mutated gene to offspring in the case of autosomal dominant diseases in heterozygous individuals. Nonetheless, the risk is elevated when compared with individuals with a normal range of CAG repeats.

In addition, individuals with a mutant form of the *huntingtin* gene also may pass on a gene with even more CAG repeats to their offspring. This means the affected children of individuals with Huntington's disease often have more severe manifestations and an earlier age of onset than their parents. This phenomenon of increasing severity of disease manifestations with each successive generation is called **genetic anticipation**. Genetic anticipation is a characteristic of Huntington's disease and other trinucleotide repeat expansion disorders.

DIAGNOSIS AND GENETIC SCREENING

Proteins carry out a wide array of different cellular functions (see Table 3.3) and the manifestation of any specific genetic disease depends on the gene that is mutated. This means that there are no general signs and symptoms associated with genetic diseases, but rather the signs and symptoms are specific for each disease according to the nature of the defective protein. In addition, many genes are only expressed in specific tissues or organs and the manifestations may therefore be organ specific. Disease manifestations may be obvious at birth or they may develop later in life.

Laboratory tests are available for the diagnosis of many genetic diseases

Specific manifestations and a family history of a particular disease can suggest a diagnosis. Laboratory tests are generally needed to confirm the diagnosis. Such laboratory tests are based on the specific gene that is mutated. For example, in the case of cystic fibrosis the mutation is in a chloride transporter and people with cystic fibrosis are characterized by having excessive levels of salt in their sweat. A common diagnostic test for cystic fibrosis is called the sweat test, in which the amount of salt in sweat is determined. Similarly, in many diseases involving defective enzymes the excess levels of a particular metabolite can be used to diagnose the disease.

In some cases, diagnosis can involve tests for the defective protein. For example, the mutation that causes sickle cell anemia results in a change in the overall charge of the hemoglobin protein which can be detected by gel electrophoresis (see Box 1.5). Therefore, it is relatively simple to make an accurate diagnosis of sickle cell anemia by analyzing a small amount of blood (FIGURE 11.12). Furthermore, such a test also identifies individuals who are heterozygous for the sickle cell trait, since they express both the normal and mutant hemoglobins. Advances in biotechnology also allow for the identification of specific mutations at the DNA level. This process enables an extremely precise diagnosis that does not require the expression of a defective protein or the manifestations of disease traits. DNA-based tests also detect recessive alleles in heterozygous individuals.

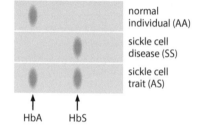

normal individual (AA)

sickle cell disease (SS)

sickle cell trait (AS)

HbA   HbS

FIGURE 11.12 Hemoglobin analysis. Due to a change of a charged amino acid (glutamate) to an uncharged amino acid (valine), normal hemoglobin (HbA) can be distinguished from the mutant hemoglobin (HbS) associated with sickle cell disease by a technique called electrophoresis (Box 1.5). Analysis of blood can determine if a person has normal hemoglobin (AA), sickle cell disease (SS), or the sickle cell trait (AS).

Prenatal diagnosis and genetic screening improve management options and better define risks

Biochemical and molecular tests can also be used to diagnose genetic diseases prenatally. Fetal cells are obtained by amniocentesis, which involves collecting fetal cells from the amniotic fluid and analyzing those cells. Early diagnosis may allow for better disease management after birth. Genetic screening can include detecting recessive alleles of genes in heterozygous individuals. Knowledge about the specific alleles of prospective parents can precisely define the risk of having children with a particular genetic disorder. For example, if both parents have a mutated allele for the same gene, then there is a one in four chance their children would inherit both alleles and have the disorder (see Figure 11.2). Without the screening neither parent may realize that they have a recessive allele of a genetic disorder.

SUMMARY OF KEY CONCEPTS

- Many diseases are due to a defect in a single gene

- Disorders due to defects in single genes generally exhibit autosomal recessive, autosomal dominant, or sex-linked inheritance

- Alteration of DNA associated with genetic diseases includes point mutations, frameshift mutations, the introduction of premature stop codons, and copy number variants

- Disease manifestations are due to DNA mutations that result in defects in the structure or function of the encoded proteins

- Generally, mutations associated with genetic diseases result in a loss of function of the affected protein, or the affected protein has toxic effects on cellular physiology

- Molecular and biochemical tests allow for prenatal testing and genetic screening to diagnose or assess the risks of genetic disorders

KEY TERMS

dominant	An allele or trait that masks other alleles and determines the phenotype in heterozygous individuals.
frameshift mutation	An insertion or deletion of nucleotides in non-multiples of three that cause the wrong codons to be translated subsequent to the point of insertion or deletion.
genetic anticipation	A phenomenon in which the signs and symptoms of some genetic disorders tend to become more severe or appear at an earlier age in subsequent generations.
haploid sufficiency	A situation in which a single functional copy of a gene is enough to maintain the normal physiology and function of an organism or cell.
point mutation	The change in a single nucleotide of the gene sequence that adversely affects the function of the gene or protein.
recessive	An allele or trait that is masked by a dominant allele. Recessive alleles are expressed as the phenotype in homozygous individuals.

REVIEW QUESTIONS

1. Why are defects in genes of enzymes usually recessive and defects in structural proteins usually dominant?

2. Describe the inheritance patterns associated with autosomal dominant, autosomal recessive diseases, X-linked recessive diseases, and Y-linked recessive diseases.

3. What are SNPs, indels, and copy number variants and how might they cause disease?

4. What is the molecular basis of Huntington's disease and the genetic anticipation observed in Huntington's disease?

ADDITIONAL QUESTIONS

In addition to the Review Questions provided above, there is a range of free online questions designed to enable students to further test their understanding of the chapter material. In order to access these interactive questions, please visit the book's website: https://routledge.com/cw/wiser

Nutritional Disorders

12

It has long been recognized that the correct balance between nutrition and exercise is a large component of good health. Nutrients are substances needed by an organism to survive, grow, and reproduce, and nutrition is the science that studies the relationship between diet and good health. Obviously if nutrients are needed for survival, any nutritional deficiency adversely affects health. Nutritional deficiencies are particularly problematic in children and adversely affect growth and development.

Overnutrition, especially an excess of calories in the diet, can also adversely affect health. Overeating can lead to weight gain and eventually to obesity. Obesity is a growing problem worldwide and is a major risk factor for the development of cardiovascular disease and diabetes. The etiology of obesity is complex and involves genetic, behavioral, cultural, and environmental factors. These various factors result in a positive energy balance that gets converted into adipose tissue (fat). Adipose tissue functions as an energy store and is regulated by several hormonal pathways. In addition to serving as an energy store, adipose tissue can be viewed as an endocrine organ and plays a role in chronic inflammation.

NUTRIENTS

UNDERNUTRITION

OBESITY

ADIPOSE TISSUE

REGULATION OF ENERGY BALANCE

NUTRIENTS

Nutrients refer to substances obtained from the environment that are ingested as food. Many nutrients are substances that our bodies cannot produce in sufficient quantities, and therefore must be obtained through ingestion. Nutrients that our bodies cannot produce are often called essential nutrients. Six major classes of nutrients are defined (TABLE 12.1). Some of these nutrients are needed in large quantities and are referred to as **macronutrients**. Included among the macronutrients are carbohydrates, fats, and proteins. **Micronutrients** are substances needed in smaller quantities and include vitamins and minerals. The simplest and most essential nutrient is water. Water allows the transportation of

Table 12.1	Six Classes of Nutrients
Carbohydrates	Sugars and starches; energy yielding (4.1 kcal/g); includes dietary fiber
Lipids	Fats and oils; energy storage (9.3 kcal/g); cell membranes; hormone synthesis
Proteins	Energy yielding (4.3 kcal/g); source of amino acids for protein synthesis
Vitamins	Organic molecules; enzyme co-factors and cell regulation; fat or water soluble
Minerals	Inorganic molecules; play structural and regulatory roles; electrolytes
Water	Most essential. Hydration and circulation

other nutrients throughout the body, as well as the removal of waste. In addition to its role in hydration, water also plays a key role in regulating body temperature.

Macronutrients are needed to generate energy and provide building blocks for synthesis of cellular components

Dietary carbohydrates, fats, and proteins are broken down into smaller molecules that are absorbed and transported throughout the body (FIGURE 12.1). Complex carbohydrates are converted primarily into glucose, fats into fatty acids, and proteins into amino acids. Glucose, fatty acids, and amino acids are energy yielding since their metabolism can produce cellular energy. Energy released from the catabolism of these substrates is captured and used to synthesize ATP. Amino acids are also needed for the synthesis of proteins, and similarly, fatty acids are needed for the synthesis of lipids that make up membranes and some hormones. Excess glucose and fatty acids can be converted into glycogen and triglycerides, respectively. Glycogen and triglycerides are fuel storage molecules that can be utilized later in the event of future deficiencies in energy-yielding macronutrients.

SIGNPOST

The production of ATP through metabolism is described in Chapter 3, pages 59–61.

FIGURE 12.1 Metabolism of macronutrients. Ingested protein, carbohydrates, and fats are broken down into substrates that can be absorbed and transported throughout the body. These substrates are taken up by cells and used as metabolic fuel or as building blocks in the synthesis of macromolecules.

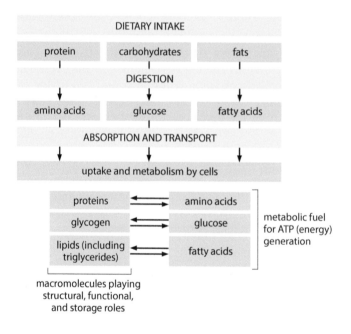

Calories are a measure of food energy

Energy-yielding macronutrients supply structural components for the biosynthetic needs of cells and tissues as well as energy. Food energy refers to the amount of energy that is potentially released through metabolism and is measured in calories. Calories can be a confusing term since the definition of calories in chemistry and physics is the amount of energy needed to raise one gram of water one degree Celsius. In addition, a food calorie is equal to one kilocalorie (1 kcal), or 1000 calories. Furthermore, the chemical definition of calorie represents the total amount of energy that is stored in a compound. Not all that potential energy is captured in the process of digestion, absorption, and metabolism. Therefore, food calories are less than the total chemical calories and are determined from standardized chemical tests or an analysis of the recipe to estimate the digestible constituents (protein, carbohydrate, and fat). Standardized tables are used to estimate the number of calories found in food. On average carbohydrates and protein provide approximately four calories per gram and fats provide approximately nine calories per gram.

A wide range of physiological functions require minerals

Minerals are elemental atoms or ions of those elements. They are found in both animal and plant food sources and the amount of specific minerals in plants can be influenced by the local concentration of those minerals in the soil. Depending on the amount needed by the body, minerals are classified as major or trace. Major minerals include sodium, potassium, chloride, calcium, phosphorous, and magnesium. These major minerals have roles in maintaining fluid balance, bone formation, enzymatic reactions, conduction of nerve impulses, and regulating blood pressure (TABLE 12.2). Deficiencies or excesses in these minerals can adversely affect health. Even though they are needed in small amounts, trace minerals are just as important for health as major minerals. Trace minerals include iron, zinc, iodine, copper, cobalt, selenium, fluoride, manganese, chromium, and molybdenum. In general, trace minerals function as co-factors for a wide range of enzymes and proteins.

Table 12.2 Major Minerals and Their Roles

Mineral	Major roles
Sodium	Electrolyte, blood pressure
Potassium	Electrolyte, muscle contraction, nerve impulses
Chloride	Electrolyte, nerve impulses
Calcium	Bone formation, muscle contraction, nerve impulses
Phosphorous	Bone formation, DNA synthesis
Magnesium	Enzyme reactions, DNA synthesis, blood clotting, ATP production

Vitamins are needed in small amounts to carry out a variety of specific functions

Vitamins are organic molecules that play a wide variety of roles in physiology and are generally classified as either fat soluble or water soluble (TABLE 12.3). The solubility properties of vitamins determine how they are absorbed and stored in the body. Fat-soluble vitamins are

Table 12.3 Function of Vitamins

Vitamin	Major functions
Fat-soluble vitamins	
A	Maturation of epithelial layers; synthesis of vision pigments
D	Regulating serum calcium levels (functions more like a hormone)
E	Antioxidant, especially in fat tissue
K	Blood coagulation; bone formation
Water-soluble vitamins	
C	Collagen synthesis; antioxidant
Thiamine (B_1)	Co-factor involved in energy metabolism and neural function
Riboflavin (B_2)	Co-factor involved in energy metabolism
Niacin (B_3)	Co-factor in oxidation/reduction reactions
Pyridoxine (B_6)	Co-factor in amino acid metabolism
Folate (B_9)	DNA synthesis
Cobalamin (B_{12})	Red blood cell formation and metabolism

absorbed and transported with the assistance of lipid molecules. Fat-soluble vitamins can also be stored and therefore reserves can be built up. However, an excess of some fat-soluble vitamins can have toxic effects. Water-soluble vitamins are easily absorbed and transported throughout the body. In contrast with fat-soluble vitamins, they are eliminated quickly and therefore need to be consumed more frequently. Toxicity due to excess water-soluble vitamins is rare.

Some non-essential nutrients also provide healthful benefits

Many foods are deemed healthy beyond their caloric value and their source of essential minerals and vitamins. For example, many plants contain bioactive food components that have antioxidant or anti-inflammatory activities or provide dietary fiber. Although these substances are not absolutely required in our diet, they do have health benefits such as reducing the risk of cancer or cardiovascular disease. Dietary fiber refers to the indigestible complex carbohydrates in plant foods that serve important functions in good health. Dietary fiber adds bulk to stools and helps to maintain regularity. In addition, fiber contributes to the sense of satiety and fullness without providing calories and thus decreases caloric intake. Dietary fiber also slows the absorption of simple sugars and cholesterol.

UNDERNUTRITION

Undernutrition, previously called malnutrition, is a deficiency in nutrients. This can be due to a general deficiency in energy-yielding nutrients and vitamins and minerals, or deficiencies in specific nutrients. The most common cause of undernutrition is poverty. However, undernutrition can also be due to ignorance or self-induced starvation. Furthermore, some diseases can lead to malabsorption. For example, diarrheal diseases (Chapter 21) also contribute to undernutrition. Undernutrition is a major problem in resource-poor countries and food insufficiency is a chronic problem. Food insecurity is often used to describe the situation in which the supply of adequate, safe, and acceptable food is limited or uncertain.

Children, including adolescents, are particularly susceptible to nutritional deficiencies since their growing bodies require substantial energy and the basic molecular building blocks supplied by protein and fats, as well as micronutrients. In general, undernutrition in children is characterized by stunted growth, slowed mental processes, and delayed puberty. Pregnancy is another situation involving an increased demand for nutrients and undernutrition during pregnancy can have adverse effects on the fetus that continue with the newborn and into infancy. In addition, more nutrients are temporarily needed if the body is undergoing repair processes following traumatic injuries, infections, or surgery.

Chronic protein–energy malnutrition has severe consequences

Undernutrition results in a deficiency in calories supplied by protein, carbohydrates, and fats. In addition, protein is needed for the growth and maintenance of the body. Diets deficient in protein are also generally deficient in carbohydrates and fats and, likewise, diets deficient in carbohydrates and fats are usually deficient in protein. Such deficiencies are referred to as **protein–energy malnutrition** (PEM). Severe protein–energy undernutrition that is associated with extreme poverty is divided into two forms: marasmus and kwashiorkor.

Marasmus is due to extreme deficiencies in both calories and protein and is characterized by wasting. In fact, the word marasmus is derived from a Greek word meaning withering. A primary manifestation of marasmus is skeletal thinness, especially in the limbs (FIGURE 12.2). This is due to a loss of subcutaneous and muscle-associated fat. Initially, the liver and other visceral organs are spared from the loss of fat tissue and a bulging abdomen is often present despite the skeletal thinness. Eventually the body starts to break down protein to supply energy needs and the wasting condition is exacerbated. Treatment with a high-calorie protein-rich diet can reverse the process. However, marasmus can progress to the point of no return and the body is no longer able to synthesize protein.

FIGURE 12.2 Marasmus. Picture of child exhibiting extreme skeletal thinness typically associated with marasmus. [Courtesy of CDC/Dr. Edward Brink.]

Extreme protein deficiency with some calories derived from carbohydrates can lead to **kwashiorkor**. This is often associated with weaning and the word kwashiorkor is derived from the Ghana language and roughly translates as the disease of weaning. Breast milk contains proteins and amino acids that are needed for growth and development. Replacing breast milk with a diet deficient in protein, but that has some calories derived primarily from carbohydrates, affects the ability of the liver to synthesize blood proteins such as albumin. Low plasma levels of albumin and other serum proteins result in the retention of water in the tissues. Thus, there is swelling, called edema, in the tissues. This edema gives patients a puffy appearance that can mask the underlying skeletal thinness. Often children with kwashiorkor have puffy faces and swelling primarily in the feet and ankles (FIGURE 12.3), in addition to skeletal thinness and a bulging abdomen. Kwashiorkor is treated by the gradual re-introduction of protein into the diet.

Micronutrient deficiencies tend to have specific physiological manifestations

Food insecurity can also lead to deficiencies in vitamins and minerals. Notable vitamin deficiencies are vitamin A, thiamine, riboflavin, niacin,

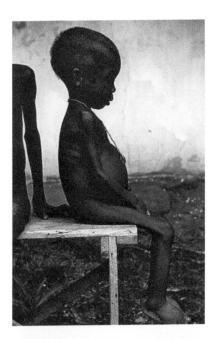

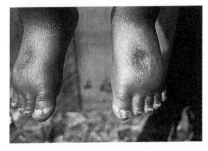

FIGURE 12.3 Kwashiorkor. Picture of child exhibiting skeletal thinness and protruding abdomen typically associated with kwashiorkor. Kwashiorkor is also associated with edema throughout the body. Lower panel shows pitting edema in the feet of the same child as demonstrated by the remnant of depression following the application of pressure. [Courtesy of CRC/Dr. Lyle Conrad.]

and folic acid. Historically deficiencies in vitamins C and D were somewhat common. Now vitamin C deficiency is rare, except in diets lacking citrus fruits and vegetables for prolonged periods. Fortification of milk with vitamin D has essentially eliminated vitamin D deficiency and the associated disease called rickets. Similarly, iodized salt has eliminated iodine deficiencies in inland and mountainous areas where the soil content of iodine is low. In general, mineral deficiencies are quite rare. One potential exception is iron deficiency. However, iron deficiency is usually associated with an increased demand for iron such as during pregnancy, infancy, or after blood loss combined with a dietary deficiency.

Since certain foods may be better sources of a particular micronutrient it is possible to have a micronutrient deficiency in the absence of protein–energy malnutrition. In this case, a specific micronutrient is lacking, and the manifestations are determined by the physiological role of that specific vitamin or mineral. For example, many vitamins function as co-factors of enzymes (see Table 12.3) and, therefore, those specific enzymes and associated metabolic pathways are affected by such a deficiency. Furthermore, a particular tissue or organ may be affected by a particular vitamin or mineral deficiency (TABLE 12.4).

Table 12.4	Effects of Micronutrient Deficiencies	
Nutrient	**Deficiency effect**	**Good sources**
Vitamin A	Impaired immunity; vision problems	Orange vegetables
Folate	Anemia	Green leafy vegetables
Thiamine	Neurological disorders	Whole grains
Niacin	Dermatitis	Enriched grains
Iron	Anemia; impaired immunity	Meat
Zinc	Impaired immunity	Seafood, meat, beans, whole grains

Malnutrition adversely affects immunity

A major consequence of undernutrition is an increased susceptibility to infections and increased severity of infectious diseases. In fact, undernourished children die primarily of infections. Almost all nutrients play a role in optimal immunity (Chapter 9). Protein–energy malnutrition obviously affects immunity since the immune system requires a lot of energy. Immune effector cells are continuously renewed through cellular replication and the rate of cellular replication is elevated during infections. In particular, the function of the thymus is impaired and there is a decrease in T cells in individuals affected by protein–energy malnutrition. Atrophy of the thymus also affects the development of secondary lymphatic organs and tissues.

Deficiencies in several micronutrients also increase the severity of infectious diseases. Furthermore, supplementation of the diet with these micronutrients decreases the severity of infectious diseases. The most notable among these are vitamin A, zinc, and iron. Vitamin A is needed for proper T-cell function as well as the growth and maturation of epithelial tissue, which functions as a barrier to pathogens. Zinc is a co-factor for several enzymes and other proteins that are involved in various physiological functions and metabolism. Iron deficiency is associated with decreased T-cell function and iron supplementation restores T cells to normal levels.

OBESITY

Obesity is a major problem worldwide and particularly in developed countries. This is more than a cosmetic concern in that being overweight is a risk factor for many chronic diseases (FIGURE 12.4). In particular, obesity is a major risk factor for developing diabetes (Chapter 13), cardiovascular disease (Chapter 14), and some types of cancer (Chapter 16). In addition, excess weight causes mechanical stress on the body and leads to osteoarthritis, lower back pain, shortness of breath, and sleep apnea. Low-grade chronic inflammation and hormonal imbalances are also associated with obesity. Obesity is the most notable and obvious example of dietary excess and is due in large part to an excess of energy-yielding macronutrients in the diet. However, excesses in micronutrients, particularly fat-soluble vitamins, can also be toxic. In addition, an excess of dietary sodium can lead to hypertension (Chapter 14).

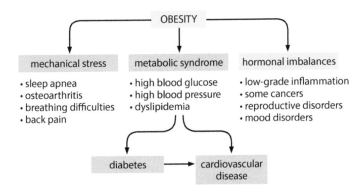

FIGURE 12.4 Pathological consequences of obesity. Obesity causes excess mechanical stress on the body, imbalances in several hormones, and a phenomenon called the metabolic syndrome. Metabolic syndrome and chronic inflammation increase the risk of developing diseases such as type 2 diabetes and cardiovascular disease. Hormonal imbalances and mechanical stress also contribute to disease states.

Excess body fat can be estimated from the body mass index

The body responds to overeating by converting the excess dietary calories into adipose tissue (fat). Accordingly, obesity is defined as having an excess of adipose tissue. However, directly measuring adipose tissue is difficult and expensive. A simple substitute is the ratio of weight to height known as the **body mass index** (BMI). BMI is defined as the weight in kilograms divided by the square of the height in meters. Ideally, the BMI should be between 18.5 and 25 and obesity is defined as a BMI greater than 30 (TABLE 12.5).

Table 12.5	BMI Categories
<18.5	Underweight
18.5–24.9	Normal
25–29.9	Overweight
30–34.9	Obese
>35	Extreme obesity

BMI is not a direct measure of body fat and other factors may need to be considered in determining ideal body weight. For example, BMI does not account for differences in body composition, and athletes may have a high BMI without having an excess of adipose tissue. This is because muscle tissue has a greater mass than adipose tissue. Similarly, loss of bone density in older individuals results in a lower BMI. In addition, not all fat is equal, and abdominal body fat is considered unhealthier than adipose tissue in other parts of the body. Therefore, the waist circumference or

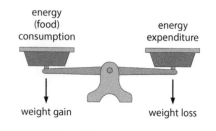

FIGURE 12.5 Weight reflects a balance between energy input and output. Weight is maintained due to a balance between food consumption and energy expenditure.

FIGURE 12.6 Factors influencing obesity. A wide range of biological, environmental, and behavioral factors affect weight gain and weight loss.

the ratio of the waist circumference and hip circumference can be used as simple anthropometric measurements to approximate health risks associated with being overweight.

The etiology of obesity is complex and involves many biological and behavioral factors

Weight gain and weight loss are quite simple. It is a matter of the balance between calories consumed and calories expended (FIGURE 12.5). A positive energy balance between calories consumed and calories expended typically results in weight gain and vice versa. However, the factors that influence food consumption and the subsequent metabolism of food into energy are numerous (FIGURE 12.6). On the biological side, genetics plays a big role in metabolism. Furthermore, certain diseases or medications may affect appetite as well as metabolism. Volition and other psychological factors also play a role in food consumption. Similarly, the types of food eaten have a strong cultural element. And the environment, in terms of the availability of food, influences consumption.

One biological factor that plays a role in weight gain is the **basal metabolic rate** (BMR). The BMR is the amount of energy needed to maintain essential activities and organ function. This is the energy expended beyond physical activity. Up to two-thirds of the energy expended by an individual can be attributed to BMR. The BMR is not a constant value and varies between individuals and changes over time and circumstances within an individual (TABLE 12.6). For example, the BMR decreases with age and that accounts for the weight gain that typically occurs with age. Gender, body composition, health status, and stress levels also influence BMR. The act of eating increases the BMR since energy is needed to digest food. Conversely, fasting lowers the BMR. Thus, decreasing the calories consumed also decreases the calories expended and the amount of weight lost due to simply decreasing calories is not proportional to the reduction in calorie consumption.

Table 12.6 Factors Influencing BMR	
Factor	**Effect**
Gender	BMR higher in males
Body composition	Muscle consumes more energy than fat
Age	BMR declines with age
Health	Disease can either increase or decrease the BMR
Stress	Epinephrine increases BMR
Food intake	Eating increases BMR and fasting decreases BMR
Genetics	BMR varies between individuals

Genetics has a major influence on weight gain and obesity

An individual's weight gain and risk of becoming obese have a strong genetic component. The heritability of obesity is estimated to range from 40% to 70%. This means that about half of obesity can be accounted for by an individual's genes. However, obesity does not exhibit a Mendelian type of inheritance. Rather, genetic predisposition influences how susceptible an individual is to weight gain. For example, some individuals are quite susceptible to weight gain, whereas others are more resistant to weight gain. This means that some individuals have a genetic predisposition to gain weight and become obese.

However, environmental factors often influence the manifestation of genetic traits, and this appears to be especially true in the case of obesity. The availability and free access to high-calorie foods and the low level of physical activity associated with modern lifestyles combined with a genetic predisposition to gain weight are the foundation for the current epidemic of obesity. The genetic predisposition to gain weight is possibly a holdover from the evolutionary past of humans when food was not always readily available and considerable amounts of energy were expended to procure food (BOX 12.1).

The Thrifty Gene Hypothesis

Obesity is common in the human population despite its detriment to health. Furthermore, obesity has a high degree of heritability. This raises questions since generally gene variants that are detrimental to health are normally lost or decrease in frequency during evolution, and therefore are not found at a high frequency in the human population. One possible explanation for this discrepancy is that gene variants that give rise to a genetic predisposition for obesity may have been beneficial in the past. For example, human evolutionary history has been influenced by periods of an abundance of food alternating with periods of food scarcity. The ability to convert excess calories to fat during periods of food abundance and then use the stored fat during periods of food scarcity would be a selective advantage during such a feast-or-famine scenario. In other words, genes that favored weight gain may indeed have been a selective advantage in the past instead of a selective disadvantage as they appear to be now. The idea that genes that promote the efficient conversion of excess calories into fat may have previously been a selective advantage is often called the thrifty gene hypothesis.

However, there are not much data that support the thrifty gene hypothesis. In fact, most available data tend to argue against the hypothesis. Furthermore, no so-called thrifty genes have been identified. Nonetheless, there has been a rapid increase in the incidence and prevalence of obesity and this rapid increase seems to be predominantly associated with a dramatic change in lifestyle during the last century involving a high intake of calorie-rich food and a decrease in physical activity. Obesity is clearly a combination of genetics and environment and the epidemic of obesity may be a product of a genetic predisposition toward energy storage in the context of the modern lifestyle.

BOX 12.1

Attempts have been made to identify genes that contribute to energy balance and weight gain. Genome-wide association studies (GWASs) have revealed many genes that have small effects on obesity. In other words, obesity is a polygenic trait. This is not surprising since metabolism, appetite regulation, and adipose tissue formation are all complex processes involving numerous genes. However, the obesity-related gene variants identified in GWASs explain less than 10% of the variation between individuals. Furthermore, many of the variants are in non-coding regions of the genome suggesting that specific gene expression in

GWASs are described in Chapter 6, pages 124–130.

SIGNPOST

SIGNPOST

Epigenetics is described in Chapter 5, pages 100–101.

particular tissues may contribute to the development of obesity. Similarly, there could be an epigenetic component. Epigenetics could constitute a link between genetic susceptibility and environmental factors.

Obesity is a key aspect of metabolic syndrome

Abnormal glucose metabolism, abnormal plasma lipid levels, and high blood pressure are often observed in overweight and obese individuals. The grouping of these factors combined with abdominal obesity is often referred to as a **metabolic syndrome**. Abdominal obesity as measured by waist circumference is the predominant risk factor for developing metabolic syndrome. The exact definition of metabolic syndrome can vary according to agency or institute but generally involves having three or more of the specific diagnostic criteria (TABLE 12.7). The development of metabolic syndrome is generally associated with a high carbohydrate diet and a sedentary lifestyle. Those individuals with metabolic syndrome have an enhanced risk of developing cardiovascular disease and diabetes.

Table 12.7 Metabolic Syndrome	
Risk factor	**Specific criteria**
Abdominal obesity	• Waist circumference >40 inches (100 cm) in men or >35 inches (88 cm) in women
Abnormal glucose metabolism	• Fasting blood glucose levels >110 mg/dL
Abnormal plasma lipids (dyslipidemia)	• Triglyceride levels >150 mg/dL • HDL ('good') cholesterol <40 mg/dL in men or <50 mg/dL in women
High blood pressure	• >140/90 mmHg
Metabolic syndrome is generally defined as having three or more of the five specific criteria.	

ADIPOSE TISSUE

Adipose tissue is described as loose connective tissue composed primarily of fat cells, called adipocytes. There are two major types of adipose tissue: brown and white. Brown adipose tissue primarily functions to generate heat. The adipocytes of brown adipose tissue contain numerous mitochondria, which are important in the generation of ATP from metabolic fuel. To generate heat, the synthesis of ATP is decoupled from the oxidation of fatty acids and triglycerides and the energy released from this catabolism of metabolic fuel is converted into heat. This process is known as non-shivering thermogenesis.

A major function of white adipose tissue is to store energy in the form of lipids and, in particular, triglycerides. In addition to serving as a major energy reserve, white adipose tissue also provides some mechanical protection to organs and serves as insulation. In humans, white adipose tissue is primarily located beneath the skin (subcutaneous fat), around the internal organs (abdominal fat), in the bone marrow, between muscles, and in breast tissue. Abdominal fat is also called visceral fat, and this form of fat is correlated with adverse health effects. The abdominal fat is packed between the organs of the abdominal cavity in specific locations called depots. These depots provide some cushioning for the organs, but the main function is to serve as a reserve of triglycerides.

Weight gain is generally associated with an increase in adipocyte size and not an increase in adipocyte number

Adipocytes contain a single large vacuole that stores triglycerides and other lipids. As the adipocyte takes up and synthesizes triglycerides, the size of the vacuole and the size of the cell increase (FIGURE 12.7). When these energy reserves are needed, the triglycerides are converted into fatty acids and released into the blood. Consequently, the vacuole and the cell decrease in size. Thus, weight gain and weight loss are not associated with a change in the number of adipocytes, but rather a change in adipocyte size. This means that the mass of the adipose depot increases primarily through the hypertrophy of adipocytes and only through limited adipocyte hyperplasia.

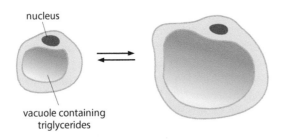

nucleus

vacuole containing triglycerides

FIGURE 12.7 Adipocyte. Adipocytes are characterized by having a single large vacuole that occupies most of the cytoplasm of the cell. Triglycerides and other lipids are stored in the vacuole. The nucleus is found at the cell periphery. The adipocyte increases or decreases in size according to the amount of lipid that is stored.

Several studies have suggested that adults have a constant number of adipocytes. However, adipocytes do increase in number during childhood and adolescence. Once reaching adulthood the older adipocytes are replaced by mesenchymal stem cells (see Box 8.1). Individuals exhibit differences in the number of adipocytes with lean individuals having fewer adipocytes than obese individuals (FIGURE 12.8). This suggests that a component of the genetic predisposition to gain weight may be due in part to the number of adipocytes an individual has. However, the exact factors regulating adipocyte number are not known.

Adipose tissue plays a role in the regulation of metabolism

As more is learned about adipose tissue, it is becoming clear that adipose tissue also has endocrine functions and regulates metabolism. In fact, the absence of adipocytes is metabolically detrimental. Adipose tissue is now known to secrete numerous proteins, called **adipokines**, which control various metabolic functions. The best characterized among these adipokines are adiponectin and leptin. Obesity and metabolic syndrome are associated with high levels of leptin and low levels of adiponectin (FIGURE 12.9).

Both leptin and adiponectin are proteins primarily synthesized and secreted by adipose tissue. Circulating levels of leptin are proportional to the mass of adipose tissue, whereas adiponectin levels are inversely correlated with adipose mass. Both of these proteins function as hormones that regulate energy balance. Leptin binds to receptors in the hypothalamus and this binding leads to decreased energy uptake and increased energy expenditure. This subsequently reduces adipose mass and subsequently lowers leptin levels. In contrast, adiponectin promotes

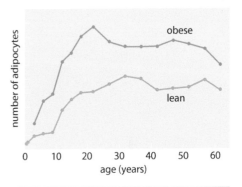

FIGURE 12.8 Comparison of adipocyte number in obese and lean individuals. Adipocytes increase in number during childhood and adolescence and remain somewhat constant during adulthood. Obese people tend to have more adipocytes than lean people. [Adapted from Spalding KL, Arner E, Westermark PO et al. (2008) Dynamics of fat cell turnover in humans. *Nature* 453: 783–787. doi: 10.1038/nature06902. With permission from Springer Nature.]

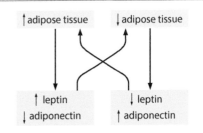

FIGURE 12.9 Homeostasis of adipose tissue by adipokines. Increasing the mass of adipose tissue results in higher levels of leptin and lower levels of adiponectin. These changes in leptin and adiponectin levels cause a decrease in adipose tissue that then lowers leptin levels and increases adiponectin levels. This decrease in leptin and increase in adiponectin cause an increase in the amount of adipose tissue. The opposing roles of leptin and adiponectin, plus their correlations with the amount of adipose tissue, maintain adipose tissue at somewhat constant levels.

lipid storage and increases the amount of adipose tissue. The normal function of these adipokines is to maintain the amount of adipose tissue at a constant level. Thus, there appears to be a dysfunction in the endocrine role of adipose tissue in obese individuals since weight is gained despite the high leptin and low adiponectin levels. This hormonal dysfunction probably contributes to the progression of obesity.

Chronic low-grade inflammation is also associated with obesity

Inflammation is an immunological process by which the body responds to injury. Macrophages are a key cellular element in the inflammatory response and the number of macrophages in adipose tissue increase with obesity. Presumably these macrophages function to clean up dying and dysfunctional adipocytes. Activated macrophages also secrete proinflammatory cytokines, which are hormones produced by immune effector cells that stimulate inflammation. In addition, some adipokines, such as leptin, are proinflammatory, whereas adiponectin is anti-inflammatory. The increase in adipose tissue mass associated with abdominal obesity increases the levels of proinflammatory adipokines and decreases the levels of anti-inflammatory adipokines, resulting in low-grade chronic inflammation (FIGURE 12.10). This chronic inflammation contributes to the development of metabolic syndrome associated with an increased risk of diabetes and cardiovascular disease (see Figure 12.4).

SIGNPOST Macrophages and inflammation are described in Chapter 9, pages 175–177.

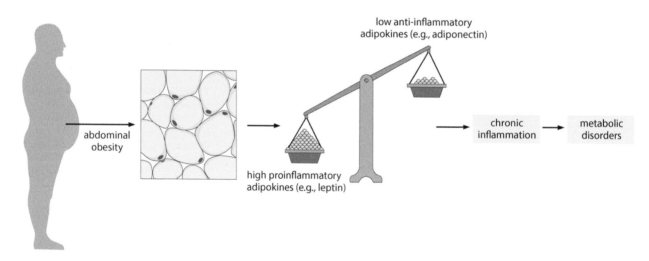

FIGURE 12.10 Abdominal obesity and inflammation. The increased mass of adipose tissue associated with abdominal obesity results in higher levels of proinflammatory adipokines and lower levels of anti-inflammatory adipokines. This results in chronic low-grade inflammation that contributes to the metabolic syndrome of insulin resistance, dyslipidemia, and hypertension.

REGULATION OF ENERGY BALANCE

Several distinct hormonal systems are involved in the regulation of metabolism as it relates to energy balance. Probably the best known are insulin and glucagon, which regulate glucose levels in the blood (Chapter 13). Other hormones that affect metabolism are glucocorticoids. Glucocorticoids are synthesized by the adrenal gland and are released in response to stress. Glucocorticoid receptors are found on most cells and control metabolism, immunity, and development. Regarding metabolism, glucocorticoids increase the levels of glucose, amino acids, and fatty acids in the blood. Glucocorticoids also suppress the inflammatory response.

Other hormones involved in the regulation of energy balance are adipokines and ghrelin. Both insulin and glucocorticoids play important roles in the homeostasis of metabolites, but do not have direct roles in adipose tissue gain or loss. The adipokines leptin and adiponectin play a direct role in the regulation of adipose tissue (see Figure 12.9).

Leptin plays a key role in energy balance and fat storage

Leptin was the first adipokine described and its role in adipose tissue homeostasis is relatively well characterized. The gene for leptin was discovered in mutant obese mice and was initially called the obese gene (BOX 12.2). Many factors can regulate the leptin levels in the blood, but in general the leptin levels correlate with the amount of adipose tissue. Individuals with a high mass of adipose tissue have high levels of leptin and, conversely, low levels of leptin are found in individuals with low amounts of adipose tissue. Therefore, leptin levels vary with weight gain and weight loss due to changes in body fat composition.

The *Obese* Gene and the Discovery of Leptin

A strain of mice that ate voraciously and gained excessive amounts of weight was discovered by chance. These mice start out at birth with a normal size, but typically weigh three times more than normal by adulthood (**BOX 12.2 FIGURE 1**). Genetic studies revealed that a single gene was responsible for this trait and the hypothetical gene was named the *obese* (*Ob*) gene. Subsequent studies mapped the gene and eventually the gene was cloned and sequenced. Characterization of the protein encoded by the gene revealed it to be a protein hormone expressed by adipose tissue that plays a role in energy balance. The protein was named leptin from the Greek word *lepto* meaning thin. Leptin was also the first adipokine to be characterized.

The mutation in the *Ob* gene is recessive and mice with two mutant copies exhibit the trait. These mice do not have any functional leptin. Humans also have the *leptin* gene. However, obesity due to

missing or defective leptin in humans is exceedingly rare and has only been identified in a few families worldwide. In fact, leptin levels are generally higher than normal in obese individuals. Thus, obesity in humans is associated with leptin resistance.

BOX 12.2 FIGURE 1 Mutation in the *Ob* gene results in obesity. The mouse on the right is normal and the mouse on the left has two defective copies of the *Ob* gene. (With permission from Oak Ridge National Laboratory/US Department of Energy/Science Photo Library.)

BOX 12.2

Leptin receptors are primarily located in the hypothalamus. Binding of leptin to its receptor causes the hypothalamus to send a signal that increases thermogenesis (FIGURE 12.11). This results in increased energy expenditure as glucose and lipid catabolism is used to generate heat instead of ATP. At the same time the hypothalamus sends a signal to suppress appetite. Therefore, energy uptake decreases at the same time energy expenditure is increased. This means that excess adipose tissue and high leptin levels slow the further accruement of adipose tissue or even cause a loss of adipose tissue. When adipose tissue decreases the leptin levels fall. The hypothalamus senses this and sends out signals to decrease thermogenesis and increase appetite. The increase in energy uptake and decrease in energy expenditure result in a positive energy balance. The resulting excess energy can then be converted into triglycerides and stored in the adipocytes causing an increase in adipose tissue and weight gain.

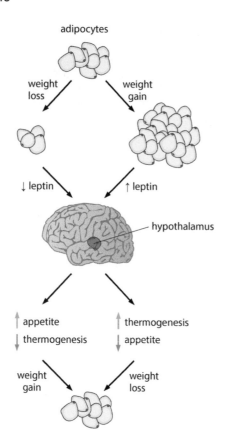

FIGURE 12.11 Leptin action. Leptin is secreted by adipocytes and as adipocytes increase in mass during weight gain more leptin is secreted. High leptin levels cause an increase in thermogenesis and a decrease in appetite via the hypothalamus resulting in a negative energy balance. Conversely weight loss leads to less leptin and an increase in appetite and a decrease in energy expenditure via thermogenesis. This positive energy balance can be used to generate more adipose tissue.

Obesity is associated with leptin resistance

Obese individuals exhibit higher circulating levels of leptin than normal-weight individuals due to their higher amounts of body fat. Therefore, obesity is generally not due to a deficiency in leptin. However, the higher levels of leptin are not resulting in increased energy expenditure or decreased energy uptake as expected. This failure of leptin action is known as leptin resistance. The cause of leptin resistance is not known. Leptin receptors appear normal and the downstream effectors in the signaling pathways appear intact. Some evidence suggests that leptin is not reaching its target site in the hypothalamus. Leptin resistance explains the failure of leptin to treat obesity (BOX 12.3).

Leptin Is Not a Cure for Obesity

The discovery that leptin decreases appetite and increases thermogenesis raised some hope that leptin could be used as a therapy against obesity. However, several clinical trials did not demonstrate any significant weight loss following the administration of leptin. The exception to this was in families in whom obesity was associated with a lack of leptin. In these cases, the administration of leptin did result in weight loss. However, obesity associated with missing leptin is extremely rare and most obesity is associated with leptin resistance. In the case of leptin resistance, the administration of more leptin does not significantly reduce weight.

Ghrelin is another regulator of appetite

Ghrelin is a protein hormone produced primarily by specialized cells in the stomach. When the stomach is empty, these cells secrete ghrelin into the circulatory system. Ghrelin receptors are found in the hypothalamus and binding of the hormone to the receptor results in signals to increase appetite (FIGURE 12.12). A full stomach causes less ghrelin to be secreted and this decreases appetite. Therefore, in some respects, ghrelin opposes leptin in that high levels of leptin increase appetite and low levels of leptin suppress appetite. However, the two hormones operate on different time scales. Ghrelin functions in a relatively short time span and regulates hunger between meals. Therefore, it is sometimes called the hunger hormone. Leptin functions on a longer time scale and is involved in the homeostasis of adipose tissue.

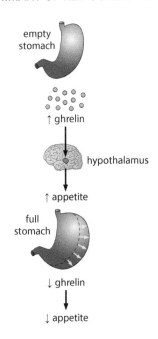

FIGURE 12.12 Ghrelin action. An empty stomach stimulates the secretion of the hormone ghrelin, which in turn increases appetite and stimulates eating. As the stomach fills less ghrelin is secreted leading to decreased ghrelin secretion and suppression of hunger.

SUMMARY OF KEY CONCEPTS

- Essential nutrients are substances obtained from the diet that the body cannot synthesize
- Macronutrients, such as carbohydrates, fats, and protein, provide energy and basic building blocks for anabolic purposes
- Micronutrients, such as many vitamins and minerals, are needed in small amounts and provide a wide range of physiological functions
- Protein–energy malnutrition is particularly adverse for children and leads to stunting and poor physical and mental development as well as decreased immunity
- An excess of body fat, particularly in the abdomen, can have adverse effects on a person's health and is a risk factor for developing diabetes or cardiovascular disease
- The etiology of obesity is complex and involves genetic, other biological factors, cultural practices, behavior, and environmental influences
- The genetics of obesity is a genetic predisposition that involves many genes
- A positive energy balance is converted into triglycerides and other lipids and stored in adipocytes
- Obesity adversely affects the endocrine functions of adipose tissue and promotes chronic systemic inflammation
- Leptin is a hormone secreted by adipocytes that plays a key role in regulating energy balance and the formation of adipose tissue
- Obesity is generally associated with leptin resistance

KEY TERMS

adipocyte	A specialized cell of adipose tissue that stores excess energy in the form of triglycerides and other lipids and regulates energy balance through adipokines.
adipokine	Any of the cytokines (hormones) produced by adipose tissue that affect metabolism and energy balance. Leptin and adiponectin are two well-characterized adipokines.
adipose	A loose connective tissue composed primarily of adipocytes and commonly referred to as fat tissue.
basal metabolic rate (BMR)	A measure of the minimum number of calories that are needed to maintain organs and perform necessary bodily functions excluding physical activity.
body mass index (BMI)	An indirect measure of adipose tissue defined by weight in kilograms divided by the height in meters squared.
kwashiorkor	A condition caused by a diet low in protein but containing some calories in the form of carbohydrates.
leptin	A peptide hormone (adipokine) synthesized by adipocytes that regulates thermogenesis and appetite.
macronutrients	Dietary constituents that are needed in large quantities to supply energy and support cell replication and repair. Includes carbohydrates, proteins, lipids, and water.
marasmus	A condition caused by deficiency of both protein and calories derived from carbohydrates or fat characterized by extreme emaciation.
metabolic syndrome	A syndrome characterized by abdominal obesity, high blood pressure, hyperglycemia, and dyslipidemia.
micronutrients	Essential dietary constituents, such as vitamins and minerals, that are required in small amounts and needed for metabolic processes.
protein–energy malnutrition (PEM)	A deficiency in dietary protein and other energy-yielding macronutrients that leads to suboptimal growth, weight loss, and immune deficiency. Also called protein–energy undernutrition.

REVIEW QUESTIONS

1. Identify and define the six classes of nutrients. For each class indicate whether they contain macronutrients, micronutrients, or both.

2. What are the two major types of protein–energy undernutrition and the differences between them?

3. What are adipokines and how do they influence obesity? What are the specific roles of leptin and adiponectin?

4. Discuss the genetics of obesity.

5. List and briefly describe hormones that are involved in the regulation of energy balance.

ADDITIONAL QUESTIONS

In addition to the Review Questions provided above, there is a range of free online questions designed to enable students to further test their understanding of the chapter material. In order to access these interactive questions, please visit the book's website: https://routledge.com/cw/wiser

Diabetes 13

Diabetes is a major public health problem that is increasing in prevalence throughout the world. Significant morbidity and mortality are associated with diabetes and all organs and tissues can be affected by diabetes. Diabetes is defined by persistently high levels of glucose in the blood, called **hyperglycemia**. One consequence of hyperglycemia is frequent urination, and the word diabetes is derived from a Greek word essentially meaning frequent urination. There are two distinct types of diabetes: diabetes insipidus and diabetes mellitus. Diabetes insipidus is rare and due to endocrine dysfunction associated with the pituitary gland. Diabetes mellitus is common and is generally simply referred to as diabetes. Incidentally, insipidus and mellitus are derived from Latin words and respectively mean bland or tasteless and honeyed or sweet. The sweetness of urine associated with diabetes mellitus is due to the excess glucose in the blood being removed by the kidneys.

The persistent hyperglycemia associated with diabetes is due either to a deficiency in insulin production or to a diminished effect of insulin. Diabetes due to insulin deficiency is classified as type 1 diabetes, whereas type 2 diabetes is characterized by a lack of response to insulin. The two types of diabetes have distinct and complex etiologies. Despite the different etiologies, the pathophysiology of the two types of diabetes is similar. The disruption of glucose homeostasis has an adverse effect on physiology, and the pathology associated with diabetes primarily involves damage to the vascular system. There is no cure for diabetes and treatment involves the management of hyperglycemia and the prevention of vascular damage.

GLUCOSE HOMEOSTASIS

Blood levels of glucose need to be maintained in a relatively narrow range. Glucose levels that are too low cannot supply the energy requirements needed for the normal function of organs and physical activity. However, high glucose levels stress the endothelial cells lining the blood vessels and affect many organs. Several distinct hormone systems play roles in regulating blood glucose levels. The most important of these with regard to rapidly regulating blood glucose levels are glucagon and insulin. Glucagon and insulin are opposing peptide hormones that maintain blood glucose levels (FIGURE 13.1) between 70 and 110 milligrams per deciliter. The liver, skeletal muscle, and adipose tissue play predominant roles in glucose homeostasis.

FIGURE 13.1 Regulation of blood glucose levels by glucagon and insulin. Glucagon is secreted by the pancreas in response to low blood glucose levels. Glucagon stimulates cells to secrete glucose and thereby raise blood glucose levels. High blood glucose levels stimulate the pancreas to secrete insulin, which stimulates cells to take up glucose.

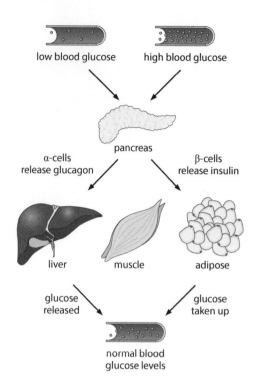

SIGNPOST

Hormones, second messengers, and protein phosphorylation are described in Chapter 3, pages 62–66

Glucagon stimulates glucogenesis

Glucagon is a peptide hormone, produced by alpha-cells of the islets of Langerhans within the pancreas. Low glucose levels stimulate the synthesis and secretion of glucagon. Glucagon receptors are most abundant in liver cells and binding of glucagon to the receptor activates second messenger and protein kinase cascades that ultimately result in the phosphorylation of two enzymes called glycogen phosphorylase and glycogen synthase. Glycogen phosphorylase is an enzyme that converts glycogen to glucose and is activated by phosphorylation. Glycogen synthase is an enzyme that converts glucose to glycogen and is deactivated by phosphorylation. Therefore, glucagon binding to receptors generates glucose. This generation of glucose is called **glucogenesis**. The newly formed glucose is secreted from cells resulting in an increase in blood glucose levels (see Figure 13.1). When the glucose level returns to normal, the alpha-cells discontinue glucagon synthesis resulting in a decrease in glucogenesis and glucose secretion by cells.

Insulin stimulates glucose uptake and metabolism

Blood glucose levels rise following a meal, and if the levels surpass the optimal range, cells in the pancreas called **beta (β)-cells** of the islets of Langerhans detect the high glucose and secrete **insulin**. Most cells in the body have insulin receptors and respond to insulin by activating glucose transporters on the membrane to take up glucose from the blood and thereby lowering blood glucose levels (FIGURE 13.2). The glucose taken up by the cells is metabolized according to the type of cell. For example, in muscle and liver cells excess glucose is converted to glycogen, the storage form of glucose. In adipose tissue the excess glucose is metabolized to pyruvate and the pyruvate is used to synthesize fatty acids and triglycerides.

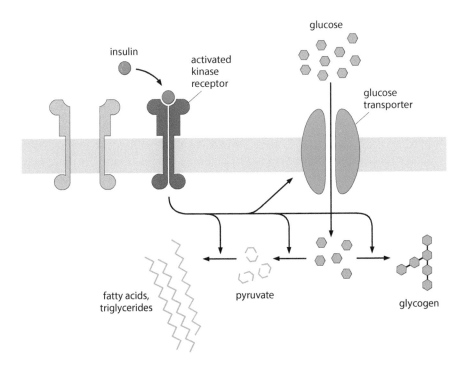

FIGURE 13.2 Insulin action. Binding of insulin to the receptor on the cell surface activates a protein kinase that phosphorylates several target proteins, resulting in the activation of several biochemical processes. This includes increased uptake of glucose via the glucose transporter, conversion of glucose to glycogen, breakdown of glucose to pyruvate, and the synthesis of fatty acids and triglycerides. The net result of insulin action is the removal of glucose from the blood and the conversion of glucose to either glycogen or lipids.

ETIOLOGY OF DIABETES

Diabetes is associated with insulin dysfunction. There are two types of diabetes called type 1 and type 2. In **type 1 diabetes** insulin is not produced. **Type 2 diabetes** is associated with a lack of response to insulin, called insulin resistance. Both situations lead to persistently high levels of blood glucose since insulin dysfunction means that excess glucose is not removed from the blood. The underlying causes for the dysfunction in insulin homeostasis are distinct for the two types of diabetes.

Type 1 diabetes is due to autoimmunity

The autoimmune destruction of beta-cells in the pancreas leads to type 1 diabetes (FIGURE 13.3). The underlying etiology of autoimmune diseases is a combination of genetic predisposition, environmental triggers, and other factors (Chapter 17). Certain alleles of genes called human leukocyte antigens (HLA) increase the risk of type 1 diabetes. Possible triggers for the autoimmune destruction of beta-cells include viral infections or damage to the beta-cells. There is some inflammation and fibrosis in the pancreas associated with this autoimmunity, but, overall, the pathology in the pancreas is minimal. The destruction of beta-cells means that insulin is not produced in response to hyperglycemia. It is possible to manage this hyperglycemia through the administration of insulin. Patients with type 1

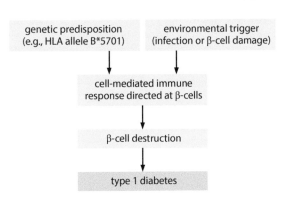

FIGURE 13.3 Etiology of type 1 diabetes. An environmental trigger combined with a genetic predisposition can lead to the autoimmune destruction of beta (β)-cells resulting in type 1 diabetes.

diabetes monitor their blood glucose after meals and, if the glucose levels are too high, they inject insulin, which then lowers blood glucose levels. For this reason, type 1 diabetes has also been called insulin-dependent diabetes.

Obesity is a major predisposing factor for type 2 diabetes

The etiology of type 2 diabetes is complex and involves numerous predisposing and risk factors (TABLE 13.1). The biggest predisposing factor for the development of type 2 diabetes is obesity, and the increasing incidence of diabetes coincides with the high rates of obesity observed in the human population. Likewise, features of the metabolic syndrome (see Table 12.6), such as dyslipidemia and high blood pressure are factors that contribute to the development of type 2 diabetes. Other factors linked with diabetes are pregnancy, stress, and corticosteroid therapy. About 1–3% of women develop temporary hyperglycemia during pregnancy that is referred to as gestational diabetes. This diabetic condition reverses itself after birth of the baby. However, women who experience gestational diabetes are at increased risk of developing type 2 diabetes. Corticosteroids, often used as anti-inflammatory drugs, antagonize the effects of insulin and can lead to hyperglycemia. Similarly, persistent stress leads to elevated levels of natural corticosteroid hormones, such as glucocorticoids, which can lead to persistent hyperglycemia.

Table 13.1 Predisposing and Risk Factors for Type 2 Diabetes
> 45 years old
Obesity (>120% of desirable body weight)
Dyslipidemia
Hypertension
History of impaired glucose homeostasis (e.g., gestational diabetes)
Long-term corticosteroid therapy (leads to hyperglycemia)
Family history of type 2 diabetes in a first-degree relative (i.e., parent or sibling)
Hispanic, Native American, African American, Asian American, or Pacific Islander descent

Diabetes has a strong genetic component

Diabetes is a complex disease involving both genetic and environmental factors. However, neither type 1 nor type 2 diabetes exhibits a Mendelian inheritance pattern and specific genes associated with diabetes have not been identified. Type 2 diabetes has a higher degree of hereditability than type 1 diabetes and exhibits a strong family predisposition. Having a primary relative, such as a parent or sibling, with type 2 diabetes increases the lifetime risk of developing type 2 diabetes by 5- to 10-fold.

The prevalence of type 2 diabetes is also higher in certain groups with shared ancestry, which is also consistent with a genetic component in the etiology. However, genetic predisposition does not necessarily lead to diabetes and other factors such as diet and lifestyle play a large role in the development of diabetes. A good example of this combination of genes and lifestyle is the Pima Indians (BOX 13.1). A high-calorie diet and sedentary lifestyle combined with a genetic predisposition greatly increase the likelihood of developing type 2 diabetes, whereas a proper diet and physical activity can overcome the genetic predisposition to develop diabetes.

Pima Indians and Diabetes

Some of the highest prevalences of type 2 diabetes are found in the Pima Indians and other people of Native American ancestry. Genetic studies have demonstrated a strong genetic predisposition for the development of diabetes among the Pima Indians but have failed to identify specific genes involved in this genetic predisposition. In addition, genetics is not the entire answer since before the 1940s diabetes among the Pima Indians was essentially non-existent. The high prevalence did not become apparent until the 1960s and beyond. This increase in prevalence was associated with a change in diet and lifestyle. The increasing prevalence of obesity and type 2 diabetes in such populations has sometimes been attributed to the effects of the thrifty gene hypothesis (see Box 12.1).

The Pima Indians were an agrarian culture living along the Gila River in southwestern United States. Starting in the 1920s and 1930s the rivers were dammed, and water was diverted away from the Pima Indians and essentially ended their farming lifestyle and eliminated their food supply. In response to the starvation of these people the government began to supply commodity foods consisting primarily of processed foods high in sugar. Thus, the diet changed from a high-fiber diet consisting of grains and vegetables to a diet primarily of processed foods. In addition, the lifestyle changed from a labor-intensive agricultural society to a more sedentary routine. As a consequence of the change in diet and decrease in physical activity there was an epidemic of obesity among the Pima Indians followed by an epidemic of type 2 diabetes.

PATHOGENESIS OF DIABETES

In addition to different etiologies, type 1 and type 2 diabetes progress differently (TABLE 13.2). The onset of type 1 diabetes is usually abrupt and often occurs during adolescence. Therefore, type 1 diabetes was previously called juvenile-onset diabetes. Type 2 diabetes, conversely, has a slow and subtle onset and generally develops later in life. Therefore, historically type 2 diabetes has been called adult-onset diabetes. However, coinciding with the increase in obesity among children and adolescents, type 2 diabetes is now being seen more frequently at younger ages.

Table 13.2   Comparison of Type 1 and Type 2 Diabetes

Criterion	Type 1	Type 2
Percentage of cases	~10%	~90%
Age of onset	Usually teenagers	Usually adults
Onset of disease	Sudden and intense	Slow and subtle
Body weight	Normal to underweight	Overweight or obese
Primary relatives with disease	< 20%	> 60%
Autoimmunity	Yes	No
Pancreas pathology	Minor inflammation or fibrosis	Amyloid deposits (late stage)
Islet beta-cells	Marked decrease	Near normal
Insulin levels	Low or absent	Normal or increased → decreased
Management	Insulin and diet	Diet, exercise, oral medication

Insulin resistance is a precursor to type 2 diabetes

The development of type 2 diabetes is associated with a mix of genetic and environmental factors (FIGURE 13.4). Obesity, a sedentary lifestyle, aging, and other factors combined with a genetic predisposition lead to persistently high blood glucose levels. Initially the pancreas responds to the high glucose levels by secreting insulin and blood glucose levels are maintained at near-normal levels. This state in which blood glucose is at normal or slightly elevated levels in the presence of higher-than-normal insulin levels is pre-diabetes. Over time, the persistence of this pre-diabetic state leads to insulin resistance and cells become less responsive to insulin.

FIGURE 13.4 Progression of type 2 diabetes. A mixture of genetic, environmental, and behavioral factors can result in persistent hyperglycemia, which induces insulin secretion to maintain glucose at normal levels. The combination of normal or slightly elevated glucose levels and high insulin levels is pre-diabetes. Over time the beta (β)-cells become exhausted and type 2 diabetes develops.

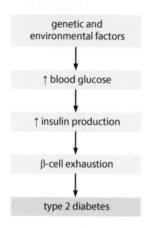

genetic and environmental factors

↓

↑ blood glucose

↓

↑ insulin production

↓

β-cell exhaustion

↓

type 2 diabetes

Muscle, liver, and adipose tissues play the largest roles in regulating glucose homeostasis. Insulin resistance means that muscle cells do not take up glucose, liver cells continue with glucogenesis and glucose secretion, and adipocytes synthesize and secrete fatty acids despite the presence of insulin (**FIGURE 13.5**). Thus, hyperglycemia persists, and even more insulin is needed to maintain proper glucose levels. The early stages of insulin resistance do not present with clinical manifestations. However, eventually, the beta-cells become exhausted, and the secretion of insulin is diminished. This decrease in insulin production leads to persistently high glucose levels and full-blown diabetes develops.

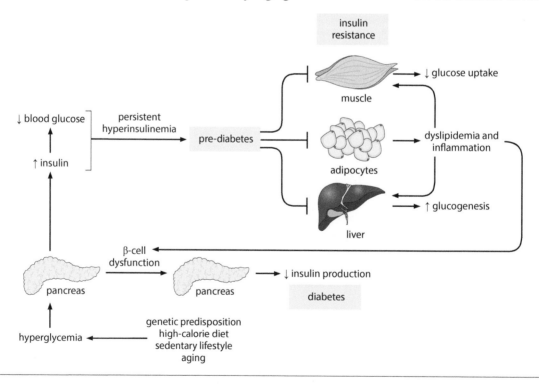

FIGURE 13.5 Pathogenesis of type 2 diabetes. Factors causing persistent hyperglycemia result in a state of hyperinsulinemia and nearly normal or slightly elevated blood glucose levels called pre-diabetes. Persistent hyperinsulinemia leads to insulin resistance characterized by decreased glucose uptake by muscle cells, increased glucogenesis and glucose secretion by liver cells, and increased fatty acid synthesis and secretion by adipocytes. Dyslipidemia and associated inflammation possibly exacerbate insulin resistance. Over time hyperinsulinemia and dyslipidemia lead to beta-cell dysfunction and decreased insulin production characteristic of diabetes.

Obesity, especially visceral fat in the abdomen, is clearly related to the development of insulin resistance in the tissues. However, the exact role obesity plays in the development of insulin resistance is not clear. One possibility is that a dysfunction in fatty acid metabolism and a high-calorie diet can affect the cellular physiology of peripheral tissues, so that the cells become less responsive to insulin. In addition, obesity is also associated with increased inflammation (Chapter 12) which can also have an adverse effect on peripheral tissues, including the pancreas. Obesity and the associated inflammation may contribute to beta-cell exhaustion.

PATHOPHYSIOLOGY OF DIABETES

The pathological consequences of type 1 and type 2 diabetes are similar despite the differences in etiology and pathogenesis. Pathology associated with diabetes is due to vascular damage caused by persistent hyperglycemia and the disruption of normal glucose homeostasis. Low insulin or insulin resistance both result in extracellular hyperglycemia and intracellular hypoglycemia (FIGURE 13.6). This means that glucose levels are too high outside of cells, whereas glucose levels are too low inside cells. Without the insulin signal, whether it is due to insulin deficiency or insulin resistance, cells do not take up enough glucose to meet their energy needs and glucose levels remain high in blood and the extracellular fluid. In some respects, cells are starving even though there is an overabundance of extracellular glucose. Therefore, cells need to synthesize glucose from other metabolites.

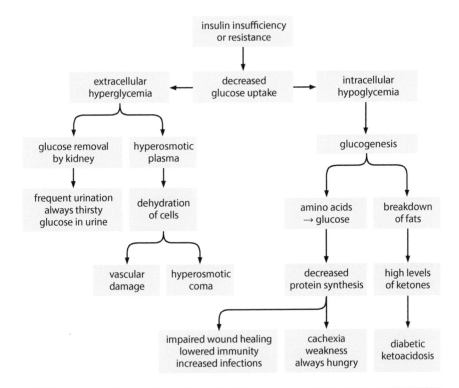

FIGURE 13.6 **Pathophysiology of diabetes.** The lack of insulin or unresponsiveness to insulin results in extracellular hyperglycemia and intracellular hypoglycemia. Intracellular hypoglycemia stimulates the synthesis of glucose (glucogenesis) by cells through the metabolism of fatty acids and amino acids. The by-products of this metabolism are the production of ketones and decreased protein synthesis, which results in some of the manifestations associated with diabetes. The hyperosmotic plasma caused by extracellular hyperglycemia results in cellular dehydration, which can lead to coma or damage to the endothelial cells lining the blood vessels. Excess glucose is removed by the kidneys resulting in frequent urination and excessive thirst.

Diabetes causes cachexia and lowers immunity

Intracellular hypoglycemia results in glucogenesis as the cells convert glycogen, amino acids, and fatty acids to glucose. This lowers the pool of available amino acids and impairs protein synthesis. Decreased protein synthesis leads to wasting (cachexia) and weakness. Thus, people with diabetes are often hungry. Impairment of protein synthesis also has an impact on wound healing and immunity since both processes are highly dependent on protein synthesis. For example, the synthesis and secretion of antibodies by B cells is an important element of the immune response against pathogens. In addition, the immune response requires a lot of energy since polyclonal expansion of lymphocytes is part of the adaptive immune response (Chapter 9). Thus, diabetes also lowers immunity and people with diabetes experience frequent and persistent infections.

Coma can result from diabetic ketoacidosis

A by-product of the metabolism of amino acids and fatty acids is the production of small organic molecules called ketones. High levels of ketones in the blood cause the blood to become acidic and this condition is referred to as ketoacidosis. Acidosis combined with the dehydration that accompanies diabetes can lead to a medical emergency referred to as **diabetic ketoacidosis**. Nausea and vomiting are often associated with acidosis, thus exacerbating dehydration. Severe acidosis and dehydration can lead to coma. Diabetic ketoacidosis is more common in type 1 diabetes than type 2 diabetes and it is sometimes associated with a failure to inject insulin in a timely fashion.

Dehydration is a potential pathological consequence of diabetes

High levels of glucose in the blood create a **hyperosmotic** state. This means that there are higher concentrations of solutes outside the cell than inside the cell and, therefore, water flows out of cells by simple diffusion (Chapter 2). This causes cells to become dehydrated and this dehydration damages the cells. Brain cells are particularly susceptible to dehydration and dehydration can lead to a hyperosmotic coma. Hyperosmotic coma is more common in type 2 diabetes than in type 1 diabetes. The kidneys are the primary organ involved in regulating osmotic concentration and removing the excess glucose from the blood and glucose is excreted with urine. As part of removing excess glucose from the blood, water is also excreted leading to frequent urination and excessive thirst.

Damage to the vascular system leads to long-term complications

The endothelial cells that line the blood vessels are stressed and damaged as a result of the persistent hyperosmotic conditions associated with diabetes. This damage to the vascular system is largely responsible for the long-term complications associated with diabetes (TABLE 13.3). For example, damage to large blood vessels speeds up atherosclerosis (Chapter 14), leading to a higher risk of heart attacks, strokes, and peripheral artery disease in people with diabetes. Therefore, diabetics die at younger ages than non-diabetics. Similarly, damage to the microvascular can lead to an increased risk of kidney disease (Chapter 15) and blindness. Diabetes is the most common cause of kidney failure and the leading cause of blindness in people under 60 years old. Poor circulation also contributes to sexual dysfunction, slow wound healing, and the development of gangrene in the toes and feet.

Table 13.3 Long-term Complications Associated with Diabetes

Atherosclerosis	Damage to large blood vessels increasing risk of strokes, heart attacks, and premature death
Kidney disease	Damage to microvascular tissue of the nephrons results in kidney dysfunction
Eye disease	Damage to microvascular tissue of retina leads to vision loss and blindness
Diabetic neuropathy	Damage to sensory neurons initially results in diabetic pain and later loss of sensation
Diabetic foot	A combination of neuropathy and poor circulation results in injuries and infections possibly leading to amputation

The peripheral nervous system is also affected by diabetes

Sensory neurons are affected more than motor neurons and initially many diabetic patients suffer from irritation, pain, and abnormal sensations. But as the disease progresses there is a loss of fine touch and pain sensations. As a result, patients are often unaware of irritation or injuries, particularly in the feet. This combined with poor circulation and the propensity for infections to develop into gangrene lead to a syndrome called a diabetic foot. The slow healing process and lowered immunity may allow for infections to progress to the point at which amputation is necessary. People with diabetes are much more likely to need lower limb amputations than those who are not diabetic.

DIAGNOSIS

Ten symptoms are generally associated with diabetes (TABLE 13.4). These symptoms are related to the pathological consequences of extracellular hyperglycemia and intracellular hypoglycemia. Frequent urination and excessive thirst are key symptoms resulting from the removal of excess glucose by the kidneys. The starvation of cells despite the high extracellular glucose results in tiredness, slow wound healing, frequent infections due to impairment of immunity, sudden unexplained weight loss, and a state of always being hungry. Damage to the vascular system affects vision and sexual function. In the case of type 1 diabetes the appearance of the symptoms is usually sudden and quite noticeable. However, in the case of type 2 diabetes the onset of symptoms is generally gradual and insidious. The gradual appearance of the symptoms is often not noticed by the patient until the disease is quite advanced.

Table 13.4 Ten Symptoms Commonly Associated with Diabetes

Frequent urination	Due to the osmotic diuresis associated with hyperglycemia
Always thirsty	Replacement of water due to excessive urination
Always tired	Weakness associated with cachexia and decreased protein synthesis
Always hungry	Due to glucogenesis from amino acids
Sudden unexplained weight loss	Result of cachexia
Slow wound healing	Associated with decreased protein synthesis
Frequent infections	Due to a lowered immunity
Numb or tingling feet or hands	Sensory nerve dysfunction
Blurred vision	Damage to microvascular tissue of the retina
Sexual dysfunction	Due to vascular damage

The clinical definition of diabetes is persistent hyperglycemia and laboratory tests are used to make the diagnosis in combination with the symptoms. The most common test is the fasting plasma glucose test in which blood glucose is measured after fasting for at least 8 hours. The glucose tolerance test provides information about the body's ability to control glucose levels. In the glucose tolerance test a specific amount of glucose is ingested and 2 hours later blood glucose levels are measured. Another test for diabetes is glycated hemoglobin, called HbA1c. Glycated hemoglobin is formed as a result of glucose chemically combining with the hemoglobin and this can serve as a biomarker for the average blood glucose levels during the preceding 2–3 months. For all three tests values below a specified value are defined as normal and values above a specified value are diagnostic of diabetes (TABLE 13.5). Test results between the upper and lower extremes are diagnostic of pre-diabetes. Pre-diabetes is a major risk factor for developing diabetes.

Table 13.5 Diabetes Diagnostic Criteria			
Condition	**FPG (mg/dL)**	**Oral GTT (mg/dL)**	**HbA1c (%)**
Normal	<110	<140	<6.0
Impaired fasting glucose levels	110–125	–	6.0–6.4
Impaired glucose tolerance	–	140–199	6.0–6.4
Diabetes	≥126	≥200	≥6.5
FPG, fasting plasma glucose; GTT, glucose tolerance test.			

MANAGING DIABETES

Diabetes is a chronic disease with no cure. Therefore, treatment involves regulating blood glucose levels to prevent the pathology associated with hyperglycemia. The more time blood glucose levels are maintained at near-normal levels, the less damage is done. This is especially important in terms of preventing long-term complications. Monitoring blood glucose levels is an important aspect of managing diabetes (FIGURE 13.7). Devices that can quickly and accurately determine blood glucose levels are available. If blood glucose levels are high, it is possible in the case of type 1 diabetes and some late-stage type 2 diabetes to administer insulin. Injecting insulin quickly lowers blood glucose levels and this is a means to maintain blood glucose levels at normal levels.

Weight reduction through diet and exercise is another means to maintain normal blood glucose levels for both type 1 and type 2 diabetes. In addition, diet and exercise can be used to prevent the progression of pre-diabetes to type 2 diabetes. Diets that are high in fiber and low in fat and sugar are recommended. In cases in which diet and exercise are insufficient to maintain blood glucose levels there are several drugs that can be used to lower blood glucose (TABLE 13.6). These drugs are primarily used for type 2 diabetes.

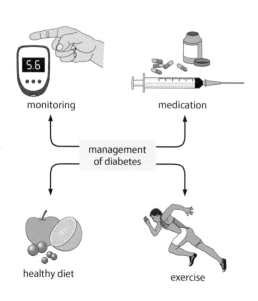

monitoring

medication

management of diabetes

healthy diet

exercise

FIGURE 13.7 Management of diabetes. Through monitoring, exercise, diet, and medications it is possible to maintain normal glucose levels and avoid some of the pathology associated with diabetes and the progression of pre-diabetes to type 2 diabetes.

Table 13.6 Oral Hypoglycemic Drugs	
Drug	**Mechanism of action**
Sulfonylureas	Stimulate secretion of insulin by the pancreas
Metformin	Increase the sensitivity of target organs to insulin
Thiazolidinediones	Increase glucose utilization and diminish glucogenesis
α-Glucosidase inhibitors	Inhibit conversion of starch to simple sugars

In addition to monitoring and maintaining normal glucose levels, care needs to be taken to treat and avoid complications associated with diabetes. Serious acute complications associated with ketoacidosis and dehydration require medical therapy and intervention. Similarly, special attention to chronic long-term complications such as cardiovascular and kidney disease are also needed. This requires regular visits to the doctor and special attention to cholesterol levels, blood pressure, smoking, kidney function, the feet, and the eyes. The goal of managing diabetes is to avoid the acute and chronic pathology of diabetes.

SUMMARY OF KEY CONCEPTS

- Glucagon and insulin play key roles in regulating blood glucose levels
- Diabetes is defined as persistently high blood glucose levels associated with a dysfunction in insulin
- Type 1 diabetes is due to insufficient insulin production by the pancreas and is caused by an autoimmune attack on the beta-cells of the pancreas
- Type 2 diabetes is due to unresponsiveness of tissues to insulin
- Obesity combined with a genetic predisposition are the major risk factors for developing type 2 diabetes
- The onset of type 1 diabetes is usually sudden, intense, and typically occurs in adolescence or young adulthood
- Type 2 diabetes tends to progress slowly and subtly with symptoms appearing during middle age
- Diabetes is characterized by high extracellular levels of glucose and low intracellular levels of glucose
- The low intracellular glucose levels lead to wasting, fatigue, always feeling hungry, lower immunity, and slow wound healing
- The metabolism of lipids and fatty acids to produce energy can lead to diabetic ketoacidosis, a severe and acute medical emergency
- Excess extracellular glucose is removed by the kidneys resulting in glucose in the urine, frequent urination, and always feeling thirsty
- Hyperglycemia causes a hyperosmotic state that damages endothelial cells of the vascular system and can lead to hyperosmotic coma
- Diabetes also damages the peripheral nervous system, initially causing irritation and pain that later develops into loss of sensation
- Long-term complications of diabetes include cardiovascular disease, kidney disease, blindness, and diabetic foot
- Diagnosis of diabetes is confirmed through laboratory tests measuring blood glucose levels
- Type 1 diabetes is managed by monitoring blood glucose levels and using insulin to maintain blood glucose at normal levels
- Weight reduction through lifestyle changes is used to manage type 2 diabetes and prevent the development of type 2 diabetes
- Drugs that lower blood glucose can also be used to manage blood glucose levels

KEY TERMS

beta-cell	A type of cell found in a region of the pancreas called islets of Langerhans that synthesizes and secretes insulin
diabetic ketoacidosis	A severe complication of diabetes that can lead to coma and death due to acidosis caused by the metabolism of fatty acids to glucose.
glucagon	A peptide hormone produced by alpha-cells of the islets of Langerhans of the pancreas that raises blood glucose levels by activating glucogenesis and the secretion of glucose into the blood.
glucogenesis	The synthesis of glucose from other metabolites such as glycogen, fatty acids, or amino acids.
hyperglycemia	A condition in which blood glucose levels are above normal. Persistent hyperglycemia is the primary feature of diabetes mellitus.
hyperosmotic	A condition in which the extracellular fluid (e.g., blood) has a higher concentration of osmotically active compounds (i.e., solutes that can cross membranes) than the cytoplasm of the cells.
insulin	A peptide hormone produced by beta-cells of the islets of Langerhans of the pancreas that lowers blood glucose levels by stimulating cells to take up glucose and convert glucose into glycogen or fatty acids.
type 1 diabetes	Diabetes due to insulin deficiency resulting from a failure of the pancreas to produce insulin. Also called insulin-dependent diabetes, or, previously, juvenile-onset diabetes.
type 2 diabetes	Due to a failure of cells to respond properly to insulin (insulin resistance). Also called non-insulin-dependent diabetes or previously adult-onset diabetes.

REVIEW QUESTIONS

1. Discuss the differences in the etiology between type 1 and type 2 diabetes.
2. List and describe the symptoms associated with diabetes.
3. What are the clinical manifestations and pathologies associated with the hyperosmotic state caused by diabetes?
4. What are the medical consequences associated with the intracellular hypoglycemia experienced during diabetes?
5. Discuss the diagnosis and management of diabetes.

ADDITIONAL QUESTIONS

In addition to the Review Questions provided above, there is a range of free online questions designed to enable students to further test their understanding of the chapter material. In order to access these interactive questions, please visit the book's website: https://routledge.com/cw/wiser

Cardiovascular Disease

14

Cardiovascular disease is the number one cause of death worldwide. The high rate of mortality associated with diseases of the circulatory system is not surprising since the circulatory system functions to supply the entire body with nutrients and oxygen as well as to remove metabolic wastes. The circulatory system consists of the heart, blood vessels, and blood. Cardiovascular disease includes damage to the heart and blood vessels, as well as damage to other organs that result from damage to the blood vessels (TABLE 14.1). The causes and manifestations of these various cardiovascular diseases are interconnected, and one cardiovascular disease can lead to the development of other cardiovascular manifestations.

ATHEROSCLEROSIS

HYPERTENSION

CORONARY ARTERY DISEASE

STROKES

Table 14.1 Cardiovascular Disease

Disease	Manifestations
Atherosclerosis	Blocked arteries
Hypertension	High blood pressure
Coronary artery disease	Chest pain, heart attack
Cerebrovascular events	Stroke
Peripheral artery disease	Poor circulation
Congestive heart failure	Weakening of the heart
Chronic kidney disease	Kidney failure

Atherosclerosis and **hypertension** are the underlying causes of most other heart disease. In particular, damage to the arteries due to atherosclerosis and hypertension can have profound effects on the heart and brain, and thus are associated with high mortality. For example, the accumulation of cholesterol and inflammation in the walls of arteries associated with atherosclerosis can obstruct blood flow. Blocked arteries can lead to coronary heart diseases, such as heart attacks, and cardiovascular events such as strokes. The persistent high blood pressure associated with hypertension can initiate atherosclerosis, weaken the heart and lead to congestive heart failure, or damage the microvascular tissue of the kidneys leading to chronic kidney disease.

ATHEROSCLEROSIS

Atherosclerosis is a thickening of the arterial wall due to the formation of an **atheroma**, more commonly referred to as a plaque. An atheroma is an accumulation of cholesterol and scar tissue in the wall of large- and medium-sized arteries that can obstruct the lumen of the blood vessel (FIGURE 14.1). In addition, over time the atheroma lowers the flexibility of the blood vessel which is often referred to as hardening of the arteries. The atheroma can also weaken the wall of the artery and result in an aneurysm. An aneurysm is a dilation of the artery due to a thinning of the artery wall. The weakened artery wall at the aneurysm is at a higher risk of rupturing, which can lead to a life-threatening hemorrhage. Blood clots can also be associated with atheromas, and these blood clots can lead to a sudden and acute blockage of a blood vessel.

FIGURE 14.1 Narrowing of an artery due to atherosclerosis. The formation of plaques, also called atheromas, in the walls of arteries can restrict blood flow, leading to manifestations of cardiovascular disease.

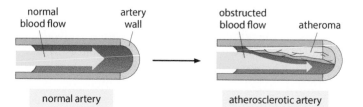

The clinical manifestations of atherosclerosis include coronary artery disease, stroke, and peripheral artery disease. These clinical conditions are primarily characterized by obstruction of blood flow. Obstruction of cerebral and coronary arteries can have severe consequences and therefore receive the greatest amount of attention. However, atherosclerosis potentially affects arteries in all organs and tissues, even though the consequences may not be as dramatic or life threatening as a heart attack or a stroke. Peripheral artery disease most commonly affects the legs, and the classic symptom is pain while walking that resolves with rest. Poor circulation can also result in necrosis and infections. Infections can progress to the point of gangrene and amputation may be necessary.

Dyslipidemia is a major risk factor for developing atherosclerosis

The etiology of atherosclerosis is complex and involves both behavioral and genetic factors (TABLE 14.2). Some of these factors such as genetic predisposition, age, and gender cannot be controlled. However, the major risk factors are associated with diet and activity levels. Diet and activity levels are also risk factors for obesity (Chapter 12) and type 2 diabetes (Chapter 13). Consequently, obesity and type 2 diabetes are also risk factors for developing cardiovascular disease, and in particular atherosclerosis. The single biggest risk factor for developing atherosclerosis is **dyslipidemia**. Dyslipidemia refers to abnormal levels of lipids in the blood, especially triglycerides and cholesterol. There is a clear correlation between the levels of triglycerides in the blood and the risk of developing cardiovascular disease (TABLE 14.3).

Table 14.2 Atherosclerosis Risk Factors	
Major	**Minor**
Dyslipidemia	Age ($\uparrow$)
Hypertension	Family history
Smoking	Gender (male > female)
Diabetes	Activity level
Obesity	Psychological stress

Table 14.3 Lipidemia and Cardiovascular Disease Risk

	TG	LDL-C	HDL-C
Normal (low risk)	<150	<100	>60
Slightly above normal	150–199	100–129	
Modest risk	200–499	130–159	
High risk	>500	>160	<40

HDL, cholesterol associated with high-density lipoprotein; LDL-C, cholesterol associated with low-density lipoprotein; TG, triglyceride. Concentrations in mg/dL.

Lipid levels are determined in part by genetics due to differences in the metabolism of lipids. However, blood lipid levels, especially triglycerides, are primarily determined by the amount of dietary fat consumed. Dietary sources of fat include vegetable oils, animal fats, and trans fats. Vegetable oils are more unsaturated than animal fats. Saturation refers to the number of double bonds in the fatty acid chains making up the triglycerides (see Figure 1.15). A completely saturated fatty acid has no double bonds. It was previously thought that unsaturated fats were less of a risk factor for developing cardiovascular disease than saturated fats. However, more recent data have questioned whether this is true. Regardless, diets high in trans fats are a risk factor for developing cardiovascular disease. Trans fats are artificially prepared by hydrogenating unsaturated fats into saturated fats, which are widely used in processed foods and the restaurant industry (BOX 14.1).

Trans Fats and Cardiovascular Disease

Trans fatty acids, commonly called trans fats, are prepared by chemically hydrogenating liquid vegetable oil. Trans fats have been widely used in baked and fried foods. Partially hydrogenated oil is less likely to spoil, so foods made with it have a longer shelf-life. Some restaurants use partially hydrogenated vegetable oil in their deep fryers because it does not have to be changed as often.

It is widely accepted that an excess of artificially prepared trans fats in the diet significantly raises the risk of overall mortality and mortality due to cardiovascular disease. Many regulatory agencies now regulate the amount of trans fats that can be added to food. Trans fats raise the levels of 'bad' cholesterol (LDL) and lower the 'good' cholesterol (HDL) levels.

BOX 14.1

'Good' and 'bad' cholesterol refer to the lipoproteins that transport cholesterol

High cholesterol is also associated with an increased risk of developing atherosclerosis, but the situation is not straightforward. Lipids in the blood are associated with large complexes of proteins and lipids called **lipoproteins** that transport lipids through the blood and lymph (BOX 14.2). Cholesterol is primarily transported by two lipoproteins called **low-density lipoprotein** (LDL) and **high-density lipoprotein** (HDL). LDL transports cholesterol from the liver, where it is synthesized, to the tissues. HDL has the opposite function and transports cholesterol from the tissues back to the liver where it can be metabolized (FIGURE 14.2). Therefore, abnormally high levels of cholesterol associated with LDL can lead to an accumulation of cholesterol in the arteries within tissues. Cholesterol is a key element of the plaques and, thus, high levels of LDL increase the risk of developing atherosclerosis. Conversely, HDL can remove the excess cholesterol from the tissues and therefore high levels of HDL have a protective effect against developing atherosclerosis (see Table 14.3). Because of this protective effect, HDL is commonly called the 'good' cholesterol and, since high levels of LDL can promote atherosclerosis, it is called the 'bad' cholesterol.

BOX 14.2

Lipoproteins

Lipids are rather insoluble in aqueous solutions. Therefore, lipids are incorporated with proteins into complexes known as lipoproteins so that they can be transported to cells throughout the body. Lipoproteins are spherical structures with phospholipids making up the surface. The polar portions of the phospholipids are oriented outward and therefore look similar to large micelles (see Figure 2.1). Like micelles, lipoproteins are soluble in water. Proteins, called apolipoproteins, are embedded in the outer phospholipid layer and triglycerides and cholesterol are found in the interior of the lipoprotein (**BOX 14.2 FIGURE 1**). The apolipoproteins bind to receptors on the surface of cells and the lipoproteins are taken up by receptor-mediated endocytosis (see Figure 2.9). The lipids within the lipoproteins become available to the cell for biosynthetic purposes or for storage in the case of adipocytes.

Four major lipoproteins are responsible for the transport of lipids (**BOX 14.1 TABLE 1**). Chylomicrons (CMs) are assembled in the intestinal epithelial cells with lipids absorbed from the diet and CMs transport lipids from the intestines to the tissues, particularly the liver and adipose. The liver is a major site for the metabolism and synthesis of triglycerides and cholesterol, whereas adipose tissue functions to store triglycerides. Transport of lipids from the liver involves two different lipoproteins called very-low-density lipoprotein (VLDL) and low-density lipoprotein (LDL). VLDL primarily transports lipids from the liver to adipose tissue and LDL transports lipids from the liver to other tissues. High-density lipoprotein (HDL) functions to transport lipids from the peripheral tissues back to the liver for re-utilization or excretion in the form of bile. In addition, the various lipoproteins differ as to the type of lipid transported. CMs and VLDL transport primarily triglycerides, whereas LDL and HDL are the major transporters of cholesterol.

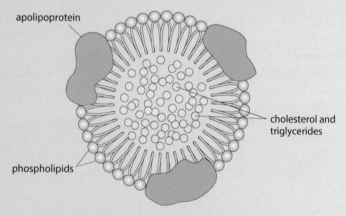

BOX 14.2 FIGURE 1 Schematic representation of a lipoprotein. Lipoproteins have an outer shell of phospholipids with embedded apolipoproteins that bind to receptors. The interior of the lipoprotein contains triglycerides and cholesterol.

Box 14.2	Table 1 Lipoproteins	
Type*	**TG/CH****	**Transport function**
CM	83%/5%	Intestines → tissues
VLDL	50%/20%	Liver → adipose tissue
LDL	4%/20%	Liver → tissues
HDL	2%/25%	Tissues → liver

*CM, chylomicron; HDL, high-density lipoprotein; LDL, low-density lipoprotein; VLDL, very-low-density lipoprotein.
**Approximate percentages of lipoprotein mass composed of triglyceride (TG) and cholesterol (CH). The remainder of the mass is made up of phospholipids and protein.

FIGURE 14.2 Transport functions of HDL and LDL. Low-density lipoproteins (LDLs) transport cholesterol and other lipids from the liver to the peripheral tissues. High-density lipoproteins (HDLs) transport cholesterol and other lipids from the peripheral tissues to the liver.

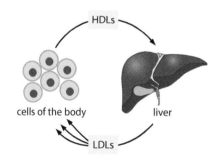

Damage to the endothelium initiates atheroma formation

Atheroma formation is a multistep process (FIGURE 14.3). The first step generally involves some type of damage to the endothelial cells lining the blood vessels. Things that can damage the endothelium include hyperlipidemia, hypertension, diabetes, toxins in tobacco smoke, inflammation, and infections. The injuries to the endothelial cells cause the endothelium to become more permeable. This allows cholesterol associated with LDL to accumulate in the intima layer of the artery wall that is located between the endothelial cells and the smooth muscle layer. The cholesterol oxidizes within the intima and the oxidized cholesterol can further damage the endothelium.

Some macrophages have receptors for oxidized cholesterol and migrate to the areas of cholesterol accumulation and begin ingesting the oxidized cholesterol (FIGURE 14.4). These macrophages take up large amounts of cholesterol and are called 'foam' cells since they have a foamy appearance due to a large number of cholesterol-filled vesicles within their cytoplasm. If the accumulation of cholesterol is minimal or transient, the macrophages can remove the cholesterol and prevent the formation of the atheroma. However, if the cholesterol continues to accumulate and is exacerbated by high levels of LDL and low levels of HDL the capacity of the macrophages to clean up the excess cholesterol is overwhelmed.

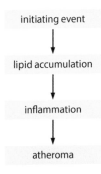

FIGURE 14.3 Steps in atheroma development. An injury to the endothelium leads to the accumulation of lipids in the artery wall. Subsequent inflammation leads to the formation of atheroma.

Macrophages are described in Chapter 9, pages 176–177.

SIGNPOST

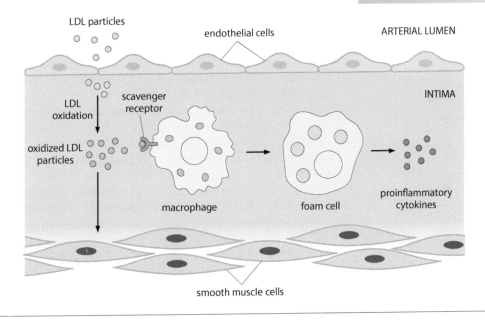

FIGURE 14.4 Early steps in atheroma development. Damage to the endothelial cells allows for increased permeability of LDL-cholesterol into the intima. The intima is the space between the endothelial cells and the muscle layer. The oxidized cholesterol is taken up by macrophages via the scavenger receptor and phagocytosis. Macrophages filled with cholesterol are often called foam cells and secrete inflammatory cytokines.

Vascular inflammation and scarring are components of atheroma development

The plaque enlarges as more cholesterol accumulates and additional macrophages migrate to the lesion. The smooth muscle cells within the artery wall also begin to migrate and rearrange in response to this inflammation and fibrocytes are attracted to the lesion. Fibrocytes are cells that secrete extracellular matrix proteins such as collagen which promotes fibrosis (i.e., scarring). The smooth muscle cells and collagen form a fibrous cap on the atheroma (FIGURE 14.5). As the plaque enlarges the foam cells in the center of the atheroma die and leave behind necrotic material and crystallized cholesterol. This results in a soft fatty mass surrounded by inflammatory cells, smooth muscle cells, and extracellular matrix (e.g., collagen) within the wall of the artery. This mass reduces the diameter of the lumen of the artery and partially obstructs blood flow.

FIGURE 14.5 Atherosclerotic plaque. Schematic drawing showing components of an atheroma. Smooth muscle cells migrate, and collagen is deposited to form a cap on the atherosclerotic lesion. Foam cells in the center of the lesion die and release cholesterol that crystallizes. Immune effector cells such as lymphocytes also migrate to the lesion.

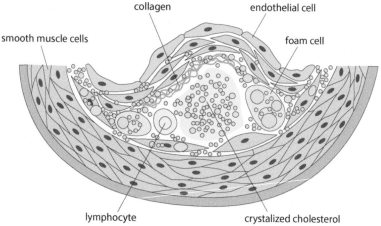

Aging of the atheroma is associated with hardening and other complications

The scarring and inflammation within the atheroma also cause calcification. Calcification is the deposition and accumulation of calcium salts in the lesion. This accumulation of calcium causes the walls of the artery to become less flexible, often described as hardening of the arteries. In fact, sclerosis refers to the stiffening of a structure. Plaques progressively enlarge and the amount of scarring and calcification increases. This results in large atheromas that substantially impede blood flow (FIGURE 14.6). The fibrous cap on the atheroma can also erode and lead to ulceration of the endothelium. This ulceration attracts platelets and leads to the formation of a blood clot, also called a **thrombus**. In addition, the thrombus or pieces of the thrombus can break off and become a moving clot, called an **embolism**. Thrombi and embolisms are especially dangerous since they can result in a sudden and total blockage of a blood vessel. Furthermore, the atheroma can weaken the wall of the artery and increase the risk of an **aneurysm** forming and a subsequent rupture of the aneurysm.

FIGURE 14.6 Enlargement and complications of an atheroma. As the atheroma ages it becomes fibrotic and increases in size to the point that the lumen of the artery is substantially reduced. In some cases, there can be an ulceration of the endothelium leading to clot (thrombus) formation. These clots can obstruct blood flow or break off to become emboli.

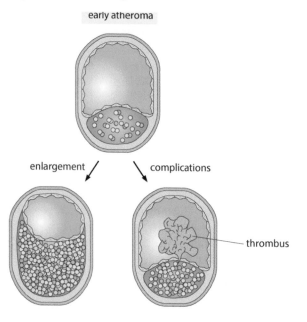

Atheroma development proceeds slowly over decades

The initiation of atheromas often starts during childhood or adolescence. In the beginning the lesions are benign fatty streaks in the walls of the artery. Over time these fatty streaks increase in size due to the accumulation of foam cells, cholesterol, and scar tissue. A fibrous cap forms in the large atheromas and fibrosis continues as well as calcification. This process takes decades and during this time there are no clinical manifestations (FIGURE 14.7). Thus, the disease has been clinically silent and progressively worsening for decades. Eventually the cap erodes and there are complications such as thrombosis, weakening of the artery wall, and hardening of the arteries. Generally, the clinical manifestations of atherosclerosis are not seen until 50 years of age and older. Common manifestations include heart attacks and strokes due to thrombi or emboli that block coronary and cerebral arteries, respectively. Similarly, obstruction of arteries in other parts of the body can lead to peripheral artery disease which can result in gangrene due to poor circulation. Aneurysms are also associated with the weakening of the artery wall and can rupture leading to hemorrhaging.

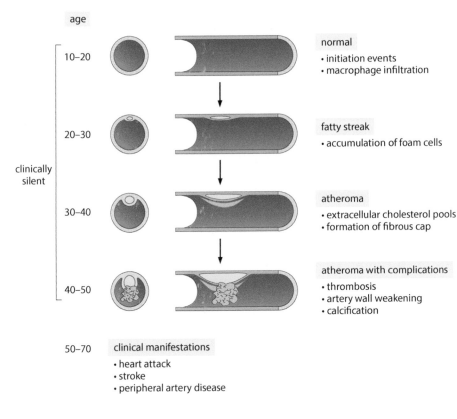

FIGURE 14.7 The progression of atherosclerosis. Atheromas slowly progress from fatty streaks, which can appear in childhood, to larger lesions with complications. Generally, atheromas are clinically silent until they become large or are associated with complications. Manifestations of the clinical disease include heart attacks, strokes, gangrene due to peripheral artery disease, and less frequently ruptured aneurysms due to weakened arteries.

Atherosclerosis can be prevented through lifestyle changes and medication

Lifestyle changes that prevent and slow the progression of atherosclerosis include: a low-fat diet, increased exercise, maintaining a healthy weight, and not smoking. These measures lower blood cholesterol and triglyceride levels, as well as lowering blood pressure. High blood pressure, hyperlipidemia, and toxins in cigarette smoke are all factors associated with the initiation of the fatty streaks that develop into atheromas. High levels of LDL-cholesterol increase the rate of atheroma progression. Therefore, cholesterol levels should be monitored. Other medical interventions include monitoring blood pressure and managing diabetes.

Medications can also be used to lower cholesterol and triglyceride levels in the blood. Several classes of drugs modulate lipid levels in the blood (TABLE 14.4). The most common and effective among these are statins. Statins inhibit the first enzyme in the cholesterol biosynthesis pathway. Bile sequestrants inhibit the reabsorption of bile. This forces the body to use available triglycerides and cholesterol for the synthesis and secretion of bile. Niacin and fibric acids affect lipid metabolism.

Table 14.4   Drugs Affecting Blood Lipid Levels

Class	Mechanism	Effect on blood lipid levels*		
		LDL-C	HDL-C	TG
Statins	HGM-CoA reductase inhibitor	↓18–55%	↑5–15%	↓7–30%
Bile sequestrants	Inhibit reabsorption of bile	↓15–30%	↑3–5%	No change
Niacin	↓ Triglyceride (TG) synthesis	↓5–25%	↑15–35%	↓20–50%
Fibric acids	Activate receptors regulating glucose and lipid metabolism	↓5–20%	↑10–20%	↓20–50%

*Typical effects of various classes of drugs on cholesterol associated with LDL and HDL and TG levels. HGM-CoA, 3-hydroxy-3-methylglutaryl coenzyme A.

HYPERTENSION

Hypertension refers to sustained high blood pressure. Normal fluctuations in blood pressure, such as those associated with increased physical activity or stress, are not considered hypertension. For example, epinephrine released as part of the flight-or-fight response rapidly increases the heart rate and cardiac output to meet the increased demand for nutrients and oxygen, thereby transiently increasing blood pressure. There are no specific symptoms associated with high blood pressure and the clinical manifestations of hypertension are due to damage to the cardiovascular system and kidneys. As with atherosclerosis, the damage caused by hypertension slowly develops and is clinically silent for decades, with the clinical symptoms manifesting later in life.

Blood pressure measurements consist of two numbers. The first or higher number is the systolic blood pressure that occurs during the contraction of the heart. As the heart relaxes the blood pressure lowers and this is called diastolic blood pressure. Optimal blood pressure is defined as less than 120/80 mm mercury (mmHg). Generally low blood pressure is not considered a problem unless it is associated with symptoms such as dizziness, fainting, or fatigue. Extremely low blood pressure, or hypotension, can result in shock and be life threatening. Blood pressure above 120/80 mmHg is defined as elevated blood pressure, stage 1 hypertension, stage 2 hypertension, and hypertensive crisis (TABLE 14.5). Generally medical interventions are not started until stage 2 hypertension. Blood pressure above 180/120 mmHg is considered an emergency situation and requires immediate medical attention.

Table 14.5   Guidelines for Defining Hypertension

Category	Systolic (mmHg)		Diastolic (mmHg)
Low blood pressure	<90	and	<60
Normal	<120	and	<80
Elevated blood pressure	120–129	and	<80
Hypertension stage 1	130–139	or	80–89
Hypertension stage 2	>140	or	>90
Hypertensive crisis	>180	or	>120

The regulation of blood pressure is complex and involves many factors

Blood pressure is the force that blood exerts on the walls of the arteries due to the pumping action of the heart. This pressure is necessary to drive the blood through the circulatory system. The physics of blood pressure is relatively simple and is determined by the volume of blood being pumped and the resistance created by the blood vessels (i.e., pressure = flow × resistance). The flow component is the cardiac output and is determined by the heart rate and the force of the heart contractions. The resistance component is the friction of the blood flowing through the vessels and is determined by the diameter of the arteries, with resistance increasing as the vessel diameter decreases. In summary, blood pressure is determined by the heart rate, heart contraction force, blood volume, and constriction or dilation of blood vessels (FIGURE 14.8).

Accordingly, blood pressure is regulated by controlling cardiac output via changing the heart rate and contraction force, causing arteries to dilate or constrict, or adjusting blood volume. The kidneys, and in particular the renin–angiotensin–aldosterone system, play a key role in regulating blood pressure. The regulation of blood pressure is complex and involves many different hormones and neurotransmitters. For example, there are neurons that can measure oxygen and carbon dioxide levels in the blood and increase or decrease blood flow accordingly. In addition, there are also sensory neurons that measure arterial blood pressure and alert the brain if blood pressure is too high or too low, which is followed by corrective measures. Similarly, imbalances in blood electrolytes affect blood volume, and therefore the physiological mechanisms that control water and sodium retention and excretion also influence blood pressure. Through the action of various hormones and neurotransmitters blood pressure can be adjusted to meet physiological needs, as well as maintaining optimal levels.

Most hypertension has a complex etiology involving genetic and behavioral components

Hypertension is defined as being either primary hypertension or secondary hypertension. In primary hypertension, also called essential hypertension, there is no known cause, and this represents about 90% of the cases of hypertension. In the remaining 10% of hypertension cases an underlying cause can be identified. As one would expect, considering the major role the kidneys play in blood pressure, kidney disease is a major cause of secondary hypertension. Other causes of hypertension could be tumors in endocrine glands, such as the adrenal or pituitary, which can affect hormones involved in blood pressure homeostasis. Pregnancy and some medications can also raise blood pressure. Identifying a specific cause of hypertension provides a possible means to treat or cure hypertension, rather than simply treat the symptoms as is the case in essential hypertension.

Essential hypertension is a multifactorial disease defined by risk factors rather than etiological agents. Some of these factors are genetic in that hypertension tends to run in families and rates of hypertension vary with ancestry (TABLE 14.6). Many of the risk factors are behavioral. For example, a high-sodium diet is probably the biggest single risk factor. High sodium intake increases water retention, which then increases blood volume and results in increased blood pressure. Other major risk factors are related to obesity and a sedentary lifestyle. It is believed that obesity causes the heart to work harder to move the blood through the

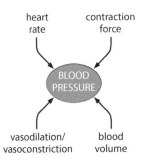

FIGURE 14.8 Factors contributing to blood pressure. Major factors that influence blood pressure include heart rate, the strength of the heart contraction, total blood volume, and the dilation or constriction of blood vessels.

The renin–angiotensin–aldosterone system is described in detail in Chapter 15, pages 287–288.

SIGNPOST

circulatory system, and thus cardiac output is persistently high. Similarly, the nicotine in cigarette smoke increases the heart rate and therefore smokers have persistently high blood pressure.

Table 14.6 Predisposing and Risk Factors for Essential Hypertension	
Factor	**Comments**
Age	Blood pressure tends to increase throughout life
Hereditary	Family history of hypertension
Ancestry	Hypertension is more common in some races
Gender	Males > Females
High salt intake	Sodium levels affect blood volume
Obesity	Heart pumps harder to push blood through vessels
Sedentary lifestyle	Probably related to hyperlipidemia and weight gain
Smoking	Nicotine raises epinephrine, which stimulates the heart
Persistent emotional distress	Stress activates elements of nervous system that elevate blood pressure
Diabetes	People with diabetes have higher incidence of hypertension for unknown reasons
High alcohol consumption	Mechanism unknown

Hypertension and atherosclerosis are often comorbidities associated with other cardiovascular manifestations

Persistent high blood pressure damages the cardiovascular system. For example, damage to endothelial cells due to persistent high blood pressure can initiate atherosclerosis and speed up the progress of atherosclerosis (FIGURE 14.9). A consequence of atherosclerosis is a narrowing of the arteries, which in turn increases blood pressure. Furthermore, the hardening of the arteries reduces their flexibility, and this decreased flexibility does not allow for the blood vessels to expand, thereby increasing blood flow resistance and blood pressure. This creates a situation in which hypertension increases the progression of atherosclerosis and atherosclerosis increases the progression of hypertension.

FIGURE 14.9 Relationship between hypertension and atherosclerosis. Hypertension can initiate and increase the progression of atherosclerosis. Atherosclerosis can increase the progression of hypertension.

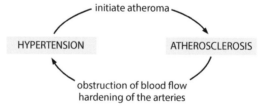

The obstruction of medium and large arteries associated with hypertension and atherosclerosis increases the risk of coronary artery disease, stroke, and peripheral artery disease (FIGURE 14.10). In addition, high blood pressure increases the risk of an aneurysm rupture in arteries damaged by atherosclerosis. Hypertension also affects the microvascular tissue. Both the retina and kidneys have extensive microvascular tissue that can be damaged by high blood pressure. Chronic kidney disease is a common clinical manifestation of hypertension (Chapter 15). Hypertension also damages the microvascular tissue of the retina causing hypertensive retinopathy. Some patients with hypertensive retinopathy may experience blurred vision or headaches, but quite often there are no symptoms. Damage to the retina is often detected in routine eye examinations. Hypertensive retinopathy can also be a predictor of future morbidity and mortality associated with heart disease, stroke, and kidney disease.

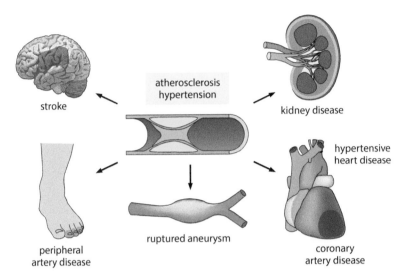

FIGURE 14.10 **Consequences of atherosclerosis and hypertension.** Atherosclerosis and hypertension adversely affect many aspects of the cardiovascular system and kidneys.

Hypertensive heart disease leads to congestive heart failure

The increased workload for the heart due to high blood pressure causes the muscle to grow larger and results in a thickening of the heart wall (FIGURE 14.11). A thickening of the heart wall results in smaller chamber sizes for the heart and that means less blood is pumped per contraction. This hypertrophy of the heart due to hypertension is called hypertensive heart disease. Hypertensive heart disease can lead to congestive heart failure. **Congestive heart disease** is a condition in which the heart is unable to pump blood efficiently through the circulatory system. This leads to blood backing up into the veins and causes leakage of fluid into the tissues, called edema. Major clinical manifestations of congestive heart disease are shortness of breath due to pulmonary edema, swelling of the legs, and tiredness. Congestive specifically refers to pulmonary edema and the fluid building up in the lungs.

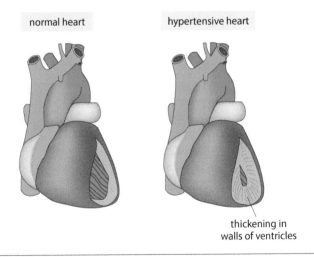

FIGURE 14.11 **Hypertensive heart.** Persistent high blood pressure can lead to hypertrophy of the heart muscle and decrease the ventricle volume. This results in less blood being pumped by the heart and an accumulation of fluid in the tissues.

Lifestyle changes and medications can be used to manage hypertension

Many of the risk factors for developing essential hypertension are modifiable (see Table 14.6). Therefore, it is possible to reduce blood pressure and prevent the development and progression of hypertension through changes in lifestyle (TABLE 14.7). These include a change in diet, a reduction in sodium intake, regular exercise, losing weight, and drinking alcohol in moderation. A diet rich in whole grains, fruits, vegetables, and low in fat is often referred to as the dietary approach to stop hypertension (DASH) eating plan.

Table 14.7	Lifestyle Modifications to Manage Hypertension	
Modification	**Recommendation**	**BP reduction***
Reduce dietary sodium	Daily sodium intake should be less than 2.3 grams/day and ideally less than 1.5 grams/day	2–8
DASH eating plan	Diet rich in fruits and vegetables and low in fat	8–14
Weight reduction	Possible to lower BP by 1 mmHg per kilogram lost	5–20
Increase physical activity	At least 30 minutes of aerobic physical activity per day	4–9
Moderate alcohol consumption	Limit drinking to two drinks per day for men and one drink per day for women	2–4

*Approximate reduction is systolic blood pressure (BP) measured in mmHg. DASH, Dietary Approaches to Stop Hypertension.

There are also drugs that can lower blood pressure which are used when lifestyle changes alone do not sufficiently lower blood pressure. Different classes of drugs affect the various factors that are involved in regulating blood pressure (TABLE 14.8). The choice of drug class depends on age, race, and the presence of diabetes or chronic kidney disease. Diuretics are generally the most often used antihypertensive drug. Diuretics are drugs that promote **diuresis**, which results in the excretion of water and electrolytes in the urine. This lowers blood volume and therefore may lower blood pressure. Other classes of drugs affect heart rate, heart output, or vasodilation as a means to lower blood pressure. The hormone angiotensin II, which plays a key role in regulating blood pressure (Chapter 15), is also the target of some antihypertension drugs. These various drugs do not cure hypertension but maintain blood pressure in an acceptable range.

Table 14.8	Medications to Lower Blood Pressure
Class	**Mechanism of action**
Diuretics	Increase sodium and water excretion and thereby lower blood volume and blood pressure
β-Blockers	Block adrenaline receptors to slow the heart rate
Ca^{2+}-channel blockers	Promote vasodilation and decrease contraction force and heart rate
ACE inhibitors	Block angiotensin-converting enzyme (ACE) which inhibits the formation of angiotensin II and results in vasodilation and diuresis
A-II receptor blockers	Block the receptors of angiotensin II resulting in vasodilation and diuresis

CORONARY ARTERY DISEASE

The arteries that supply blood to the heart are called coronary arteries. The heart plays a central role in the cardiovascular system and the heart is continuously working. Therefore, anything that affects blood flow to the heart is going to have dire consequences. In fact, compromising the blood flow to the heart due to atherosclerosis is the most common cause of death and disability. The deprivation of blood flow to a tissue or organ is called **ischemia** and ischemia deprives cells of oxygen and glucose. Therefore, coronary artery disease is also called ischemic heart disease. Generally, four clinical syndromes are associated with coronary artery disease: angina pectoralis, myocardial infarction, sudden cardiac death, and congestive heart failure due to chronic ischemic heart disease (TABLE 14.9).

Table 14.9	Syndromes of Coronary Artery Disease
Clinical syndrome	**Description**
Angina pectoralis	Also called cardiac chest pain. A distinctive and frightening sensation of chest discomfort due to ischemia
Myocardial infarction	Also called heart attack. Necrosis of the heart muscle due to ischemia
Sudden cardiac death	Natural death from cardiac causes occurring within one hour of the appearance of symptoms
Chronic myocardial ischemia	Gradual weakening of the heart muscle due to the long-term effects of ischemia and can lead to congestive heart failure

Heart attacks are due to blockage of coronary arteries

The clinical syndrome associated with coronary artery disease primarily depends on the degree and nature of the obstruction. Obviously, a complete obstruction of a blood vessel does more damage than a partial obstruction. Similarly, obstruction of a larger blood vessel does more damage than obstruction of a smaller blood vessel. The obstruction of an artery means that all cells being supplied by that artery are deprived of oxygen and nutrients. A complete blockage results in those cells dying and causing an area of necrosis called an **infarct** (FIGURE 14.12). The sudden occlusion of a coronary artery is commonly called a heart attack. Heart attacks are generally associated with complete and sudden occlusion of coronary arteries and are typically due to thrombotic or embolic events associated with atheroma (see Figure 14.6). Blockage of a large blood vessel can result in sudden death due to a substantial weakening of the heart muscle, which renders it incapable of pumping blood.

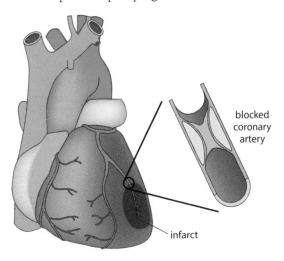

blocked coronary artery

infarct

FIGURE 14.12 Myocardial infarct. Blockage of a coronary artery results in an area of necrosis called an infarct.

Chronic ischemia can lead to congestive heart failure

The heart has little regeneration potential and those heart muscle cells that die as a result of ischemia are not replaced. Repair of the damaged heart is primarily fibrosis which results in a weaker heart. This means that ischemic damage to the heart is permanent and the heart is not able to function to its full capacity thereafter. In addition, the heart can also be damaged by partial occlusion of coronary arteries. Some of the occlusions may develop slowly and therefore not result in myocardial infarction. In addition, new collateral blood vessels may form around the occlusion and re-route the flow of blood. However, this formation of new blood vessels does not completely restore the normal blood flow and the clinical manifestation is often chest pain, called angina, rather than a heart attack. Nonetheless, this decreased blood flow associated with chronic ischemia damages heart muscle.

The damage caused by chronic ischemia, repeated anginal attacks, and small or large infarcts accumulates. This accumulated damage weakens the heart muscle and is called ischemic **cardiomyopathy**. Cardiomyopathy is any disease of the heart muscle that affects the ability of the heart to pump blood effectively. Ischemic cardiomyopathy generally manifests in older patients as congestive heart failure. In this case, the heart muscles are too weak to pump blood through the circulatory system. Congestive heart failure caused by a hypertensive heart is due to not enough blood filling the chambers because of the thickening of the heart walls.

Interventions for artery disease due to atherosclerosis include balloon angioplasty, stents, and bypass surgery

There are no treatments for diseases associated with late-stage atherosclerosis. Interventions focus on overcoming the obstruction of the coronary arteries. One such example is balloon angioplasty. In this procedure, a deflated balloon attached to a catheter is guided to a narrowed vessel and then inflated. The balloon forces the expansion of the blood vessel and the surrounding artery wall. The balloon is then deflated and withdrawn, leading to an expanded blood vessel that allows for improved blood flow. A stent may also be inserted at the time of ballooning to ensure the vessel remains open (FIGURE 14.13). The stent is a tube-shaped mesh device that can be expanded to the diameter of the artery. After the deflation of the balloon the stent maintains its shape and holds the artery open. One problem with stents is that they can induce an

FIGURE 14.13 Balloon angioplasty and stenting. A deflated balloon is guided to the arterial occlusion with a catheter and inflated. This expands the stent to hold open the artery and increases blood flow.

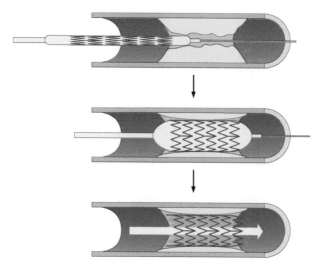

inflammatory response in the artery wall, which can result in a re-closing of the artery.

Another procedure for overcoming arterial occlusion is bypass surgery. In bypass surgery a vein is taken from the arm or leg and transplanted to the heart to redirect some blood from the aorta to a coronary artery downstream of the occlusion (FIGURE 14.14). Bypass surgery can relieve angina in cases in which stents are not possible and possibly prevent a future heart attack.

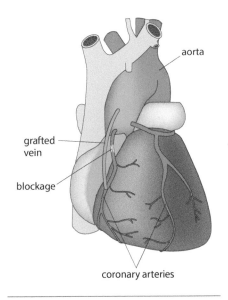

FIGURE 14.14 Coronary artery bypass graft. A vein is removed from the arm or leg and grafted from the aorta to a coronary artery to bypass a blockage.

STROKES

The brain is an active organ that requires copious amounts of glucose and oxygen, and thus is also extremely sensitive to disruptions in blood flow. A loss of blood flow to the brain results in a sudden loss of neurological function and this is called a stroke. There are two major types of strokes referred to as ischemic strokes and hemorrhagic strokes (FIGURE 14.15). **Ischemic strokes** are due to obstructed blood vessels and **hemorrhagic strokes** are due to ruptured blood vessels. Either event results in a loss of blood flow to a region of the brain and causes an infarct, analogous to the myocardial infarcts seen in coronary artery disease. Thus, strokes are sometimes called brain attacks.

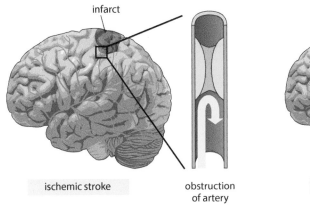
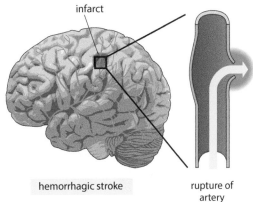

FIGURE 14.15 Two major types of strokes. Strokes result from obstruction of blood vessels in the brain (ischemic) or bleeding in or around the brain (hemorrhagic).

Atherosclerosis is a major risk factor for ischemic strokes

Ischemic strokes are more common and account for 80–85% of strokes. Most ischemic strokes are associated with atherosclerosis and are due to thrombi associated with the atheroma or an embolism. The occlusion of the artery results in the death of the tissue (i.e., an infarct) supplied by the blocked blood vessel. A brief loss of neurological function caused by a temporary decrease in blood flow is called a transient ischemic attack. Transient ischemic attacks exhibit the same symptoms as strokes, but the symptoms only last minutes or hours. In addition, there is no infarct associated with the transient ischemic attack. A transient ischemic attack, however, does increase the risk of subsequently having a stroke and is sometimes called a warning stroke.

Hemorrhagic strokes are due to the rupture of arteries in or around the brain

Arteries that are weakened by atherosclerosis can rupture and the likelihood of rupturing increases with hypertension. Bleeding within the brain results in the tissues distal to the rupture not being supplied oxygen and glucose. In addition, the accumulation of blood within the brain tissue can exert pressure on the brain, leading to neurological dysfunction. This bleeding within the brain is called intracerebral hemorrhage. Bleeding between the brain and the tissues surrounding the brain, called a subarachnoid hemorrhage, can also cause a stroke. Subarachnoid hemorrhages are often associated with head trauma and are not necessarily correlated with hypertension or atherosclerosis.

Symptoms of a stroke are sudden loss of neurological function

The symptoms of ischemic and hemorrhagic strokes are quite similar and imaging techniques are needed to distinguish the two types of strokes. Both ischemic and hemorrhagic strokes result in rapid death or dysfunction of neurons in a localized area. This necrosis occurs within minutes to hours. The exact symptoms depend on the specific region of the brain infarct and the neurological functions controlled by that region of the brain. However, there is one commonality in that the onset of symptoms is almost always sudden. Symptoms can include loss of motor functions, loss of vision, confusion or difficulty speaking, loss of balance or coordination, or severe headache (FIGURE 14.16). Quite often only one side of the body is affected since the lesion is confined to one side of the brain.

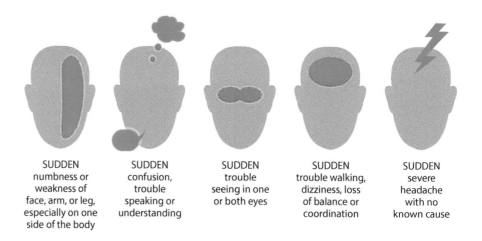

SUDDEN numbness or weakness of face, arm, or leg, especially on one side of the body

SUDDEN confusion, trouble speaking or understanding

SUDDEN trouble seeing in one or both eyes

SUDDEN trouble walking, dizziness, loss of balance or coordination

SUDDEN severe headache with no known cause

FIGURE 14.16 Symptoms of stroke. The rapid appearance of neurological dysfunction is indicative of a stroke. The region of the brain affected determines the exact symptoms.

Table 14.10   Stroke Outcomes
• Death within 2 weeks
• Recovery with profound disability
• Recovery with minor disability
• Complete recovery with no deficit

Stroke can result in a wide range of outcomes from essentially a complete recovery to death (TABLE 14.10). The outcome depends largely on the size of the occluded or ruptured blood vessel and the corresponding size of the infarct. Obviously, larger infarcts have more severe consequences. The brain has little regenerative capacity and the lost neurons are not replaced. However, through physical therapy it may be possible to retrain some of the remaining neurons to compensate for the loss of cells. Thus, it is possible to recover some of the lost functions through physical therapy.

Treatment and intervention for strokes include a clot-busting drug, endovascular procedures, and surgery

Due to the rapid death of neurons following an obstruction or aneurysm rupture, treatment and interventions are most effective if initiated within a few hours of the symptoms. The only available drug for the treatment of stroke is tissue plasminogen activator (t-PA), which is a protease that converts plasminogen to active plasmin (FIGURE 14.17). Plasmin is a protease that breaks down fibrin, which is the major protein found in blood clots. Therefore, t-PA initiates the breakdown of blood clots and is often called a clot-busting drug. Administration of t-PA within a few hours after the appearance of symptoms generally results in a good prognosis for the patient. However, t-PA is contraindicated in hemorrhagic strokes since it prevents the formation of clots.

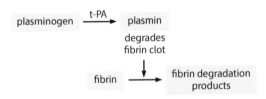

FIGURE 14.17 Tissue plasminogen activator. Tissue plasminogen activator (t-PA) is a protease that converts plasminogen to plasmin. Activated plasmin is a protease that breaks down the protein (fibrin) that forms blood clots.

Endovascular procedures involving the use of catheters to reach the clot or ruptured blood vessels are also available. For example, there are devices that can be used to remove a clot from the blocked artery. Similarly, devices to close off a ruptured blood vessel or promote clotting at the rupture are also available for hemorrhagic strokes. Surgery to repair the ruptured aneurysm is another option.

SUMMARY OF KEY CONCEPTS

- Atherosclerosis is an accumulation of cholesterol and scar tissue in the walls of arteries that can obstruct the blood flow
- Dyslipidemia, and especially high LDL and low HDL, is a major factor in the development of atherosclerosis and cardiovascular disease
- Manifestations of atherosclerosis include heart attacks, strokes, and peripheral artery disease
- Lifestyle changes and cholesterol-lowering drugs are the primary interventions in the management of atherosclerosis
- Hypertension is a sustained high blood pressure
- The etiology of primary hypertension is complex and involves both genetic and behavioral risk factors with the single largest risk factor being a high-sodium diet
- Hypertension damages endothelial cells and increases the initiation and progression of atherosclerosis
- The increased workload of the heart during hypertension can cause hypertrophy of the heart and lead to congestive heart failure
- Lifestyle changes and medications to lower blood pressure are used to manage hypertension

- Heart attacks are due to sudden and complete blockage of coronary arteries, primarily from blood clots associated with atherosclerosis

- Accumulated damage to the heart due to ischemic events weakens the heart and leads to congestive heart failure

- Strokes can be caused by blockage (ischemic stroke) or rupture of arteries (hemorrhagic stroke) supplying the brain

- Death of neurons due to lack of blood supply results in sudden neurological manifestations

---

KEY TERMS

aneurysm	An abnormal bulge of a blood vessel, especially an artery, that results from a weakening of the vessel wall.
atherosclerosis	A disease of the arteries characterized by the formation of plaques of fatty material (i.e., atheroma) within the artery wall.
atheroma	A thickening of an artery wall, commonly called a plaque, which consists of fatty deposits, smooth muscle cells, macrophages and other inflammatory cells, and fibrosis.
cardiomyopathy	Disease of the heart muscle that affects the ability of the heart to effectively pump blood.
congestive heart disease	A condition in which the heart is unable to pump sufficient blood through the circulatory system resulting in pulmonary edema.
diuresis	An increased excretion of urine. Diuretic drugs are sometimes used to lower blood pressure.
dyslipidemia	An abnormality in, or abnormal amounts of, lipids or lipoproteins in the blood.
embolism/embolus	An obstruction in a blood vessel due to a blood clot or other foreign matter that gets stuck while traveling through the bloodstream (plural = emboli).
endovascular (procedure)	Refers to procedures carried out inside blood vessels through the use of catheters.
hemorrhagic stroke	A loss of blood flow to a region of the brain due to a ruptured blood vessel.
high-density lipoprotein (HDL)	A lipoprotein that primarily functions in the transport of cholesterol from the tissues to the liver.
hypertension	A persistent, abnormally high blood pressure.
infarct	An area of necrosis in a tissue or organ resulting from obstruction, such as a thrombus or embolus, of the local circulation.
ischemia	A deficiency of blood being supplied to a tissue or organ due to either a constriction or a blockage of a blood vessel.
ischemic stroke	A localized infarct due to an occlusion in a blood vessel supplying a region of the brain.
lipoprotein	A complex of lipids and specific proteins that functions to transport lipids throughout the body.
low-density lipoprotein (LDL)	A lipoprotein that functions primarily in the transport of cholesterol from the liver to the tissues.
thrombus	A stationary blood clot on the wall of a blood vessel, frequently causing vascular obstruction (plural = thrombi).
tissue plasminogen activator (t-PA)	An enzyme (protease) that activates plasmin (another protease) that breaks down the fibrin in blood clots. Can be used therapeutically to treat strokes or heart attacks.

REVIEW QUESTIONS

1. Describe the initiation and progression of atheroma formation and the time frame in which this occurs.

2. What are the risk factors for developing hypertension and how do these factors lead to high blood pressure?

3. Identify and describe the two conditions that can lead to congestive heart failure. What are the clinical manifestations associated with congestive heart failure?

4. Discuss the similarities and differences between ischemic and hemorrhagic strokes.

5. What are the four clinical manifestations often associated with coronary artery disease?

6. What are endovascular procedures and give some examples for the treatment of coronary artery disease and stroke?

ADDITIONAL QUESTIONS

In addition to the Review Questions provided above, there is a range of free online questions designed to enable students to further test their understanding of the chapter material. In order to access these interactive questions, please visit the book's website: https://routledge.com/cw/wiser

Chronic Kidney Disease and Cardiorenal Syndrome

15

Chronic kidney disease is a slow and progressive loss of kidney function over a period of several years that eventually leads to permanent kidney failure. Hypertension (Chapter 14) and diabetes (Chapter 13) are two major predisposing factors in the development of chronic kidney disease. As diabetes and hypertension are increasing, the prevalence of chronic kidney disease is also on the rise. It is predicted that chronic kidney disease may become the fifth leading cause of death by 2040. The progression of chronic kidney disease is usually insidious since symptoms generally do not appear until kidney function is lower than 25% of normal. Furthermore, chronic kidney disease can initiate or accelerate the development of heart disease, and heart disease can initiate or accelerate kidney disease. This combination of kidney and heart disease is referred to as a cardiorenal syndrome. Treatment and control of chronic kidney disease and cardiorenal syndrome are aimed at stopping or slowing down the progression of the disease.

KIDNEY FUNCTION

CHRONIC KIDNEY DISEASE

CARDIORENAL SYNDROME

KIDNEY FUNCTION

The kidneys are paired organs that are located on the back side of the abdominal cavity. Blood that needs to be filtered enters via the renal artery and the filtered blood exits via the renal vein (FIGURE 15.1). Soluble wastes and excess fluids containing electrolytes and other solutes are carried from the kidneys to the urinary bladder by the ureter. The filtration process is carried out by an arrangement of blood vessels and tubules known as the **nephron**. Each kidney contains approximately one million nephrons.

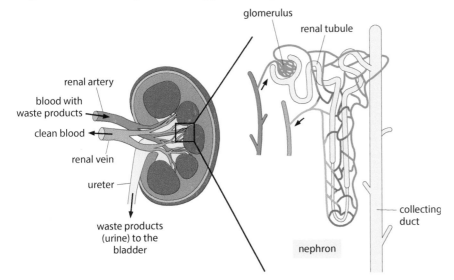

FIGURE 15.1 Basic kidney anatomy and nephron structure. Blood enters the kidney via the renal artery and is filtered by approximately one million interconnected nephrons that make up the kidney. The nephron consists of the glomerulus and a looping renal tubule that is surrounded by peritubular capillaries. The blood is filtered by the nephrons and clean blood is returned to the circulatory system via the renal vein. The filtrate moves into the collecting duct, which connects to the ureter and is passed to the urinary bladder.

The nephron functions as a filter to remove waste products from the blood

Each nephron contains an intertwined mass of capillaries called the **glomerulus**. The glomerulus is surrounded by a funnel-like structure called Bowman's capsule. At the base of Bowman's capsule is a renal tubule that loops within the nephron and is surrounded by a capillary bed. The fluid phase of the blood exits the glomerulus and flows down the tubule (FIGURE 15.2). As the first step in the filtration process, water and small-molecular-mass solutes exit the glomerulus, whereas proteins and cells are retained within the capillaries that surround the renal tubules.

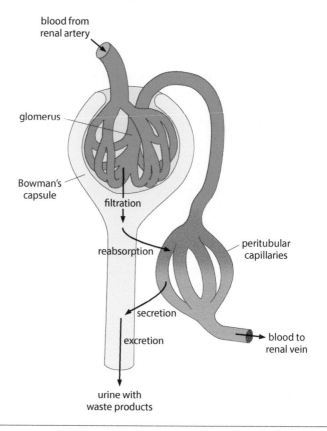

FIGURE 15.2 Nephron function. The fluid phase of blood exits the glomerulus into Bowman's capsule. Water and solutes, such as glucose and amino acids, are reabsorbed as the fluid progresses down the tubule. Metabolic waste and toxins are secreted from the blood into the tubules and carried to the bladder as urine. (Adapted from Madhero88, published under CC BY 3.0.)

Water and solutes that are to be retained by the body are reabsorbed by the capillaries surrounding the renal tubule. Waste and excess solutes are retained in the tubule and are ultimately transported as urine to the bladder via the ureter. The primary metabolic waste products are hydrogen ions (i.e., acid), urea, and creatine. In addition, toxins and drug metabolites are also excreted by the kidneys. Obviously, metabolites such as glucose and amino acids need to be reabsorbed into the blood, unless they are in excess. As the fluid passes through the tubule there is a mix of absorption and secretion of solutes between the capillaries and the tubules to maximize the excretion of waste products and the retention of needed metabolites and electrolytes. Numerous transporters are found in the tubules to carry out the absorption and secretion of solutes. These absorption and secretion activities are largely controlled by osmotic phenomena. However, various hormones also regulate the absorption and secretion of specific solutes.

Kidneys also play a major role in sodium and water balance

To maintain the correct sodium concentration in the blood and other extracellular fluids, the amount of sodium excreted in urine needs to be adjusted according to the amount of sodium in the diet. Hormones such as atrial natriuretic peptide, angiotensin II, and aldosterone regulate the amount of sodium that is reabsorbed by the nephron (TABLE 15.1). Similarly, the kidneys adjust the amount of water excreted or retained in response to changes in the extracellular fluid volume. The amount of water lost in urine is regulated by antidiuretic hormone, also known as vasopressin.

Table 15.1	Examples of Hormones Regulating Water Balance
Atrial natriuretic peptide	Secreted by heart muscle in response to excess stretching due to increased blood volume. Reduces extracellular fluid volume by increasing sodium and water excretion by the kidneys
Angiotensin II	A peptide hormone that increases antidiuretic hormone production, aldosterone secretion, sodium retention in the kidneys, and vasoconstriction
Aldosterone	A steroid hormone produced by the adrenal glands that increases sodium retention in the kidneys and causes vasoconstriction
Antidiuretic hormone	A peptide hormone secreted by the pituitary gland that regulates water retention by the kidneys

Sodium and water balance are important in the regulation of blood pressure since blood volume is an important component of blood pressure (Chapter 14). Dehydration, sodium loss, or blood loss can reduce the extracellular fluid volume and lower blood pressure. In response to the reduction of extracellular fluid volume, the kidneys retain water and electrolytes, leading to an increase in the extracellular fluid volume. Conversely, if the extracellular fluid volume increases or blood pressure goes up, the kidneys excrete more water and electrolytes. Both the cardiovascular system and the kidneys respond to changes in blood pressure or blood volume in a way that maintains homeostasis. Maintaining homeostasis prevents damage to the kidneys, as well as the cardiovascular system.

The renin–angiotensin–aldosterone system plays a central role in the regulation of blood volume and blood pressure

The kidneys play a major role in regulating blood pressure via a protein called renin. Phenomena such as low blood pressure, low sodium concentration, or stimulation of the fight-or-flight response cause the kidneys to secrete renin. Renin is an enzyme that converts a protein called angiotensinogen into angiotensin I. Angiotensin I is then converted into angiotensin II by angiotensin-converting enzyme (ACE). Angiotensin II is a peptide hormone that affects several organ systems involved in the regulation of blood pressure (FIGURE 15.3). The net effect of the actions of angiotensin II is to increase the heart rate, induce vasoconstriction, and increase blood volume, all of which raise blood pressure.

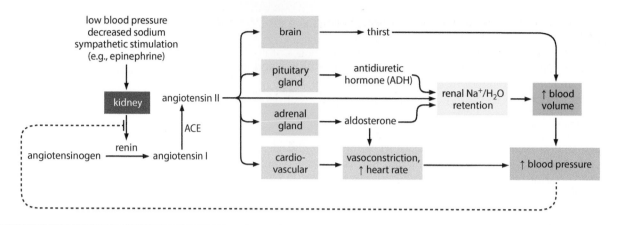

FIGURE 15.3 Regulation of blood pressure by renin. Renin is secreted by the kidneys in response to low blood pressure, low sodium levels, or stimulation of the sympathetic nervous system (e.g., epinephrine). Renin is a protease that converts angiotensinogen into angiotensin I, which is subsequently converted into angiotensin II by a peptidase called angiotensin-converting enzyme (ACE). Angiotensin II is a peptide hormone that affects several organs and glands that work together to increase blood volume, constrict arteries, and increase the heart rate, resulting in an increase in blood pressure. High blood pressure blocks the synthesis and secretion of renin, which helps to maintain blood pressure at an optimal level.

Specific actions of angiotensin II include stimulating thirst, increasing water and sodium retention in the kidneys, increasing the intrinsic heart rate, causing vasoconstriction, and inducing the secretion of two other hormones called antidiuretic hormone (ADH) and aldosterone. ADH and aldosterone also cause the kidney to retain sodium and water and thus increase blood volume. Aldosterone also increases vasoconstriction. All these actions increase blood pressure. High blood pressure feeds back to the kidney and blocks the action of renin. The combination of these various factors is called the **renin–angiotensin–aldosterone system (RAAS)**, and RAAS plays a central role in the homeostasis of blood pressure.

CHRONIC KIDNEY DISEASE

Damage to the glomerulus or other parts of the nephron can compromise the ability of the kidney to filter blood, and if enough damage is done this can result in acute or chronic kidney failure. Acute renal failure is often due to circulatory collapse associated with shock. Other causes of acute renal failure include toxic damage to the nephron associated with drugs, toxins, or severe hemolysis. Chronic kidney disease, conversely, is a slowly progressing disease that can take years or decades to manifest clinical symptoms. It is well known that, even if one kidney stops functioning, the remaining kidney is sufficient to carry out normal functions. Therefore, clinical manifestations associated with kidney disease are usually not apparent until the damage is extensive.

Diabetes and hypertension are two common predisposing factors for chronic kidney disease

As with most chronic diseases, both environmental and genetic factors contribute to kidney disease initiation and progression. In general, the risk factors for developing chronic kidney disease are similar to the risk factors for developing cardiovascular disease. There is also a genetic predisposition to develop chronic kidney disease and therefore having a family history of chronic kidney disease is a risk factor. Kidney function naturally decreases with age and therefore older people are more prone to develop chronic kidney disease. Chronic kidney disease is more frequent in males than females, as well as being more frequent in some races. Modifiable risk factors include obesity and smoking, both of which are also associated with diabetes and hypertension.

Diabetes is the leading cause of chronic kidney disease. It is believed that the hyperfiltration associated with removing excess glucose may damage the nephrons. Many patients diagnosed with type 2 diabetes have protein in the urine at the time of diagnosis. Protein in the urine is an indicator of damage to the nephrons, called **nephropathy**. Many patients with type 2 diabetes develop nephropathy and approximately 10% of these patients experience a gradual loss of kidney function.

Hypertension has been long recognized as a major risk factor for chronic kidney disease and hypertensive end-stage renal disease, and hypertension may account for one-quarter of kidney failures. Overt kidney dysfunction is often observed in patients following years of uncontrolled hypertension. Systemic hypertension increases the intraglomerular pressure and can damage the microvascular of the nephrons. Furthermore, a hardening of the glomerulus, similar to atherosclerosis (Chapter 14), can also occur. Glomerulosclerosis decreases filtration capacity and ultimately causes loss of kidney function.

Gradual loss of nephrons slowly accumulates until overt symptoms manifest

Nephropathy can be viewed as a vicious cycle, in that damage to some nephrons leads to more nephrons subsequently being damaged (FIGURE 15.4). Several factors can damage nephrons. The two most common are hyperfiltration associated with diabetes and glomerulosclerosis associated with hypertension. Other factors that can damage nephrons include exposure to toxins, deposition of antibody complexes, and autoimmune attack (Chapter 17). Loss of nephron function causes the remaining nephrons to work harder. This hyperfiltration then causes more damage and a further loss of nephron function. The decreased nephron function also increases blood pressure, which causes even more damage. Since there is limited regeneration of nephrons, the damage accumulates as the disease progresses and the disease is irreversible.

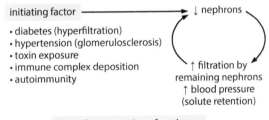

FIGURE 15.4 Vicious cycle of nephropathy. Factors that lead to loss of nephron function cause the remaining nephrons to accommodate for the decreased filtration capacity. The resulting hyperfiltration causes more damage. In addition, decreased filtration capacity increases water and solute retention causing additional glomerulosclerosis due to an increase in blood pressure, which further decreases nephron function.

Initially there are no overt symptoms associated with nephropathy. Since chronic kidney disease rarely exhibits symptoms until the later stages, screening is recommended for those who are at risk. Kidney function can be assessed by determining the **glomerular filtration rate** (BOX 15.1). Chronic kidney disease is divided into five stages based on the percentage of remaining kidney function (TABLE 15.2). During stage 1 chronic kidney disease there are generally no indicators of nephropathy other than a decrease in the glomerular filtration rate. Stage 2 kidney disease exhibits a mild decrease in normal kidney function as noted by protein in the urine and an increase in blood pressure. However, there are no overt clinical symptoms during stage 2 kidney disease. Similarly, in stage 3 there are generally no overt clinical symptoms other than possible signs of anemia. This is likely due to the role kidneys play in the production of new red blood cells via the synthesis of erythropoietin, a hormone that stimulates the production of red blood cells.

BOX 15.1

Glomerular Filtration Rate

Glomerular filtration rate (GFR) is an estimate of how much blood passes through the glomeruli per minute. The GFR can be viewed as a measure of how efficiently the kidneys are filtering blood. A traditional GFR calculation requires an injection into the bloodstream of a substance that is later measured in urine. For example, inulin, a synthetic non-metabolizable sugar, passes completely from the blood into the glomerular filtrate and is not reabsorbed by the tubules. Thus, the amount of inulin found in the urine after a specified time following injection is the amount of inulin removed from the blood during that time period. Simple calculations can convert the amount of inulin in the urine to the GFR.

Creatine clearance is another method to estimate the GFR that does not involve injection and subsequent urine collection, and therefore is more convenient to carry out and more widely used. Creatinine is produced naturally by the body and is especially abundant as creatine phosphate in skeletal muscle. Creatine is a waste product from the normal breakdown of muscle and is filtered by the glomerulus. The GFR can be estimated by measuring the amount of creatine found in either urine or blood. Higher-than-normal levels of creatine in the blood or lower-than-normal creatine levels in the urine are indicative of decreased kidney function.

Table 15.2 Stages of Chronic Kidney Disease

	Normal GFR	Description	Manifestations
1	>90%	Normal function	No symptoms
2	60–89%	Mildly decreased function	Protein in urine
3	30–59%	Moderately decreased function	Some anemia and loss of bone density
4	15–29%	Severely decreased function	Fatigue, swelling, nausea
5	<15%	End-stage kidney disease	Kidney failure requiring dialysis or transplantation

Generally overt clinical symptoms are not observed until stage 4 chronic kidney disease when the kidneys are functioning at less than 30% of the normal levels. Some of the overt symptoms of chronic kidney disease are due to the fluid retention associated with the loss of kidney function (TABLE 15.3). In addition, the deficiencies in filtration and excretion lead to the accumulation of toxins that result in the loss of appetite, nausea, and vomiting. Kidney function below 15% of normal levels is considered kidney failure and is called stage 5 chronic kidney disease or **end-stage renal disease**. At this stage patients either need to be on dialysis or receive a kidney transplant.

Table 15.3 Clinical Features Associated with Chronic Kidney Disease

- Often silent until stages 4 and 5
- Changes in urination
- High blood pressure
- Edema (swelling of the legs and puffiness around the eyes)
- Shortness of breath from fluid accumulation in the lungs (i.e., pulmonary edema)
- Fatigue and weakness (due to anemia and accumulation of waste products)
- Itching, easy bruising, and pale skin (due to anemia)
- Loss of appetite, nausea, and vomiting
- Headaches, numbness in the feet or hands, disturbed sleep, altered mental status

Managing chronic kidney disease involves controlling hypertension and diabetes

There is no direct treatment for kidney disease and there are no drugs that can directly improve the glomerular filtration rate. Therefore, treatment and management are directed at the underlying causes of kidney disease (TABLE 15.4). Furthermore, damage to the kidney is largely irreversible and therefore management is focused on preserving kidney function and slowing the progression of the disease. Since hypertension and diabetes are the two most common underlying causes of chronic kidney disease, management of chronic kidney disease is directed at managing hypertension or diabetes. For example, drugs to lower blood pressure are often prescribed. In addition, drugs to lower cholesterol to manage atherosclerosis may also be prescribed. For kidney disease associated with diabetes it is important to maintain optimal blood glucose levels to prevent further kidney damage.

Table 15.4 Managing Chronic Kidney Disease
• Maintain optimal glucose levels
• Lower blood pressure
• Lower cholesterol
• Erythropoietin ± iron supplementation
• Calcium and vitamin D supplementation
• Low-protein diet

Anemia is often associated with chronic kidney disease and, thus, supplementation with the hormone erythropoietin and sometimes iron is often included in the management of chronic kidney disease. Similarly, loss of bone density is another common comorbidity associated with chronic kidney disease and disease management may include supplementation with vitamin D and calcium. A diet low in protein may also help with the manifestations that are associated with kidney disease. A major metabolic waste product removed from the blood by the kidneys is urea, which is the waste product from the breakdown of protein and amino acids. Therefore, lowering the amount of protein in the diet reduces the stress on the kidneys.

CARDIORENAL SYNDROME

Combined cardiac and renal dysfunction is increasingly being recognized and is often referred to as **cardiorenal syndrome**. Currently there is no precise definition for cardiorenal syndrome and cardiorenal syndrome generally includes a wide range of acute and chronic conditions in which the failure of either the heart or the kidney initiates or accelerates failure of the other. Some attempts to better define cardiorenal syndrome have been made by developing five classes of cardiorenal syndrome based on which organ initiated the failures and whether the failures are acute or chronic.

Cardiac and renal failures have similar predisposing factors

Shared predisposing factors contribute in part to the comorbidities of cardiac and renal dysfunctions (FIGURE 15.5). Specifically, old age, diabetes, and hypertension are major predisposing factors for cardiorenal syndrome, as well as being major predisposing factors for developing either heart failure or kidney failure alone. Both diabetes and hypertension damage the kidneys and lead to nephron loss. Similarly, diabetes and hypertension damage the cardiovascular system, including the heart. Furthermore, diabetes and hypertension promote the development of atherosclerosis, which damages both the kidney and the heart.

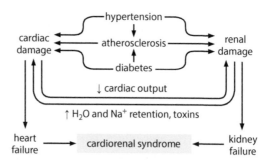

FIGURE 15.5 Cardiorenal syndrome. Hypertension, atherosclerosis, and diabetes can damage the heart and kidneys. Damage to the heart leads to decreased cardiac output, which damages the kidneys. The fluid retention and toxins associated with kidney disease damage the heart. This comorbidity of cardiovascular disease and chronic kidney disease is often referred to as cardiorenal syndrome.

Connections between the heart and kidney promote cardiorenal syndrome

The high prevalence of cardiorenal syndrome cannot be explained solely by shared predisposing factors and there are other connections between the heart and kidney. For example, acute or chronic heart failure may affect the kidneys' ability to carry out filtration, regulate fluid and electrolytes, and clear waste. Similarly, acute kidney injury or chronic kidney disease can affect cardiac performance due to failures in maintaining proper blood volume and electrolyte balance. In addition, higher levels of toxins in the blood associated with kidney dysfunction can affect heart function. Therefore, cardiovascular disease increases the incidence and severity of chronic kidney disease and chronic kidney disease increases the incidence and severity of cardiovascular disease (TABLE 15.5).

Table 15.5   Comorbidities of Cardiovascular and Kidney Diseases
• Shared predisposing and risk factors
• Decreased cardiac output causes kidney dysfunction
• Kidney dysfunction increases blood pressure and toxins in the blood, leading to heart damage

The maintenance of blood flow is a shared responsibility of the heart and kidney. A consequence of heart disease and heart failure is decreased cardiac output. This decreased cardiac output activates the sympathetic nervous system and the renin–angiotensin–aldosterone system (RAAS). Activation of the RAAS results in the retention of sodium and water, as well as increasing the heart rate and vasoconstriction. This attempt to compensate for decreased cardiac output raises blood pressure and may further damage an already damaged heart. At the same time, the decreased cardiac output adversely affects kidney function leading to less filtration and an accumulation of toxins that may further damage the heart. However, cardiorenal syndrome cannot be completely explained by the cycling of low cardiac output and kidney dysfunction. It is likely that other factors, such as inflammation and other neurohormones, play crucial roles in the development and progression of cardiorenal syndrome.

Diuretics and drugs to improve cardiac output are often used to manage cardiorenal syndrome

The management and treatment of cardiorenal syndrome can be complicated, because cardiorenal syndrome is somewhat of an umbrella term to describe various disorders involving both heart and kidney dysfunction. Therefore, the exact treatment depends somewhat on which organ is primarily or secondarily affected and whether the dysfunctions are acute or chronic. Furthermore, there are no drugs that directly improve the glomerular filtration rate. However, improving cardiac output does improve the glomerular filtration rate. Therefore, much of the management of cardiorenal syndrome is focused on improving cardiac output. For example, drugs that increase the heart rate and contraction force may be used. In addition, fluid retention is a common manifestation of cardiorenal syndrome and diuretics are frequently used to reduce the extracellular fluid volume.

SUMMARY OF KEY CONCEPTS

- Kidneys function to remove waste products from the blood and maintain extracellular fluid volume and electrolyte balance
- The nephron is the functional filtration unit of the kidney
- The two most common predisposing factors leading to the development of chronic kidney disease are hypertension and diabetes
- A loss of nephron function compromises the ability of the kidneys to filter blood and maintain the optimal extracellular fluid volume
- Loss of some nephron function promotes further loss of nephrons, leading to a progressive loss of kidney function
- End-stage renal disease is characterized by extensive loss of normal kidney capacity to carry out filtration and requires dialysis or a kidney transplantation to sustain life
- Cardiorenal syndrome refers to the co-occurrence of cardiovascular disease and chronic kidney disease
- Cardiovascular disease initiates and accelerates kidney disease and chronic kidney disease initiates and accelerates cardiovascular disease
- There are no treatments to cure kidney disease and improve the glomerular filtration rate
- Management of chronic kidney disease is primarily directed at the underlying causes, such as hypertension and diabetes

KEY TERMS

cardiorenal syndrome	A wide range of acute and chronic conditions in which the failure of either the heart or the kidney initiates or accelerates the failure of the other.
end-stage renal disease	Stage 5 of chronic kidney disease is characterized by kidney failure and requires dialysis or transplantation to maintain life.
glomerular filtration rate (GFR)	An estimate of the amount of blood that passes through the glomeruli per minute used to assess kidney function.
glomerulus	A network of capillaries within the nephron that perform the first step of filtering blood.
nephron	A vascular structure in the kidney that filters blood and forms urine.
nephropathy	Refers to damage to the kidney that can lead to loss of kidney function.
renin–angiotensin–aldosterone system (RAAS)	A combination of hormones and proteins that regulates blood pressure and fluid and electrolyte balance.

REVIEW QUESTIONS

1. Identify four hormones involved in regulating water and electrolyte balance and describe how they function?

2. What are the clinical manifestations and pathologies associated with the five stages of chronic kidney disease?

3. What is meant by cardiorenal syndrome?

4. Discuss the management of chronic kidney disease and cardiorenal syndrome.

ADDITIONAL QUESTIONS

In addition to the Review Questions provided above, there is a range of free online questions designed to enable students to further test their understanding of the chapter material. In order to access these interactive questions, please visit the book's website: https://routledge.com/cw/wiser

Cancer

16

Cancer is a common disease, and some cancers exhibit a high rate of mortality, making cancer a major cause of death worldwide. Cancer can be simply defined as uncontrolled cell division. Such uncontrolled cell division often has pathological consequences. The etiology of cancer involves both hereditary components and environmental components, but ultimately involves changes in genes that regulate cell division and cell stability. This loss of regulation of cell division can result in tumors and in some cases these tumor cells can become malignant and spread to other parts of the body.

Genes play a major role in the development of cancer, even though there are relatively few hereditary syndromes associated with cancer. Mutations in genes that function in cell division and cell survival lead to cells becoming cancerous. The damage to genes can be due to external factors, such as chemicals and radiation, or can be due to spontaneous mutations that occur during normal cell division. Hundreds of different genes are potentially involved as normal cells progress to cancer cells, and it is the accumulation of mutations in multiple genes that causes cells to become cancerous.

Cancer can essentially afflict any tissue or organ, and therefore can manifest as a large number of rather distinct diseases. The most common cancers resulting in the most deaths are lung cancer, colon cancer, breast cancer, and prostate cancer, and these four cancers are discussed in greater detail. Diagnosis of cancer is difficult due to manifestations of cancers being somewhat non-specific and these manifestations usually do not occur until the disease is in the late stages. This inability to diagnose cancer early greatly compromises the ability to treat cancer. Current cancer treatments rely heavily upon surgery, radiation therapy, and chemotherapy. All these approaches are beset with adverse side effects. Future approaches to cancer treatment may be more targeted at the cancer cells with fewer adverse side effects.

NEOPLASMS

Neoplasia refers to abnormal cell replication. A neoplasm is a mass of cells produced from abnormal replication and neoplasms are more commonly known as tumors. Neoplasia does not refer to normal cellular replication from stem cells (Chapter 8). Neoplasms can range from **benign** to **malignant**. Benign tumors generally exhibit a slow replication rate and are often referred to as **hyperplasia** (TABLE 16.1). In addition, cellular structure and function are only slightly altered during hyperplasia. Many benign tumors are also encapsulated and the cells making up the tumor generally remain localized in a cellular mass. Examples of benign tumors include skin moles and polyps in the colon. Benign tumors rarely cause health problems and have an extremely low mortality rate. An exception is when the tumor presses against something important and interferes with some normal physiological function. For example, most brain tumors are benign, but nonetheless can result in death due to their location and ability to interfere with brain function.

Table 16.1 Benign vs. Malignant Tumors	
Benign	**Malignant**

• Hyperplasia
• Slow growth rate
• Slightly altered cell structure and function
• Cell mass often encapsulated
• Remains localized
• Generally not lethal
	• Hyperplasia → dysplasia
	• Rapid growth rate
	• Loss of cytoplasm, larger and irregular nucleus, chromosomal abnormalities
	• Dysplasia → malignancy
	  ○ loss of adherence and infiltration
	  ○ metastasis via blood and lymphatic system
	  ○ ± angiogenesis
	• Can cause serious disease and death

Progression to malignancy involves changes in cellular structure and function

Benign tumors can evolve into malignant tumors. This is not to say that all benign tumors develop into malignant tumors. Accordingly, many benign tumors remain benign and cause no health problems. The progression from benign to malignant is generally a slow process. Hyperplastic cells may slowly change their appearance and start to lose their normal functions. This pre-malignant state is often referred to as **dysplasia**. Dysplasia can revert back to a more normal state or dysplasia can further progress to malignancy. As cells progress to malignancy, they generally replicate at higher rates, resulting in an increase in the size of the tumor. The malignant cells also have notable changes in their morphology. In particular, the ratio of the size of the nucleus to the size of the cytoplasm increases and the nucleus exhibits an altered shape and staining properties. Furthermore, as cells progress from dysplasia to malignancy these cells often lose their differentiated phenotype and become poorly differentiated. Poorly differentiated cells have a different appearance when examined microscopically and are not organized in their usual patterns (FIGURE 16.1).

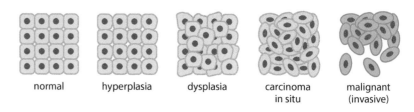

normal   hyperplasia   dysplasia   carcinoma in situ   malignant (invasive)

FIGURE 16.1 Progression of cells to malignancy. As cells become cancerous there are changes in their morphology that can be noted by a pathologist. Benign hyperplastic tumors only have a slightly altered morphology. Dysplasia is characterized by increased alterations in cellular structure. Precancerous cells that have not invaded adjacent tissues are referred to as carcinoma in situ. These cells have also lost many of the characteristics of differentiated cells. Malignancy is characterized by loss of adherence to adjacent cells and invasion of neighboring tissues.

Malignancy is also characterized by loss of adherence to other cells and an ability to invade other tissues. Normal cells are generally tightly connected to their neighboring cells to form tissues. As cells become more malignant, they tend to lose this adherent phenotype. These malignant cells are now free to move and invade neighboring tissues. Furthermore, malignant cells can invade the circulatory and lymphatic systems, allowing them to spread to other parts of the body. This spread throughout the body is called **metastasis**. The metastasizing tumor cells can establish secondary tumors in other organs. There is a high level of mortality associated with metastatic cancer.

Cancers are defined by their tissue origin

Most cell types are capable of becoming malignant and the specific names of cancers are in reference to the type of tissue from which the tumor develops (TABLE 16.2). The major types of cancer are carcinoma, sarcoma, leukemia, lymphoma, and melanoma. Carcinomas are by far the most common and are derived from epithelial tissue. Common carcinomas include skin cancer, lung cancer, colon cancer, breast cancer, and prostate cancer. Sarcomas are derived from mesenchymal tissue, which includes bone, cartilage, fat, muscle, and blood vessels. Cancers derived from white blood cells are called either leukemia or lymphomas. If the cancer is found circulating in the blood and lymph it is leukemia, and if the cancer is a mass in the lymphatics or other organs it is called a lymphoma. Melanomas refer to cancers originating from melanocytes, the pigment-producing cells in the skin.

The basic tissue types are described in Chapter 8, pages 156–158.

SIGNPOST

Table 16.2 Nomenclature of Cancer	
Cancer	**Tissue of origin**
Carcinoma	Epithelial tissue
Sarcoma	Mesenchymal tissue
Leukemia	White blood cells in the blood
Lymphoma	White blood cells as a mass
Melanoma	Melanocytes

The degree of malignancy is defined by four stages

In addition to their cellular origin, cancers are also characterized by their degree of malignancy. Generally, four stages and a precancerous stage are recognized based on the degree of invasiveness of the cancerous cells (TABLE 16.3). For example, stage 0 is defined as a tumor that has not yet invaded neighboring tissues. This is often referred to as carcinoma in situ, which means cancer in place. In addition, stage 0 is often considered precancerous and there is rarely, if ever, mortality associated with this stage. The spread of tumor cells into adjacent tissues marks the beginning of stage 1. Stage 1 tumors are small and, as they increase in size and expand into additional tissues within the same organ, they become stage 2. Stage 3 is characterized by invasion of the lymphatics within the same organ and stage 4 is characterized by metastasis to other organs. However, the exact description of the stages varies according to the specific type of cancer.

Table 16.3	Stages of Cancer
Stage	**Definition**
Stage 0	A small tumor with no invasion of adjacent tissue (precancerous)
Stage 1	A small tumor with invasion into adjacent tissue
Stage 2	A larger tumor with more invasion into the surrounding tissues of the afflicted organ
Stage 3	Invasion of the lymphatics within the afflicted organ
Stage 4	Metastasis to other organs

ETIOLOGY OF CANCER

The most fundamental basis of all cancer is damaged DNA and there are various agents that can damage DNA and increase the risk of cancer (FIGURE 16.2). Many chemicals are known to damage DNA and these chemicals are called **carcinogens**. The list of chemical carcinogens is quite long. Most notable are chemicals called benzopyrenes in cigarette smoke associated with lung and other cancers. Similarly, ionizing radiation damages DNA and causes mutations. Some viruses integrate into the genome of the host cell, and this integration can potentially produce mutations that contribute to carcinogenesis. In addition, some viruses and other infectious agents may stimulate host cell division as part of their life cycle, and this stimulated cell division may promote carcinogenesis. An external source of DNA damage is not necessary to cause carcinogenesis since mutations can also spontaneously occur during normal cell division.

FIGURE 16.2 Examples of agents increasing the risk of cancer. Exposure to chemical carcinogens, radiation, and infectious agents can increase the risk of developing cancer.

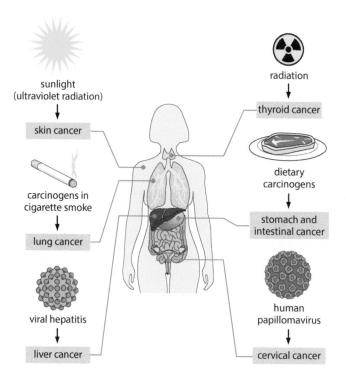

The accumulation of mutations in somatic cells results in cancer

Damage to DNA implies that genes are affected and the damage may alter gene function or expression. In addition, cancer is polygenic since multiple genes are affected during tumorigenesis. This accumulation of mutations in multiple genes can take years or decades to develop. Therefore, over time exposure to carcinogens and ionizing radiation increases the risk of developing cancer by potentially increasing the number of accumulated mutations. The mutations associated with the development of cancer are present in somatic cells, and the cells of a tumor are clonal in nature, since a single cell may initially start the carcinogenesis process. Therefore, the degree of malignancy increases as mutations accumulate in subsequent generations of the tumor cells (FIGURE 16.3).

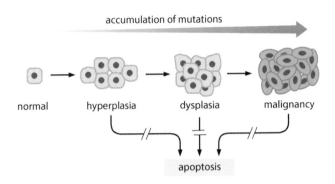

FIGURE 16.3 Mutations and the progression of cancer. Mutations accumulate in somatic cells during carcinogenesis leading to increasing malignancy. Mutations also occur in genes that interfere with apoptosis, and this decreases the removal of abnormal cells.

As tumor cells evolve the integrity of cell division diminishes and this results in chromosomal abnormalities. Therefore, tumor cells often no longer have two complete copies of all the chromosomes, may have extra copies of some chromosomes, or abnormal chromosomes due to breakage and translocation events. This state of chromosomal abnormality is referred to as aneuploidy and accounts for the large and darkly stained nuclei in malignant cells. Generally, cells that have mutated DNA and chromosomal abnormalities are induced to undergo apoptosis. **Apoptosis** is a form of cell suicide in which defective cells are eliminated (BOX 16.1). Therefore, carcinogenesis and the development of cancer also involve a failure of apoptosis to eliminate these abnormal cells.

Apoptosis

In contrast to necrosis, which is a form of traumatic cell death due to injury, apoptosis is a programmed cell death that is a normal part of development and the maintenance of cellular stability. Through apoptosis, cells that are no longer needed, or that have been damaged beyond repair, are removed in an orderly fashion with a minimum of damage to other cells. Cells undergoing apoptosis fragment into apoptotic bodies that can be phagocytosed by macrophages. This process prevents cell lysis and the release of cellular contents that can damage other cells and cause inflammation.

There are two pathways by which apoptosis is initiated: an intrinsic pathway and an extrinsic pathway. The intrinsic pathway is initiated by intracellular signals in response to cell stress or cell damage. This pathway involves the release of mitochondrial proteins and is sometimes referred to as the mitochondrial pathway. The extrinsic pathway is mediated by extracellular signals that bind to receptors on the surface of cells. Both initiation pathways lead to the activation of proteases called caspases that degrade intracellular proteins. During apoptosis cells shrink, the DNA fragments, and the cell forms apoptotic bodies that are phagocytosed.

BOX 16.1

HEREDITARY CANCER SYNDROME

The fact that mutated genes are the basis of cancer implies that there is a strong genetic element in the etiology of cancer. However, there are relatively few examples of single genes being both sufficient and necessary to cause cancer. A heavily cited example is familiar retinoblastoma. Retinoblastoma is a rare cancer of the retina that is inherited in an autosomal dominant manner and exhibits 90% penetrance. The inherited form of retinoblastoma is found almost exclusively in young children and children with this syndrome are prone to develop other tumors later in life.

Cancers that exhibit a monogenic inheritance pattern, such as retinoblastoma, are referred to as hereditary cancer syndromes (TABLE 16.4). The tumors associated with hereditary cancers tend to develop at a relatively early age and multiple independent tumors may arise. In addition, as many hereditary cancers are autosomal dominant, the same syndrome is often seen in first-degree relatives, such as parents, children, and siblings. However, many of these syndromes exhibit a low penetrance and may not be observed in every generation. The low penetrance means that not everyone with the mutated allele will develop cancer and, thus, the mutated alleles can be viewed as predisposing factors that increase the risk of developing specific cancer. Hereditary cancer syndromes are estimated to account for approximately 5% of all cancer.

Table 16.4   Common Hereditary Cancer Syndromes

Syndrome (genes)	Gene function	Incidence	Penetrance	Common tumor types
HNPCC* (*MSH2, MLH1, MSH6, PMS2*)	DNA repair	1:500	80–90%	Colon cancer and others
Hereditary breast and ovarian cancer (*BRCA1, BRCA2*)	DNA repair	1:500–1:1000	30–80%	Breast, ovarian, and prostate cancer
Neurofibromatosis type 1 (*NF1*)	Tumor suppressor	1:3000	?	Benign neurofibroma
Familial retinoblastoma (*RB1*)	Tumor suppressor	1:15,000–1:30,000	90%	Bilateral retinoblastoma in childhood
Familial adenomatous polyposis (*APC*)	Tumor suppressor	1:33,000	100%	Numerous (>100) colon polyps

*HNPCC, hereditary non-polyposis colorectal cancer.

GENES OF CANCER

Cancer is associated with DNA damage leading to the malfunction of genes. This malfunction could be a mutation in the gene that affects its function, or a mutation that affects gene expression. Genes associated with cancer are often categorized as either **proto-oncogenes** or **tumor suppressor** genes. Proto-oncogenes promote cell division, whereas tumor suppressor genes inhibit cell division. Mutations in proto-oncogenes that result in an increase in activity, or a gain of function, may lead to inappropriate cell division. In contrast, mutations in tumor suppressor genes that result in a loss of function tend to lead to inappropriate cell division. Cell division is highly regulated, and it generally takes several mutations in multiple proto-oncogenes and tumor suppressor genes for cells to progress to a malignant state.

Proto-oncogenes are regulatory proteins

Proto-oncogenes generally encode proteins that are a part of pathways that stimulate cells to divide (FIGURE 16.4). For example, there are many hormones that activate cell division, and these hormones are generally referred to as growth factors. The overexpression of such growth factors can promote the development of cancer. Hormones bind to receptors and activate signal transduction pathways that often include protein kinases. Thus, mutations anywhere between the receptor and final protein target may lead to inappropriate cell division. For example, a mutation in a protein that is part of the signal transduction pathway, such as a protein kinase, may cause that protein to be active even when a hormone is not present. Thus, the cell thinks that the hormone is present even when it is not. Some proto-oncogenes are transcription factors that regulate the expression of genes which promote cell division. In this case mutations in transcription factors may lead to the inappropriate expression of genes and activation of cell replication.

> **Protein kinases and signal transduction pathways are described in Chapter 3, pages 63–66.**
>
> SIGNPOST

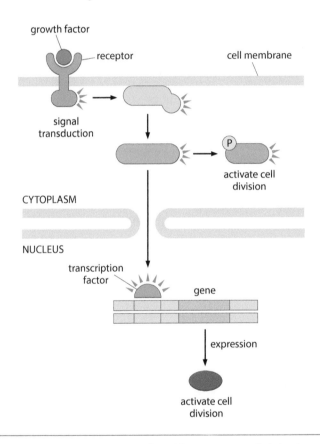

FIGURE 16.4 Potential proto-oncogenes. Potential proto-oncogenes include a large number of regulatory proteins in which a gain of function could result in the activation of cell division. These include growth factors, growth factor receptors, components of the signal transduction pathway, and transcription factors

Increased expression of a gene is also considered a gain of function and mutations that lead to higher-than-normal levels of expression of the proto-oncogene may promote inappropriate cell division. Gain-of-function mutations lead to proto-oncogenes becoming oncogenes and oncogenes facilitate the conversion of normal cells into cancer cells. In other words, a mutated proto-oncogene that is more active than normal, or expressed at higher levels than normal, is an oncogene (FIGURE 16.5).

FIGURE 16.5 Conversion of proto-oncogene to oncogene. Mutations in proto-oncogenes that result in a gain of function can promote tumorigenesis. For example, some mutations may result in an overly active protein, or other mutations can increase the expression of the proto-oncogene.

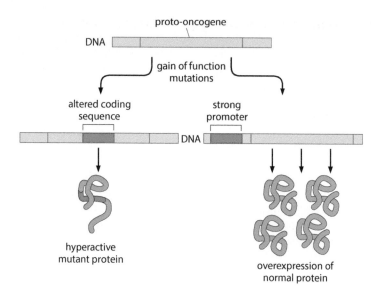

Tumor suppressors function as anti-oncogenes

In addition to proteins that activate cell division there are many proteins that inhibit cell division. Those proteins that block cell division are called tumor suppressors. Therefore, tumor suppressor genes are sometimes referred to as anti-oncogenes since they have the opposite effect as proto-oncogenes. The expression of tumor suppressors blocks cell division and mutations that lead to a loss of function promote cellular replication (FIGURE 16.6).

FIGURE 16.6 Tumor suppressors. Genes classified as tumor suppressors carry out various functions related to cell replication and cell survival. Loss of function of tumor suppressor genes increases the risk of cells becoming cancerous.

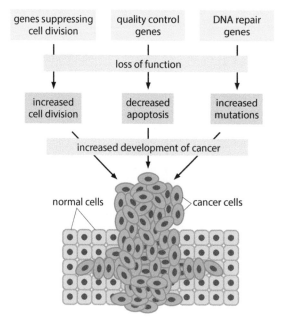

Some tumor suppressors function by blocking the expression or action of genes that regulate cell division. Thus, if the tumor suppressor is not expressed, or its function is compromised, inappropriate cell division may occur. Some tumor suppressors function in the quality control of DNA and cellular replication. If chromosomal or cellular damage is detected, the tumor suppressor blocks the continuation of the cell division process and if the damage is beyond repair the tumor suppressor induces apoptosis (see Box 16.1). A loss of function in these genes allows the survival of cancerous cells. DNA repair genes are also considered to be tumor suppressors. A diminished capacity to repair damaged DNA increases the risk of acquiring mutations that potentially lead to cancer. Tumor suppressor genes are generally recessive and require the loss of both alleles to promote the development of cancer.

Numerous cellular processes associated with cell survival are involved in tumorigenesis

Genes affecting the development of cancer can generally be classified as proto-oncogenes or tumor suppressor genes based on gain-of-function or loss-of-function phenotypes, respectively. However, this does not speak about the normal cellular functions of proto-oncogenes and tumor suppressor genes. Genes that are associated with tumorigenesis are generally involved in three cellular processes: cell fate, cell subsistence, and genome maintenance (TABLE 16.5). Such genes are sometimes called **cancer driver genes** since the abnormal function of these genes promotes the development of cancer.

Table 16.5   Cellular Processes Associated with Tumorigenesis

Cell fate	Cell subsistence	Genome maintenance
• Balance of stem cells and differentiated cells • Epigenetic chromatin remodeling	• Growth factor signaling pathways • Angiogenesis	• DNA repair • G2 quality control (aneuploidy)

It has long been recognized that cancer cells are in a de-differentiated state and somewhat similar to stem cells. Therefore, it is not surprising that many of the mutations associated with cancer are in genes that regulate the differentiation process. This would also include genes involved in chromatin-modifying activities that are associated with the epigenetic modification of chromatin. Mutations that block or reverse the differentiation process of cells can lead to a cell becoming cancerous.

Cancer cells are often able to replicate under conditions of suboptimal nutrients. Mutations that allow cancer cells to proliferate under limited nutrients include genes for growth hormones, their receptors, and the associated signal transduction mechanisms. Such mutations allow cancer cells to override the normal controls that would prevent normal cells from proliferating under suboptimal conditions. Similarly, cancer cells can often override the signals of apoptosis. Thus, the survival of cancer cells is enhanced and cancer cells persist in situations when normal cells would not.

Another process that increases the survival of cancer cells is the stimulation of angiogenesis. Tumor cells require nourishment and as the tumor mass increases the cells in the inner portion of the mass are deprived of nutrients. Therefore, many tumors induce the formation of blood vessels to supply nutrients to the entire tumor. This process of blood vessel formation is called angiogenesis. Furthermore, the stimulation of angiogenesis is also associated with increased malignancy and metastasis.

Many genes associated with cancer are involved in DNA repair. Obviously, any mutations adversely affecting the DNA repair process increase the risk of developing cancer. A less efficient DNA repair process means that there will be more mutations and a greater chance of a mutation in a cancer-driving gene. Tumor cells also tend to be aneuploidy. Aneuploidy refers

Cellular differentiation and epigenetics are discussed in Chapter 8, pages 161–162.

SIGNPOST

to a state in which the chromosomes have an abnormal morphology, and the normal diploid state is not maintained. Cells that have chromosomal abnormalities are usually eliminated during the gap 2 (G2) period of the cell cycle (see Box 2.5). This means that there has been a breakdown in quality control mechanisms in cancer cells.

TUMOR PROTEIN 53

There are hundreds of genes that can drive cells to become cancerous. Among these cancer drivers, the tumor protein 53 gene (*TP53*) is particularly noteworthy since mutations in *TP53* are associated with about half of all cancers. The *TP53* gene encodes for a protein designated as p53. This designation is in reference to p53 being a protein (p) of 53 kilodaltons in mass. The p53 has cellular roles in genome stability, DNA repair, apoptosis, and cell metabolism. Because of its role in genome stability and DNA repair, p53 is sometimes called the 'guardian of the genome.' In its guardian-of-the-genome role p53 functions as a tumor suppressor via the induction of cell cycle arrest, senescence, or apoptosis in response to a variety of cellular stress signals.

Cellular stress increases p53 levels within cells and activates p53 as a transcription factor

The p53 serves as a sensor of cellular stress and responds to signals associated with DNA or chromosome damage, hypoxia, nutrient deprivation, osmotic shock, and ribosome dysfunction (FIGURE 16.7). These stressors increase the levels of p53 within the cell by stabilizing p53. Normally p53 is kept at low levels in unstressed cells by continuous degradation of the protein. The various stress signals activate protein kinases that phosphorylate p53 and render p53 resistant to the normal degradation process. Thus, the levels of p53 increase in response to stress signals. Furthermore, phosphorylation also activates p53 as a transcription factor that then initiates the expression of other genes.

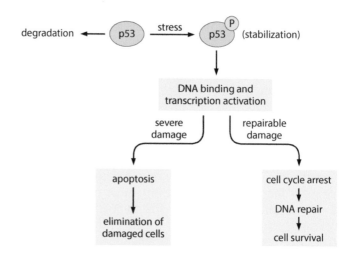

FIGURE 16.7 **Activation and function of p53.** Various cellular stresses activate the phosphorylation of p53. This phosphorylation of p53 leads to higher p53 levels and activates p53 as a transcription factor. Depending on the cellular context the activated p53 initiates a specific gene expression program. For example, p53 activates the genes associated with apoptosis in severely damaged cells. If the damage is reparable, then p53 activates genes associated with cell cycle arrest and DNA repair.

The p53 plays key roles in regulating cell division and apoptosis

The cell type, cellular environment, and nature of the stress determine the specific genes that are turned on by p53. For example, in response to sustained or severe stress, genes associated with either apoptosis or senescence are activated. The p53-activated apoptosis results in the expression of genes associated with both the intrinsic and extrinsic apoptosis pathways (see Box 16.1). In other situations, p53 may activate genes promoting senescence, which leads to a state of permanent cell division arrest. Thus, apoptosis and senescence permanently suppress potential cancer cells by eliminating the cells or blocking their replication, and thus suppress tumor formation.

If cellular repair is possible, p53 activates genes that temporarily arrest cells at the G1/S boundary of the cell cycle (see Box 2.5). In addition, p53 activates DNA repair genes and the temporary arrest of cell division allows for the repair of damaged DNA. This cell division arrest and DNA repair prevent the errors in the DNA sequence from being propagated in future cell generations. Thus, the accumulation of mutants necessary for tumorigenesis is decreased. As p53 plays a major role in maintaining the genome as well as suppressing unregulated cell division, it is not surprising to find that mutations in the *TP53* gene are associated with approximately half of all cancers.

LUNG CANCER

Among the various cancers, lung cancer causes the highest number of deaths. The number of deaths due to lung cancer is greater than the combined deaths due to colon, breast, and prostate cancers. Lung cancer rates began to rise in the 1920s following the increase in the popularity of smoking tobacco (FIGURE 16.8). Approximately 85% of lung cancer is directly linked to smoking tobacco and smoking tobacco increases the risk of developing lung cancer 10-fold. Causes of lung cancer in non-smokers include exposure to radon gas, asbestos, air pollution, and second-hand smoke. Smoking increases the risk of developing other cancers as well as increasing the risk of developing cardiovascular disease (Chapter 14). Thus, smoking is often referred to as the number one preventable cause of death.

Benzopyrenes found in tobacco smoke damage DNA and increase the risk of developing cancer

Dozens of known or suspected carcinogens are found in tobacco smoke as well as toxic metals and formaldehyde. The most notable among the carcinogens found in tobacco smoke is benzo[a]pyrene. Benzo[a]pyrene results from the incomplete combustion of organic matter. After being absorbed, the benzo[a]pyrene is oxidized to benzopyrene diol epoxide (FIGURE 16.9). Benzopyrenes have a flat structure and can intercalate between the strands of a DNA molecule. The oxygen of the epoxide group can then form an adduct with the guanine bases of DNA. This disrupts normal DNA replication and increases the probability of a mutation occurring. In particular, guanine is often converted to thymidine as a result of the DNA–benzopyrene adduct. Such mutations may affect the function or expression of proto-oncogenes or tumor suppressor genes and promote the development of cancer.

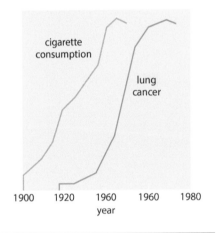

FIGURE 16.8 Correlation with cigarette smoking and lung cancer. Lung cancer rates began to rise approximately 20 years after the increase in cigarette smoking and increased proportionately to cigarette consumption.

FIGURE 16.9 Role of benzopyrenes in carcinogenesis. Benzo[a]pyrene is oxidized to an epoxide by the cellular protein P450. The flat benzopyrene can intercalate between the strands of a DNA molecule (red circle) and the highly reactive epoxide group (blue) forms a chemical adduct with guanine. These DNA–benzopyrene adducts can result in mutations that affect the function or expression of genes that may possibly lead to neoplasia.

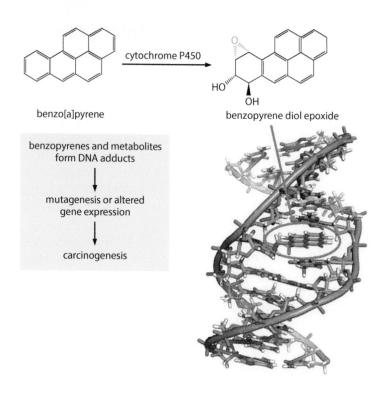

benzo[a]pyrene

cytochrome P450

benzopyrene diol epoxide

benzopyrenes and metabolites form DNA adducts

↓

mutagenesis or altered gene expression

↓

carcinogenesis

The type and stage of lung cancer influence prognosis

Lung cancers are initially asymptomatic and generally not diagnosed until stage 4. Therefore, the prognosis and survival among patients with lung cancer are rather low. In addition, there are different types of lung cancer that influence development and prognosis. Approximately 95% of lung cancers are carcinomas derived from epithelial cells. Most of these lung carcinomas fall into four distinct histological types based on their microscopic appearances (TABLE 16.6). Small cell carcinomas arise from specialized neuroendocrine cells of the bronchial epithelium and have the strongest relationship to cigarette smoking. Small cell carcinomas grow rapidly and metastasize early. Adenocarcinomas are the most common type of lung cancer. They are derived from secretory cells of the epithelium in the small bronchi and tend to grow more slowly and metastasize later than other lung carcinomas. Squamous cell carcinomas tend to arise in the main bronchi and differentiate into the types of cells found in the epithelium of the skin. Large cell carcinomas are large round cells that lack differentiation and probably represent squamous carcinomas or adenocarcinomas that are too undifferentiated to be classified. They are rapidly dividing cells that metastasize early.

Table 16.6 Most Common Types of Lung Cancer			
Type	**Percentage**	**Cell origin**	**Characteristics**
Small cell carcinoma	15–20	Neuroendocrine cells	Grows and spreads rapidly
Adenocarcinoma	30–40	Mucus-secreting cells	More slowly growing
Squamous cell carcinoma	25–30	Cells lining bronchi	More slowly growing
Large cell carcinoma	10–15	Undifferentiated (unknown)	Grows and spreads rapidly

COLORECTAL CANCER

Colorectal cancer, or more commonly colon cancer, is the third most common form of cancer, making up approximately 10% of all cancer. Most colon cancer is associated with older age and lifestyle factors. Lifestyle risk factors include diet, obesity, smoking, and lack of physical activity. The dietary factors that increase the risk of colon cancer include the high consumption of red meat, processed meat, and alcohol. People with inflammatory bowel disease are also at increased risk of developing colon cancer. In addition, there are two hereditary syndromes associated with colon cancer (see Table 16.4). These hereditary cancers (familial adenomatous polyposis and hereditary non-polyposis colon cancer) represent less than 5% of total colon cancer.

Inflammatory bowel disease is described in Chapter 17, page 332.

SIGNPOST

More than 95% of colon cancer arise from benign tumors called polyps. More specifically these benign tumors are usually adenomas since they arise from glandular or secretory cells of the colonic epithelium. These adenomatous polyps are considered to be precancerous tumors that may develop into carcinoma. As with all cancer, this development from a benign tumor to a malignant cancer is due to the sequential accumulation of mutations in genes regulating normal growth and development. This accumulation of mutations in the progression of cancer has been particularly well studied in colorectal cancer (BOX 16.2).

Progression of Colon Cancer

Tumors evolve from benign to malignant through the sequential accumulation of mutations and this process has been particularly well studied in colorectal cancer. The first mutation leading to the initial development of an adenoma is often in the adenomatous polyposis coli (*APC*) gene. The *APC* gene is also the affected gene in familial adenomatous polyposis syndrome (see Table 16.4) and variations in the *APC* gene lead to the formation of hundreds to thousands of polyps. Loss of function of this gene also leads to the formation of a slowly growing and small adenoma, and thus, the *APC* gene is classified as a tumor suppressor. A subsequent mutation in the *RAS* gene leads to additional growth and the small adenoma develops into a larger adenoma (**BOX 16.2 FIGURE 1**). The K-ras protein plays a role in signal transduction and increased expression of this gene promotes cell growth; thus it is considered a proto-oncogene. Additional mutations in other genes, and especially *TP53*, lead to the development of malignant tumor cells that can invade the lymph nodes and metastasize to other organs.

BOX 16.2

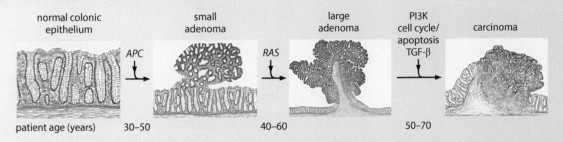

BOX 16.2 FIGURE 1 Gene alterations associated with the progression of colon cancer. Sequential mutations correlate with the progression from benign adenomas to metastatic carcinomas. PI3K, phosphatidylinositol 3-kinase; TGF, transforming growth factor. [From Vogelstein B, Papadopoulos N, Velculescu VE et al. (2013) Cancer genome landscapes. *Science* 339: 1546–1558. doi: 10.1126/science.1235122. With permission from AAAS.]

Development from the polyp to metastatic stage 4 cancer is defined by the sequential invasion of the tumor into the submucosa, the muscle wall, and lymph nodes, and finally becoming metastatic (FIGURE 16.10). During stages 1 and 2 the tumor directly extends into the wall of the colon and thus it is still relatively localized. During stage 3 the tumor cells invade the lymph nodes, and the spread to other organs marks the beginning of stage 4. The most common sites for secondary colon tumors are the liver and lungs. The high level of secondary tumors in the liver is likely due to the portal circulatory system that drains blood from the intestines directly to the liver. This development from a benign polyp to stage 4 carcinoma is rather slow and can take decades. Unfortunately, the early stages of colon cancer are almost always asymptomatic, and diagnosis is usually made after cancer has become metastatic when the five-year survival rate is only around 5%. Therefore, starting at age 50, screening for polyps is recommended. Polyps detected by colonoscopy are removed at the time of detection.

FIGURE 16.10 Stages of colorectal cancer. Precancerous polyps are restricted to the epithelial cells of the mucosa. Spread of cancer cells to the submucosa is designated as stage 1 and penetration of the muscle layer is stage 2. Detection of cancer cells in intestinal lymph nodes marks the beginning of stage 3 and spread to other organs is stage 4.

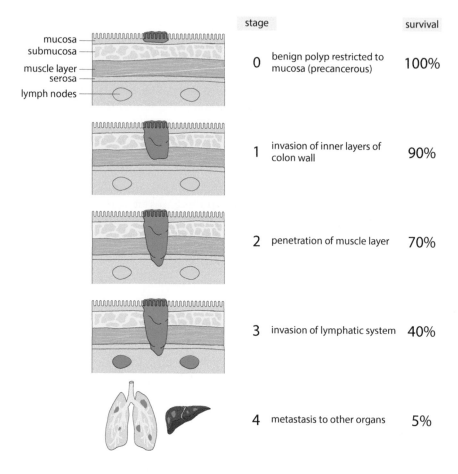

	stage		survival
	0	benign polyp restricted to mucosa (precancerous)	100%
	1	invasion of inner layers of colon wall	90%
	2	penetration of muscle layer	70%
	3	invasion of lymphatic system	40%
	4	metastasis to other organs	5%

BREAST CANCER

Breast cancer is the most common cancer of women and only in 1986 did lung cancer pass breast cancer as the most common cause of cancer death in women. Furthermore, it is a highly feared cancer in that, in addition to the associated mortality, breast cancer threatens disfigurement, self-esteem, and relationships more than any other type of cancer. Most breast cancers are carcinomas derived from epithelial cells of the breast ducts or the epithelial cells of the milk glands. Ductal carcinoma accounts for 75–80% of breast cancer and lobular carcinoma (derived from the milk glands) accounts for 10–15% of breast cancer. Sarcomas derived from fibrous tissue, blood vessels, adipose, or nervous tissue are relatively rare and generally behave similarly to malignant tumors derived from those tissues found in other parts of the body.

The etiology of breast cancer is a complex mix of genetic, hormonal, and environmental factors

The largest risk factor for developing breast cancer is being a woman, and women have a 100-fold higher risk of developing breast cancer than men (TABLE 16.7). This is largely due to having more breast tissue and in part due to increased exposure to estrogen. Early onset of menses, childlessness, and delayed childbearing all increase exposure to endogenous estrogen, as well as increasing the risk of developing breast cancer. Similarly, women with naturally high estrogen and progesterone levels are more likely to develop breast cancer. Furthermore, exposure to exogenous estrogen from oral contraceptives and estrogen replacement therapy slightly increases the risks of developing breast cancer. This also includes exposure to diethylstilbestrol (DES), which is a synthetic estrogen prescribed to pregnant women between 1940 and 1971 to prevent miscarriage, premature labor, and related complications during pregnancy. Age, smoking, alcohol consumption, and obesity also contribute to breast cancer risk. Breast cancer also has a strong genetic component and breast cancer in female first-degree relatives, such as mothers, sisters, or daughters, greatly increases the risk.

Table 16.7	Major Risk Factors Associated with Breast Cancer
Sex	Women are at 100× greater risk than men
Family history	Having a first-degree female relative with breast cancer doubles the risk
Estrogen	Prolonged exposure to endogenous or exogenous estrogen increases the risk
Age	Most breast cancer appears after age 55
Lifestyle factors	Cigarette smoking, alcohol consumption, lack of physical activity, and obesity increase risk

Some breast cancer is associated with hereditary breast–ovarian cancer syndrome

The genetic predisposition to develop breast cancer is well documented and 5–10% of breast cancer is associated with a hereditary breast–ovarian cancer syndrome (see Table 16.4). Gene mapping studies led to the identification of two genes associated with the increased risk of developing breast cancer and these genes were named *BRCA1* and *BRCA2*. BRCA stands for breast cancer and is pronounced bra-ca. Some variations in the *BRCA* genes increase the risk of developing breast or ovarian cancer. Both *BRCA1* and *BRCA2* are involved in DNA repair and are classified as tumor suppressor genes. The presumption is that these variations in the *BRCA* genes decrease the functionality of the BRCA proteins, possibly leading to an increase in errors in other genes, thus increasing the risk of developing cancer. Women with deleterious mutations in either *BRCA1* or *BRCA2* have a risk of breast cancer that is approximately five times the normal risk, and a risk of ovarian cancer that is 10–30 times higher than normal. These mutations also increase the risk of other cancers, but not to the extent of breast and ovarian cancer.

PROSTATE CANCER

The prostate is an exocrine gland that surrounds the urethra as it emerges from the bladder. A major function of the prostate is to secrete fluid into the semen. Most prostate cancers are derived from the secretory cells producing the prostatic fluid, and thus are adenocarcinomas. As with most cancer, the initial stages of prostate cancer are generally asymptomatic. It is possible to screen asymptomatic men for prostate cancer by detecting prostate-specific antigen (PSA) in the blood or the digital rectal exam. However, there is some controversy regarding prostate cancer screening (BOX 16.3).

BOX 16.3

Prostate Cancer Screening

The digital rectal examination (DRE) accesses the prostate manually through the wall of the rectum. The examiner slides a gloved finger into the rectum and presses against the prostate to check its size and detect any lumps. However, there is debate on the efficacy of the DRE in detecting prostate cancer and there is no clear evidence that DRE reduces prostate cancer mortality. Another potential screen for prostate cancer is measuring prostate-specific antigen (PSA) in the blood. PSA is a protease produced by the prostate gland that liquefies semen and allows sperm to swim freely following ejaculation. Normally PSA levels are extremely low in the blood. Damage to the prostate increases the level of PSA in the blood and, thus, PSA can be a biomarker for prostate cancer. However, PSA is not a unique indicator of prostate cancer, but it may also detect inflammation of the prostate or benign prostatic hyperplasia. Thus, PSA screening has a high false-positive rate.

Due to the low sensitivity and specificity of the DRE and PSA test, prostate cancer screening is controversial. PSA screening has undoubtedly resulted in the early diagnosis of prostate cancer and saved some lives. However, overdiagnosis can lead to unnecessary biopsies, which carry some risk of harm. Furthermore, most prostate cancer develops rather slowly and overdiagnosis could lead to unnecessary treatment for a disease that would not have advanced to a stage that causes serious manifestations or premature mortality. Most recommendations suggest screening after a discussion between the patient and health-care provider about the uncertainties, risks, and potential benefits associated with prostate cancer screening.

Prostate cancer is the most common cancer in men. It is not as aggressive as lung or colon cancer and therefore the mortality associated with prostate cancer is less than those cancers. Prostate cancer is generally associated with older age and the slow growth of most prostate cancer has resulted in the adage that one is more likely to die 'with' prostate cancer than die 'of' it. Nonetheless, some prostate cancer can be aggressive and produce a metastatic disease with high mortality. Malignant prostate cancer cells have a strong tendency to metastasize to the lymph nodes and bones, particularly of the pelvis and lower spine.

Androgens, such as testosterone, also play a role in prostate cancer. However, it is unlikely that androgens drive prostatic epithelial cells to become tumorigenic. Rather testosterone is a growth factor for prostatic epithelial cells, and the epithelial cells retain testosterone receptors after they become malignant. Thus, prostatic cancer cells often require testosterone for their growth.

Age, family history, race, and hormones play a role in the development of prostate cancer

Nearly 99% of prostate cancer is diagnosed after the age of 50 and the average age at the time of diagnosis is 70. Men with a first-degree relative (i.e., brother or father) with prostate cancer have a two- to threefold increased risk of developing prostate cancer compared with men without prostate cancer in the family. Mutations in *BRCA2*, one of the breast cancer genes, are also associated with an increased risk of prostate cancer. There is also a strong racial variation in the incidence of prostate cancer. The incidence in white men is two-thirds the incidence in black men, and the incidence in Asian men is about half of that in white men. Currently there is no explanation to explain those racial differences. However, there is also an environmental component as Asian men moving permanently to the USA or Europe have a higher incidence of prostate cancer compared to men living in China or Japan.

DIAGNOSIS AND SCREENING

There are no tests that can accurately diagnose all cancers. Cancers are often defined by the organ in which they originate and there may be organ-specific symptoms associated with a specific cancer. Furthermore, most cancers do not produce symptoms until the cancer has reached a rather advanced stage. Therefore, cancer can be very difficult to diagnose in a timely fashion which allows for a good prognosis. There is a great need for biomarkers that accurately diagnose cancer in the early stages. There are some clinical manifestations that suggest cancer. For example, local manifestations can include an abnormal mass, pain, obstruction, hemorrhage, and bone fracture (TABLE 16.8). Systemic manifestations can include cachexia, anemia, infection, and hormonal imbalances. However, manifestations may not appear at all and, if they do, they usually manifest after malignancy and metastasis. Quite often cancer is diagnosed in older people during doctor visits for other issues.

Table 16.8	Clinical Manifestations Associated with Cancer
Manifestation	**Comment**
Abnormal mass	Depending on location and size, a tumor may be visible or palpable. Imaging techniques are also useful for detecting some tumors
Pain	Cancer can destroy tissues, invade nervous tissue, or cause inflammation, which can result in localized pain
Obstruction	Body passageways can become obstructed by a tumor growing in the lumen of the passageway or by compression of the passageway by adjacent tumors
Hemorrhage	Cancers of mucosal surfaces or the skin may ulcerate and bleed
Fractures	Primary bone cancers or cancers that metastasize to the bones can weaken the bones resulting in fractures during minor injuries
Cachexia	Persistent fatigue and generalized wasting are often seen in late-stage cancer. This may be due to a combination of loss of appetite, increased nutritional demands of the growing tumor, and increased inflammation associated with the tumor
Anemia	Anemia is one of the most frequent manifestations associated with cancer. This can be due to blood loss. In addition, anemia is also associated with decreased erythropoiesis due to the side effects of chemotherapy
Infection	Infection is a common complication associated with cancer and is a common cause of death in patients with cancer. Increased infections are associated with tissue erosion, obstructions, cachexia, and decreased immunity caused by the cancer
Hormone production	Benign tumors of endocrine glands commonly cause an overproduction of hormones that can lead to clinical manifestations

Biopsies are often used to confirm the diagnosis of cancer

Since the manifestations associated with cancer tend to be non-specific, a confirmed diagnosis of cancer generally requires cytological examination. For example, a biopsy of the tumor or cells collected from the affected tissue can be examined microscopically. The pathologist looks for characteristics that define the cells as being dysplastic or malignant. As cells progress from normal to malignant, there are changes in the cell structure that can be seen with microscopy (see Figure 16.1). In particular, changes in the structure of the nucleus and the ratio of the nucleus/

cytoplasm sizes increases as cells become malignant (FIGURE 16.11). In addition, the stage of cancer (see Table 16.3) and the involvement of lymph nodes can also be determined from a biopsy. The stage of the cancer is important regarding the types of treatment and the prognosis of the disease.

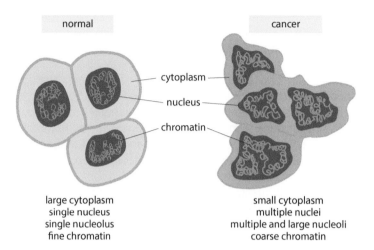

FIGURE 16.11 Comparison of normal and cancer cells. As cells become malignant the relative size of the nucleus compared with the cytoplasm increases and there are changes in the structure of the nucleus. (Adapted from the National Cancer Institute of the National Institutes of Health of the United States.)

Screening for early cancer detection is limited

It is widely accepted that the prognosis is much better and mortality much lower when the cancer is detected in the early stages. However, cancer is usually asymptomatic in the early stages and difficult to detect. Ideally, screening tests that detect early stage cancer would lower the mortality associated with cancer. Unfortunately, relatively few such screening tests exist. In addition, in cases in which screening tests exist, most are fraught with high false-positive rates that may lead to unnecessary procedures, such as biopsies, and possibly unnecessary treatment. Therefore, screening for many cancers is limited to people in high-risk groups, or who have family histories of a specific cancer.

One screening test that is generally viewed as successful is the Papanicolaou (Pap) smear for cervical cancer (TABLE 16.9). The Pap smear may reduce cervical cancer deaths by up to 80%. In this test, cells from the epithelium of the cervix are collected and examined for signs of dysplasia. Abnormal cells can then be removed to prevent progression to cervical cancer. Similarly, colon cancer screening appears effective since most colon cancer arises from pre-existing polyps. These polyps can be removed during a colonoscopy. Other screening tests are somewhat controversial in that screening may not reduce overall mortality and subject some individuals to unnecessary medical procedures. For example, mammography is now questioned for women under the age of 50 and is generally not recommended for screening except in high-risk groups, such as those individuals expressing the *BRCA* mutations or previously having had breast cancer. Similarly, prostate cancer screening is controversial due to its low sensitivity and specificity (see Box 16.3).

Table 16.9 Examples of Cancer Screening Tests	
Test	**Recommendation**
Pap smear for cervical cancer	Every 3 years in ages 21–65
Colonoscopy for colon cancer	Every 10 years from age 50 to 75
Mammogram for breast cancer	Every 2 years beginning at age 50
PSA test for prostate cancer	Controversial (Box 16.3)

TREATMENT

Surgery, radiation therapy, and chemotherapy are the three major approaches to treating cancer. Other approaches include hormonal therapy and targeted therapies based on biological features of specific cancers (TABLE 16.10). In addition, combinations of these therapeutic approaches are often used. For example, the tendency of cancer to invade neighboring tissues and metastasize to other locations often limits the efficacy of surgery. Therefore, chemotherapy and/or radiation therapy is often used as a follow-up or adjunct therapy to the surgery. The location, characteristics, and stage of the tumor, as well as the overall well-being of the patient, determine the therapeutic approach. The primary goal of the therapy is the complete removal of cancer with minimal damage to normal tissue and minimal adverse effects of the treatment. Other therapeutic goals can include suppressing cancer to a subclinical state and maintaining the quality of life, or palliative care to mitigate suffering.

Table 16.10 **Cancer Treatment Approaches**		
Approach	**Process**	**Limitations**
Surgery	Removal of the tumor	Works best for small, localized tumors in accessible tissue. May miss some cancer cells
Radiation therapy	Expose tumor cells to radiation, which causes breaks in DNA that lead to apoptosis	Normal cells are also damaged, leading to adverse side effects
Chemotherapy	Use of cytotoxic drugs to inhibit rapidly dividing cells	Normal cells are also damaged, leading to adverse side effects
Hormonal therapy	Block the synthesis of hormones required for tumor growth or use hormone antagonists	Only a limited number of cancers are dependent on hormones for growth
Targeted therapy	Design drugs that target specific changes in cancer cells and therefore are less toxic	Very few drugs have been developed and those drugs only effective against a few types of cancer
Immunotherapy	Stimulates the immune system to attack and kill cancer cells more efficiently	In the early stages of development and relatively few immunotherapies are available

Surgery is most useful for the removal of small localized tumors

In theory, surgery could be used to remove any tumor except for cancers of the blood such as leukemia. However, most often cancers have invaded neighboring tissues or metastasized to distal organs at the time of surgery. Nonetheless, removing the primary tumor is usually beneficial. Common examples of surgical procedures for cancer include mastectomy for breast cancer, prostatectomy for prostate cancer, and lung cancer surgery. The surgical intent can be either the removal of only the tumor or the entire organ, as is the case of mastectomy and prostatectomy. Surgery is sometimes needed to determine the stage of the cancer and the extent to which the cancer has invaded other tissues such as the lymph nodes, and whether the cancer has metastasized to other tissues. Occasionally surgery is necessary to control the symptoms such as bowel obstruction in colon cancer, or compression of another tissue such as nervous tissue due to the mass of the tumor.

Radiation therapy is the use of ionizing radiation to kill cancer cells and shrink tumors

Radiation either directly damages DNA or generates free radicals that damage DNA. In particular, double-stranded DNA breaks are often induced. Normal cells are also affected, but dividing cells, and especially rapidly dividing cancer cells, are more sensitive to radiation. The damage to normal cells, however, does result in a lot of adverse side effects associated with radiation therapy. To minimize these side effects, radiation is focused on the tumor as much as possible. Despite the side effects, radiation therapy is successful in treating many cancers. Radiation therapy is often used in conjunction with chemotherapy since the two approaches often act synergistically. Similarly, there are some drugs that enhance the sensitivity of cancer cells to radiation. Radiation therapy is also sometimes used as an adjunct therapy following the surgical removal of a tumor. Some leukemia and other blood cancers can also be treated via radiation therapy in conjunction with bone marrow transplantation. Bone marrow transplantation replaces the hematopoietic stem cells that were destroyed during radiation.

Anti-cancer drugs interfere with cell division

Chemotherapy refers to treatment with cytotoxic drugs that kill or inhibit the replication of cancer cells. Generally, chemotherapy kills rapidly replicating cells and therefore does not necessarily target specific cancer cells. As with all forms of cancer treatment, the efficacy of chemotherapy depends largely on the type of cancer and its stage. As expected, the cytotoxic drugs used for chemotherapy exhibit adverse side effects. The side effects are typically associated with rapidly dividing cells such as hematopoietic stem cells, cells of the intestinal epithelium, and hair follicles. Thus, immunosuppression, anemia, gastrointestinal distress, inflammation, and hair loss are common side effects of chemotherapy.

Several distinct classes of drugs based on their mechanism of action are used in cancer chemotherapy (TABLE 16.11). For example, alkylating agents form chemical adducts with DNA and some agents cross-link the two strands of DNA. Inhibitors of topoisomerase, an enzyme involved in unwinding the DNA helix during replication, generates single-stranded and double-stranded breaks in the DNA molecule. Some nucleotide analogs can incorporate into the replicating DNA molecule. In all these cases, the damage to the DNA induces apoptosis (see Box 16.1). Some nucleotide analogs block the synthesis of DNA and thus block cell division. Similarly, anti-folates interfere with folate metabolism, which indirectly interferes with DNA synthesis. Folates are crucial in many biosynthetic activities and most notable is the synthesis of the nucleotide thymidine. Blocking the synthesis of thymidine leads to a deficiency and indirectly inhibits DNA synthesis. Other anti-cancer drugs are anti-microtubule agents. Microtubules are cytoskeletal elements that form the mitotic spindle, which is responsible for separating the sister chromatids during mitosis (see Box 2.4).

Table 16.11	Common Anti-cancer Drugs Used in Chemotherapy
Drug class	**Mechanism of action**
Alkylating agents	Forms adducts with DNA molecules and cross-links DNA strands
Nucleotide analogs	Blocks DNA synthesis or damages DNA when incorporated into nascent DNA strand
Anti-folates	Interferes with folate metabolism and blocks synthesis of nucleotides, particularly thymidine
Topoisomerase inhibitors	Generates single- and double-stranded DNA breaks during DNA replication
Anti-microtubule agents	Inhibits the formation of the mitotic spindle and prevents the completion of cell division

Some cancers are responsive to hormone therapy

Hormone therapy is used to treat some cancers derived from hormonally responsive tissues. In such cases, the tumor cells retain hormone receptors and are dependent on the hormone for continued cell division. For example, some forms of breast and prostate cancer are dependent on estrogen and androgens, respectively, for their growth. Generally, hormone dependency involves nuclear-receptor hormones, also called steroid hormones, which regulate gene expression. Treatment of such cancers can be accomplished through drugs that inhibit the synthesis of the hormone, or antagonists that block the action of the hormone. For example, estrogen blocks the androgen receptor and can be used to treat some forms of prostate cancer. One limitation of hormonal therapy is that some cancer cells may lose their growth dependence on hormones and thus the drug becomes ineffective.

Nuclear receptors are described in Chapter 4, pages 78–81.

SIGNPOST

Development of future cancer treatments is targeted at specific cancers

The toxicity and adverse side effects associated with chemotherapy and radiation therapy highlight the need for less toxic drugs which are more specific for cancer cells. Ideally anti-cancer drugs should target some biological feature that is unique to the cancer cell, rather than targeting replicating cells in general. Substantial research is directed at this goal and such therapy is often called targeted therapy. Targeted therapy is expected to be less harmful to normal cells and therefore more effective in the treatment of cancer. For example, the specific mutated protein, or other genetic alteration associated with cancer, would be a unique target that is specific for a specific type of cancer. Many of the drugs developed for targeted therapy inhibit protein kinases and other signaling molecules that regulate cell division.

Stimulation of the immune system may improve the natural ability of the immune system to eliminate cancer cells

Cancer immunotherapy refers to diverse strategies aimed at stimulating the patient's own immune system to recognize and eliminate cancer cells. For example, cytokines are sometimes used in the treatment of cancer. Cytokines are hormones that function in the regulation of the immune system (Chapter 9). The rationale for using cytokines is to stimulate the immune system to assist in the elimination of cancer cells.

The use of monoclonal antibodies (see Box 9.2) is another type of immunotherapy. Monoclonal antibodies are antibodies that recognize a single epitope and can be produced as drugs. These antibodies recognize proteins that are specific for certain tumor cells. When administered, the monoclonal antibodies bind to the cancer cells and recruit other elements of the immune system to eliminate the cancer cells. In addition, the monoclonal antibody can be coupled with a toxin that specifically kills the cancer cell. In other words, the monoclonal antibody functions like a magic bullet that specifically targets the cancer cell and does not affect normal cells. It is also possible through genetic engineering (Chapter 7) to create immune cells with tumor-specific receptors that can be returned to the patient (BOX 16.4).

BOX 16.4

Gene Therapy for Cancer Treatment

Cancer has a strong genetic component and there are substantial efforts being made to develop gene therapy (Chapter 7) approaches for cancer treatment. Several treatments targeting B-cell tumors based on chimeric antigen receptor T cells (CAR-T cells) have been approved by the FDA since 2017 (**BOX 16.4 TABLE 1**). The approach is to isolate T cells from the patient and introduce a gene for a T-cell receptor that specifically binds to antigens on the cancer cells. The receptors are chimeric because they combine the antigen binding of antibody molecules with T-cell-activating functions into a single receptor. These genetically modified T cells are then returned to the patient and the engineered T cells attack and kill the tumor cells.

Another genetic engineering approach has been to develop an oncolytic virus to specifically attack and kill tumor cells. IMLYGIC® is a modified herpes virus that is injected directly into melanoma tumors. The virus is modified so that it does not replicate in normal cells but does replicate in cancer cells. Therefore, the virus is preferentially toxic to cancer cells. The virus also contains a cytokine gene and stimulates a strong immune response against cancer cells. Since many cancers are due to defective *TP53* genes, another approach has been to create an adenovirus vehicle containing the wild-type *TP53* gene (Gendicine®). Gendicine® is only approved in China.

Box 16.4 Table 1   Examples of Gene Therapy Approaches to Cancer

Treatment	Cancer	Approach
CAR-T cells	B-cell lymphomas and B-cell leukemias	Genetically engineered T cells express chimeric antigen receptor which recognizes the cancer cells
IMLYGIC®	Melanoma	A modified herpes virus to specifically kill cancer cells and activate immunological attack on the cancer cells
Gendicine®	Squamous cell carcinoma	An adenovirus recombined with *TP53* to replace damaged *TP53* in cancer cells

SUMMARY OF KEY CONCEPTS

- Cancer is defined as abnormal cellular replication

- Cancers can be benign and of little medical consequence, or malignant with a high level of morbidity and mortality

- The underlying etiology of cancer is the accumulation of damage in genes that play a role in the regulation of cell division

- Genes associated with cancer are generally classified as proto-oncogenes or tumor suppressor genes based on either a gain-of-function or loss-of-function phenotype, respectively

- Cellular processes associated with tumorigenesis include genes that regulate cell fate, cell survival, and maintenance of genome integrity

- Tumor protein 53 (*TP53*) gene, encoding the p53 protein, plays key roles in genome stability, DNA repair, apoptosis, and cell metabolism, and mutations in this gene are associated with about half of all cancers

- Benzopyrenes are carcinogens found in tobacco smoke that form adducts with DNA and are well documented to be the major risk factor in developing lung and other cancers

- Most colon cancer arises from benign tumors called polyps and early removal of the polyps may avert the development of colon cancer

- Genes linked with a hereditary form of breast cancer are involved in DNA repair

- Prostate cancer is common in men over 70 and although it can produce a severe metastatic disease is generally a slowly progressing cancer

- Cancer is difficult to diagnose due to manifestations being non-specific and occurring in the later stages of the disease

- Currently, surgery, radiation therapy, and chemotherapy are the three most common approaches that are used to treat cancer

- Future prospects for cancer treatment include a more cancer-specific targeted therapy and immunotherapy

KEY TERMS

apoptosis	A programmed cell death involving an ordered sequence of molecular events that serves as a mechanism to remove cells that are no longer needed or to remove cancerous, infected, or damaged cells.
benign	A mild disease condition that does not threaten health or life and especially in reference to cancer.
cancer driver genes	Genes involved in cell fate, cell subsistence, or genome maintenance that can promote the development of cancer if they fail to function normally.
carcinogen	A chemical or substance that damages DNA and can lead to cancer.
dysplasia	An abnormal growth or development of cells within tissues or organs that may precede the development of cancer.
hyperplasia	An abnormal increase in the number of cells or the enlargement of an organ or tissue due to increased numbers of cells.
malignant	Refers to cancer cells that exhibit uncontrolled growth, invade nearby tissues, or spread to other parts of the body.
metastasis	The process by which disease-bearing cells (e.g., cancer or pathogen) spread from one part of the body to another via the circulatory or lymphatic systems.
neoplasia	Any abnormal multiplication of cells.
proto-oncogene	A regulatory gene that promotes cell division and converts to an oncogene during the development of cancer.
TP53	The gene encoding a transcription factor called p53 that regulates cell division and apoptosis and plays a role in the maintenance of the genome.
tumor suppressor	A regulatory gene that inhibits the cell division process, promotes apoptosis, or is involved in genome maintenance.

REVIEW QUESTIONS

1. Describe the five types of cancer based on the tissue of origin.

2. What are the general features of the four stages (1–4) of cancer and pre-cancer?

3. What is meant by hereditary cancer syndrome and how does it differ from other cancers? Discuss some examples of cancers due to hereditary cancer syndrome.

4. What is *TP53* and describe its role in the development of cancer.

5. What are the normal functions of 'cancer driver' genes?

6. Discuss the various strategies in cancer therapy and some future prospects of cancer therapy.

ADDITIONAL QUESTIONS

In addition to the Review Questions provided above, there is a range of free online questions designed to enable students to further test their understanding of the chapter material. In order to access these interactive questions, please visit the book's website:
https://routledge.com/cw/wiser

Immunological Disorders

17

The immune system (Chapter 9) obviously plays a central role in health and disease. A major function of the immune system is to recognize and eliminate pathogens and promote the healing process. Thus, disorders that compromise the immune system can have profound effects on health. As expected, individuals with a compromised immune system are more susceptible to infections and experience more severe disease due to pathogens.

Obviously, a compromised immune system is not a good thing. However, inappropriate and excessive immune responses can also cause disease. Examples, of inappropriate immune responses are chronic inflammation, allergy, autoimmunity, and oral intolerance. These disorders can be viewed as hyperresponsiveness of the immune system. The inflammation associated with this hyperactivity causes pathology that ranges from mild to severe. Over the past several decades there has been an increase in these inflammatory disorders in industrialized countries. This increase is correlated with a decrease in infectious diseases leading to the hygiene hypothesis as an explanation for the increasing prevalence of inflammatory diseases. The hygiene hypothesis proposes that limited exposure to pathogens and other microbes affects the normal development of the immune system and promotes inappropriate and excessive inflammation.

IMMUNE DEFICIENCIES

CHRONIC INFLAMMATION

HYPERSENSITIVITY

AUTOIMMUNITY

ORAL TOLERANCE

HYGIENE HYPOTHESIS

IMMUNE DEFICIENCIES

Immunodeficiency refers to a state in which the immune system's ability to effectively deal with infectious disease is compromised. This deficiency can be either intrinsic or extrinsic. Intrinsic, also called primary, immunodeficiency refers to genetic defects that affect components of the immune system. More than 300 genetic immunodeficiency conditions have been described. They are generally grouped into humoral immunity disorders, T-cell and B-cell disorders, phagocytic disorders, and complement disorders. In rare cases, the immune system is essentially totally compromised, resulting in severe combined immunodeficiency (SCID). Treatment or management of immunodeficiency disorders generally involves reduced exposure to pathogens and perhaps prophylactic drugs.

Most immunodeficiency is related to environmental factors. For example, acquired immunedeficiency syndrome (AIDS) due to the human immunodeficiency virus (HIV) is the result of an infection (Chapter 22). Immunodeficiency associated with malnutrition is also well known.

A type of SCID due to an enzyme deficiency is described in Chapter 11, page 228.

SIGNPOST

SIGNPOST

The effects of malnutrition on immunity are discussed in Chapter 12, page 242.

The immune response requires substantial energy and protein–energy malnutrition compromises the ability of the immune system to fight infections. In addition, some micronutrients are needed for the optimal function of the immune system. Other factors that can suppress the immune system are drugs and toxins. In some cases, the desired effect of the drug is to suppress the immune system. However, in many cases, immunosuppression is a side effect of the drug and is undesirable. For example, many anti-cancer drugs suppress the immune system and leave patients more vulnerable to infectious diseases.

CHRONIC INFLAMMATION

The immune system responds to infections and other injuries with an inflammatory response. At the onset of an infection or injury, immune effector cells secrete cytokines that are responsible for the clinical manifestations associated with inflammation such as swelling, redness, fever, and pain (Chapter 9). Inflammation is a process by which fluid, biomolecules, and cells are brought to the injured site to limit the extent of the injury, remove necrotic debris, and prepare for the healing process. Generally, inflammation is acute and resolves in a few days as the healing process commences. If the infection or injury is not resolved, a persistent, or chronic, inflammation may develop. Generally, the magnitude of the manifestations associated with chronic inflammation is noticeably less than those of acute inflammation (TABLE 17.1). Nonetheless, there is damage associated with chronic inflammation and chronic inflammation is associated with the development of many other diseases.

Table 17.1	**Comparisons between Acute and Chronic Inflammation**	
	Acute	**Chronic**
Duration	Short-lived	Long term and non-resolving
Magnitude	High with overt symptoms and localized	Low and systemic with vague symptoms
Initiation	Infection, trauma, cellular stress	Tissue damage, metabolic dysfunction
Outcome	Healing	Slowly progressing diseases
Biomarkers	Inflammatory cytokines, C-reactive protein	None identified yet

The etiology of chronic inflammation is complex and not well understood

Acute inflammation can develop into chronic inflammation if the cause of the inflammation persists. For example, the persistence of a low-level infection, continued exposure to an irritant, or an untreated injury can lead to chronic inflammation. Increasing age, smoking, obesity, excessive alcohol consumption, or chronic stress also contribute to the development of chronic inflammation. Although chronic infections can contribute to chronic inflammation, infections are generally not the drivers of chronic inflammation. Often chronic inflammation occurs in the absence of infection with metabolic dysfunction or tissue damage frequently being the initiator of chronic inflammation. Similarly, dysbiosis, an imbalance in the microbiota of the gut (see Box 6.2), also contributes to chronic inflammation.

Diets high in refined grains, alcohol, and highly processed foods combined with low physical activity are also risk factors for developing chronic

SIGNPOST

Obesity and inflammation are discussed in Chapter 12, page 248.

inflammation. Urbanization and persistent exposure to environmental and industrial toxins, including smoking, also increase the risk of chronic inflammation. In other words, chronic inflammation may be a product of the so-called western lifestyle (FIGURE 17.1). Furthermore, obesity and increased adipose tissue associated with western diets and lack of physical activity increase inflammation. The increased chronic inflammation associated with aging is thought to be due in part to cellular senescence. A prominent feature of cellular senescence is the increased secretion of proinflammatory cytokines and other inflammatory molecules. However, aging is a conundrum in that aging is also associated with a weakened immune response.

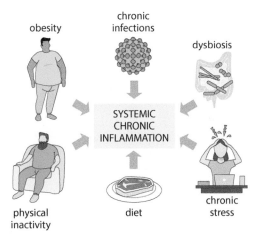

FIGURE 17.1 Contributing factors to systemic chronic inflammation. The etiology of chronic inflammation is complex and involves biological, social, and behavioral, factors.

Chronic inflammation is a contributing factor to many other diseases

Although inflammation is generally viewed as being protective and associated with healing, inflammation does have damaging effects. One problem with unresolved inflammation is the accrual of cellular and tissue damage – albeit minimal – over time. This accrual of damage plays a role in the development of many other diseases, such as cardiovascular disease (FIGURE 17.2). For example, high levels of circulating C-reactive protein, a biomarker for systemic inflammation, are associated with an increased risk of developing coronary artery disease and increased mortality associated with cardiovascular disease. In addition, the treatment of patients with drugs that inhibit proinflammatory cytokines reduces the rates of cardiovascular events and mortality.

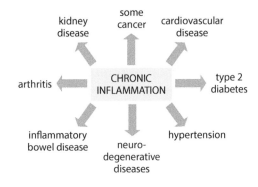

FIGURE 17.2 Possible consequences of chronic inflammation. Systemic chronic inflammation is a contributing factor to the development of many other diseases.

HYPERSENSITIVITY

Immune responses are generally helpful and protective. However, recognition of an innocuous substance by the adaptive immune response is not helpful since this induces an inflammatory response that is not needed. Allergies are an example of an unnecessary immune response against an innocuous substance. The innocuous substance is referred to as an **allergen**. Allergens are antigens recognized by elements of the adaptive immune response and these allergens have no detectable effect on non-allergic individuals. Allergies are also called **hypersensitivity** reactions. The symptoms associated with most allergies are relatively mild and are of no major medical consequence. However, some severe allergic reactions can be associated with mortality.

Hypersensitivity reactions can be either humoral or cell mediated

There are four types of hypersensitivity reactions based on the elements that recognize the allergen and the effectors that respond to this recognition (TABLE 17.2). The first three types are based on the humoral immune response and the fourth is a cell-mediated response.

Table 17.2 Types of Hypersensitivity Reactions

	Antigen	Reactant	Effector	Example
Type 1	Allergen	IgE	Mast cells	Allergic rhinitis
Type 2	Modified human component	IgG	Complement, phagocytosis	Some drug allergies
Type 3	Soluble immune complex	IgG	Complement, cytotoxicity	Serum sickness
Type 4	Allergen	T cells	Cytotoxicity	Contact dermatitis

Type 2 and 3 hypersensitivity reactions are similar, and both involve IgG. In type 2 hypersensitivity the allergen is on the surface of cells. In many cases the allergen is produced by the binding of a drug or other small molecule to cell membrane proteins. Small molecules that are normally not antigenic, but that can become antigenic when conjugated to proteins, are called haptens. IgG binding to the allergen/hapten results in complement activation or phagocytosis. In the case of type 3 hypersensitivity reactions, the allergen and antibodies form a complex that is initially soluble. This complex is then deposited on the walls of blood vessels, thus triggering an inflammatory response involving complement and various leukocytes. This can also include the release of vasoactive substances. An example of type 3 hypersensitivity is serum sickness caused by the administration of antibodies from another species to treat a disease via passive immunization. The foreign antibodies are essentially allergens that are recognized by human antibodies and form antibody–allergen complexes that are deposited on blood vessel walls.

Type 4 hypersensitivity reactions are mediated by T cells. Since activation of T cells requires antigen presentation the response is generally slower than the antibody-based responses and, thus, is also called delayed hypersensitivity. The activated T cells can cause a localized inflammatory response due to the secretion of cytokines by helper T cells or the direct action of cytotoxic T cells. A common example of a type 4 hypersensitivity reaction is dermatitis, experienced by some people after contact with poison ivy (BOX 17.1).

SIGNPOST

Passive immunization is described in Chapter 9, pages 197–198.

Poison Ivy and Urushiol

Most individuals develop contact dermatitis when they touch poison ivy or related plants. This contact dermatitis is due to urushiol, which is an oily mixture of compounds containing catechol with long-chain hydrocarbons (**BOX 17.1 FIGURE 1**). Urushiol is readily absorbed by the skin and forms chemical adducts with proteins. Normally small molecules like those making up urushiol do not act as antigens. However, urushiol covalently bound to protein can be processed by dendritic cells or other professional antigen-presenting cells (Chapter 9). These dendritic cells then migrate to the lymph nodes and present the urushiol-conjugated peptides to T cells. This activates T cells, and these T cells migrate back to the skin where the original contact occurred. These activated T cells secrete cytokines or directly cause cytotoxic damage in response to the urushiol-conjugated proteins in the skin. The pathology manifests as itching, rashes, blisters, or oozing lesions (**BOX 17.1 FIGURE 2**). A similar type of contact sensitivity develops in some people to coins, jewelry, or metals containing nickel. In this case the nickel chelates with histidine residues in some proteins and forms a T-cell epitope. Small molecules that function as allergens after being associated with proteins are called haptens.

BOX 17.1 FIGURE 1 Urushiol. The R group of the catechol consists of long hydrocarbon chains ranging from 15 carbon atoms to 17 carbon atoms and varying degrees of unsaturation (i.e., double bonds).

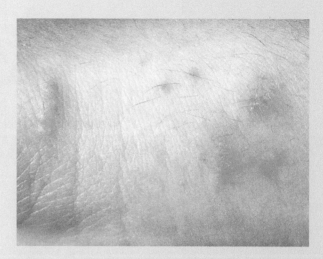

BOX 17.1 FIGURE 2 Contact dermatitis due to poison ivy. (From Wikimedia Commons; courtesy of Britannic124, published under CC BY-SA 3.0.)

Many common allergies involve IgE, mast cells, and histamine

Type 1 hypersensitivity reactions are the most common type of hypersensitivity and involve allergens binding to IgE (see Table 17.2). The IgE is generated from a previous exposure, called sensitization, of B cells that recognize the antigen (FIGURE 17.3). This initial sensitization does not produce symptoms but sets the stage for subsequent exposures to the allergen. As with other antigens (see Figure 9.10), the allergen stimulates allergen-specific B cells to undergo clonal proliferation and develop into plasma cells. These plasma cells secrete IgE that recognizes the allergen. **Mast cells** have Fc receptors on their surface that bind IgE. This binding of IgE to the receptors sensitizes the mast cell to subsequent exposure to the allergen. Mast cells are full of vesicles that primarily contain histamine. On subsequent exposure, the allergen binds to the IgE on the mast cell surface and this stimulates the release of the histamine and other inflammatory molecules.

FIGURE 17.3 Type 1 hypersensitivity. Exposure to an allergen activates B cells with receptors that recognize the allergen to proliferate and produce IgE. The IgE binds to Fc receptors on mast cells and sensitizes mast cells. Subsequent exposure to the allergen causes mast cells to release histamine and other inflammatory molecules contained in the vesicles.

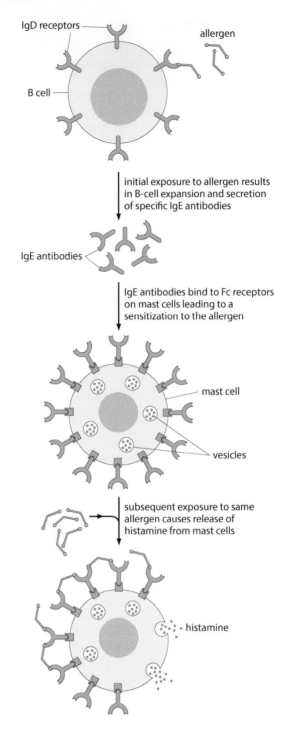

Histamine is a small organic molecule derived from the amino acid histidine by an enzymatic reaction called decarboxylation. Various types of cells have histamine receptors and account for the various symptoms associated with allergies (FIGURE 17.4). For example, histamine receptors on some nerves are responsible for the itching associated with many allergies. Other nerves stimulate sneezing. Histamine is also vasoactive and causes blood vessels to dilate and become more permeable, which causes swelling and redness in tissues. In the nasal passages histamine enhances the secretion of mucus, which results in a runny nose. In the lungs histamine causes the bronchi to constrict and that combined with increased mucus secretion results in the shortness of breath associated with asthma. Antihistamines are drugs that block the actions of histamine and thus relieve allergy symptoms.

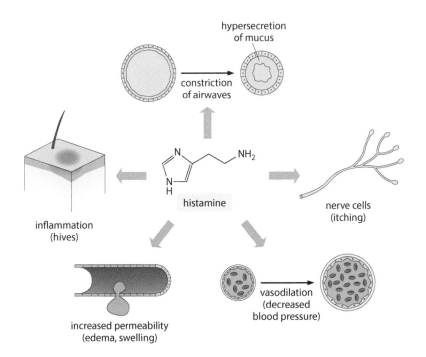

FIGURE 17.4 Mechanisms of action of histamine. Histamine binds to receptors on various types of cells, resulting in the inflammation associated with allergies such as redness, swelling, itching, constriction of bronchi, hypersecretion of mucus and other fluids, vasodilation, and increased capillary permeability.

Type 1 hypersensitivity reactions can be localized or systemic

Seasonal hay fever is a common and well-known allergic reaction. In this case, pollen or other allergens are recognized by the IgE and this activates a localized inflammatory response in the eyes and nose, including itching, sneezing, and a runny nose. Other parts of the body can be affected and, in general, the specific symptoms of the allergy are largely dependent on the route of exposure to the allergen (TABLE 17.3). For example, if the allergen reaches the lungs, asthma symptoms may result due to the constriction of the airways and increased mucus production. Contact of the allergen with the skin can produce hives. Hives are raised red skin due to localized blood flow and increased permeability of the blood vessels. Localized stimulation of the nerves produces itching at the hives. Ingestion of an allergen results in vomiting and diarrhea, common symptoms of food allergies. Ingested allergens can also be absorbed and transported throughout the body. This systemic spread of the allergen can result in hives throughout the body. A particularly severe form of systemic type 1 hypersensitivity reaction is anaphylaxis.

Table 17.3	IgE-mediated Allergic Syndromes		
Syndrome	**Common allergens**	**Route of entry**	**Symptoms**
Hay fever and seasonal allergies	Pollens, dust mite feces, pet dander	Inhalation	Running nose, sneezing, itchy eyes
Asthma	Pollens, dust mite feces, pet dander	Inhalation	Constriction of airways, increased mucus production
Rash or hives	Animal hair, insect bits	Skin contact	Raised/red skin due to local increase in blood flow and permeability, itching
Food allergy	Nuts, shellfish, milk, eggs	Oral	Vomiting and diarrhea, and itching and hives if systemic
Anaphylaxis	Drugs, insect venoms, food (rarely)	Intravenous, oral absorption	Systemic vasodilation and edema, death

Anaphylactic shock is a severe type 1 hypersensitivity reaction that can be mortal

Although most type 1 hypersensitivity reactions are localized near the entry point of the allergen, it is possible to have a systemic hypersensitivity reaction known as anaphylaxis. Systemic vasodilation results in whole-body swelling. This edema, due to dilated blood vessels, is also associated with a substantial fall in blood pressure. An individual can go into shock and possibly die if the blood pressure drops to the point at which the heart can no longer pump blood throughout the body. Other symptoms associated with anaphylaxis include difficulty breathing due to constricted airways and stomach cramps due to smooth muscle spasms.

The route of entry of allergens associated with anaphylaxis is generally intravenous or oral absorption. However, food allergies rarely result in anaphylaxis. The most common allergens associated with anaphylaxis are insect venoms or drugs. Since type 1 hypersensitivity reactions occur rapidly after exposure to the allergen, a person who is hypersensitive to insect venoms can rapidly go into anaphylactic shock and die after being stung by a bee or wasp. Such people are advised to carry epinephrine in the form of an autoinjector with them (BOX 17.2). Injection of the epinephrine after an insect sting reverses the action of the histamine by causing blood vessels to constrict and maintaining blood pressure at an adequate level.

EpiPens and Anaphylactic Shock

An EpiPen is the common name for a medical device that automatically injects a measured amount of epinephrine (**BOX 17.2 FIGURE 1**). This epinephrine autoinjector contains a spring-loaded needle that exits the tip or edge of the device and penetrates the recipient's skin to deliver epinephrine via intramuscular injection. The EpiPen is designed as an emergency treatment for severe anaphylactic reaction. People with severe allergies are prescribed EpiPens and carry them. Upon exposure to the allergen, such as a bee sting, the individual can rapidly and precisely inject epinephrine, which prevents them going into anaphylactic shock.

BOX 17.2 FIGURE 1 EpiPen. (From Wikimedia Commons; courtesy of Tokyogirl79, published under CC BY-SA 3.0.)

AUTOIMMUNITY

Another disease associated with an overly active immune system is autoimmunity. Autoimmunity is a situation in which elements of adaptive immunity recognize antigens that are normal cellular components and not foreign substances. This recognition of **autoantigens** results in chronic inflammation and destruction of normal tissue in the absence of pathogens. There are three types of autoimmune responses that correspond to types 2–4 hypersensitivity reactions (see Table 17.2). Types 2 and 3 are mediated by autoantibodies and type 4 is mediated by autoimmune T cells. Autoimmune disease is not associated with IgE, the source of type 1 hypersensitivity.

More than 80 autoimmune disorders have been described and such disorders may affect up to 7% of the population in industrialized countries. Autoimmune disorders affect a wide range of organs and tissues (TABLE 17.4). Blood vessels and connective tissue are often either the direct or the indirect target of the autoimmune reaction. Many autoimmune disorders affect a specific organ and the disease manifestations are due to the dysfunction of that organ. For example, type 1 diabetes is due to the autoimmune destruction of the β-cells in the pancreas. The destruction of the β-cells results in a loss of insulin production and causes diabetes. Similarly, in multiple sclerosis the autoantigen is myelin basic protein, found on the myelin sheath of nerve cells. The demyelination of nerve cells has neurological manifestations such as weakness, impaired vision, and incoordination. Autoimmune disorders can also be systemic, especially in situations in which antibody–autoantigen complexes are being deposited throughout the vascular system, as is the case in systemic lupus erythematosus.

Type 1 diabetes is described in detail in Chapter 13, pages 255–256.

SIGNPOST

Table 17.4   Examples of Autoimmune Disorders

Disease	Type	Autoantigen	Consequence	HLA allele*
Rheumatic fever	2	Streptococcal cell wall antigens, M protein	Myocarditis, arthritis	DRB1*07
Systemic lupus erythematosus	3	Nuclear (e.g., chromatin or ribonuclear particles)	Vasculitis, glomerulo-nephritis, arthritis	DRB1*0301
Type 1 diabetes	4	Pancreatic β-cell proteins	Loss of insulin production due to loss of β-cells	B*5701
Multiple sclerosis	4	Myelin basic protein	Neurological degeneration due to degradation of the myelin sheath	DRB1*15:01
Celiac disease	4	Modified gluten, transglutaminase	Malabsorption and intestinal atrophy	DQA1*05 + DQB1*02

*HLA allele(s) that are strongly associated with developing autoimmune disease.

Autoimmunity is due to loss of self-tolerance

Precursors to B cells and T cells mature in the bone marrow and thymus, respectively. This maturation involves the expression of proteins, including surface markers that are specific for the different types of lymphocytes, as well as the generation of the IgD receptor or T-cell receptor repertoires on B cells and T cells, respectively. The DNA rearrangements involved in generating antibodies or T-cell receptors could possibly generate autoreactive lymphocytes. In addition, an individual's lymphocytes are continuously changing due to their experience with infections, vaccinations, and other immunogenic challenges. Thus, potentially autoreactive lymphocytes are continuously being generated. Mechanisms to eliminate or suppress autoreactive lymphocytes lead to a state of **self-tolerance**. Thus, autoimmunity is a disorder in which these mechanisms of self-tolerance have failed.

The maturation of B cells and T cells in the primary lymphoid tissues (bone marrow and thymus) can also be viewed as an 'education' in that self-reactive lymphocytes are eliminated via apoptosis (FIGURE 17.5). B cells can avoid apoptosis by undergoing receptor editing in which the DNA sequence of the IgD receptor is rearranged to produce an antibody molecule that no longer reacts with self. This elimination of autoreactive lymphocytes and receptor editing is called central tolerance since it occurs in the primary lymphoid tissues. Autoreactive lymphocytes that are not eliminated and make it to the secondary lymphoid tissues are induced to go into an anergic state. In this anergic state the lymphocytes do not respond to antigens and are unable to undergo clonal proliferation. This peripheral tolerance can also involve apoptosis and the elimination of autoreactive lymphocytes.

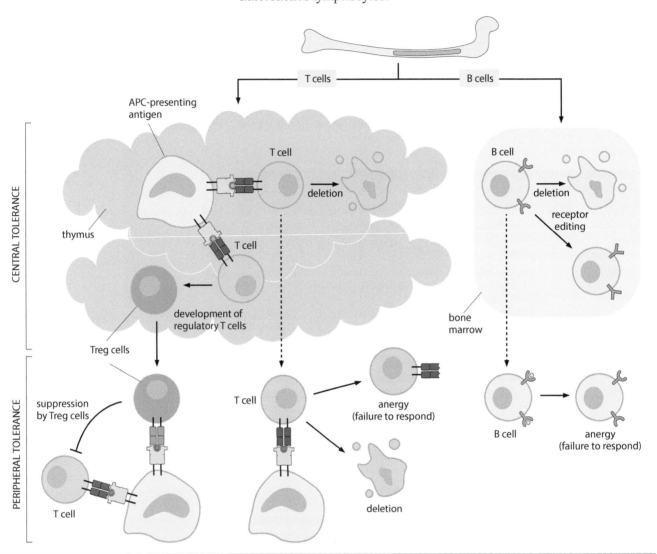

FIGURE 17.5 Mechanism of self-tolerance. Immature B cells and T cells that encounter autoantigen in the primary lymphoid tissues (bone marrow and thymus) are induced to undergo apoptosis. B cells can avoid apoptosis by editing the IgD receptor so that it no longer recognizes self. Some T cells that recognize self-antigens develop into regulatory T cells (Treg cells) instead of undergoing apoptosis. Treg cells suppress other autoreactive T cells that recognize the same antigen. Mature lymphocytes that recognize self-antigen and migrate to secondary lymphoid tissues (dashed arrows) either become unresponsive (anergy) or undergo apoptosis. [Adapted from Kumar V, Abbas AK & Aster JC (2012) *Robbins Basic Pathology*, 9th edn. Saunders. With permission from Elsevier.]

Another process to suppress autoreactivity involves regulatory T cells (Treg cells). Treg cells are a subpopulation of CD4+ T cells (i.e., helper T cells) that suppress the activation and proliferation of effector T cells in an antigen-dependent manner. Some of the immature T cells that encounter self-antigen in the thymus convert into Treg cells instead of undergoing apoptosis (see Figure 17.5). These Treg cells then migrate to the peripheral lymphoid tissues or circulate throughout the body. Upon encountering antigen, the Treg cells suppress the development of naive T cells to effector T cells and secrete anti-inflammatory cytokines. In addition, Treg cells can also induce apoptosis in effector T cells that recognize the autoantigen. The exact nature of the dysfunctions in central and peripheral tolerance that are associated with autoimmune disorders is not known.

Autoimmunity has a complex etiology involving genetic and environmental factors

The factors leading to the loss of self-tolerance are complex and not well understood. There is a genetic predisposition and autoimmunity tends to run in families. However, few genetic determinants have been identified and environmental and other factors also play a large role in the development of autoimmune disease (FIGURE 17.6). Autoimmune diseases are often triggered by environmental factors such as infection, drugs, toxins, sunlight, or stress. Gender is also a predisposing factor since many autoimmune diseases are more prevalent in women than men. Obviously, an individual's antibody or T-cell receptor repertoire could contribute to the development of autoimmunity. In addition, diet, nutrition, gut health, and microbiota may also influence autoimmunity.

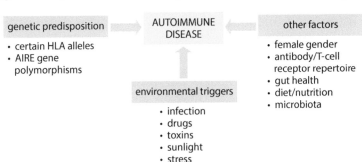

FIGURE 17.6 Etiology of autoimmune disease. Autoimmunity has a complex etiology involving a genetic predisposition, environmental triggers, and other factors. AIRE, autoimmune regulator.

Autoimmunity is associated with polymorphisms in major histocompatibility complex (MHC) genes

MHC proteins bind to peptide epitopes and present these epitopes to T cells. These MHC proteins are divided into two classes and there are several genes in each class (BOX 17.3). For each of these MHC genes there is an incredible amount of diversity found in the human population and for the major MHC genes there are thousands of known alleles for each gene. Some of these alleles are associated with an increased risk of developing autoimmune disease (see Table 17.4). Other MHC alleles have a protective effect against developing a specific autoimmune disease. As MHC proteins play a role in presenting antigens to T cells, it could be that certain alleles are more likely to present peptide epitopes that result in an aberrant autoimmune reaction.

Antigen presentation is described in Chapter 9, pages 190–192.

SIGNPOST

Major Histocompatibility Complex

The major histocompatibility complex (MHC) is a large genetic locus containing a set of linked genes that code for proteins involved in antigen presentation (Chapter 9). These genes were initially discovered due to their role in transplant rejection and matching donors and recipients for organ transplantation involves these MHC genes. In humans the genes are also known as human leukocyte antigen (HLA) genes. The MHC genes are divided into class I and class II based on their overall structure and specific function in antigen presentation. There are three major and three minor MHC class I proteins designated as HLA-A, HLA-B, and HLA-C, and HLA-E, HLA-F, and HLA-

G, respectively. There are three major and two minor MHC class II proteins designated as HLA-DP, HLA-DQ, and HLA-DR, and HLA-DM and HLA-DO, respectively. The MHC class II proteins are dimers consisting of alpha- and beta-subunits and thus there are two genes associated with each HLA protein (e.g., HLA-DRA and HLA-DRB). HLA-DR is further complicated in that there are four different B genes designated as HLA-DRB1, HLA-DRB3, HLA-DRB4, and HLA-DRB5. The MHC genes are highly variable in the human population and there are thousands of different alleles for all the major HLA genes.

The autoimmune regulator gene plays a key role in lymphocyte education

Another gene with a strong association with autoimmune disease has been named the autoimmune regulator (AIRE). The protein encoded by this gene is involved in the presentation of self-antigens to developing T cells in the thymus. Specifically, the AIRE protein is a transcription factor that drives the transcription of a wide range of genes normally expressed in other organs. This more or less creates a reflection of self in the thymus. Nascent T cells that recognize self are eliminated by apoptosis in the thymus. Disruption of the normal function of the *AIRE* gene results in T cells not being properly 'educated' in regard to distinguishing self from non-self and this leads to the development of a range of autoimmune disorders.

Infections and other factors can trigger autoimmune disease

Autoimmunity has a strong genetic predisposition. However, less than 20% of the individuals with predisposing *AIRE* polymorphisms or predisposing HLA alleles develop autoimmune diseases. This means that other factors play a central role in developing autoimmune disease. In addition, autoimmunity is often triggered by external factors. For example, sunlight triggers systemic lupus erythematosus and makes the symptoms worse (FIGURE 17.7). The ultraviolet radiation associated with sunlight damages skin cells and can lead to apoptosis. This damage potentially releases self-antigens, such as chromatin, and other damage-associated molecular patterns (DAMPs). The exposure of the immune system to DAMPs and self-antigens may initiate processes that lead to the development of autoimmunity in people with a genetic predisposition. Similarly, drugs, toxins, or other haptens can cause type 2 hypersensitivity reactions, which may initiate the development of autoimmunity.

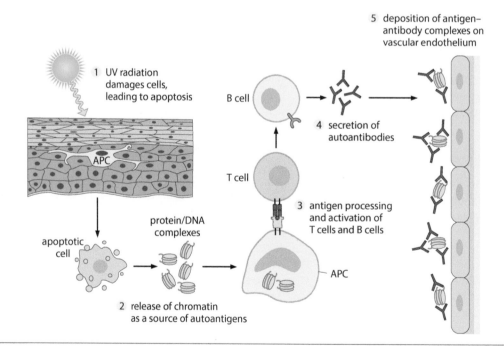

FIGURE 17.7 Role of sunlight in initiating and exacerbating systemic lupus erythematosus. Ultraviolet (UV) radiation from sunlight damages skin cells and can lead to apoptosis. Failure to completely clear the apoptotic cells can result in the release of chromatin and other nuclear components that normally are not exposed to the immune system. In genetically predisposed individuals this might result in the development of antibodies against these components. The antibody–antigen complexes can be deposited on the vascular endothelium resulting in a type 3 hypersensitivity reaction.

Other triggers for autoimmunity are infections. A well-known example is rheumatic fever. Rheumatic fever typically develops 2–4 weeks after a streptococcal throat infection. About half of the cases involve inflammation of the heart, but joints, skin, and brain are also commonly affected. The disease is less common now than historically, as antibiotic treatment can prevent disease initiation and progression. Up to 3% of streptococcal infections that are not treated can result in rheumatic fever. It is believed that antibodies against *Streptococcus pyogenes* cross-react with host antigens. In particular, a virulence factor in the cell wall of the bacterium called M protein is highly immunogenic. The antibodies generated against M protein cross-react with proteins of the heart and connective tissue (FIGURE 17.8). This cross-reactivity of anti-pathogen antibodies with host molecules is called molecular mimicry.

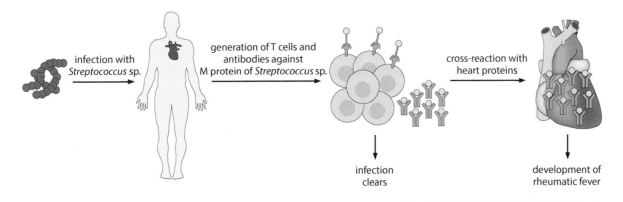

FIGURE 17.8 Molecular mimicry as a potential cause of rheumatic fever. Infection with *Streptococcus sp.* results in antibodies and T cells that recognize the M protein of *Streptococcus* sp. These antibodies can also bind to host proteins such as myosin and connective tissue. After clearance of the infection the inflammation persists, resulting in heart damage and the development of rheumatic fever.

ORAL TOLERANCE

Secondary lymphoid tissue called mucosa-associated lymphoid tissue (MALT) is found throughout the digestive and respiratory tracts. The MALT of the gastrointestinal tract can distinguish commensal organisms and foodstuff from pathogens, despite the molecular similarities of pathogens, commensals, and food. All are composed of foreign molecules consisting of complex carbohydrates and proteins. This lack of immune reactivity to the microbiota and foodstuff is called **oral tolerance**. The mechanisms of oral tolerance are not well understood, but may involve regulatory T cells, which suppress other T cells. A loss of oral tolerance leads to chronic inflammation in the gut that has features of hypersensitivity and autoimmunity. Inflammatory bowel disease and celiac disease are two common examples of diseases associated with a loss of oral tolerance (TABLE 17.5).

Table 17.5	Diseases Associated with Loss of Oral Tolerance
Crohn's disease	An inflammatory bowel disease that most commonly affects the lower portion of the small intestine and the upper portion of the large intestine
Ulcerative colitis	Inflammation of the colon and rectum with ulcers not due to an infection
IgE-mediated food allergies	A type 1 hypersensitivity reaction against specific foods
Celiac disease	Inflammation of the small intestine triggered by dietary gluten

Loss of oral tolerance can lead to inflammatory bowel disease

Inflammatory bowel disease is a group of diseases characterized by chronic inflammation of the small or large intestines. Crohn's disease and ulcerative colitis are the two main types of inflammatory bowel disease. Crohn's disease can affect the entire gastrointestinal tract, whereas ulcerative colitis is restricted to the colon and rectum. Abdominal pain, cramping, and diarrhea are common clinical manifestations of both diseases. A hyperreactive immune response damaging the intestinal mucosa is believed to be the main component of the pathophysiology. The etiology of inflammatory bowel disease is not understood. There is a strong genetic predisposition for both Crohn's disease and ulcerative colitis involving multiple genes. Another possible predisposing factor is a change in the microbiota called intestinal dysbiosis (see Box 6.2). Currently, neither disease is believed to be an autoimmune disease involving autoantigens or autoreactive T cells. Both diseases appear to be a result of an overresponse to natural substances of the intestinal milieu that may be due to dysregulation of oral tolerance.

Gluten triggers celiac disease in people with a strong genetic predisposition

Celiac disease was previously considered to be a food allergy. However, it is now known that the disease is also associated with autoantibodies and specific destruction of epithelial cells in the small intestine. The disease etiology and pathogenic mechanisms are relatively well understood in celiac disease. There is a strong genetic predisposition, and this predisposition is primarily associated with specific HLA alleles. Over 95% of people with celiac disease have either the DQ2 or DQ8 variants of the HLA-DQ protein. In addition, most patients with celiac disease have a specific haplotype of the alpha (A1) and beta (B1) subunits of HLA-DQ called DQ2.5 consisting of two specific alleles (DQA1*05 + DQB1*02). Individuals who are homozygous for DQ2.5 have more severe celiac disease.

Another key factor in celiac disease is gluten. Gluten refers to a group of proteins found in the seeds of many grains and especially wheat. This protein is rich in proline and glutamine, which is not completely digested into amino acids. The resulting peptides are absorbed by the small intestine and an enzyme called transglutaminase removes the amine groups of the glutamine residues, converting them to glutamate residues (FIGURE 17.9). The HLA-DQ2.5 protein binds to this deaminated gluten and presents it to helper T cells. A localized inflammatory response is initiated if there are T cells and B cells that recognize the modified gluten peptide. Since gluten is a necessary trigger to induce this inflammation, a gluten-free diet alleviates the symptoms associated with celiac disease.

In addition to the inflammatory response triggered by gluten, there are also antibodies produced against gluten and transglutaminase in patients with celiac disease. The antibodies against the transglutaminase are often used as a biomarker for the diagnosis of celiac disease. It is not clear why antibodies form against gluten and transglutaminase, nor what role the antibodies play in celiac disease pathology.

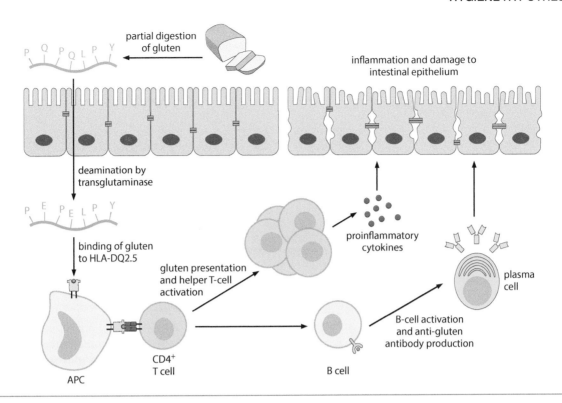

FIGURE 17.9 Pathogenesis of celiac disease. Digestion of wheat and other cereal grains results in gluten peptides that are proline (P) and glutamine (Q) rich. These peptides are absorbed across the intestinal epithelium and transglutaminase converts the glutamines to glutamates (E). These modified peptides bind to HLA proteins in antigen-presenting cells (APCs) and are presented to helper T cells (CD4$^+$). The activated helper T cells stimulate the secretion of proinflammatory cytokines and the production of anti-gluten antibodies. The cytokines and antibodies cause inflammation and damage to the intestinal epithelium.

HYGIENE HYPOTHESIS

The adaptive immune response is a powerful armament in the body's fight against pathogens. However, there are some potential negative consequences of this adaptive immune response including organ rejection associated with transplantation, hypersensitivity, and autoimmunity, and gut dysfunction associated with the loss of oral tolerance. Over the last several decades there has been an increasing incidence of immunological disorders (FIGURE 17.10). This increase has largely been in more industrialized countries. Corresponding to this increase in immunological disorders there has been a decrease in infectious disease incidence.

Decreasing infectious disease frequency correlates with increases in the frequency of hypersensitivity diseases

The increase of immunological disorders in resource-rich countries, compared with resource-poor countries, implies an environmental or behavioral component in the development of immunological disorders. There are several sociological differences between high-income and low-income countries (TABLE 17.6). People in resource-poor countries are more likely to live in rural areas, have large families, be in contact with livestock, and have less access to antibiotics and antiseptics, whereas in richer countries people are more likely to be urban, have small families, and have more access to antibiotics and antiseptics. This means that people in resource-poor environments are more likely to be exposed to infectious diseases than those in resource-rich environments. Indeed, infectious diseases are generally more prevalent in countries with low

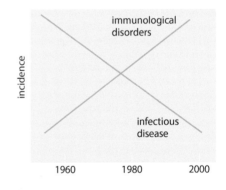

FIGURE 17.10 Incidence of infectious disease and immunological disorders. In the last half of the 20th century there has been a decrease in infectious disease (blue) and an increase in hypersensitivity disease and autoimmunity (green), especially in industrialized countries.

socioeconomic status. There is an especially high burden of intestinal infections (Chapter 21) and parasitic worm infections (Chapter 24) in resource-poor areas. In addition, people from resource-poor areas tend to have a more diverse microbiota than people from resource-rich areas. A reduction in gut microbiota diversity (i.e., dysbiosis) is commonly observed in individuals with autoimmune and hypersensitivity disorders.

Table 17.6 Differences between Resource-poor and Resource-rich Countries	
Resource poor	**Resource rich**
• Large family size • Rural, contact with livestock • Diverse and changing intestinal microbiota • Low antibiotic/antiseptic use • Poor sanitation, high fecal–oral burden • High helminth burden	• Small family size • Urban • Stable and less diverse intestinal microbiota • High antibiotic/antiseptic use • Good sanitation, low fecal–oral burden • Low or absent helminth burden

These differences between resource poor and resource rich indicate that people from resource-poor areas are exposed more frequently to pathogens and exposed to a wider diversity of microbes. The correlation of increased immunological disorders to a decreased exposure to microbes and parasites is a basic tenant of the **hygiene hypothesis**. This lack of exposure to infectious agents and microbes affects the natural development of the immune system and, in particular, results in an immune system that tends to overrespond. It appears that the decreased exposure to pathogens and other microbes predisposes the immune system to start attacking self or innocuous agents. Furthermore, this lack of exposure to pathogens and other microbes results in the immune system being more prone to an inflammatory response mediated by type 1 helper T cells.

Some researchers believe that the microbiota is more crucial than pathogens in guiding the development of the immune system and have refined the hygiene hypothesis to an 'old friends' hypothesis. In this hypothesis, early and regular exposure to harmless microorganisms (i.e., old friends) that have been present throughout human evolution train the immune system to react appropriately to threats and not to react to non-threats. In other words, children from industrialized countries are exposed to less diverse microbes in general, and this lower exposure results in an incomplete education of the immune system.

Altering the microbiota or exposure to pathogens may alleviate hyperactive immunological disorders

The epidemiological evidence for the hygiene hypothesis is rather strong and there is experimental evidence as well. However, the specific mechanisms by which exposure to microbes diminishes the development of hyperactive immunological disorders are not clear. Nonetheless, microbe exposure in a genetically predisposed individual or an induvial who already has a hyperactivity disease may be beneficial. In this regard, some studies have shown that treatment with probiotics has a beneficial outcome in the manifestation of hyperactive immunological disorders. Probiotics are a mixture of live microorganisms that have health benefits when consumed or applied to the body by altering the microbiota. Altering the microbiota may help modulate the immune system and increase oral tolerance.

Another line of research that has been explored in the treatment or prevention of hyperactive immune disorders is to intentionally infect a person with parasitic worms. Some of the rationales for using parasitic worms is that helminth infections produce a polarized type 2 helper T-cell response that is much less inflammatory than the type 1 response. In addition, strong type 2 responses are usually associated with weak type 1 responses and vice versa. Intestinal worms, which were widespread throughout the world up to a few decades ago, have been more or less eliminated in industrialized countries. However, intestinal worms are still a major public health problem in the rest of the world and especially in resource-poor countries. Although, helminth infections exert a strong immunomodulatory effect, clinical trials using helminth infections to treat hyperactive immune disorders have been ambiguous (BOX 17.4).

Helper T cells and their subtypes are described in Chapter 9, pages 193–194.

SIGNPOST

Helminth Therapy to Alleviate Intestinal Inflammation

A common helminth used in the treatment of hyperactive immunological disorders is the pig whipworm, *Trichuris suis*. *T. suis* is related to the human whipworm *T. trichiura*, but cannot establish an active infection in humans. Therefore, *T. suis* does not pose a public health risk as patients intentionally infected with the worm do not excrete eggs in their feces. Furthermore, although the worm establishes a temporary infection, it does not cause pathology in humans. However, since it cannot establish a long-term infection in humans, if treatment is successful, it would likely require periodic re-infections.

In initial studies, patients with ulcerative colitis or Crohn's disease were given viable *T. suis* eggs. The treatment was well tolerated and patients exhibited significant disease remission. Even though the beneficial effect was temporary, repeated doses with *T. suis* eggs sustained the clinical improvement and suggested a potential new therapy for inflammatory bowel disease. However, subsequent clinical trials have not been as successful in reducing the clinical manifestations associated with inflammatory bowel disease. Similarly, treatment with *T. suis* eggs has not been successful at alleviating symptoms associated with allergic rhinitis or asthma. Regardless of the disappointing results thus far, there is still interest in exploring the immunomodulatory effects of helminths as a potential treatment for hyperactive immunological disorders. As immunomodulation is the result of secreted helminth products (see Box 24.1) it may be possible to design therapeutics that do not involve the intentional infection with parasitic worms.

Helmby H. (2015) Human helminth therapy to treat inflammatory disorders – where do we stand? *BMC Immunology* 16: 12. doi: 10.1186/s12865-015-0074-3.

BOX 17.4

SUMMARY OF KEY CONCEPTS

- Dysfunction of the immune system is associated with many diseases

- Immune deficiencies result in increased susceptibility to infectious disease and more severe clinical outcomes associated with infectious disease

- Chronic inflammation, allergies, and autoimmunity are diseases due to inappropriate immune responses

- Chronic inflammation contributes to the pathophysiology of many common non-infectious diseases

- Allergies involve an inflammatory response against innocuous substances called allergens

- The most common allergies are type 1 hypersensitivity reactions and involve IgE, mast cells, and histamine

- Anaphylaxis is a severe systemic type 1 hypersensitivity reaction that can be fatal

- Autoimmunity is a disease characterized by a loss of self-tolerance and an immunological attack on normal tissues and organs

- Pathology associated with autoimmunity can be mediated by antibodies, T cells, or both

- The development of autoimmunity involves a genetic predisposition, environmental triggers, and other factors such as gender, diet, and the microbiota

- Oral tolerance refers to the ability of the immune system to be unresponsive toward non-pathogenic commensals and foodstuff

- Inflammatory bowel disease and celiac disease are examples of a loss of oral tolerance

- As infectious diseases decline, there has been a corresponding increase in autoimmune and hypersensitivity diseases leading to the development of the hygiene hypothesis

KEY TERMS

allergen	An antigen that elicits hypersensitivity or allergic reaction.
autoantigen	A normal constituent of cells or tissue that is recognized by antibodies or T cells and contributes to the development of an autoimmune disease.
celiac disease	An autoimmune disorder in genetically predisposed individuals in which gluten triggers inflammation in the small intestine.
histamine	A small organic molecule released from mast cells and other immune effector cells that has inflammatory as well as vasoactive properties and mediates the symptoms associated with type 1 hypersensitivity reactions.
hygiene hypothesis	An explanation for the increase in hypersensitivity and autoimmune disorders in industrialized countries due to less exposure to microbes and pathogens.
hypersensitivity	Immune responses to innocuous antigens that cause local or systemic allergic reactions.
inflammatory bowel disease	A term used to describe intestinal disorders associated with chronic inflammation such as Crohn's disease and ulcerative colitis.
mast cell	Immune effector cells found throughout the body that have a high affinity for IgE and play a major role in type 1 hypersensitivity reactions and immunity against parasitic worms.
oral tolerance	The suppression of immune responses against antigens associated with food or the intestinal microbiota.
self-tolerance	The lack of an immune response, primarily by B cells or T cells, directed against autoantigens.

REVIEW QUESTIONS

1. Describe the various symptoms associated with type 1 hypersensitivity reactions and how these symptoms relate to the route of exposure.

2. Discuss the various genetic and environmental factors associated with the development of autoimmune disorders.

3. Describe the etiology and pathophysiology of celiac disease.

4. Discuss why hypersensitivity and autoimmune disorders may be more common in industrialized countries than resource-poor countries.

ADDITIONAL QUESTIONS

In addition to the Review Questions provided above, there is a range of free online questions designed to enable students to further test their understanding of the chapter material. In order to access these interactive questions, please visit the book's website:
https://routledge.com/cw/wiser

Drug Misuse – Associated Disorders

18

Drug addiction and substance use disorders are major problems that can adversely affect health. These effects can be biological with clinical manifestations due to acute or chronic toxicity. Substance abuse can also increase the risk of developing other diseases such as cancer, cardiovascular disease, AIDS, or mental disorders (FIGURE 18.1). These increased risks can be due to the toxic effects of the substance being abused or to behavioral changes associated with drug misuse. Substance abuse also contributes to social problems such as homelessness, violence, and crime. These medical and social problems also have a huge economic impact in terms of increased health-care costs, decreased productivity, and accidents. The continued use of such drugs despite the adverse outcomes is due to their addictive properties. Addiction is correlated with changes in brain chemistry and physiology.

MENTAL DISEASE

SUBSTANCE USE DISORDER

MECHANISMS OF TOLERANCE AND DEPENDENCE

DISEASE ASSOCIATED WITH TOBACCO USE

DISEASE ASSOCIATED WITH ALCOHOL USE

OTHER COMMONLY MISUSED DRUGS

FIGURE 18.1 Impacts and consequences of substance abuse. The misuse of drugs has a large impact on both the individual and society. AIDS, acquired immune deficiency syndrome; CVD, cardiovascular disease.

Tobacco and alcohol are the two most commonly abused substances. Nicotine, the addictive substance in tobacco, is not overtly toxic at doses normally inhaled during smoking. However, other components in tobacco smoke can increase cancer, cardiovascular disease, and chronic obstructive pulmonary disease. Similarly, alcohol has addictive properties and chronic heavy drinking can damage the liver and brain. Other drugs that are routinely misused include opioids, cannabis, stimulants such as cocaine and amphetamines, depressants, and hallucinogens. Chronic use of these drugs can also lead to physical disease, as well as causing behavioral and societal problems.

MENTAL DISEASE

Normally we think of disease as being a physical illness that can be described in terms of pathology associated with organs and tissues. Mental illness is a functional disorder that affects behavior, emotion, and cognition, and often there is no overt pathology. Obviously, mental illness is related to problems with the brain and, even though there is no overt pathology, there can be aberrations in brain chemistry. Disorders of the brain and nervous system that involve structural, biochemical, or electrical abnormalities are called neurological disorders. Symptoms associated with neurological disorders include paralysis, muscle weakness, poor coordination, loss of sensation, seizures, confusion, pain, and altered consciousness.

Mental disorders are defined primarily by behaviors that impair a person's ability to function and interact with others. Psychiatry refers to the diagnosis and treatment of mental disorders. Like physical diseases, mental diseases have both genetic and environmental etiologies. Life experiences, such as stress, trauma, lack of sleep, and nutrition, can impact brain function and contribute to mental disorders. In addition, mental and physical health are fundamentally linked. People with mental illness are more likely to develop physical illnesses, and people with chronic physical illnesses are at risk of developing mental illness. In general, the biological bases of mental disorders are not well understood.

A wide range of behavioral abnormalities are defined as mental illness

Human behavior and personality are complex and, accordingly, many types of abnormal behavior fall into the realm of mental disorders (TABLE 18.1). These can include extreme psychotic episodes characterized by delusions and hallucinations, depression, impulse control, anxiousness, or substance abuse. The disorders can be transient or long-lasting and manifestations can range from mild to disabling. Disability in the context of mental disease usually refers to an inability to carry out daily functions associated with living, form and maintain interpersonal relationships, or acquire and keep a job. These various mental disorders are also often interconnected. For example, severe sleep deprivation can lead to psychotic episodes or severe depression can lead to sleep disorders.

Table 18.1   Mental Disorders

Type	Examples	Description
Psychosis	Schizophrenia	Characterized by delusions and hallucinations
Mood disorders	Depression, bipolar disorder	Unusually intense and sustained sadness, melancholia, or despair
Anxiety disorders	Phobias, OCD, PTSD	Anxiety or fear that interferes with normal functioning
Personality disorders	Paranoid, narcissistic, dependent	Harmful patterns of behavior that differ from social norms
Eating disorders	Anorexia nervosa, bulimia nervosa,	Disproportionate concern in matters of food and weight
Sleep disorders	Insomnia	Disruption of normal sleep patterns
Impulse control	Kleptomania, pyromania, gambling addiction	Inability to resist certain urges or impulses that could be harmful to themselves or others
Developmental	Autism, ADHD	Psychiatric conditions originating in childhood that involve substantial impairment
Substance use disorder	Drug dependence, drug abuse	Persistent use of drugs despite harm and adverse consequences

ADHD, attention deficit hyperactivity disorder; OCD, obsessive–compulsive disorder; PTSD, post-traumatic stress disorder.

Substance use disorder and mental illness are often observed as comorbidities

Comorbidity is two or more disorders observed in the same person. The disorders can occur simultaneously or sequentially in that one disorder may follow the other. Comorbidity also implies an interaction between the disorders that can worsen the outcomes of both disorders. Approximately half of the people with either substance use disorders or mental disorders also have the other. Three main pathways can contribute to the comorbidity between substance use disorders and mental illnesses (FIGURE 18.2). For example, there are shared risk factors between mental illness and substance use. Mental illness then potentially predisposes an individual toward substance use and addiction, and substance use potentially contributes to the development and exacerbation of mental illness.

Shared risk factors of substance use and mental disorders include a genetic predisposition, the involvement of similar neurotransmitters and regions of the brain, and extrinsic factors such as stress, trauma, and childhood experiences. Regarding the association between substance use disorder and mental illness, it is difficult to establish causality and directionality. For example, mental illness during adolescence may be mild and go undiagnosed. Substance abuse later in life may then exacerbate the mental illness leading to a perception that the drug abuse caused the mental illness, when in fact the reverse might be true.

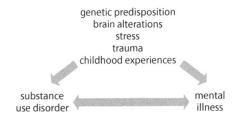

FIGURE 18.2 Comorbidity of substance use disorder and mental illness. Substance use disorder and mental illness have similar risk and predisposing factors. In addition, substance use disorder and mental illness can exacerbate each other.

SUBSTANCE USE DISORDER

Substance use disorder is the continued use of drugs despite substantial harm and adverse consequences. The disorder is characterized by an inability to reduce or stop consuming the drug despite repeated attempts and is often accompanied by **withdrawal** symptoms. Withdrawal symptoms refer to physiological manifestations that occur if drug use is ceased. Other symptoms associated with substance abuse include substantial time and effort spent on obtaining and using drugs, disengagement from social and professional activities or obligations as a result of drug use, and high-risk behaviors while using the drugs.

Persistent drug use leads to tolerance and dependence

The body adapts to repeated exposures to a specific drug and this adaptation can lead to some of the characteristics of substance use disorder (FIGURE 18.3). For example, persistent drug use often leads to a state in which more drug is needed to achieve the same effect. This diminishing effect of a drug due to repeated use is called **tolerance**. One sign of drug abuse is the need to take the drug more frequently or at higher doses to achieve the same effect. In addition, cessation of drug use may be accompanied by withdrawal symptoms. The association of withdrawal symptoms upon drug use cessation is called **dependence**. This dependence can be physical with symptoms such as fatigue or tremors, or psychological with emotional symptoms and lack of motivation. In other words, dependence is a state in which it becomes necessary to take the drug to feel normal. Tolerance and dependence are due to physiological changes in response to persistent drug use. **Addiction** is about behavior and is often used to describe a strong compulsion to take the drug despite experiencing negative consequences.

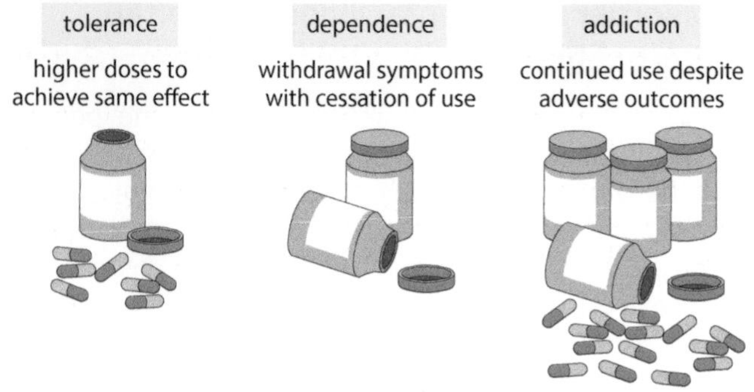

FIGURE 18.3 Qualities associated with drug misuse. Misuse of drugs can lead to tolerance, dependence, and addiction.

tolerance	dependence	addiction
higher doses to achieve same effect	withdrawal symptoms with cessation of use	continued use despite adverse outcomes

Psychoactive drugs are often misused

Psychoactive drugs are substances that affect nervous system function and result in alterations of perception, mood, consciousness, cognition, or behavior. These types of drugs are often misused, because many produce a state of relaxation or euphoria. Psychoactive drugs are further categorized as stimulants, depressants, and hallucinogens. Stimulants, such as nicotine, cocaine, and amphetamines, increase the activity of the central nervous system and may induce euphoria. Depressants, such as alcoholic beverages, opioids, cannabis, barbiturates, and benzodiazepines, do the opposite and have calmative, sleep-inducing, or anxiety-reducing effects. Hallucinogens alter perception and may distort reality.

Psychoactive drugs bind to targets associated within synapses

The mechanism of action of psychoactive drugs involves binding to specific brain receptors or transporters (TABLE 18.2). Binding of these drugs to their targets affects brain chemistry and function, and specifically affect neurotransmission at the level of the synapse (BOX 18.1). Drugs can either increase or decrease the activity of specific neurotransmission systems. In other words, drugs can function as agonists or antagonists. Some psychoactive drugs bind to receptors and either mimic the normal **neurotransmitter** or block the action of the normal neurotransmitter. Other drugs bind to neurotransmitter transporters and block the reuptake of the neurotransmitter. Blocking neurotransmitter uptake raises neurotransmitter levels. In summary, binding of psychoactive drugs to their targets interferes with brain chemistry and neurotransmission.

SIGNPOST

Mechanisms of drug action are described in Chapter 3, pages 66–69.

Table 18.2   Examples of Misused Psychoactive Drugs and Their Effects

Drug	Primary target	Primary effects
Nicotine/tobacco	Nicotinic acetylcholine receptor	Acetylcholine agonist that raises dopamine levels
Ethanol (i.e., alcoholic beverages)	Ligand-gated ion channels, especially GABA receptor	Activates the mesolimbic dopamine system and suppresses inhibition control
Opioids (e.g., heroin, morphine, oxycodone)	Opioid receptors	Block release of neurotransmitters involved in the sensation of pain, stimulates the dopamine reward system
Cannabis (THC)	Cannabinoid receptors	Agonist of anandamide that may be involved in appetite, pain sensation, mood, and memory
Benzodiazepines (e.g., Valium) and other barbiturates	GABA receptors	Increases effectiveness of GABA and leads to a decrease in the excitability of neurons, resulting in a calming effect
Hallucinogens such as LSD, mescaline, and psilocybin	Serotonin receptors	Serotonin is implicated in cognition, mood, and anxiety. Serotonergic hallucinogens are not particularly dopaminergic
Cocaine	Monoamine transporter	Blocks reuptake of dopamine
Ecstasy (MDMA) and other amphetamines	Serotonin and norepinephrine transporters	Blocks reuptake of serotonin, norepinephrine, and especially dopamine

GABA, gamma-aminobutyric acid; LSD, lysergic acid diethylamide; MDMA, 3,4-methylenedioxymethamphetamine; THC, tetrahydrocannabinol.

The Synapse, Neurotransmitters, and Psychoactive Drug Action

The nervous system is primarily composed of numerous interconnected cells called neurons. Neurons typically consist of a cell body, projections called dendrites, and a cable-like projection called the axon (**BOX 18.1 FIGURE 1**). The axon can be quite long and has branches. The termini of the axons contact other neurons and other cells, such as muscle cells. Signals are transmitted at these contact points called synapses. The cell sending the signal is the pre-synaptic cell and the cell receiving the signal is the post-synaptic cell. Signaling is partly electrical and partly chemical. An electrical pulse travels along the axon and activates the synaptic connections. In some types of synapses, called electrical synapses, the electrical pulse crosses the synapse via cell junctions. Chemical synapses involve the release of neurotransmitters into the synapse to relay the signal to the post-synaptic neuron.

In a chemical synapse, the electrical pulse traveling along the axon activates voltage-gated calcium channels at the axon termini and this activation causes an influx of calcium ions (Ca^{2+}) which triggers the fusion of vesicles containing neurotransmitters with the plasma membrane of the pre-synaptic cell. These secreted neurotransmitters bind to receptors located in the plasma membrane of the post-synaptic cell. The specific cellular response depends on the type of neurotransmitter involved and the nature of the receptor. Some receptors are coupled to ion channels, which transmit electrical signals down the axon, whereas other receptors are coupled with signal transduction and second messenger pathways.

Binding of the neurotransmitter to ligand-gated ion channels opens the channels and the influx of cations initiates an electrical pulse that is transmitted to other neurons. Many neurotransmitter receptors are similar to hormone receptors that are coupled to G-proteins and secondary messenger cascades. In this case, binding of the ligand activates a second messenger like cAMP and protein kinases. The resulting protein phosphorylation regulates cellular processes. Chemical synapses are often classified according to the neurotransmitter released and use names such as cholinergic (binds acetylcholine), dopaminergic (binds dopamine), or serotonergic (binds serotonin). The released neurotransmitters are rapidly removed from the synapse by either enzymes that break down the neurotransmitter or transporters that take the neurotransmitter back into the neurons.

Psychoactive drugs bind to protein targets found in the synapse. Many drugs mimic the natural ligands of the receptors and function as agonists or antagonists by binding to the receptor. Other drugs bind to ligand-gated channels at sites distinct from the neurotransmitter-binding site and either open or close these channels. Another target of psychoactive drugs is the reuptake transporters that remove neurotransmitters from the synapse. Inhibiting the transporters results in the persistence of the neurotransmitter in the synapse.

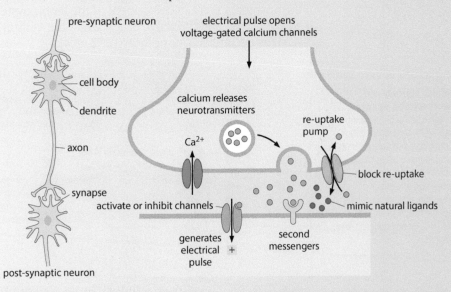

BOX 18.1 FIGURE 1 Neurons and synapses. Nerve cells communicate via electrical pulses that travel down the axons and release neurotransmitters at contact points between cells called synapses. Neurotransmitters (green circles) are analogous to hormones and bind to receptors that affect the post-synaptic cell. Psychoactive drugs (purple circles) interfere with normal neurotransmission by mimicking natural neurotransmitters (i.e., ligands), affecting ion channels, or blocking transporters that remove the neurotransmitters from the synapse.

Many psychoactive drugs mimic natural neurotransmitters and neural pathways

Opioids are drugs that bind to opioid receptors. The name is derived from opium, which is a drug prepared from the poppy plant. Opioids include drugs isolated from plants, called opiates, as well as synthetic drugs that bind opioid receptors. The body produces natural opioids, called endorphins, that bind to opioid receptors. Binding of agonists – whether natural or artificial – to opioid receptors blocks the release of neurotransmitters involved in the sensation of pain. Therefore, opioids suppress pain and are widely used as analgesics for the treatment of pain. Opioid antagonists bind to the receptors, but do not activate the receptors and therefore do not have psychoactive effects. Some opioid antagonists are used to treat opioid overdoses since blocking the binding of agonists to receptors inhibits toxicity.

Similarly, tetrahydrocannabinol (THC), the primary psychoactive component of cannabis, appears to mimic a compound produced naturally by the body called anandamide. Anandamide and THC bind to cannabinoid receptors that are part of the endocannabinoid system. The endocannabinoid system is involved in a variety of physiological processes that modulate appetite, pain sensation, mood, and memory. Serotonin is another neurotransmitter involved in cognition, mood, and anxiety and some psychoactive drugs target serotonin receptors and transporters.

A common feature of many psychoactive drugs is the activation of dopamine and the mesolimbic pathway

Although the various drugs that are prone to misuse have different molecular targets, most converge in a common mode of action involving **dopamine** and the **mesolimbic pathway**. The mesolimbic pathway is a dopaminergic pathway of the brain that originates in a part of the brain called the ventral tegmental area. Dopamine is transported from the ventral tegmental area to the nucleus accumbens and the amygdala. The nucleus accumbens is believed to play a role in reward and desire, and the amygdala is a key component of the limbic system and is associated with emotion. Psychoactive drugs target the ventral tegmental area, the nucleus accumbens, or both (FIGURE 18.4).

FIGURE 18.4 Mesolimbic pathway and psychoactive drugs. The mesolimbic pathway originates primarily in a region of the brain called the ventral tegmental area (VTA). The nerves in the VTA contain a variety of receptors and activation of these receptors results in dopamine being released in another region of the brain called the nucleus accumbens (NAc). This may lead to feelings of well-being, mild euphoria, relaxation, or excitation. Most psychoactive drugs target the VTA, NAc, or both. (EtOH, ethanol or alcohol; THC, tetrahydrocannabinol; an active ingredient in cannabis.)

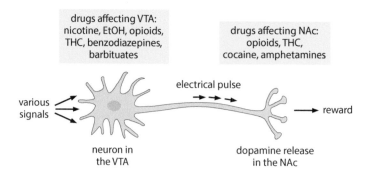

The mesolimbic pathway controls an individual's responses to natural rewards, such as food, sex, and social interactions, and thus incentivizes repetition to get that reward. For example, palatable foods activate the neurons of the mesolimbic pathway and increase dopamine levels. The higher levels of dopamine create a pleasurable experience (i.e., reward). Dopamine has been called the 'pleasure' or 'feel-good' hormone. This good experience motivates a person to repeat the activity. Obviously, activities such as eating or sex are important for the survival and continuation of a species. Drugs that activate the mesolimbic pathway can also motivate individuals to repeatedly consume the drug. This leads to a corruption of the reward mechanisms and drugs progressively replace natural rewards in determining behavior.

MECHANISMS OF TOLERANCE AND DEPENDENCE

Repeated use of a drug leads to tolerance toward that drug and possibly the development of dependence. Tolerance can be described on many levels (TABLE 18.3). One mechanism of tolerance is an increase in the enzymes or processes that degrade the drug or prevent the drug from reaching its target. Many of these processes shorten the drug's half-life. Drug half-life is the time it takes for the levels of the drug in the body to decrease by half. Increased degradation of the drug is referred to as metabolic tolerance. Metabolic tolerance is part of pharmacokinetic tolerance. Pharmacokinetics refers to the absorption, distribution, metabolism, and excretion of drugs and causes less drug to reach the target.

Table 18.3 Types of Drug Tolerance	
Metabolic or pharmacokinetic	Changes in the absorption, distribution, excretion, and especially metabolism of drugs causing less drug to reach the target
Physiological or pharmacodynamic	Changes in the brain that cause the drug to be less effective
Behavioral	Learning how to compensate for the effects of the drugs or adapting behavior accordingly
Cross-tolerance	Tolerance to a related drug due to repeated use of another drug
Reverse tolerance or sensitization	An unusual occurrence of becoming more sensitive to drugs after repeated use

Changes in brain chemistry play a major role in drug tolerance and dependence

Repeated use of a psychoactive drug can also alter the brain chemistry, so that higher drug concentrations are needed to achieve the same effect. This is called pharmacodynamic or physiological tolerance. Physiological tolerance can be associated with changes in the number of receptors or desensitization of the receptors, as well as other changes in brain chemistry. These changes could be at the level of the drug target or there could be changes in the mesolimbic dopamine pathway. Little is known about the specific mechanisms of pharmacodynamic tolerance. Pharmacodynamic tolerance is also likely to be involved in the development of dependence.

In addition to requiring more drug for the same effect, substance use disorder is also accompanied by dependence. Dependence is an adaptation of the brain in which drugs need to be present to avoid withdrawal symptoms. The potential of developing dependence is a property of the drug, as well as the individual. Drug dependence is more likely to occur in people with a genetic predisposition and these individuals may have abnormal neuronal activity in the mesolimbic dopamine pathway. Like tolerance, mechanisms of drug dependency are poorly understood.

Dependency may involve stable changes in gene expression

Effects of repeated drug use on receptors, endogenous ligands, and other drug targets have been demonstrated. However, it is difficult to reconcile dependency solely in the context of the upregulation or downregulation of specific receptors or neural pathways. Dependency could also be associated with changes in neuron gene expression. Perturbations of synaptic transmissions by drugs are known to affect transcription factors and alter gene expression. Two transcription factors that have garnered considerable attention in this respect are the cAMP response element-binding protein (CREB) and ΔFosB (**FIGURE 18.5**). CREB is a transcription factor that is activated by cAMP-dependent protein kinases and many neurotransmitter receptors, including dopamine receptors, are coupled with the cAMP signal transduction pathway. However, the precise role of CREB and ΔFosB in tolerance and dependence is not known.

SIGNPOST Details of signal transduction and second messengers are described in Chapter 3, pages 63–66.

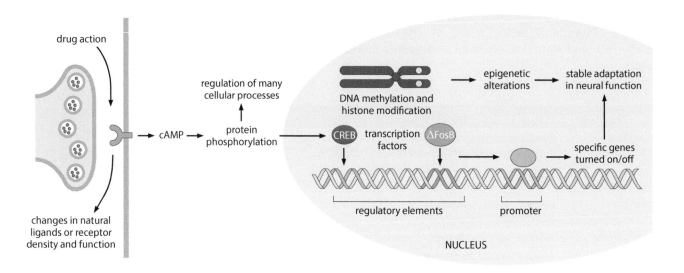

FIGURE 18.5 Possible mechanisms of tolerance and dependency. Psychoactive drugs act at the synapse and repeated drug use can lead to changes in brain chemistry and neural function. For example, the natural ligands, number of receptors, or functional state of receptors can change in response to repeated drug use, resulting in pharmacodynamic tolerance. Some neurotransmitters are coupled to second messengers such as cAMP and protein phosphorylation. In addition to regulating many cellular processes, cAMP can regulate gene expression via a protein called the cAMP response element-binding protein (CREB). CREB and other transcription factors such as ΔFosB bind to regulatory elements of genes and change the transcriptional profile of the cell. The transcriptional profile can also be altered by epigenetic changes in chromatin structure. Such phenotypic changes represent a stable adaptation of neurons to repeated drug use and might partially explain dependency.

SIGNPOST Chromatin remodeling and epigenetics are described in Chapter 5, pages 100–101.

Epigenetic changes such as DNA methylation and histone modification have also been implicated in dependency. Changes in chromatin structure repress or derepress the ability of genes to be transcribed. Changes in neuronal gene expression, whether due to changes in chromatin structure or transcription factors, can produce long-term adaptive changes in the brain that may be associated with the dependency phenotype.

DISEASES ASSOCIATED WITH TOBACCO USE

Cigarette smoking is a leading cause of preventable disease and premature death, and smoking is a factor in 20% of deaths. It is estimated that smoking reduces a person's lifespan by approximately 10 years. Smoking is a major factor in the development of cancer, cardiovascular disease, and pulmonary disease. Diseases due to smoking are caused by toxins in tobacco smoke. In particular, a class of compounds called benzopyrenes are carcinogens and greatly increase the risk of developing cancer. Similarly, components in tobacco smoke affect the cardiovascular system and are a major element in cardiovascular disease (TABLE 18.4). Nicotine is a psychoactive drug found in tobacco and has proven to be rather addictive. Withdrawal symptoms, such as irritability, a depressed mood, restlessness, and anxiety, are among the symptoms associated with the cessation of smoking. Therefore, people continue to smoke despite adverse health outcomes.

> Carcinogenesis due to benzopyrenes is described in Chapter 16, pages 305–306.
>
> **SIGNPOST**

Table 18.4	**Adverse Effects of Smoking on the Cardiovascular System**

- Nicotine causes immediate and long-term increases in blood pressure and heart rate promoting the development of hypertension
- Toxins in cigarette smoke damage blood vessels and possibly initiate atherosclerosis
- Smoking causes dyslipidemia and accelerates plaque formation in atherosclerosis
- Increased inflammation due to toxins in smoke increase the intravascular clotting phenomenon
- Carbon monoxide in smoke binds to hemoglobin and leads to hypoxia in the tissues

Nicotine binds primarily to acetylcholine receptors and activates the mesolimbic dopamine pathway

Nicotine constitutes 0.6–3.0% of the dry weight of tobacco (*Nicotiana tabacum*). During the act of smoking, nicotine is readily absorbed by the lungs, rapidly enters the circulation, and reaches the brain within seconds. The primary target of nicotine is a class of receptors called the nicotinic acetylcholine receptor (nAChR). Receptors that bind acetylcholine are called cholinergic receptors, and there are many types of cholinergic receptors that carry out a wide variety of functions. Cholinergic receptors are ion-gated channels and binding of acetylcholine propagates an electrical signal through the neuron.

Like many psychoactive drugs, nicotine primarily targets neurons in the ventral tegmental area (see Figure 18.3). Nicotine functions as an acetylcholine agonist and as an agonist, binding of nicotine to the nAChR opens the channel and ions flow into the neuron. This creates an electrical pulse that is transmitted along the axon and results in the release of dopamine into the nucleus accumbens. Dopamine release induces pleasure and reduces anxiety and stress. In addition to dopamine, nicotine affects other neurotransmitter pathways (TABLE 18.5). At lower blood concentrations, nicotine functions primarily as a stimulant via acetylcholine and epinephrine pathways. As nicotine concentrations rise, serotonin and endorphin pathways are activated, and this results in a relaxant effect.

Table 18.5	**Neurotransmitter Pathways Affected by Nicotine**
Neurotransmitter	**Primary functions**
Dopamine	Pleasure, suppress appetite
Acetylcholine	Arousal, cognitive enhancement
Epinephrine	Arousal, suppress appetite
Vasopressin	Memory improvement
Serotonin	Mood modulation
β-Endorphin	Reduce anxiety and tension

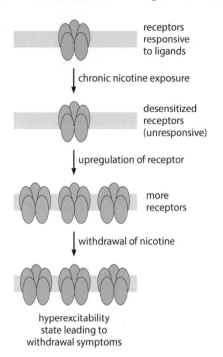

Chronic nicotine exposure causes receptor desensitization and upregulation

Cigarette smoking, in amounts typical of daily smokers, maintains the nAChR in a nearly saturated state, meaning most receptors are continuously bound with nicotine. The persistence of ligand results in **desensitization** of the receptor and in this case the ion-gated channel remains closed despite ligand binding. Desensitization is a general feature associated with most receptors after prolonged exposure to ligands. In addition, chronic nicotine exposure causes an increase in the number of nAChRs (FIGURE 18.6). This upregulation in nAChR expression is probably in response to the prolonged desensitization.

Desensitization and upregulation of nAChRs probably contribute to the development of tolerance and dependence. For example, the receptors return to a normal responsive state in the absence of nicotine. As upregulation has produced more receptors, the return of the receptors to the responsive state may result in a hyperexcitable state. This hyperexcitable state in the absence of nicotine may lead to cravings and withdrawal symptoms. The cravings and withdrawal symptoms may be especially acute during long periods of abstinence, such as overnight sleep. Nicotine binding to the receptors alleviates the cravings and withdrawal symptoms, as well as promoting pleasurable effects. Alterations in the downstream components of the mesolimbic dopamine pathway are also likely to contribute to tolerance and dependence.

FIGURE 18.6 Possible mechanisms of nicotine tolerance and dependence. Chronic exposure of acetylcholine receptors to nicotine causes desensitization and the receptors are temporarily unresponsive to ligands. The desensitization causes an increase in receptor number resulting in increased tolerance. A prolonged absence of nicotine leads to receptor re-sensitization, which may lead to a state of receptor hyperexcitability and contribute to withdrawal symptoms associated with dependency.

Smoking is the leading cause of chronic obstructive pulmonary disease

Long-term exposure to smoke and other irritants results in an inflammatory response in the lung. This chronic inflammation can develop into **chronic obstructive pulmonary disease** (COPD). The most common symptoms of COPD are chronic cough and difficulty breathing. These symptoms persist and progressively worsen. Diagnosis involves measuring lung function in terms of the amount and speed of airflow during inhalation and exhalation. Lung function naturally decreases with age and smoking accelerates this process (FIGURE 18.7). Cessation of smoking can slow this progressive loss of lung function, but the lost lung function cannot be regained.

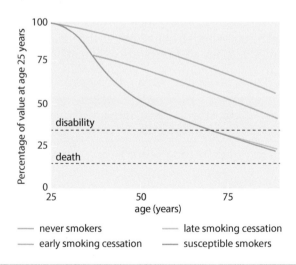

FIGURE 18.7 Lung function in smokers and non-smokers. A graphic illustrating the loss of lung function (*y* axis) with age in smokers and non-smokers.

The two primary pathological features associated with COPD are a narrowing of the air passageways and a breakdown of lung tissue. The relative contributions of these two pathological features vary between individuals. Both phenomena are due to inflammation in the lungs involving macrophages and neutrophils. Smoking increases the number of macrophages and neutrophils in the lungs and, in response to smoke and other irritants, these cells secrete inflammatory cytokines. These cytokines cause a constriction of the bronchi and bronchioles of the lungs, as well as hypersecretion of mucus (FIGURE 18.8). These two phenomena restrict airflow and produce chronic bronchitis. This results in a chronic cough that generally includes the production of sputum.

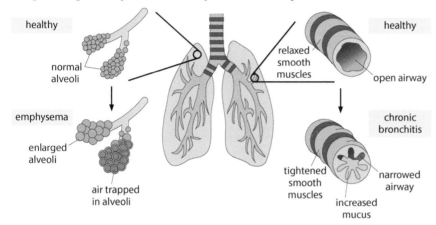

> Macrophages and neutrophils are described in Chapter 9, pages 176–177.
>
> **SIGNPOST**

FIGURE 18.8 Pathological features of COPD. Damage to the terminal air sacs, called alveoli, in the lungs results in decreased lung function and trapping of air in the lungs (emphysema). This is due to a breakdown of the alveoli walls and loss of elasticity. Inflammatory cytokines cause constriction of the bronchi and hypersecretion of mucus. The resulting bronchitis further obstructs airflow.

Activated macrophages and neutrophils also secrete hydrolytic enzymes such as elastase and collagenase. These enzymes degrade proteins of the extracellular matrix and lead to a breakdown of the walls of the alveoli. The loss of extracellular matrix proteins, such as collagen and elastin, decreases the ability of the lungs to expand and contract during breathing. This causes air to be trapped in the air sacs. Trapping of air in the tissues is called emphysema. Thus, COPD is a combination of bronchitis and emphysema.

DISEASE ASSOCIATED WITH ALCOHOL USE

The consumption of alcoholic beverages is well accepted in most cultures and alcohol use per se is not necessarily harmful. In fact, there may be benefits to the moderate use of alcohol such as lowering the risk of developing cardiovascular disease. This lowering of the risk of developing cardiovascular disease may be related to raising HDL (high-density lipoprotein) levels, the so-called good cholesterol, which protects against developing atherosclerosis. In addition, moderate alcohol consumption may be linked to improved well-being through better social interactions and networks. Moderate drinking is usually defined as 1–2 typical alcoholic beverages per day.

Consumption of more than moderate amounts of alcohol is often associated with adverse effects. At high doses alcohol can be acutely toxic and fatal. Other acute effects of high alcohol doses include accidents and injuries associated with mental and physical impairment due to intoxication. Chronic effects of long-term heavy alcohol consumption include liver and brain damage. Exposure in utero can lead to fetal alcohol spectrum disorder (BOX 18.2). Long-term alcohol abuse also tends to have detrimental effects on intrapersonal interactions with family, friends, and coworkers. Societal problems associated with alcohol abuse include driving accidents and fatalities, accidental injuries, sexual assaults, domestic abuse, and violent crime.

BOX 18.2

Fetal Alcohol Spectrum Disorder

Consumption of alcohol during pregnancy can have adverse effects on the developing fetus that persist after birth and throughout life. The various manifestations due to exposure to alcohol in utero are called fetal alcohol spectrum disorders. Manifestations can include abnormal facial appearance, short height, low body weight, small head size, poor coordination, behavior problems, learning difficulties, and problems with hearing or sight. Fetal alcohol syndrome is the most severe form of the disorder and is characterized by height and weight below the tenth percentile, distinctive facial features, and significant neurological impairment.

Ethanol meets all the criteria of being a teratogen. Teratogens are substances that are capable of disrupting fetal growth and development and producing malformations. Ethanol readily crosses the placenta, and the fetus has not yet started to express the ethanol-detoxifying enzymes. Ethanol decreases the expression of gamma-aminobutyric acid (GABA) receptors – especially in the cerebrum – during development periods. GABA, in addition to being the major inhibitory neurotransmitter, plays a role in nerve cell migration. Exposure to ethanol during brain development may affect nerve migration and thereby distort the normal development of neural networks. These distorted neural networks may account for the manifestations associated with fetal alcohol syndrome.

Ethanol is a CNS depressant

Ethanol (CH$_3$CH$_2$OH) is a psychoactive molecule that is generically referred to as alcohol. It is rapidly absorbed by the small intestine after consumption and spreads throughout the body. The acute psychological and behavioral effects of ethanol are largely dose dependent and there is a progression from mood elevation to euphoria to mental and physical impairment to severe intoxication with increasing concentrations of alcohol in the blood (TABLE 18.6). Blood alcohol content (BAC) is defined as the percentage of alcohol in the blood. A concentration of 0.4% (400 mg per liter) will kill about half of those affected. The legal definition of intoxication is typically 0.08% BAC.

Table 18.6	Concentration–Effect Relationship of Ethanol
BAC (%)	**Effects**
0.02–0.03	Mood elevation. Slight muscle relaxation
0.05–0.06	Relaxation and warmth. Increased reaction time. Decreased fine muscle coordination
0.08–0.09	Impaired balance, speech, vision, hearing, muscle coordination. Euphoria
0.14–0.15	Gross impairment of physical and mental control
0.20–0.30	Severely intoxicated. Very little control of mind or body
0.40–0.50	Unconscious. Deep coma. Death from respiratory depression

BAC, blood alcohol content expressed as a percentage.

The primary target of ethanol appears to be the γ-aminobutyric acid (GABA) receptor. GABA is the major inhibitory neurotransmitter in the central nervous system and plays a role in virtually all brain functions. Benzodiazepines, barbiturates, and many anesthetics also target the GABA receptor. Inhibitory neurotransmitters suppress neuron activity and GABA agonists, such as ethanol, facilitate the suppression of the central nervous system (CNS). The apparent stimulatory effects of ethanol at low doses may be due to selective depression of inhibitory control mechanisms. Inhibitory control is the ability to control impulsive or automatic responses and respond through reasoning. Therefore, a person becomes less inhibited and more compulsive after alcohol consumption. In addition, ethanol also increases dopamine levels in the mesolimbic reward pathway, leading to mood elevation and euphoria.

Alcohol withdrawal can be associated with severe manifestations

Alcohol use disorder, more commonly called alcoholism, is characterized by tolerance and dependence. Signs of alcoholism include consumption of increasing amounts of alcohol, frequent intoxication, and preoccupation with drinking to the exclusion of other activities. Like other dependencies, cessation of alcohol use can result in physical withdrawal symptoms (TABLE 18.7). People with long-term high alcohol intake may experience severe withdrawal symptoms called delirium tremens. Delirium tremens generally follows a long period of heavy drinking that is abruptly stopped. Physical manifestations include severe shaking, irregular heartbeat, and sweating. Mental manifestations include confusion, severe anxiety, and even hallucinations. Seizures and death are also possible. Benzodiazepines can be used to treat severe alcohol withdrawal and hospitalization is sometimes needed.

Table 18.7 Alcohol Withdrawal Symptoms	
Minor to moderate	**Delirium tremens (or DTs)**
• Shaky hands • Sweating • Mild anxiety • Nausea, vomiting • Headache • Insomnia	• Severe tremors • Disorientation, confusion, and severe anxiety • Hallucinations • Seizures • Profuse sweating • High blood pressure • Racing and irregular heartbeat • Fever

Alcohol is metabolized primarily by the liver

Chronic heavy drinking adversely affects all organs and especially the liver and brain. The adverse effects of alcohol on the liver are primarily related to the detoxification and metabolism of ethanol by the liver. The liver, as a major organ of the digestive system, metabolizes nutrients and detoxifies toxins that are absorbed by the intestines. In total, 80% of the absorbed ethanol passes through the liver. The detoxification of ethanol is a two-step process involving alcohol dehydrogenase and acetaldehyde dehydrogenase (FIGURE 18.9). Alcohol dehydrogenase converts ethanol to acetaldehyde, which is then converted to acetate by acetaldehyde dehydrogenase. Acetate is a normal metabolite that can either enter the tricarboxylic acid cycle and be metabolized to carbon dioxide as part of energy production or serve as a substrate in fatty acid synthesis.

> Metabolism and the tricarboxylic acid cycle are described in Chapter 3, pages 60–61.
>
> SIGNPOST

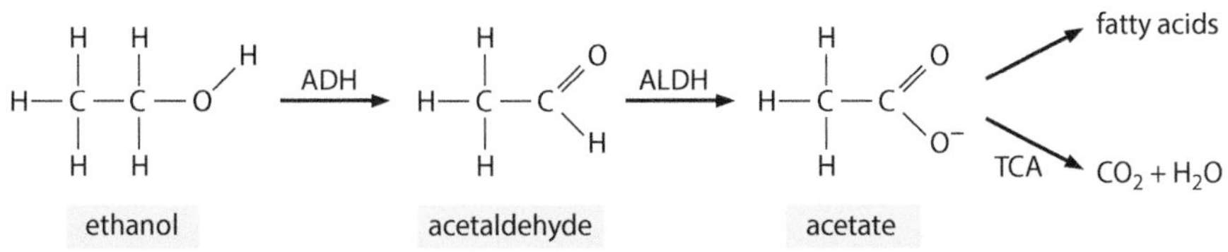

FIGURE 18.9 Metabolism and detoxification of ethanol. Alcohol dehydrogenase (ADH) converts ethanol into acetaldehyde and acetaldehyde dehydrogenase (ALDH) converts acetaldehyde into acetate. Acetate can be metabolized in the tricarboxylic acid (TCA) cycle into carbon dioxide and water, or acetate can be a substrate in the synthesis of fatty acids.

During the detoxification of ethanol, the first step catalyzed by alcohol dehydrogenase is faster than the acetaldehyde dehydrogenase reaction. Therefore, acetaldehyde accumulates following the ingestion of a large amount of alcohol. Acetaldehyde exhibits some acute toxicity and this, along with dehydration, contributes to the ill-effects – often called a 'hangover' – that sometimes occur after a night of heavy drinking. In addition, these two alcohol-detoxifying enzymes are inducible and people who drink regularly have higher levels of these enzymes and, therefore, are more capable of detoxifying alcohol. This is an example of metabolic tolerance since the higher levels of detoxifying enzymes mean that an individual needs to consume more alcohol to have the same effect. Natural variations in levels of these enzymes between individuals also explain why some people are more susceptible to the effects of alcohol than others.

Chronic heavy drinking can extensively damage the liver

The metabolism of excessive amounts of ethanol associated with chronic heavy drinking ultimately results in liver pathology. Clinical manifestations associated with alcoholism are a fatty liver, alcoholic hepatitis, and cirrhosis of the liver (TABLE 18.8). Fatty liver disease, called hepatic steatosis, is due to the accumulation of fatty acids and triglycerides in the liver. Initially, fatty liver disease is generally asymptomatic. Over time, though, there is increased inflammation due to the accumulation of fat and the toxicity of acetaldehyde. Acetaldehyde also causes inflammation and produces some oxidative stress that results in inflammation of the liver called hepatitis.

Table 18.8 Primary Hepatic Manifestations of Alcoholism	
Fatty liver	Alcohol metabolism leads to an increase in fatty acids and triglycerides (fat accumulation called steatosis)
Alcoholic hepatitis	Associated with acetaldehyde toxicity (inflammation, oxidative stress)
Cirrhosis	Chronic inflammation leads to fibrosis (scarring) and eventually liver failure

Over time, the accumulation of fat and persistent inflammation leads to fibrosis, more commonly known as scarring. The accumulation of scar tissue compromises liver function and eventually **cirrhosis** of the liver develops in 10–20% of heavy drinkers. The extensive inflammation and scarring associated with cirrhosis cause a loss of liver function and can lead to liver failure and death. The liver does have substantial regeneration abilities and if drinking is stopped soon enough the liver can recover (FIGURE 18.10). However, once cirrhosis develops, it is irreversible and therefore considered an end-stage disease requiring transplantation.

FIGURE 18.10 Liver damage associated with alcohol abuse. Chronic heavy drinking can cause the accumulation of fat in the liver (hepatic steatosis) and inflammation (alcoholic hepatitis). Continued alcohol consumption may lead to cirrhosis. The damage to the liver can be reversed if drinking is stopped soon enough. Cirrhosis cannot be reversed and can result in liver failure.

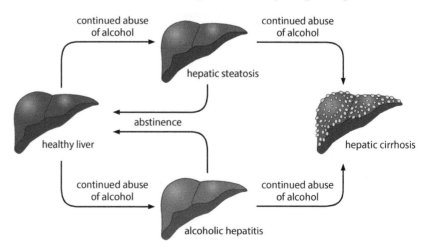

Dementia and other neurological effects are associated with alcohol abuse

Chronic heavy drinking also severely affects the brain and may lead to cerebellar degeneration (TABLE 18.9). The cerebellum is responsible for coordinating voluntary movements and motor skills such as balance, coordination, and posture. Thus, even when sober, alcoholics may exhibit a lack of coordination and stumbling. In addition, there may be a deterioration in intellectual function that affects planning, thinking, and judgment.

Table 18.9   Common Symptoms of Alcoholic Dementia

- Loss of muscle coordination
- Confusion
- Inability to form new memories
- Memory loss
- Confabulation
- Abnormal eye movement
- Double vision
- Hallucinations

Other effects on the brain are attributable to nutritional deficiencies and especially thiamine (vitamin B_1). Alcohol supplies calories but is devoid of other nutrients and 80% of alcoholics exhibit thiamine deficiency. Notable pathologies associated with thiamine deficiency in alcoholics are Wernicke's encephalopathy and Korsakoff's psychosis. Both often occur together, called Wernicke–Korsakoff syndrome, and generally Wernicke's encephalopathy precedes Korsakoff's psychosis.

Thiamine is a co-factor in glucose metabolism and deficiency leads to nerve death. Eye movement disorders combined with incoordination and confusion are key attributes of Wernicke's encephalopathy. Korsakoff's psychosis is characterized by amnesia and confabulation. Confabulation is a memory error in which a person creates false memories, and the person is confident about their recollections despite contradictory evidence. The created memories can range from subtle alterations to bizarre fabrications that are often part of rambling, albeit fluent, accounts. This syndrome can be prevented, and sometimes treated, with thiamine supplementation.

OTHER COMMONLY MISUSED DRUGS

Several other psychoactive drugs, in addition to tobacco and alcohol, are commonly misused and have adverse effects. Acute toxicity is always a health risk and drug overdoses are potentially fatal. This is especially true of CNS depressants since they can lead to respiratory failure at high doses. Chronic use can exacerbate mental disease, lead to tolerance and dependence, and ultimately addiction. In addition to affecting individual health, drug misuse also contributes to interpersonal problems, and has societal and economic issues.

Current opioid misuse is often described as an epidemic

Opium, a drug prepared from the poppy plant, has been used for thousands of years for the treatment of pain and other disorders. The availability of some powerful synthetic opioids has resulted in a substantial number of overdose deaths in the past 2–3 decades. This opioid epidemic started in part due to an increase in opioid prescriptions to treat chronic pain. As with most psychoactive drugs, long-term use results in tolerance and dependence. The mechanisms of opioid tolerance and dependence are not understood. Tolerance to some opioids may be associated with receptor downregulation via internalization. Desensitization of receptors is also observed. Protein kinases regulating gene expression are also affected and chronic opioid use may result in long-term phenotypic changes in the brain (see Figure 18.4.).

Cannabis use is the subject of numerous debates

Cannabis is probably the most used psychoactive drug in the world. There is an ongoing debate about its harmful effects versus its beneficial and medical uses. Many governmental jurisdictions have either legalized or decriminalized the possession of cannabis over the last decade. Medical uses of cannabis use include reducing nausea and vomiting during chemotherapy, improving appetite in people with HIV/AIDS, or treating chronic pain and muscle spasms. The popularity of cannabis as a recreational drug is due to its relaxing and euphoric effects. However, high doses can produce hallucinations and paranoia. Cannabis is not considered acutely toxic and there are no reported deaths due to overdoses.

Although, much less addictive than opioids or alcohol, repeated use of cannabis leads to tolerance and dependency. Like other psychoactive drugs, a downregulation of cannabinoid receptors is observed following chronic use. In addition, the mesolimbic reward pathway may also be altered with repeated use. Chronic use may also reduce attention, memory, and learning functions and affect how the brain builds connections between the areas of the brain that are necessary for these functions. This is especially true in users who started during adolescence. Repeated use of high doses sometimes leads to psychosis.

CNS stimulants include the party drugs cocaine and ecstasy

Another commonly used recreational drug is cocaine. It is a strong stimulant that also produces feelings of well-being, increased attentiveness, arousal, or euphoria. Anxiety, paranoia, and restlessness can also occur, especially as the effects of the drug wear off. Cocaine also raises the heart rate and increases blood pressure. High doses can produce a life-threatening heart rate and high blood pressure that increases the risk of heart attack or stroke. Cocaine addiction involves ΔFosB expression in the nucleus accumbens and causes an altered transcriptional profile in neurons (see Figure 18.4).

Other stimulants, such as amphetamine, methamphetamine, and ecstasy, have similar mechanisms of action and effects. Cocaine and ecstasy are often called party drugs or club drugs since they are commonly used to stay up all night at dance clubs. Long-term and high-dose use of amphetamines does produce changes in the brain and sometimes leads to psychosis.

SUMMARY OF KEY CONCEPTS

- Mental illnesses are functional disorders that affect behavior, emotion, and cognition

- Substance use disorder is a type of mental illness that is often associated with other mental illnesses

- Psychoactive drugs alter perception, mood, consciousness, cognition, or behavior and are frequently misused

- Binding of psychoactive drugs to receptors or transporters in the brain interferes with neurotransmission and brain function

- Many psychoactive drugs activate the neurons of the mesolimbic reward pathway, increase dopamine levels, and produce a feeling of pleasure

- Repeated drug use results in a state called tolerance in which more drug is needed to produce the same effects

- Pharmacodynamic changes in receptors such as upregulation, downregulation, and desensitization may contribute to tolerance

- Dependency and addiction are states in which drug use is needed to prevent withdrawal symptoms

- Nicotine addiction increases the risk of cancer, cardiovascular disease, and chronic obstructive pulmonary disease (COPD) due to long-term exposure to toxins in cigarette smoke

- Inflammation directed at irritants in cigarette smoke can cause bronchitis and emphysema associated with COPD

- Alcohol is a CNS depressant that with increasing dose progresses from mood elevation to euphoria to mental and physical impairment to severe intoxication and ultimately death

- The two organs most affected by chronic heavy drinking are the liver and brain

- Detoxification of ethanol results in the accumulation of acetaldehyde and fatty acids in the liver that leads to fatty liver disease or alcoholic hepatitis, which may progress to cirrhosis

- Degeneration of the cerebellum is often observed in heavy drinkers and both motor and cognitive skills are affected

- Other commonly abused drugs include opioids, cannabis, stimulants such as cocaine and amphetamines, depressants, and hallucinogens

KEY TERMS

addiction	A compulsive physiological or psychological need for a habit-forming substance, especially a drug or behavior, despite negative consequences.
cirrhosis (liver)	A chronic disease of the liver characterized by the replacement of normal tissue with scar tissue and the loss of functional liver cells often due to alcohol abuse.
COPD	Chronic obstructive pulmonary disease commonly caused by smoking and associated with a combination of emphysema and bronchitis.
dependence (drug)	A state in which it is necessary to take a drug to feel normal and cessation of drug use leads to withdrawal symptoms.
desensitization	The temporary loss of responsiveness of a receptor to its ligand due to a persistent presence of the ligand.
dopamine	A neurotransmitter in the central nervous system that helps regulate emotion and is often associated with pleasure.

mesolimbic pathway	A neurological pathway of the midbrain associated with dopamine receptors that influences motivation and euphoria.
neurotransmitter	Signaling molecules that are secreted by neurons (i.e., nerve cells) which affect the adjacent neurons or muscle cells at the synapse.
psychoactive (drug)	A substance that affects nervous system function and results in alterations of perception, mood, consciousness, cognition, or behavior.
tolerance	The phenomenon of a decreased effect of a drug due to substantial prior exposure, resulting in the need for higher doses to have the same effect.
withdrawal (symptoms)	Refers to the feelings of discomfort, distress, or intense craving for a substance that occurs when use of the substance is stopped.

REVIEW QUESTIONS

1. What is the mesolimbic pathway and what role does it play in psychoactive substances and their abuse?

2. Discuss tolerance, dependence, and addiction and the relationships between them and substance abuse.

3. Briefly describe the pathophysiology of COPD. What are the two lung disorders associated with COPD?

4. List and describe the various classes of psychoactive drugs.

5. Describe the pathological effects of alcohol abuse on the brain and liver.

ADDITIONAL QUESTIONS

In addition to the Review Questions provided above, there is a range of free online questions designed to enable students to further test their understanding of the chapter material. In order to access these interactive questions, please visit the book's website:
https://routledge.com/cw/wiser

Infectious Disease

Infectious diseases are an important cause of disease throughout the world with resource-poor countries being especially afflicted. The first chapter of this section gives an overview of infectious disease and describes the types of pathogens that cause disease. Also included in this chapter is an overview of modes of transmission of pathogens and a discussion of drug-resistance mechanisms. The following five chapters cover respiratory infections, diarrheal diseases, sexually transmitted infections, vector-transmitted diseases, and parasitic worms. Each of the chapters gives an overview of the type of disease being described and focuses on the most important diseases in that class of infectious disease. For example, the chapter on respiratory infections focuses on influenza, COVID-19, and tuberculosis. The chapter on diarrheal diseases describes the pathophysiology of diarrhea and how pathogens cause diarrhea, and in some cases produce an invasive disease. HIV and AIDS is the focus of the chapter on sexually transmitted infections and malaria is the focus of the chapter on vector-transmitted diseases. Although parasitic worms are not responsible for much mortality, they do cause substantial human morbidity among impoverished people living in tropical areas. As with the chapters on non-infectious diseases, the individual chapters on infectious diseases also include discussions on disease manifestations, diagnosis, treatment, and control of the diseases. Examples of host–pathogen interactions at the molecular and cellular levels are also discussed.

Infectious Diseases

19

Approximately 25% of mortality is still due to infectious diseases, despite great advances in the treatment and prevention of infectious disease throughout the 20th century. Although the greatest disease burden from infectious disease is in resource-poor countries, infectious disease is also important in developed countries. Lower respiratory infections (Chapter 20), diarrheal diseases (Chapter 21), AIDS, (Chapter 22), tuberculosis (Chapter 20), and malaria (Chapter 23) are the five leading causes of death among infectious diseases. AIDS, tuberculosis, and malaria are sometimes collectively referred to as the 'Big Three'.

Infectious diseases are caused by biological entities – ranging in complexity from infectious proteins to multicellular parasites – that are transmitted from person to person or acquired from the environment. For this reason, infectious diseases are also called communicable or contagious diseases. Generally, contagious diseases refer to those diseases that are highly transmissible between persons. Similarly, communicable disease is generally more restrictive than infectious disease in that communicable refers to diseases that must be transferred from person to person and cannot exist independently in the environment.

The entities that cause infectious disease are **pathogens** since they directly or indirectly cause damage to the host. There are also many microbes that inhabit the human body as **commensals**. Commensals are symbiotic organisms that do not cause pathology and some even benefit their host. These commensal organisms are part of the microbiota (see Box 6.2). Among the pathogens, some cause more severe diseases than others. The degree of pathology associated with an infection is known as **virulence**. **Pathogenicity** and virulence are often used interchangeably despite having different meanings (BOX 19.1).

AGENTS OF INFECTIOUS DISEASE

PRION DISEASES

VIRUSES

BACTERIA

PROTOZOA

FUNGI AND MYCOSES

TRANSMISSION

INFECTIVITY AND PATHOGENESIS

DIAGNOSIS, TREATMENT, PREVENTION AND CONTROL

DRUG RESISTANCE

Pathogenicity vs. Virulence

Pathogenicity is the ability to produce disease and is an inherent property of the pathogen. An organism is either a pathogen or not. This inherent ability to produce disease is associated with an ability to overcome host barriers and establish an infection. Virulence is the relative capacity to cause disease or the degree of pathology associated with the infection. For example, a highly virulent pathogen causes severe disease manifestations, whereas an avirulent pathogen causes little or no pathology. Virulence is generally correlated with the ability of the pathogen to replicate within the host. Furthermore, virulence can be conditional, and not necessarily an inherent property of the pathogen. This means that any specific pathogen can cause a wide range of disease manifestations ranging from asymptomatic to mortal.

BOX 19.1

AGENTS OF INFECTIOUS DISEASE

Infectious diseases are caused by a wide range of biological entities (TABLE 19.1). In general, the diseases caused by pathogens are due to the replication of the pathogen within the host. In some cases, the pathogen directly damages host tissue to cause disease. However, most of the time, the pathology associated with infectious disease is due largely to the inflammatory response directed against the pathogen.

Table 19.1 Biological Agents of Infectious Disease	
Prions	Infectious protein
Viruses	Non-cellular (nucleic acid + protein)
Bacteria	Prokaryotic microbe
Protozoa	Eukaryotic microbe
Fungi	Eukaryotic microbe
Helminths (worms)	Multicellular, macroscopic
Blood-feeding arthropods	Ectoparasites, vectors

Prions and viruses are non-cellular and are completely dependent on the host for their replication. Therefore, many do not consider them to be organisms. Nonetheless, they are composed of biomolecules and thus are biological entities. Many infectious diseases are due to single-cell microorganisms such as bacteria, protozoa, and fungi. These pathogenic microbes often have features that facilitate their transmission and the ability to replicate with the host. Many pathogenic microbes are obligate pathogens since they require the host for replication and are no longer able to survive independently in the environment as free-living organisms. However, some microbes are facultative pathogens that can replicate and survive in the environment, as well as in a host.

Parasitic worms are multicellular and generally macroscopic (Chapter 24). Blood-feeding arthropods are not infectious agents per se, but nonetheless are related to infectious disease. They can be considered ectoparasites because they attach to the host to suck blood and can cause localized inflammation. However, many consider mosquitoes and similar blood-feeding insects as predators since they do not attach to the host. Regardless of the semantics, blood-feeding arthropods are an element of infectious disease since some can serve as vectors in disease transmission (Chapter 23).

PRION DISEASES

Prions are the simplest infectious agent since they consist of misfolded proteins that are able to 'transmit' their misfolded shape onto normal variants of the same protein. This triggers a chain reaction that produces large amounts of the misfolded protein. These abnormal prions form protein aggregates called amyloid. Accumulation of amyloid in cells leads to cell death. Nerve cells are particularly affected by amyloid deposits and most prion diseases are neurodegenerative. Many other neurodegenerative disorders involve amyloid deposits (see Box 3.2). The necrosis of nervous tissue is characterized by 'holes' when examined by microscopy and this results in a spongy architecture of the affected brains. Hence prion diseases are collectively known as transmissible spongiform encephalopathies (TABLE 19.2).

Table 19.2	Examples of Transmissible Spongiform Encephalopathies	
Disease	**Host**	**Features and transmission**
Scrapie	Sheep and goats	Known since 1732 and named after the compulsive scraping against rocks, trees, or fences by the affected animals due to apparent itching. May be transmitted through urine and ingestion. PrP^{Sc} can persist in the environment for decades
Bovine spongiform encephalopathy (BSE)	Cattle	Commonly known as mad cow disease due to the associated abnormal behavior. The source of the infection is likely to be meat-and-bone meal prepared from the remains of cattle who spontaneously developed the disease or scrapie-infected sheep
Variant Creutzfeldt–Jakob disease (vCJD)	Humans	Creutzfeldt–Jakob disease can occur spontaneously (approximately 85%) or is inherited in an autosomal dominant manner (approximately 7.5%). The variant (vCJD) refers specifically to the transmissible form, which is associated with eating beef tainted with BSE
Kuru	Humans	Formerly common among the Fore people of Papua New Guinea and the name is derived from a Fore word that means 'trembling.' It is widely accepted that kuru was transmitted via funerary cannibalism. After the Fore people ended the practice, cases dropped from 200 deaths per year in 1957 to no deaths from 2010 onward

Prions are transmissible proteins

Prion diseases can be acquired, exhibit a genetic predisposition, or spontaneously arise for no known reason. Ingestion appears to be the major method of acquisition in animals and thus prion diseases are considered transmissible. In fact, the word prion is a blending of protein and infection. It is possible that the ingestion of meat or animal products from diseased animals could initiate the disease. In addition, ingestion of grass from areas where affected animals have died and decayed, urinated, or defecated could be a source of infection. The incubation periods for prion diseases are generally 5–50 years and, despite these rather long incubation periods, the disease progresses rapidly once symptoms appear. Symptoms are similar to other neurodegenerative diseases and include convulsions, dementia, loss of balance and coordination, and behavioral changes. Prion diseases are untreatable and always fatal.

Prion protein (PrP) is highly conserved in mammals and is most abundant in neurons. The protein can exist in two different structural forms. The normal form is called PrP^C and the other form is the disease-associated form, or PrP^{Sc} for scrapie. The PrP^C exhibits predominantly alpha-helix as its secondary structure and the PrP^{Sc} is predominantly beta-sheet. The disease PrP^{Sc} isoform is extremely stable and has persisted in the environment for decades. Furthermore, the PrP^{Sc} isoform is resistant to proteases and other treatments that normally destroy proteins. This stability likely contributes to its transmissibility. In addition, the PrP^{Sc} isoform promotes conversion the PrP^C isoform to the PrP^{Sc} isoform (FIGURE 19.1). This means that the normal PrP^C isoform is progressively converted to the disease form of the protein. The PrP^{Sc} isoforms associate with large fibrils that form the amyloid deposits responsible for the disease pathogenesis.

Secondary protein structure is described in Chapter 3, pages 51–53.

SIGNPOST

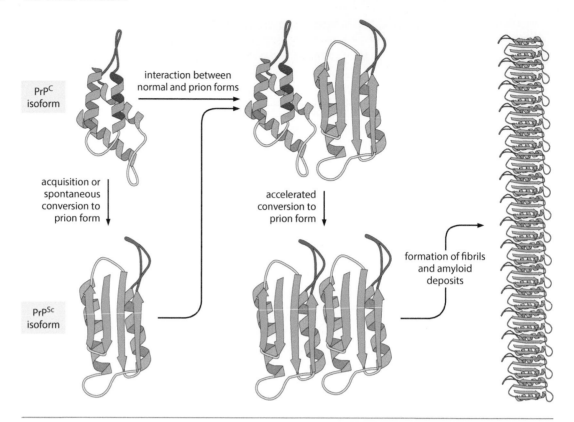

FIGURE 19.1 Prion protein and disease pathogenesis. Prion proteins exist in a normal isoform (PrP^C) and a prion disease isoform (PrP^{Sc}). The interaction of these two isoforms results in the conversion of the PrP^C isoform to the PrP^{Sc} isoform. The accumulated PrP^{Sc} isoforms form long fibrils that make up the amyloid deposits found within cells. The buildup of amyloid deposits interferes with normal cell function.

VIRUSES

Viruses are obligate pathogens that are completely dependent upon a host cell for replication. This inability to replicate independently from a host cell has raised semantic arguments about whether viruses are living organisms (see Box 2.1). Regardless of whether one defines viruses as living or not, viruses are clearly biological entities and have a similar molecular composition and structure as cells. All organisms are potentially infected by viruses. There are viruses that infect single-celled organisms, such as bacteria and eukaryotic microbes, viruses that infect plants or fungi, and viruses that infect all classes of animals. There are numerous viruses that infect humans and many of these viruses are responsible for major human diseases (TABLE 19.3). Furthermore, viruses have a huge impact on agriculture as many domestic plants and animals are affected by viral infections.

A wide range of diverse diseases are caused by viruses

There are numerous distinct types of viruses as defined by their morphological structures and genomes. Similarly, the diseases caused by viruses are diverse and can involve essentially any tissue or organ. However, most viruses do have a tropism for a specific tissue or organ, and the disease caused by a particular virus often has distinctive clinical manifestations. Clinical manifestations can range from mild to severe and highly mortal. For most viruses, the immune system effectively controls the infection and protective immunity may develop.

Table 19.3   Examples of Human Viral Diseases

Disease and/or virus	Nucleic Acid*	Env**	Comments
Common and notable viruses with significant morbidity and mortality			
HIV/AIDS	ss(+)RNA-9.7	Yes	A pandemic emerged in the late 1970s that has infected >75 million people
Influenza	ss(−)RNA-13.6	Yes	The seasonal flu causes approximately 5 million severe cases annually worldwide with an estimated 0.1% mortality rate
Dengue	ss(+)RNA-11	Yes	A common mosquito-transmitted virus that can cause severe hemorrhagic fever and dengue shock syndrome in the tropics
HPV	dsDNA-8	No	The most common sexually transmitted infection globally and is associated with cervical cancer
Norovirus	ss(+)RNA-7.5	No	Common cause of gastroenteritis and notable for outbreaks in schools and on cruise ships
Viruses that historically have been major pathogens but are now controlled through vaccination			
Smallpox	dsDNA-186	Yes	A fatal infection with a mortality rate of approximately 30% that was eradicated in 1980
Polio	ss(+)RNA-7.5	No	A virus that can infect the central nervous system and causes paralysis in about 15% of the cases
Rubella	ss(+)RNA-9.8	Yes	Previously a common childhood infection and associated with birth defects if transmitted congenitally
Mumps	ss(−)RNA-15.4	Yes	Widespread vaccination has led to a major decrease in the disease and now generally only found in resource-poor areas
Viruses associated with newsworthy outbreaks during the last decade			
COVID-19	ss(+)RNA-30	Yes	A pandemic coronavirus that emerged in late 2019 and shut down most of the world's economy
Ebola	ss(−)RNA-19	Yes	Associated with periodic outbreaks in Africa of severe hemorrhagic fever with a mortality rate ranging from 25% to 90%
Zika	ss(+)RNA-10	Yes	First spread to the Americas in 2015 and is associated with birth defects when transmitted congenitally
Viruses commonly associated with childhood infections			
Rotavirus	dsRNA-19.6	No	The most common cause of diarrhea in children with nearly every child being infected at least once by the age of 5 years
RSV	ss(−)RNA-10	Yes	A major cause of lower respiratory tract infections and hospital visits during infancy and childhood
Measles	ss(−)RNA-15.9	Yes	A highly contagious air-borne virus that infects about 20 million people a year primarily in Africa and Asia
Chickenpox	dsDNA-125	Yes	Before the introduction of the vaccine a rather common childhood disease caused by a herpesvirus
Viruses associated with long-term chronic infections			
Shingles	dsDNA-125	Yes	Caused by the same virus that causes chickenpox and is a flare-up later in life of this lifelong infection
Herpes simplex	dsDNA-152	Yes	Oral (HSV-1) and genital (HSV-2) are lifelong infections with sporadic flare-ups of blisters that affect 65–90% of adults
Hepatitis C	ss(+)RNA-9.6	Yes	A long-lived latent infection of the liver that can lead to liver disease, including cancer
Notable zoonotic viruses			
Rabies	ss(−)RNA-12	Yes	A neurotropic virus usually acquired from animal bites that is almost always fatal if not treated before reaching the CNS
West Nile virus	ss(+)RNA-11	Yes	A mosquito-transmitted virus of birds that can cause encephalitis in humans

*Single-stranded (ss) positive (+) or negative (−) strand or double-stranded (ds), DNA or RNA, and approximate size of the genome in kilobases.
**Enveloped, Yes or No.
AIDS, acquired immunedeficiency syndrome; CNS, central nervous system; HIV, human immunodeficiency virus; HPV, human papillomavirus; HSV, herpes simplex virus; RSV, respiratory syncytial virus.

The vast majority of viruses are specific to a particular host species and exhibit a defined life cycle involving a particular mode of transmission. In addition, there are zoonotic viruses. Most of the time humans are a dead-end host for zoonotic viruses, such as rabies, and subsequent human-to-human transmission does not occur. However, sometimes the virus can adapt to humans and cause an epidemic or become a new human pathogen. COVID-19, Ebola, and HIV are examples of animal viruses adapting to humans and becoming human pathogens.

Viruses consist of a nucleic acid genome and associated proteins, and some are surrounded by a membrane

Although viruses cannot replicate independently of a host cell, viruses do replicate and, like other organisms, information is passed to subsequent generations via nucleic acids. In the case of viruses, the hereditary molecule can be either DNA or RNA. No viruses have both DNA and RNA simultaneously, although some viruses do exhibit either DNA or RNA at different life cycle stages. The viral genome can be either double stranded or single stranded. In the case of viruses with single-stranded genomes, either the sense strand, also called the positive strand, or the antisense strand, also called the negative strand, can serve as the genome.

Viral genomes are associated with proteins that are encoded by the viral genome. These nucleic acid-interacting proteins participate with other viral proteins in forming a virus particle, also called a virion. The proteins involved in the formation of the virion are structural proteins. The virus genome also encodes for non-structural proteins that are generally involved in the replication or transcription of the viral genome. The assembled structural proteins and viral nucleic acid are known as the capsid (FIGURE 19.2). In some viruses the capsid is surrounded by a membrane of host cell origin. Viruses with an outer membrane covering are called enveloped. Sometimes viruses without the envelope are called naked. Viruses are grouped into taxonomic families based on the type of nucleic acid making up their genome, whether they are enveloped or not, and their overall morphological structure based on electron microscopy.

FIGURE 19.2 Generalized virus structure. The nucleic acid of a virus (genome) is enclosed within a capsid made up of viral structural proteins. The capsids of enveloped viruses are surrounded by a membrane of host cell origin. Most viruses have proteins projecting from the surface that are often called spike proteins. These spike proteins bind to host cell receptors and facilitate entry into the host cell.

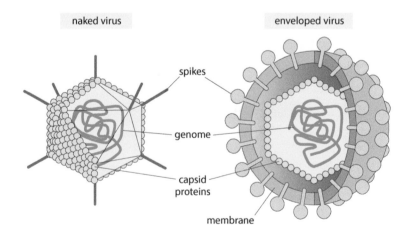

Viruses exploit the host cell machinery to express their genes and replicate their genomes

Virus replication involves virus entry into a host cell. Viruses recognize receptors on the host cell and therefore most viruses have a specific host or a limited range of potential hosts. Some viruses are even specific for specific cell types within the host. This recognition of host cells is mediated by proteins on the virus surface that are often called spike proteins since they protrude from the virus. Virus entry can involve endocytosis, fusion of the virus envelope with the host cell membrane, or injection of the viral nucleic acid. After entry, the virus capsid disassembles to release the viral nucleic acid and non-structural proteins.

After the release of the genome, non-structural proteins facilitate the transcription of viral genes and the replication of the viral genome. Generally, viruses have specialized nucleic acid polymerases to deal with the unique nature of the viral genome, such as being single stranded or RNA. Viral mRNA is translated into proteins on host cell ribosomes and the structural proteins assemble with the newly replicated viral genomes to form progeny virus. Enveloped viruses may assemble at the host cell plasma membrane and bud from the host cell. Some viruses exploit the host cell's secretory pathway. Non-enveloped viruses generally cause the host cell to lyse following viral replication and this releases the progeny viruses. Such viruses are called lytic viruses. Although all viruses exhibit this generalized life cycle (FIGURE 19.3), each specific type of virus has its own unique life cycle.

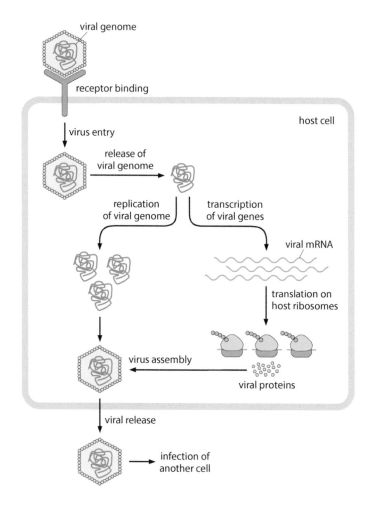

FIGURE 19.3 Generalized virus life cycle. Viruses recognize receptors on the surface of host cells. Binding to the receptor facilitates the entry of the virus into the host cell. After entry the virus particle disassembles and releases the viral genome so that it can be replicated and transcribed. Viral mRNA is translated into proteins on the host ribosomes. The replicated viral genomes and newly synthesized viral proteins are assembled into viral particles that are released from the host cell.

Viral pathogenesis is associated with cellular damage and inflammation

Viruses are clearly pathogenic since viruses kill or severely damage the cells they infect. Most viruses also induce a strong innate immune response involving inflammatory cytokines and natural killer cells. The exact manifestations of viral disease depend largely on the specific cells, tissues, or organs that are infected. Viral infections are generally characterized by acute symptoms occurring shortly after infection and these acute symptoms subside as the adaptive immune response clears the infection. Many viral infections are completely cleared by the host immune response without treatment. However, some viruses are capable of evading the immune system and establishing long-term – and even lifetime – chronic infections. HIV (Chapter 22) and herpes viruses (BOX 19.2) are examples of lifelong chronic infections.

BOX 19.2

Herpesviruses

Herpesviruses are a large family of DNA viruses that cause several human diseases. The most common herpesviruses in humans are herpes simplex virus-1 (HSV-1), herpes simplex virus-2 (HSV-2), and varicella-zoster virus (VZV). HSV-1 primarily causes oral herpes commonly known as cold sores or fever blisters, and HSV-2 primarily causes genital herpes. Herpes commonly presents as blisters around the mouth or genitalia that may subsequently ulcerate. The disease cycles between active disease with lesions and latent periods without lesions. Periodically the lesions may reappear. A person remains infected for life and there are no drugs to cure the infection.

VZV causes chickenpox and shingles. Chickenpox, also known as varicella, is a highly contagious disease characterized by small itchy blisters that spread over the entire body due to the initial infection with VZV. Like HSV, VZV goes into a latent stage after the lesions heal and the people remain infected for life. This dormancy lasts for decades and can re-emerge in older persons as shingles, also known as herpes zoster. Approximately 20% of people experiencing chickenpox during childhood develop shingles sometime after the age of 50. Shingles typically presents as a wide stripe of painful blisters across the back.

BACTERIA

Bacteria are prokaryotic cells and have several features distinct from eukaryotic cells (see Table 2.1). For example, bacteria are covered with a peptidoglycan cell wall and have no internal membranes (FIGURE 19.4). In addition, most bacteria have a single large circular DNA molecule as their genome and replicate by binary fission. In the presence of abundant nutrients most bacteria can replicate rapidly with generation times of the order of 20 minutes. During periods of low nutrients or other harsh conditions some bacteria form spores that can survive in the environment for years. The spores can revert into replicating bacteria when conditions become more favorable. Several diseases are caused by spore-forming bacteria (BOX 19.3).

SIGNPOST

Binary division is described in Chapter 2, pages 42–43.

FIGURE 19.4 Structural features of a bacterial cell. Prokaryotic cells have no nucleus, and the genomic (chromosomal) DNA is in the cytoplasm in a region called the nucleoid. Cell support is derived from a cell wall just outside of the plasma membrane. Many bacteria are encapsulated in a complex polysaccharide layer secreted by the bacteria called the capsule. Some bacteria contain hair-like structures called pili and some bacteria are motile via a flagellum, which is structurally distinct from eukaryotic flagella.

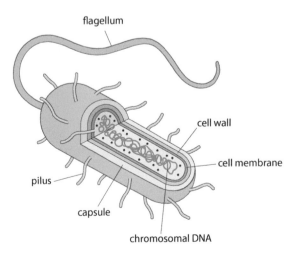

Spore-forming Bacteria and Disease

Clostridioides difficile and *Bacillus anthracis* are two examples of spore-forming bacteria that cause disease. *C. difficile* is often associated with nosocomial infections and antibiotic-associated diarrhea. Nosocomial infections are infections spread in health-care facilities. Spores are excreted in the feces of infected individuals and can contaminate surfaces. People touching these contaminated surfaces can contaminate their hands and this may lead to ingestion and initiate an infection. The hardy spores are difficult to kill and contribute to the spread of *C. difficile* in hospitals and clinics. *C. difficile* infections are quite common, but the normal intestinal microbiota protects against diarrhea caused by these bacteria.

Antibiotic treatment alters the normal microbiota and this alteration of the microbiota allows for an overgrowth of *C. difficile* that leads to antibiotic-associated diarrhea.

Anthrax is a potentially deadly disease caused by *B. anthracis*. The disease is acquired through inhalation or ingestion of the spores or the introduction of spores through a broken area in the skin. Spores can survive for decades in the soil. Anthrax spores have been used in biological warfare and bioterrorism. Both the ability to aerosolize anthrax spores, which allows dispersal over a large area, and the stability of the spores are attractive features of this potential weapon.

BOX 19.3

Bacteria are found in essentially all ecosystems and most bacteria do not cause disease. In fact, many bacteria are components of our microbiota (see Box 6.2) and play a beneficial role in human health. Other bacteria can be viewed as opportunists in that they normally do not cause disease, but under certain conditions can cause disease. For example, an injury may introduce bacteria from the environment that normally would not have access to internal organs. Once gaining access the bacterium can now become virulent. Similarly, an immunosuppressed host may allow the replication of normally avirulent microorganisms. However, there are many bacteria that are clearly pathogenic and cause numerous human diseases. Many of these pathogenic bacteria are obligate pathogens that need a host for survival and cannot exist in the environment, and thus are transmitted from person to person.

Disease-causing bacteria express virulence factors

The ability of bacteria to cause disease depends on the pathogen gaining access to and replicating within the host. The pathology associated with bacterial infections is generally due to the destruction of tissue by the pathogen, inflammation associated with the immune response against the pathogen, or toxins secreted by the pathogen. **Virulence factors** are pathogen products that are associated with any aspect of the disease process (TABLE 19.4). For example, factors that allow bacteria to gain access and replicate within the host are virulence factors. Similarly, factors associated with immune evasion are virulence factors. In some cases, the bacteria per se do not damage the host, but the damage is due to toxins that are associated with the bacteria. In the case of secreted toxins, the secretory apparatus involved in toxin secretion is also considered to be a virulence factor.

Table 19.4	Virulence Factors
Immune system avoidance	Pathogen products that interfere with or suppress any aspect of the host immune response against that pathogen (e.g., enzymes that break down antibodies)
Capsules	Some bacteria are surrounded by a polysaccharide capsule that can promote adherence, block complement, and avoid phagocytosis
Adhesins	Pathogen products that promote the adherence of pathogens to host cells or tissues resulting in damage, dysfunction, or inflammation
Invasins	Pathogen products that facilitate the entry of a pathogen into a host, organ, tissue, or cell
Nutrient acquisition	Pathogen products that increase the ability of the pathogen to acquire nutrients from the host
Hydrolytic enzymes	Pathogen enzymes that break down host tissues or extracellular components resulting in pathology. Destruction may also facilitate pathogen entry or exit, facilitate the acquisition of nutrients from the host, or destroy components of the immune system
Toxins	Pathogen products that can cause pathology and disease manifestations independent of the living pathogen. Can be an integral component of the pathogen, called endotoxin, or an exotoxin that is secreted by the pathogen
Secretory pathways	Components of the pathogen that are responsible for secreting exotoxins, destructive enzymes, or other virulence factors
Drug resistance	Polymorphisms in genes that allow pathogens to survive and replicate in the presence of antibiotics and other drugs

Adherence of microbes to host tissues plays a large role in virulence and pathogenesis. Factors that promote adherence are in general called adhesins. For example, many bacteria exhibit extensions on the cell surface called pili (see Figure 19.4) and some of these pili promote bacterial adherence and are thus virulence factors. Similarly, the capsules of many bacteria promote adherence, in addition to protecting the bacterium from the immune system. Adherence may also be coupled with the invasion of host tissues or cells and factors that promote invasion are called invasins.

Genes for virulence factors are often found on pathogenicity islands

Pathogenicity islands are clusters of several virulence factor genes found on either the chromosome or extrachromosomal elements of pathogenic organisms. These pathogenicity islands are absent in non-pathogenic organisms of the same or closely related species. These virulence factor genes can be transferred from pathogenic organisms to non-pathogenic organisms via horizontal gene transfer (BOX 19.4). This means that non-pathogenic organisms can suddenly acquire virulence factor genes and become pathogenic. These pathogenicity islands can even be transferred between different species of bacteria.

Horizontal Gene Transfer

BOX 19.4

Normally we think of genetic inheritance as being vertical, meaning the genes move from parents to offspring. However, it is possible for genes to move horizontally from one individual to another. Such horizontal gene transfer events are known to shape evolution and can quickly impart new abilities to the recipient. Horizontal gene transfer is more common in unicellular organisms, especially bacteria, but can also occur in multicellular organisms. However, to be inherited in subsequent generations the newly acquired genes would need to find their way into the germline.

There are three major mechanisms of horizontal gene transfer: transformation, transduction, and conjugation (**BOX 19.4 FIGURE 1**). Conjugation is a normal process in bacteria that involves the transfer of DNA in the form of plasmids from a donor to a recipient. This can be viewed as a type of sexual reproduction in prokaryotes. Drug resistance genes and virulence factor genes are often transferred in this manner. Cells can also take up DNA from the environment. The source of the DNA might be from other cells that have died and released their DNA into the environment. This does not happen readily, nor often, but nonetheless happens. The donor DNA taken up by the cell can be incorporated into the recipient genome and then maintained in subsequent generations. This is called transformation. Transduction is similar to transformation, except horizontal transfer involves a virus. Some DNA of the donor becomes incorporated into a viral genome. When that virus infects another cell the donor DNA can be incorporated into the recipient genome.

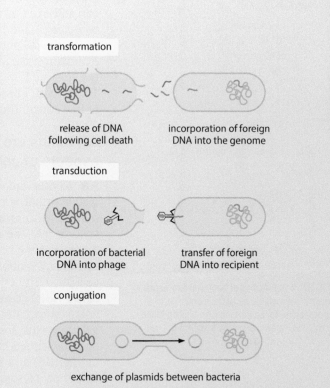

transformation

release of DNA following cell death

incorporation of foreign DNA into the genome

transduction

incorporation of bacterial DNA into phage

transfer of foreign DNA into recipient

conjugation

exchange of plasmids between bacteria

BOX 19.4 FIGURE 1 Horizontal gene transfer mechanisms. DNA released by dying or lysed cells can be taken up by a recipient cell and that DNA can be incorporated into its genome in a process called transformation. Transduction is a process in which host DNA is incorporated into a virus – or bacteriophage in the case of bacteria – and then transferred to a recipient following infection. Conjugation is a process by which bacteria exchange DNA via extrachromosomal elements called plasmids. All three mechanisms are exploited in recombinant DNA technology (Chapter 7). [Adapted from Furuya EY & Lowy FD (2006) Antimicrobial-resistant bacteria in the community setting. *Nature Reviews Microbiology* 4: 36–45. doi: 10.1038/nrmicro1325. With permission from Springer Nature.]

Bacterial exotoxins are secreted proteins or peptides that have a toxic effect on host physiology

Secreted proteins that are toxic to host cells or tissues are called **exotoxins**. A wide range of bacterial products are described as exotoxins and there are numerous mechanisms by which these toxins affect host physiology. For example, some toxins cause cell death and such toxins are called cytotoxins (TABLE 19.5). Some cytotoxins affect eukaryotic protein translation by altering ribosomes or elongation factors, resulting in the inhibition of translation. Cells that are unable to translate proteins die. Other cytotoxins lyse cells by forming pores in the plasma membrane or via the action of hydrolytic enzymes such as lipases that disrupt the plasma membrane. Some exotoxins disrupt normal cell function instead of killing host cells. For example, some exotoxins interfere with normal nerve cell signaling and are called neurotoxins. Other exotoxins interfere with the normal function of intestinal epithelial cells and are called enterotoxins (Chapter 21).

Table 19.5	Cellular Effects of Exotoxins
Cytotoxins	Kill or lyse host cells
Neurotoxins	Interfere with nerve transmission
Enterotoxins	Affect intestinal epithelium transport functions

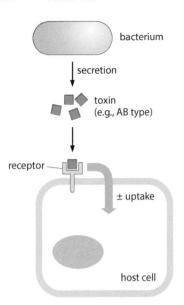

FIGURE 19.5 Action of bacterial exotoxins.
Exotoxins are secreted by bacteria and bind to host receptors. For some exotoxins, binding to the receptor facilitates entry of the toxin into the cell.

Exotoxins are generally heat labile, as they are proteinaceous in structure and cooking destroys their activity. However, a few are heat stable and can still exert a toxic effect after cooking. Generally, exotoxins bind to a receptor on the host cell (FIGURE 19.5) and binding to the host receptor facilitates the entry of some exotoxins into the host cell. Quite often these exotoxins have an enzymatic activity that modifies specific targets within the host cell and modification of the target affects cellular physiology. Toxins with enzymatic activity are quite potent and exhibit toxic effects at low concentrations.

AB toxins are a class of exotoxins that alter specific aspects of cellular physiology

AB toxins are composed of two subunits: the A subunit has some 'activity' that is responsible for the toxicity and the B (binding) subunit that binds to receptors on the host cell and facilitates the entry of the A subunit into the cell. The A subunit typically has an enzymatic activity that modifies a host cell component. For example, several bacterial toxins add an ADP-ribose group to the cellular target protein, and this causes cellular dysfunction or toxicity (FIGURE 19.6). One class of AB toxins is called AB5 because the B subunit is a pentamer of five polypeptides. The host cell receptor for AB5 toxins is the carbohydrate portions of specific glycoproteins and glycolipids. Four different classes of AB5 toxins have been described based on the receptor recognized by the B subunit, and the activity and target of the A subunit (TABLE 19.6). Like many other virulence factors, AB toxins are often associated with mobile DNA elements that can be transferred within a species or between species via horizontal gene transfer (see Box 19.4).

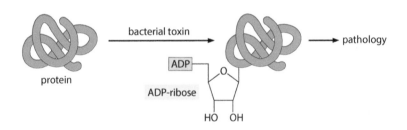

FIGURE 19.6 Mechanism of action of bacterial toxins. The A subunit of some bacterial AB5 toxins that carry out ADP-ribosylation of host cell target proteins. This ADP-ribosylation can affect protein function and lead to pathology.

Table 19.6	Four Classes of AB5 Toxins			
Toxin	**B-subunit receptor**	**A-subunit activity**	**A-subunit target**	**Cellular pathology**
Cholera toxin	Glycolipid (GM1)	ADP-ribosylation	α-subunit of a Gs protein	Stimulation (Gs) of signal transduction leading to dysfunction of ion transport
Pertussis toxin	Glycoprotein (Neu5Ac)	ADP-ribosylation	α-subunit of a Gi protein	Lack of inhibition (Gi) of signal transduction leading to disruption of cellular functions
Shiga toxin	Glycolipid (Gb3)	RNase	28S rRNA	Inhibition of protein translation leading to cell death
SubAB toxin	Glycoprotein (Neu5Gc)	Protease	ER HSP70	Decreased ER stress response leading to apoptosis

Bacterial endotoxin can cause toxic shock syndrome

The original meaning of endotoxin was a toxin tightly associated with the pathogen and not secreted from the pathogen. Now endotoxin is synonymous with **lipopolysaccharide** (LPS). LPS is a large molecule composed of a complex lipid called lipid A and a complex polysaccharide (FIGURE 19.7). The lipid A component anchors LPS into the outer cell membrane of Gram-negative bacteria and the polysaccharide makes up the outer surface of the bacterium and is part of the capsule.

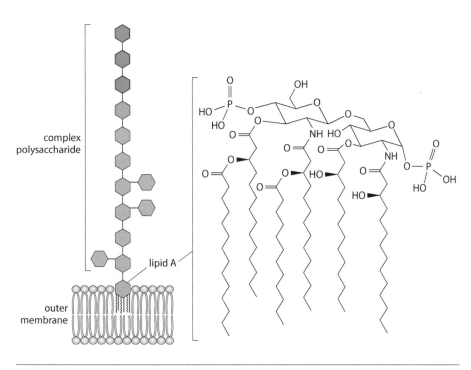

FIGURE 19.7 Structure of lipopolysaccharide and lipid A. Lipopolysaccharide (LPS) is anchored into the outer membrane of the bacterial surface by lipid A. The complex polysaccharide forms part of the bacterial capsule. Lipid A is a complex glycolipid with six fatty acid chains.

Lipid A is the toxic component of LPS and this toxicity is due to overstimulation of the innate immune system. Toll-like receptor-4 (TLR4) recognizes the lipid A component of LPS. Binding of LPS to the receptor activates an innate immune response that helps to clear the bacteria. Therefore, at low concentrations, LPS initiates an appropriate immune response against the pathogen. However, high concentrations of LPS due to high levels of Gram-negative bacteria can over-activate the innate immune response. The resulting strong proinflammatory response activates many pathways that can lead to fever, low blood pressure (hypotension), disseminated intravascular blood coagulation, and multiple organ failure (FIGURE 19.8). Damage to endothelial cells can result in low blood pressure and, if the blood pressure falls below levels needed to pump blood through the circulatory system, the person can go into shock. Shock due to high levels of Gram-negative bacteria is called **endotoxic shock** or septic shock. Disseminated intravascular coagulation (DIC) is a condition characterized by abnormal blood clots throughout the circulatory system. Both endotoxic shock and DIC are medical emergencies that can lead to organ failure and death.

Toll-like receptors and innate immunity are described in Chapter 9, pages 181–182.

SIGNPOST

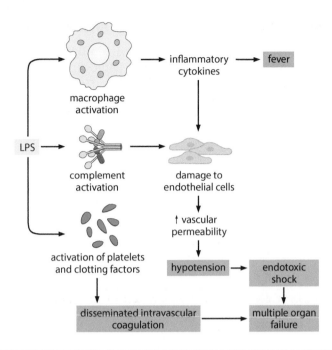

FIGURE 19.8 Effects of lipopolysaccharide. High levels of lipopolysaccharide (LPS) can activate macrophages, complement, and blood clotting factors including platelets. Activated macrophages secrete inflammatory cytokines, which can cause fever and damage endothelial cells. Similarly, activation of complement also damages endothelial cells. Damage to endothelial cells causes an increase in vascular permeability leading to hypotension (i.e., low blood pressure). Severe systemic hypotension due to endotoxin can result in endotoxic shock. Activation of platelets and blood clotting factors causes disseminated intravascular coagulation (DIC). Multiple organ failure can result from DIC and endotoxic shock.

A simple means to characterize bacteria is based on morphology and staining

Bacteria can be detected by microscopy and diagnosis often involves microscopy of clinical samples taken from patients. The sizes of most bacteria are in the range of 0.5–5 μm. A micrometer is one millionth (10^{-6} of a meter). Different species of bacteria exhibit different shapes (FIGURE 19.9). The most common shapes of bacteria are spherical (cocci), rod-shaped (bacilli), curved rods (vibrio), spiral-shaped (spirilla) or tightly coiled (spirochetes). Bacteria forming short rods are called coccobacilli. Most bacteria exist as single cells, but some bacteria aggregate into distinct patterns such as long chains (streptococci or streptobacilli) or masses reminiscent of a cluster of grapes (staphylococci).

Biological dyes, or stains, can be used to detect bacteria via microscopy. A common stain to detect bacteria is **Gram stain**. Gram staining separates bacteria into two groups: Gram positive and Gram negative. This difference is due to a fundamental difference in the cell-wall structure of bacteria (FIGURE 19.10). Gram-positive bacteria stain purple due to retention of the crystal violet dye following a de-staining step. De-staining extracts the crystal violet from Gram-negative bacteria and therefore they stain pink due to the subsequent staining with another dye.

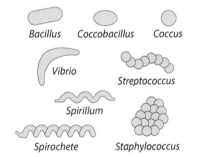

FIGURE 19.9 Common morphologies of bacteria. Bacteria have descriptive names based on their shape. Bacteria that form long chains are preceded with the prefix strepto- and bacteria that aggregate into clusters have the prefix staphylo-.

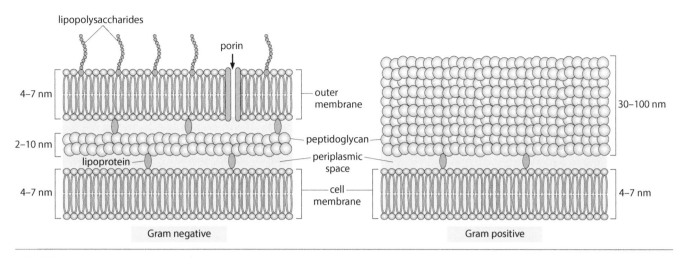

FIGURE 19.10 Gram staining distinguishes bacterial cell wall architecture. Gram-negative bacteria have a double membrane with a thin layer of peptidoglycan between the membranes. Gram-positive bacteria have a single membrane with a thick layer of peptidoglycan. Gram-positive bacteria retain more crystal violet during the staining procedure than Gram-negative bacteria and stain purple, nm = nanometers (10^{-9} meters). [Adapted from Berezin S, Aviv Y, Aviv H et al. (2017) Replacing a century old technique – modern spectroscopy can supplant Gram staining. *Scientific Reports* 7: 3810. doi: 10.1038/s41598-017-02212-2. Published under CC BY 4.0.]

PROTOZOA

Protozoa are ubiquitous eukaryotic microbes that are found in nearly all ecological niches. Although most protozoa are free-living, a large number exist in symbiotic or parasitic relationships with other organisms. Of relevance to human health, several protozoa are human pathogens and cause important human diseases (TABLE 19.7). Several protozoa infect the gastrointestinal tract and are common causes of diarrhea and dysentery (Chapter 21). Malaria (Chapter 23) is a major disease caused by a protozoan pathogen. Malaria, along with African sleeping sickness, Chagas' disease, and leishmaniasis, is transmitted by arthropod vectors. Trichomoniasis is a common sexually transmitted disease. However, it is not highly virulent. Similarly, toxoplasmosis is also rather common, but generally benign. Exceptions are if the *Toxoplasma* infection is acquired congenitally or if the host is immunocompromised. Congenital infections are often associated with vision or development problems during childhood. Toxoplasmosis and cryptosporidiosis are opportunistic diseases often associated with AIDS (see Box 22.3).

Table 19.7 Major Protozoan Pathogens Infecting Humans

Protozoan	Disease	Transmission	Importance
Trichomonas vaginalis	Inflammation of the urogenital tract	Sexually transmitted	Most common non-viral sexually transmitted infection with >140 million cases per year worldwide
Entamoeba histolytica	Amoebic dysentery	Fecal–oral	An estimated 50 million symptomatic cases with 100,000 deaths per year
Giardia duodenalis	Diarrhea	Fecal–oral	Common intestinal infection causing 180 million clinical cases per year
Cryptosporidium species	Acute watery diarrhea	Fecal–oral, water-borne outbreaks	Prevalence ranges from 1% to 10% and can be life threatening in patients with AIDS
Plasmodium falciparum	Malaria, acute febrile disease	Mosquito	Approximately 200 million clinical cases with 500,000 deaths per year
Leishmania species	Cutaneous or visceral disease	Sandfly	1.5–2 million new cases and 70,000 deaths per year
Trypanosoma cruzi	Chagas' disease	Triatomine bug	Major cause of cardiac disease in South and Central America with at least 6 million infected
Trypanosoma gambiense	African sleeping sickness	Tsetse fly	70–80,000 new cases with 30,000 deaths per year
Toxoplasma gondii	Neurological and vision defects	Ingestion (e.g., undercooked meat)	A common infection that can cause birth defects. Seroprevalence ranges from 6% to 75% depending on the region

FUNGI AND MYCOSES

Fungi include yeasts, molds, and mushrooms. Yeasts are eukaryotic single-cell organisms that replicate via mitosis. However, in many yeasts, the cell division process involves a smaller daughter cell 'budding' from the larger mother cell. Molds grow in the form of multicellular filaments called hyphae (FIGURE 19.11). The mass of interconnected hyphae is called a mycelium. Many fungi are dimorphic in that they can switch between the yeast form and mycelium form. Typically, molds and yeast secrete hydrolytic enzymes that break down complex macromolecules into simpler metabolites that can be absorbed. This is known as saprotrophic nutrition.

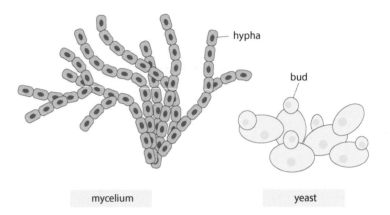

FIGURE 19.11 Yeast and mycelium forms of fungi. Yeasts grow as single cells and molds form long chains of cells called hyphae. The mycelium is a mass of hyphae. Mitosis in yeast often involves an unequal cytoplasmic division called budding.

Mycoses are generally avirulent in immunocompetent persons

Diseases caused by fungi are known as mycoses. Fungal infections usually do not cause severe disease and many infections caused by fungi are superficial and relatively non-virulent (TABLE 19.8). An exception is in the case of immunodeficiencies and immunocompromised people. Generally, fungal infections only cause severe disease in individuals with a compromised or weakened immune system. Therefore, fungal infections are often considered to be opportunistic infections. For example, *Pneumocystis jirovecii* pneumonia (PCP) is a common opportunistic infection associated with AIDS (see Box 22.3). Most fungi are not obligate pathogens and normally are free-living. Many fungal species are abundant in the environment and thousands of fungal spores may be inhaled daily without causing disease due to the immune system easily controlling or eliminating the infection.

Table 19.8 Human Mycoses		
Description	**Tissue Colonized**	**Characteristics**
Superficial	Outermost layers of the skin, nails, or hair	Skin discoloration called tinea versicolor
Cutaneous	Dead keratinized layers of the skin	Redness, itching, and scaling such as athlete's foot
Subcutaneous	Dermis and subcutaneous tissues (often introduced by skin trauma)	Localized inflammatory response at the site of infection
Systemic	Originates primarily in the lungs and subsequently spreads to other parts of the body	Fever, fatigue, and enlarged liver and spleen
Opportunistic pathogens	Skin, lungs, and dissemination to other organs	Infections of patients with immune deficiencies

Most fungal infections are associated with either the skin or the lungs. However, fungi are also commonly found in the gastrointestinal tract and urogenital tract. Many of these fungi infecting the intestinal and urinary tracts are viewed as commensals. Regarding skin infections, there are some fungi that have specifically adapted to living on the skin called dermatophytes. These fungi secrete enzymes that specialize in breaking down keratin and other skin molecules and absorbing the metabolites. Cutaneous fungal infections are known as tinea and probably the best known is athlete's foot (tinea pedis). Athlete's foot and other cutaneous fungal infections tend to occur in moist areas and skin folds. The metabolic products and disruption of the skin produced from the digestion of keratin can result in some inflammation causing redness, itching, and scaling. Rarely do these dermatophytic fungi penetrate beyond the outermost layer of skin. Generally, subcutaneous fungal infections are associated with a traumatic skin injury that introduces fungi found in soil into the deeper tissues.

Severe and systemic fungal infections are generally associated with an immunocompromised state

Another common site for fungal infections is the lungs. Fungal infections in the lungs are generally initiated by inhaling fungal spores, which can be very abundant in the environment. Most of the time, either the infections are not established or the infections are rather superficial. However, fungal infections in the lungs can produce a pneumonia-like disease similar to pathogenic bacteria and viruses (Chapter 20). Although the amount of pneumonia caused by fungi pales in comparison to bacteria and viruses, it is increasingly being recognized among immunosuppressed patients. The infections can spread from the lungs to other organs and cause systemic infections (FIGURE 19.12). Systemic infections are generally restricted to patients with an impaired immune system, such as patients with AIDS, cancer patients undergoing chemotherapy, and patients receiving immunosuppressive therapy. Systemic fungal infections can be medical emergencies with potentially high mortality rates.

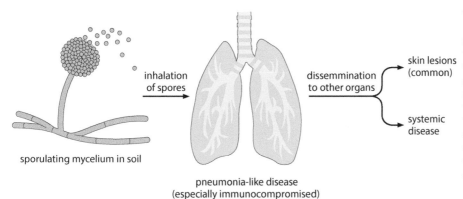

sporulating mycelium in soil

inhalation of spores

dissemination to other organs

skin lesions (common)

systemic disease

pneumonia-like disease (especially immunocompromised)

FIGURE 19.12 Pathology of endemic mycoses. A common route of fungal infection is the inhalation of spores associated with disrupting the soil or contact with the soil. This can result in a pneumonia-like disease in immunocompromised individuals due to the growth of yeast forms. From the lungs the yeast forms can spread throughout the body via the circulatory system. Skin lesions are common in many endemic mycoses. Systemic disease is characterized by fever, fatigue, and an enlarged liver and spleen.

Many of the fungi causing systemic fungal infections are often grouped together and called **endemic mycoses**. Endemic mycoses are caused by dimorphic fungi that exist in a mycelial form in the environment and grow as yeasts when causing infections. Endemic mycoses often exhibit defined geographical distributions according to the specific species causing the disease (TABLE 19.9). Most of these fungi are found in the soil and infections are acquired by inhalation of the spores associated with disruption of the soil or contact with the soil. Dissemination of the infection from the lungs to other parts of the body is often accompanied by skin lesions.

Table 19.9 Endemic Mycoses		
Disease	**Species**	**Geographical region**
Histoplasmosis	*Histoplasma capsulatum*	Worldwide and especially along the lower Ohio and Mississippi rivers in the USA
Blastomycosis	*Blastomyces dermatitidis*	Eastern USA and Canada
Coccidioidomycosis	*Coccidioides immitis, Coccidioides posadasii*	Southwestern USA and northern Mexico
Paracoccidioidomycosis	*Paracoccidioides brasiliensis, Paracoccidioides lutzii*	Central and South America
Talaromycosis	*Talaromyces marneffei*	Southeast Asia including southern China and eastern India
Emergomycosis	*Emergomyces africanus*	Southern Africa

TRANSMISSION

Infectious disease can be acquired by several mechanisms (FIGURE 19.13). In some cases, the disease may be acquired from the environment. This could involve organisms that are free-living but can become pathogenic after gaining entry into the human body. In other cases, pathogens may produce dormant forms that can survive for extended periods in the environment. Another source of pathogens is animals. Diseases that are naturally transmissible between humans and animals are called **zoonoses**. The natural animal host of a zoonotic infection is called the reservoir. The most prevalent mode of acquisition of infectious disease involves human-to-human transmission. Most pathogens are obligate and cannot survive for extended periods outside of the human host. Thus, they require human-to-human transmission to survive and are sometimes called communicable. Conversely, non-communicable pathogens are those that normally live in an abiotic environment such as soil or water. Contagious refers to those pathogens that are easily transmissible between hosts.

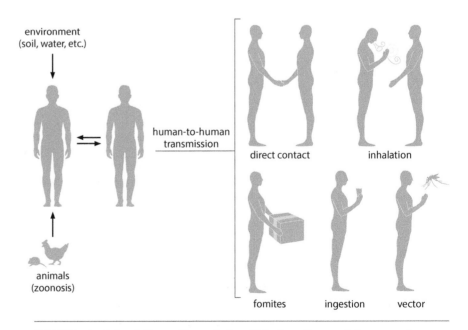

FIGURE 19.13 Acquisition and transmission of infectious diseases. Pathogens can be acquired from abiotic sources and animals. Most infectious diseases are due to human-to-human transmission involving direct contact, inhalation, ingestion, or blood-feeding vectors. Pathogens can also be spread from person to person via inanimate objects called fomites.

INFECTIVITY AND PATHOGENESIS **377**

There are multiple mechanisms for pathogen transmission

Most pathogens have evolved mechanisms to increase their transmission between hosts. There are five main routes utilized by pathogens to infect the host: direct contact, fomites, ingestion, inhalation, and biological vectors (see Figure 19.13). These routes are not mutually exclusive, and some pathogens can exploit multiple routes. Pathogens can be directly transferred from person to person through activities such as touching, kissing, and sex. This tends to be a relatively efficient process, but the pathogen needs to be located on hands or other parts of the body, or there needs to be an exchange of bodily fluids. A somewhat artificial direct transmission is through blood transfusions, organ transplants, or sharing needles. This type of transmission is referred to as blood-borne. Some pathogens can be transferred to an inanimate object, such as a doorknob, and then subsequently transferred to another person if that person touches that object. These objects that play a role in disease transmission are called **fomites**.

Other indirect means of moving from person to person or acquiring an infection are ingestion, inhalation, and biological vectors. Many of the pathogens that are acquired by ingestion inhabit the gastrointestinal tract and involve a type of transmission called fecal–oral transmission (Chapter 21). Similarly, many pathogens that are acquired by inhalation inhabit the lungs and transmission involves the formation of droplets during coughing and sneezing that contain the pathogen, or the pathogen becomes aerosolized (Chapter 20). Pathogens that do not inhabit the lungs or gastrointestinal tract and that are not amenable to direct contact or fomite transmission may rely upon a biological vector to move from person to person (Chapter 23). Biological vectors are blood-feeding arthropods, such as mosquitoes, that can acquire the infection during blood feeding and then later transmit that infection to another person when that vector takes another blood meal.

INFECTIVITY AND PATHOGENESIS

Pathogens need to be able to establish an infection once they have come in contact with the host. This ability to establish infection is infectivity (TABLE 19.10). The initial contact between the host and pathogen generally involves an epithelial layer such as the skin or linings of the digestive, respiratory, or urogenital tracts. Some pathogens adhere to these epithelial layers and establish a colony, whereas other pathogens have the ability to cross such epithelial barriers. Those that cross epithelial barriers and enter the host must contend with the host immune system and have mechanisms to avoid elimination by the immune system. A key element of infectivity is the ability of the pathogen to replicate within the host. This replication may be part of the mechanism by which the pathogen stays ahead of the immune system. In addition, the pathogen needs to produce progeny to spread to other hosts. In some cases, this spread to other hosts may involve some type of exit strategy.

Table 19.10 Infectivity	
Factor	**Requirement**
Attachment and/or entry	Adherence to the epithelium or breaking host barriers to initiate infection
Local or systemic spread	Evade host defenses and establish infection
Multiplication	Increase pathogen numbers and continue evading host defenses
Shedding or exit	Release sufficient numbers to ensure transmission
Host damage (i.e., pathogenesis)	Not strictly necessary, unless needed for exit

Many infections are asymptomatic or have mild clinical manifestations. For example, pathogens that establish superficial infections on epithelial layers normally cause rather benign diseases (FIGURE 19.14). However, those pathogens that penetrate epithelial layers tend to cause more serious disease. This is due in part to tissue destruction caused directly by the pathogen, and to the inflammatory response initiated by the host immune system. In many cases, this inflammation remains localized to the original site of infection. Some pathogens can spread to other internal organs. In some cases, the pathogens target a specific organ, whereas other pathogens can affect multiple organs and spread throughout the body. In general, systemic infections affecting internal organs are more severe than infections that remain localized.

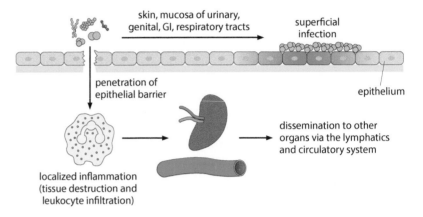

FIGURE 19.14 Superficial, localized, and disseminated infections. Pathogens can establish a colony on epithelial surfaces to cause a superficial infection. Penetration of the epithelial barrier causes localized inflammation and disease manifestations. Inflammation is characterized by the infiltration of leukocytes, also called white blood cells, and tissue destruction. Some pathogens can disseminate to other organs via the lymphatic or circulatory systems. GI, gastrointestinal.

Infectious diseases have a natural history that defines the typical progression of pathology and recovery

Upon establishing an infection, pathogens begin to replicate. Generally, the clinical manifestations associated with infectious diseases are determined largely by the abundance of the pathogen. Since infections are generally initiated with a low number of pathogen, it takes some time after the introduction of the pathogen until the pathogen reaches sufficient numbers to cause symptoms. The period of time between the initiation of the infection and the appearance of symptoms is called the **incubation period** (FIGURE 19.15). At the beginning of the symptomatic phase the symptoms may be non-specific and rather mild, and this period is called the prodromal period. Increases in pathogen abundance cause damage and inflammation, and the symptoms become more severe and specific during this acute stage of infection.

SIGNPOST

Adaptive immunity is described in Chapter 9, pages 182–183.

FIGURE 19.15 Typical course of infectious disease. Upon infection pathogens begin to replicate. Once reaching a threshold number, symptoms begin to appear after the incubation period. Initially the symptoms are non-specific (prodromal) and later become specific as the number of pathogens increases. An overwhelming infection and possibly death occur if the immune system cannot sufficiently control the infection. Otherwise, the immune system limits the propagation of the pathogen, leading to convalescence and possibly the elimination of the pathogen. If the infection cannot be completely resolved the disease enters a chronic state characterized by low numbers of pathogens.

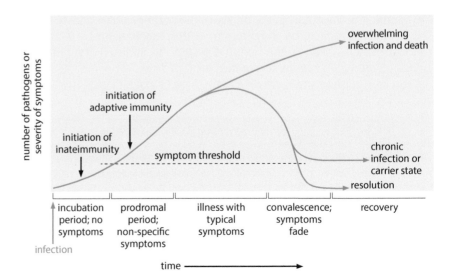

As the pathogen replicates, the immune response is also increasing and, in particular, adaptive immunity is activated. The immune response eliminates the pathogen and slows its replication, and therefore, the abundance of the pathogen starts to decrease. A failure of the immune response results in an overwhelming infection that likely ends in death. As pathogen abundance decreases the symptoms become less severe and this marks the beginning of the convalescence period. Most of the time the immune system eliminates the pathogen and there is complete recovery from the disease. Sometimes, this can even lead to protective immunity and subsequent infections only produce mild disease or no disease at all. Some pathogens can establish long-term chronic infections following convalescence. This can lead to a carrier state in which an individual is infectious even though acute symptoms are no longer present. Many pathogens have evolved strategies to maintain long-term chronic infections in order to increase their transmissibility.

DIAGNOSIS, TREATMENT, PREVENTION, AND CONTROL

Confirmation that an infectious disease is causing the clinical symptoms relies on directly or indirectly detecting the pathogen. For example, bacteria and other microorganisms can be directly observed in clinical specimens using a microscope (FIGURE 19.16). Viruses are not visible with light microscopy and electron microscopy is rather impractical for diagnosis. A major limitation of microscopy is its sensitivity, as sufficient organisms need to be present in the sample. Carrying out in vitro culture of the specimen can overcome this limitation. However, it is not possible to culture many of the disease-causing organisms.

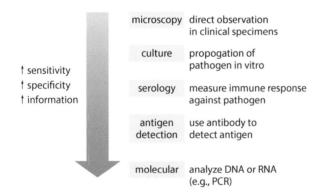

FIGURE 19.16 General diagnostic methods. Shown are various types of diagnostic methods for infectious disease in the order of increasing sensitivity and specificity. Furthermore, molecular methods may provide more information about the pathogen, such as drug sensitivity and virulence.

Molecular and other laboratory methods increase the specificity and sensitivity of the diagnosis

Another common method used in the diagnosis of infectious disease is serology. This involves the detection of antibodies that recognize the pathogen. The presence of antibodies indicates that the immune system has been exposed to the pathogen. Serology is highly sensitive and specific, but it is limited since serology cannot always distinguish between current infections and past infections. In addition, antibodies can be used for antigen detection, which detects proteins or other products of the pathogen. With the incredible advancements in nucleic acid technology, such as PCR (see Box 7.2), the detection of DNA or RNA for the diagnosis of infectious diseases is becoming more routine. In general, serology, antigen detection, and PCR tend to be much more sensitive than microscopy and culture as diagnostic methods.

In addition to the increased sensitivity, molecular- and antibody-based methods can provide more detailed information about the pathogen. For example, serotyping is a method to distinguish strains within a species based on reference antibodies. Similarly, molecular methods can distinguish closely related pathogens, as well as detect virulence factors or provide information about drug resistance. This increased specificity in the diagnosis may influence treatment and control strategies.

Drugs that exhibit selective toxicity against pathogens are often used to treat infectious diseases

The ability to diagnose infectious diseases implies treatment options. There are many available drugs to treat infectious diseases and many infectious diseases are readily curable or can be effectively managed by drug treatment. Drug action involves drugs binding to specific targets associated with the pathogen, such as enzymes, transporters, and receptors, and affecting the normal function of those proteins. Drugs also exhibit selective toxicity against the pathogen in that the drugs are more toxic for the pathogen than for the host (see Table 3.8). In addition, killing or slowing the replication of the pathogen allows the immune system to better control the pathogen and this leads to a faster recovery than without the drug.

In the case of bacterial infections, there is a wide array of bactericidal and bacteriostatic drugs to fight infections. Bactericidal drugs kill bacteria and bacteriostatic drugs inhibit their replication. There are fewer options for eukaryotic pathogens and viruses. A major reason for the greater number of effective anti-bacterial drugs is that the eukaryotic cell is fundamentally different from the prokaryotic cell, resulting in many unique drug targets and pathways in bacteria being exploited by drugs. However, pathogens are continuously evolving and developing resistance to drugs. This means that other strategies to control infectious diseases are needed.

Numerous strategies are available for the prevention and control of infectious diseases

Prevention of infectious diseases usually focuses on minimizing contact between the pathogen and host or minimizing the ability of the pathogen to establish a productive infection (TABLE 19.11). These measures can be applied at the individual level to prevent disease in those individuals, or community-based strategies can be applied to protect the population. Vaccination (Chapter 9) is probably one of the most efficacious methods of prevention and control. Many diseases that were historically major problems are now minor problems due to highly efficacious vaccines. Diseases that are currently major problems generally lack efficacious vaccines.

SIGNPOST

Mechanisms of drug action and selective toxicity are described in Chapter 3, pages 66–69.

Table 19.11 Prevention and Control Strategies	
Strategy	**Examples**
Immunization	• Numerous vaccines available to control numerous infectious diseases
Improved health and nutrition	• Proper nutrition bolsters immunity and lessens the morbidity of infectious disease
Chemoprophylaxis and chemotherapy	• Prophylactic drugs prevent disease following infection • Treating infected individuals reduces the pathogen reservoir (i.e., carriers)
Personal protection	• Handwashing • Face masks • Bed nets
Environmental improvements	• Sanitation • Water treatment • Eradicating mosquito breeding sites

Since undernutrition lowers immunity (Chapter 12), improved nutrition and health are effective to control and prevention strategies. Drugs also play a large role in the prevention and control of infectious diseases. Treatment of infected individuals reduces carriers in the population and lowers transmission. In addition, drugs can be taken for prophylaxis. Measures to lower an individual's exposure to pathogens include handwashing to minimize the direct transfer of pathogens or bed nets to prevent mosquito-transmitted diseases. Environmental changes, such as improved sanitation or water treatment, can also minimize transmission at the population level.

DRUG RESISTANCE

Drug resistance is a huge problem for many infectious diseases. Pathogens, being biological entities, are subject to evolutionary and selective forces. In the presence of a drug, most pathogens can evolve to become more tolerant to the drug and eventually resistant to the drug. This means that new drugs to combat infectious disease are continuously needed. Drug resistance is particularly problematic among the 'big 3' (AIDS, TB, and malaria) and is a contributing factor to why those three diseases are so prevalent and cause so much morbidity and mortality. Without efficacious drugs, there is much more morbidity and mortality associated with infectious diseases due to the inability to treat infections. It has been predicted that drug resistance will be a leading cause of death by 2050.

Drugs exert selective pressure on pathogens

The individual microbes making up a pathogen population exhibit diversity regarding their sensitivity to drugs. In other words, some of the individual pathogens can tolerate more drug than others. Tolerance is a condition in which higher concentrations of drugs are needed to have a therapeutic effect, such as inhibiting replication or killing the pathogen. This means that those individual pathogens with more drug tolerance survive better in the presence of drugs than those individuals completely sensitive to the drug. Drug usage selects pathogens that have increased tolerance to the drug, and this eventually evolves into drug resistance. Drug resistance is a condition in which the normal therapeutic dosage of the drug no longer eliminates the pathogen.

Drug pressure is the exposure of pathogens to drugs and drug pressure selects those individuals with the highest tolerance for the drug. This creates an evolutionary bottleneck (FIGURE 19.17) in which the pathogen population becomes less diverse, and the drug-resistance trait becomes highly prevalent in the pathogen population. Therefore, drug-resistant pathogens are more likely to be transmitted and the drug will no longer be effective and lead to more morbidity and mortality. However, drug resistance is not necessarily a biological advantage and, in fact, it is almost always a biological disadvantage in the absence of drug pressure. For example, drug-sensitive pathogens will replace drug-resistant pathogens if the use of the drug is discontinued for several years.

diverse population of pathogen

drug pressure creating
an evolutionary bottleneck

predominance of
drug-resistant variants

drug-resistant variant

FIGURE 19.17 Selection of drug resistance. Exposure to drugs selects individual pathogens with a higher tolerance or resistance to the drug. Pathogens that are drug sensitive are eliminated resulting in an evolutionary bottleneck creating a predominance of drug-resistant variants.

Mutations in genes drive the development of drug resistance

Mechanisms by which pathogens develop drug resistance include changes in the drug target, effects on drug transport or permeability, and drug inactivation (FIGURE 19.18). The most common mechanism of drug resistance is a mutation in the gene for the drug target. For example, a mutation that decreases the binding affinity of the drug without overtly affecting the normal function of the protein makes the drug less efficacious. The decreased affinity of a drug for its target protein means that higher concentrations of the drug are needed to inhibit the target protein. Furthermore, continuous exposure of pathogens to drug pressure results in the accumulation of more mutations and the pathogen progressively becomes more tolerant to the drug, until eventually the pathogen becomes completely drug resistant.

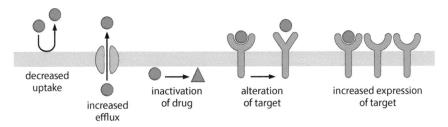

decreased
uptake

increased
efflux

inactivation
of drug

alteration
of target

increased expression
of target

FIGURE 19.18 Mechanisms of drug resistance. A common mechanism of drug resistance is a mutation in the drug target resulting in decreased affinity of the target for the drug. Increasing the level of the target through either gene expression or gene amplification is another mechanism of drug resistance. Mechanisms of drug resistance not involving the target include decreased drug uptake, increased drug efflux, or inactivation of the drug.

Another possible alteration in the drug target is an increase in the expression of the target protein. Higher levels of the target protein also require higher levels of the drug to inhibit the protein. Higher expression of the target can be accomplished by increasing the copy number of the gene, called gene amplification, or increasing the level of expression of the gene. Other common mechanisms of drug resistance include changes in membrane transporters that lead to either a decreased uptake of the drug into the cell or an increased efflux of the drug out of the cell. This change in drug permeability means that less drug reaches the target. Quite often a change in a transporter that results in drug resistance affects multiple drugs simultaneously. This is best characterized by a class of transporters called the ATP-binding cassette (ABC) transporters. In fact, ABC transporter genes were originally called multidrug resistance (MDR) genes due to their ability to confer resistance to several drugs in bacteria and cancer cells (BOX 19.5).

Multidrug Resistance and ABC Transporters

BOX 19.5

Permeability (P)-glycoprotein was discovered in the late 1970s due to its association with drug resistance in cancer cells. Subsequently, homologs of the gene were found in bacteria. In both cases the gene was associated with drug resistance to multiple drugs. Thus, the gene was named multidrug-resistant (MDR) gene. This name may have been a bit premature since it is now known the primary function of the P-glycoprotein is not drug resistance per se. P-glycoproteins are part of a large family of transporters called ATP-binding cassette (ABC) transporters. ABC transporters are one of the largest and oldest gene families and are found in all organisms from bacteria to humans. In bacteria the ABC transporters function in both influx and efflux, whereas in eukaryotes the transporters are part of efflux pathways.

Many of the ABC transporters are involved in metabolism and nutrient acquisition. Some of the transporters have a rather narrow range of substrates, whereas others have a broader range of substrates. A subset of the ABC transporters is involved in the efflux of foreign substances out of cells. These genes likely evolved to protect cells from toxic and harmful substances. In that regard, it is not surprising that pathogenic microorganisms utilize MDR genes in the development of drug resistance. Since many of the ABC transporters have a wide range of substrates, the expression of a particular ABC transporter gene may provide resistance to multiple drugs. Furthermore, mutations within an MDR gene may enhance the efflux of a particular drug and increase drug resistance.

Resistance due to drug inactivation involves specialized enzymes

Another mechanism of drug resistance not directly involving the target is the inactivation of the drug by enzymes that degrade or modify the drug. A good example of drug inactivation is the β-lactam antibiotics, such as penicillin. The β-lactam antibiotics have a similar structure and most inhibit bacterial cell wall synthesis. They are a widely used class of antibiotics that are effective against a broad spectrum of Gram-negative and Gram-positive bacteria. Bacteria that are resistant to β-lactam antibiotics secrete an enzyme called β-lactamase that hydrolyzes the antibiotic and makes it ineffective. β-lactamase is an ancient bacterial enzyme that may have evolved from proteases or RNases billions of years ago. Sometimes the expression of the β-lactamase gene is activated by the presence of the antibiotic. In addition, β-lactamase genes are often found on plasmids and can be transferred within a species, or even between species, by horizontal gene transfer (see Box 19.4).

Factors that expose pathogens to sub-therapeutic drug concentrations promote the spread of drug resistance

Drug-resistant pathogens can quickly emerge and spread. Several factors contribute to the emergence and spread of drug resistance (TABLE 19.12). Self-medication and poor compliance are situations in which the normally recommended drug treatment is not followed. For example, shortly after starting the treatment regimen a person may quit taking the drug because they are feeling better or because of adverse side effects. This interrupted treatment creates a situation in which most, but not all, pathogens are eliminated from the body. Those remaining pathogens are exposed to sub-therapeutic concentrations of drugs that are too low to completely eliminate the pathogen. This is especially true for drugs with longer half-lives. Drug half-life refers to the time it takes for half of the drug to be broken down or excreted from the body. Longer exposure of a pathogen to sub-therapeutic drug concentrations creates more opportunity for mutations that lead to increased drug tolerance to occur. And this opportunity for the accumulation of mutations is further increased by mass drug administration and high intensity of pathogen transmission.

Table 19.12 Factors Contributing to the Emergence and Spread of Drug Resistance	
Factor	**Explanation**
Self-treatment	Only taking the drug until symptoms clear or taking lower doses
Poor compliance	Not completing the full course of treatment because of side effects or the cessation of symptoms
Long drug half-life	Results in a longer exposure of the pathogen to sub-therapeutic drug concentrations
Mass administration	Exposes a larger pathogen population to the drug
Transmission intensity	Increased person-to-person spread results in possible re-infection while drugs are at sub-therapeutic levels

Drug combinations are often used to prevent the development of drug resistance

In addition to the continuous development of new drugs, there are other strategies to prevent the development and spread of drug resistance. The most widely used method is drug combinations. Using two or more drugs affecting multiple targets has a couple of advantages. One advantage is that the drugs often act synergistically. In other words, the combined effect of the two drugs is more than the sum of each drug used separately. In addition, mutations would need to occur in multiple targets nearly simultaneously for drug resistance to develop. This greatly reduces the rate of development of drug resistance. In fact, the recommended treatment of TB, AIDS, and malaria all involve drug combinations.

Other strategies to prevent the development and spread of drug resistance center around the control and regulation of the drugs to avoid their indiscriminate use. Since the development of drug resistance depends substantially on exposure to the drug, less exposure means a lower chance of developing drug resistance. Drugs that are not readily available and require prescriptions from health-care professionals can be better controlled. For example, health-care professionals can educate patients on the importance of taking the drug as prescribed and finishing the prescribed dosage. Furthermore, monitoring for the appearance of drug resistance may allow for a change to another drug. This removes the drug pressure on the previous drug and prevents further selection for drug resistance.

SUMMARY OF KEY CONCEPTS

- Infectious diseases are caused by biological entities – ranging from infectious proteins to multicellular parasites – that are acquired from the environment or animals or, most often, transmitted from person to person

- Non-cellular pathogens include prions and viruses

- Unicellular pathogens include bacteria, fungi, and protozoa

- Modes of transmission include direct or indirect contact, ingestion, inhalation, or biological vectors.

- Pathogens have the ability to overcome host barriers and replicate within the host

- Pathology associated with infectious disease is due to direct tissue or cellular damage caused by the pathogen or the inflammatory response associated with the infection

- Diagnosis of infectious disease relies on direct observation via microscopy or in vitro culture, or more sensitive serological or molecular methods

- Most infectious disease is readily treated and managed with drugs that exhibit selective toxicity toward the pathogen

- Infectious disease is prevented by minimizing host–pathogen contact or vaccination

- Drug resistance is an increasing problem in the control and management of infectious diseases

- Most pathogens can rapidly evolve drug resistance in the presence of drug pressure

KEY TERMS

AB toxin	A type of exotoxin composed of an A subunit that has toxic activity and a B subunit that binds to the host cell and facilitates entry of the A subunit into the host cell.
commensal	An organism participating in a symbiotic relationship in which the commensal derives some benefit while the host organism is unaffected. Refers to commensalism.
drug resistance	The reduction in effectiveness of a medication, such as an antimicrobial or an antineoplastic drug, in treating a disease or condition.
endemic mycoses	Infections caused by any of a diverse group of fungi that occupy specific ecologic niches in the environment. Transmission usually involves contact with soil or inhalation of spores.
endotoxic shock	A severe systemic inflammatory response induced by lipopolysaccharide on the surface of Gram-negative bacteria.
exotoxin	A secreted toxin of pathogenic bacteria that causes cell death or interferes with normal cellular function.
fomite	Inanimate objects, such as clothes, utensils, or furniture, that serve as intermediates in the transmission of infectious diseases.
Gram stain	A biological staining method that distinguishes bacteria into two classes (positive or negative) based on the structure of their cell wall.
incubation period	The time between exposure to a pathogen and the appearance of symptoms.
lipopolysaccharide (LPS)	A large glycolipid found in the cell wall of Gram-negative bacteria that can cause endotoxic shock. Also called endotoxin.
pathogen	A microbe or other biological entity that can cause disease.
pathogenicity	The ability of an organism to cause disease.

prion	An infectious and misfolded protein that can 'transmit' its misfolded shape onto normal variants of the same protein.
systemic (disease)	A disorder that can affect several organs or tissues or even the entire body.
virulence	The degree to which a pathogen can cause disease that is primarily determined by its ability to replicate within a host.
virulence factor	A pathogen product that is associated with any aspect of the disease process, including survival and replication of the pathogen.
zoonosis	A disease that can be transmitted from animals to humans.

REVIEW QUESTIONS

1. What are the six biological classes of pathogens? Briefly describe each.

2. In what situations are fungal infections, including endemic mycoses, most likely to cause a severe disease?

3. What is the difference between endotoxin and exotoxin? How do these toxins cause disease?

4. What is meant by Gram positive and Gram negative, and what is the basis of the difference between the two?

5. Discuss the various means of diagnosing infectious diseases and their relative specificities and sensitivities.

6. Discuss the mechanisms by which pathogens become resistant to drugs.

ADDITIONAL QUESTIONS

In addition to the Review Questions provided above, there is a range of free online questions designed to enable students to further test their understanding of the chapter material. In order to access these interactive questions, please visit the book's website:
https://routledge.com/cw/wiser

Respiratory Tract Infections

20

Numerous pathogens infect the respiratory tract. Transmission of respiratory tract infections primarily involves the inhalation of the pathogen or contaminated hands. Such transmission is generally rather efficient and infections can rapidly spread through a population. In that regard, there have been several notable pandemics of influenza and, most recently, COVID-19. Most respiratory tract infections are mild and self-resolving, but a few pathogens can cause death, especially in the very young and the very old. Respiratory tract infections are a major cause of childhood mortality and, like diarrheal diseases (Chapter 21), they are substantial problems in resource-poor countries.

Infections of the upper respiratory tract tend to be mild, and the most common symptoms are a runny nose and sore throat. More severe clinical manifestations are observed if infections migrate to the lower respiratory tract. The most serious manifestation of a respiratory tract infection is pneumonia. Influenza and COVID-19 are viral diseases capable of causing severe pneumonia. Bacterial pneumonia is primarily observed in older individuals and other immunocompromised people. Tuberculosis is another bacterial infection of the lungs that tends to be more severe in immunocompromised individuals.

TRANSMISSION

The most common mode of transmission of respiratory infections is through respiratory droplets (FIGURE 20.1). Respiratory droplets consist of saliva, mucus, and other matter derived from the respiratory tract and range in size from 5 μm (10^{-6} m) to 1 mm. They are formed and expelled during breathing, speaking, singing, sneezing, coughing, or vomiting. Pathogens can be enclosed within these respiratory droplets and infection occurs when respiratory droplets containing pathogens reach a susceptible mucosal surface, such as the eyes, nose, or mouth. For example, inhalation of respiratory droplets can deposit pathogens in the respiratory tract. However, respiratory droplets are large and cannot remain suspended in the air for very long and, thus, are usually only dispersed over short distances. It is generally accepted that transmission via respiratory droplets primarily occurs over distances of less than 2 meters. Settling respiratory droplets can contaminate surfaces and cause indirect transmission. Touching around the face after touching a contaminated surface (i.e., fomite) can also introduce the pathogen and initiate an infection.

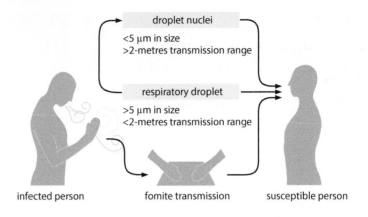

FIGURE 20.1 Transmission of respiratory infections. Infectious pathogens are expelled in phlegm, sputum, or respiratory droplets. A common mode of transmission is the inhalation of droplets. Generally, the transmission range is less than 2 meters since these droplets settle quickly. Evaporation of the smaller respiratory droplets creates droplet nuclei resulting in air-borne transmission of aerosolized pathogens, which can occur over greater distances. Objects (i.e., fomites) can also be contaminated with pathogens and in this case the disease is spread by touching the contaminated objects and then touching around the eyes, nose, or mouth.

Some respiratory infections exhibit air-borne transmission

Smaller respiratory droplets rapidly evaporate and shrink in size and form **droplet nuclei**. Droplet nuclei refer to respiratory droplets that are less than 5 µm in diameter and can become aerosolized. Transmission via aerosolized particles is called air-borne transmission. Aerosolized particles can travel longer distances and remain air-borne for longer periods, thus potentially increasing transmissibility. In addition, droplet nuclei are more likely to reach the alveoli before interacting with a mucosal surface, and thereby can establish an infection deep in the lungs sooner. However, the drying of the respiratory droplet adversely affects many pathogens, and ambient temperature and humidity can affect the survivability of pathogens. For example, viruses with a lipid envelope are generally more stable in dry air, whereas those without an envelope are more stable in moist air. In addition, pathogens typically survive longer at lower air temperatures. Thus, not all pathogens can be transmitted through the air-borne route and air-borne transmission is determined in part by the conditions.

Preventive measures minimize risk factors and involve both infected people and susceptible people

The factors involved in the transmission of respiratory infections are similar to transmission factors for infectious diseases in general, and involve infectiousness, environment, exposure, and susceptibility (TABLE 20.1). Small and enclosed spaces favor transmission by either respiratory droplets or droplet nuclei. Clearly the number of infectious pathogens being expelled by an infected person and the duration of contact between an infected person and a susceptible non-infected person contribute to the spread of infections.

Preventive measures include measures taken by infected people to avoid spreading the disease to others and measures taken by susceptible people to avoid becoming infected (TABLE 20.2). Susceptible people wishing to avoid becoming infected should minimize the time spent in close contact with people exhibiting symptoms of respiratory infections, limit touching around the nose and mouth, wash hands often, and decontaminate commonly touched surfaces. Covering the mouth and nose when coughing or sneezing limits the spread of pathogens to other people. Vaccines are also available against several respiratory tract pathogens and some of these vaccines have played a large role in substantially reducing the prevalence of some respiratory diseases.

Table 20.1	Transmission Factors
Infectiousness	Number of infectious pathogen particles expelled Survival of a pathogen in the environment
Environment	Small and enclosed spaces favor transmission
Exposure	Duration, frequency, and physical proximity (e.g., daily contact with family, friends, coworkers)
Susceptibility	Immune status of an exposed individual

Table 20.2	**Prevention of Respiratory Infections**
Measures taken by susceptible people to avoid infection	

- Minimize close contact with people who have symptoms of respiratory infection
- Avoid touching mouth and nose

Measures taken by infected people to avoid spreading disease

- Covering mouth and nose when coughing or sneezing, including wearing surgical or other masks
- Properly dispose of tissues used to contain respiratory droplets or secretions

Measures taken by both susceptible and infected people

- Frequent handwashing
- Frequent cleaning and disinfection of commonly touched surfaces
- Vaccination against common respiratory diseases

PATHOPHYSIOLOGY

The respiratory tract consists of the lungs and the various passageways that connect to the lungs. Air is inhaled through the nose and mouth and passes to the trachea via the throat. The trachea branches into two main bronchi, which enter each lung, and progressively branch into narrower secondary and tertiary bronchi. These bronchi further branch into numerous smaller tubes called bronchioles. At the end of the bronchioles are the alveoli, commonly called air sacs, where gas exchange with the blood occurs. The entire respiratory tract is lined with epithelial cells and the epithelial cells are covered with a mucus layer.

The respiratory tract is divided into the upper respiratory tract, consisting of the nose, mouth, throat, and sinuses, and the lower respiratory tract, consisting of the trachea, bronchi, bronchioles, and alveoli. Any portion of the respiratory tract can become infected. However, there is a tendency for some pathogens to colonize a specific region of the respiratory tract (FIGURE 20.2).

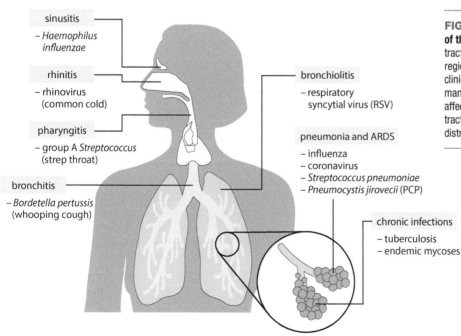

sinusitis
– *Haemophilus influenzae*

rhinitis
– rhinovirus (common cold)

pharyngitis
– group A *Streptococcus* (strep throat)

bronchitis
– *Bordetella pertussis* (whooping cough)

bronchiolitis
– respiratory syncytial virus (RSV)

pneumonia and ARDS
– influenza
– coronavirus
– *Streptococcus pneumoniae*
– *Pneumocystis jirovecii* (PCP)

chronic infections
– tuberculosis
– endemic mycoses

FIGURE 20.2 Examples of notable pathogens of the respiratory tract. All parts of the respiratory tract can be infected. Infection of the different regions of the respiratory tract exhibits different clinical manifestations (see Table 20.3). The clinical manifestations, the region of the respiratory tract affected, and a few examples of common respiratory tract pathogens are shown. ARDS, acute respiratory distress syndrome.

Infection of the respiratory tract is associated with inflammation and hypersecretion of mucus

The symptoms and clinical manifestations associated with respiratory tract infections correspond to the region(s) of the respiratory tract that is affected (TABLE 20.3). A common feature of respiratory tract inflammation is the hypersecretion of mucus. Excess mucus produced by the respiratory system in response to disease is called phlegm. Upper respiratory tract infections are often accompanied by a runny nose, which refers to excess phlegm being excreted. In lower respiratory tract infections the accumulation of phlegm is often accompanied by coughing. A cough can be dry, in which no phlegm is expelled, or productive, in which excess phlegm is expelled. Coughed-up phlegm is called sputum. **Pneumonia** is characterized by a buildup of fluid in the lungs and, if not treated, it can have a high rate of mortality.

Table 20.3	Inflammation of the Respiratory Tract and Clinical Manifestations
Sinusitis	• Thick nasal mucus, plugged nose, facial pain, headaches, poor sense of smell • Usually accompanied by rhinitis and upper respiratory tract infection • Most often of viral origin
Rhinitis	• Runny nose, sneezing, stuffy nose, and post-nasal drip • Allergy most common cause • Infections include numerous viruses (e.g., common cold) and several bacteria • Often accompanied by pharyngitis
Pharyngitis	• Sore throat, often accompanied by fever • Numerous viruses and bacteria
Bronchitis	• Productive or dry cough, wheezing, shortness of breath, and chest discomfort • More than 90% due to viruses
Bronchiolitis	• Fever, cough, runny nose, wheezing, and breathing problems • Usually only occurs in children less than 2 years of age • Viral infection and especially respiratory syncytial virus (RSV)
Pneumonia	• Productive or dry cough, chest pain, fever, and difficulty breathing • Results in approximately 4 million deaths annually • Generally viral or bacterial in origin, and occasionally fungal • Most often bacterial in older individuals and viral in young people

In general, infections of the lower respiratory tract are more severe than infections of the upper respiratory tract

Infections of the upper respiratory tract tend to be mild and self-resolving. Generally, acute symptoms only last a few days and there is no major tissue destruction. As infections spread deeper into the respiratory tract the clinical manifestations tend to become more severe. For example, mortality is almost exclusively associated with pneumonia, which is defined as inflammation of the alveoli. In some cases, pathogens that infect the upper respiratory tract are less virulent (see Box 19.1) than those pathogens that infect the lower respiratory tract. However, many pathogens of the respiratory tract can infect any part of the respiratory tract and the increased virulence is due to a more intense inflammatory response in the lower respiratory tract. As an example of balance in nature, upper respiratory tract infections tend to be more transmissible than lower respiratory tract infections.

Severe damage to the lung epithelium results in possible dissemination of the infection

Most respiratory tract pathogens only cause superficial damage to the epithelium and the pathogen is confined to the respiratory tract. However, more extensive damage to the epithelium, which diminishes its barrier function, can result in submucosal inflammation. For example, many commensal bacteria are found in the respiratory tract and these normally non-pathogenic bacteria can cause inflammation if the epithelium is disrupted. Indeed, it is rather common for secondary bacterial infections to be associated with viral infections following the epithelial damage caused by the virus. Furthermore, after the invasion of the submucosa, some pathogens may be able to enter the bloodstream and cause systemic infections.

Pneumonia is the most serious manifestation of respiratory tract infections

The epithelium of the alveoli is substantially thinner and in close contact with capillaries to facilitate the exchange of gases. This means that damage to the alveolar epithelium is likely to cause more severe disease than damage to other parts of the respiratory tract. In addition, activation of alveolar macrophages by pathogens or epithelial damage results in the secretion of inflammatory cytokines that activate and attract neutrophils from the blood (FIGURE 20.3). This inflammation also causes the capillaries to become leaky and fluid from the blood begins to fill the interstitial space and alveoli. Interstitial fluid is fluid found outside of blood vessels. This fluid in the lungs can become purulent with pus containing white and red blood cells. Cytokine production also leads to fever, chills, and fatigue. Filling of the alveoli and interstitial space with fluid is called pulmonary edema and results in cough, difficulty breathing, and chest pain.

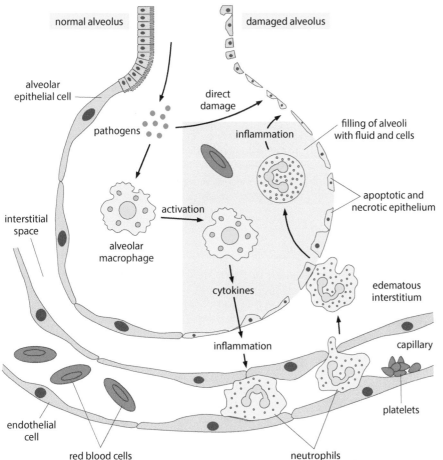

FIGURE 20.3 Pathophysiology of pneumonia and acute respiratory distress syndrome. Pathogens entering the alveoli can directly damage the epithelium of the alveoli and activate alveolar macrophages. Activated macrophages secrete inflammatory cytokines that further damage the epithelium and the cytokines attract neutrophils, leading to more inflammation and damage. Fluids and cells leaking from capillaries cause pulmonary edema due to the accumulation of fluid in the interstitial space (edematous interstitium) and alveoli.

Pneumonia can progress to **acute respiratory distress syndrome** (ARDS) which is characterized by the rapid onset of widespread inflammation in the lungs. Symptoms of ARDS include shortness of breath (dyspnea), rapid breathing (tachypnea), and bluish coloration (cyanosis). ARDS can quickly evolve into a respiratory failure and often requires mechanical ventilation to prevent death. A decrease in quality of life often occurs in those who survive ARDS.

PATHOGENS OF THE RESPIRATORY TRACT

Respiratory diseases of the lower respiratory tract are caused by several viruses and bacteria, as well as some fungi (TABLE 20.4). Most of these pathogens remain in the upper respiratory tract and resolve without causing severe disease manifestations. For example, numerous viruses infect the upper respiratory tract and generally only cause mild disease that is often referred to as the common cold. Some pathogens infect the lower respiratory tract and cause more severe disease. Respiratory infections often exhibit more severe manifestations in older individuals and children. For example, respiratory syncytial virus (RSV) generally causes a mild disease in adults but can cause severe disease in infants and children (BOX 20.1). Similarly, pneumonia is more often of viral etiology in children, whereas, in adults, pneumonia is more frequently caused by bacteria. Influenza and COVID-19 are examples of potentially severe diseases of the lower respiratory tract caused by viruses.

Table 20.4   Examples of Infectious Diseases of the Respiratory Tract

Disease	Description
Endemic mycoses	Fungal infections acquired from the soil with manifestations primarily in immunocompromised individuals
Pneumocystis jirovecii pneumonia (PCP)	A pneumonia disease caused by the opportunistic fungus *Pneumocystis jirovecii* (previously *carinii*)
Bacterial pneumonia	Historically a major cause of death before the introduction of antibiotics
Tuberculosis	Chronic infection of the lungs caused by *Mycobacterium tuberculosis*
Common cold	An upper respiratory tract infection caused by numerous species of virus characterized by relatively mild symptoms
Influenza	A potentially severe infection of the lower respiratory tract caused by the influenza virus
COVID-19	Caused by a recently emerged coronavirus (SARS-CoV-2) that can result in acute respiratory disease syndrome

BOX 20.1

Respiratory Syncytial Virus

Respiratory syncytial virus (RSV) is a common virus that usually causes mild, cold-like symptoms, especially in older children and adults. But for infants and small children it can cause severe disease that frequently results in hospitalization (**BOX 20.1 TABLE 1**). Nearly all children have experienced at least one RSV infection before reaching 4 years old. Each subsequent infection induces more immunity and, hence, decreasing severity of disease with age. However, it is being increasingly recognized that RSV infection can cause severe disease in older patients. RSV causes infected cells of the mucosa to fuse together, resulting in a large multinucleated cell called a syncytium, hence the name respiratory syncytial virus.

Box 20.1   Table 1 Clinical Consequences of RSV Infection

Age group affected	Disorder
Children <1 year old	Bronchiolitis, pneumonia, or both
Children	Febrile rhinitis and pharyngitis
Older children and adults	Common cold

The immune status of an individual also influences disease progression, since respiratory infections are generally more severe in very old, very young, or immunocompromised individuals. For example, endemic mycoses are fungal infections due to the inhalation of spores from the environment, and primarily affect immunocompromised people. Similarly, *Pneumocystis jirovecii* (previously *carinii*) is an opportunistic fungal pathogen that causes pneumonia often found in patients with AIDS and other immunocompromised people (see Box 22.3). Tuberculosis is a chronic bacterial infection that afflicts millions of people worldwide and is more severe in the immunocompromised individual.

Endemic and opportunistic mycoses are described in Chapter 19, pages 372-374.

SIGNPOST

BACTERIAL INFECTIONS AND PNEUMONIA

Several bacterial species can infect the upper respiratory tract and cause rhinitis or pharyngitis. *Streptococcus pyogenes* and related *Streptococcus* species, collectively known as group A *Streptococcus* (GAS), are common pathogens of the upper respiratory tract. GAS is responsible for a large portion of sore throats in children and the disease is commonly known as strep throat. *Bordetella pertussis*, which causes whooping cough (BOX 20.2), is another notable bacterial infection of the respiratory tract that primarily affects children.

Pertussis

Pertussis, also called whooping cough, is a highly contagious bacterial disease that primarily affects children. Initial symptoms are generally a runny nose, fever, and mild cough. The disease then progresses to severe coughing fits that can last for weeks. The name whooping cough is in reference to a high-pitched whooping sound or gasping that might occur as the person breathes in following a coughing fit. The causative agent is the bacterium *Bordetella pertussis*. Pathogenesis associated with

pertussis is largely due to an AB5 toxin (Chapter 19) called pertussis toxin. Pertussis toxin perturbs the cAMP signal transduction pathway that leads to an inhibition of phagocytosis by macrophages and a reduction in the recruitment of neutrophils. Therefore, pertussis toxin helps the bacteria avoid innate immunity and allows the infection to persist. In addition, pertussis toxin may also contribute to cough pathology.

BOX 20.2

Bacterial pneumonia can have a high rate of mortality if not treated. Before the discovery of antibiotics, bacterial pneumonia was a major cause of death in very young and very old people. Bacterial pneumonia still exhibits high mortality in resource-poor countries. Many bacteria can cause inflammation and pneumonia if they colonize the alveoli (TABLE 20.5), especially if there has been a disruption of the epithelium. Infections can include bacteria that commonly inhabit the upper respiratory tract or other parts of the body, or even bacteria normally found in the environment. In addition, there are several bacteria, including *Streptococcus pneumoniae*, that have virulence factors that make them more prone to cause pneumonia.

Table 20.5 Examples of Bacteria That Cause Pneumonia

- *Streptococcus pneumoniae*
- *Haemophilus influenzae*
- *Moraxella catarrhalis*
- *Mycoplasma pneumoniae*
- *Staphylococcus aureus*
- *Klebsiella pneumoniae*
- *Chlamydia pneumoniae*
- *Legionella pneumophila*
- *Bacillus anthracis*

The most common cause of bacterial pneumonia is *Streptococcus pneumoniae*

Streptococcus pneumoniae accounts for 15–50% of all community-acquired pneumonia. Community-acquired infections are transmitted outside of a health-care facility, in contrast with hospital-acquired diseases. *S. pneumoniae* is generally considered a part of the microbiota and normally does not cause disease. However, the bacterium may become virulent and cause pneumonia in individuals with weaker immune systems, as well as disseminate to other parts of the body and cause diseases like meningitis. In the lungs, bacteria colonize the air sacs and stimulate an inflammatory response that results in plasma, blood, and white blood cells filling the alveoli (see Figure 20.3).

S. pneumoniae has several virulence factors that contribute to the increased virulence of pneumococcal pneumonia (TABLE 20.6). These virulence factors include the polysaccharide capsule, a toxin called pneumolysin, and an IgA protease, as well as various adhesins and immunogenic cell wall components. Probably the most important virulence factor is the capsule. It has been long recognized that *S. pneumoniae* strains without capsules are far less virulent. The capsule blocks the action of complement and impairs phagocytosis. The bacterium also secretes an enzyme that breaks down the immunoglobulin IgA. IgA is a secreted antibody that plays an important role in mucosal immunity (see Table 9.10). Pneumolysin is a pore-forming protein that damages epithelial cells and can promote tissue invasion and intensify the inflammatory response.

Table 20.6	Major Virulence Factors of *Streptococcus pneumoniae*
Polysaccharide capsule	• Clinical isolates of *S. pneumoniae* causing invasive disease are encapsulated • Inhibits complement and impairs phagocytosis
Pneumolysin	• A 53-kDa pore-forming protein produced by clinical isolates • Facilitates invasion of the epithelium
IgA protease	• Bacterial enzyme that specifically cleaves human IgA1 (type of antibody) as part of avoidance of the immune system

TUBERCULOSIS

Tuberculosis is a common disease caused by the bacterium *Mycobacterium tuberculosis*. Other *Mycobacterium* species from animals can also cause human tuberculosis, but this is rather rare and is generally restricted to immunocompromised individuals. Over a billion people are currently infected worldwide and tuberculosis causes millions of deaths per year, making it one of the most common causes of death by a single organism. It is commonly grouped with malaria (Chapter 23) and AIDS (Chapter 22) as part of the 'big three.' Like most infectious diseases, tuberculosis is more prevalent in resource-poor countries.

Infections are primarily localized to the lungs and cause pulmonary tuberculosis. However, mycobacteria can also spread to other parts of the body and cause extrapulmonary disease. Most people with tuberculosis have a latent disease without symptoms and are not infectious to other people. **Active tuberculosis**, conversely, is a contagious disease with potentially severe pathology and a high rate of mortality if not treated.

Tuberculosis is an old disease that has plagued humans for millennia

It was previously believed that tuberculosis emerged coincident with the development of agriculture and that *M. tuberculosis* may have been derived from *M. bovis* of cattle. However, molecular data indicate that tuberculosis has been present in humans for at least 70,000 years and spread around the world with humans as they migrated out of Africa. It is most likely that tuberculosis in cattle and other domestic animals was derived

SIGNPOST

Early human evolution is described in chapter 5, pages 107-109.

from human tuberculosis. The agricultural revolution and urbanization, however, did create conditions for higher rates of transmission, allowing *M. tuberculosis* to become a major human pathogen.

In the 17th century, tuberculosis reached pandemic proportions and throughout the 17th and 18th centuries upward of 25% of all deaths were due to tuberculosis. The disease was referred to as the white plague due to the pallor of those infected. The term tuberculosis to describe the disease was introduced in 1839 and in 1882 Robert Koch described *Mycobacterium tuberculosis* as the etiological agent. He was later awarded a Nobel Prize for his discovery. Identification of the causative agent and its mode of transmission led to public health measures that decreased transmission. The subsequent antibiotic era further reduced its prevalence and mortality. In the mid-20th century it was believed that tuberculosis might be eradicable. However, the disease has persisted – especially in poor people – and since the 1980s tuberculosis has re-emerged as a major human pathogen due to drug resistance and the AIDS epidemic.

A distinct feature of *M. tuberculosis* is a thick lipid layer in its cell wall

Mycobacterium species are small non-motile bacilli that do not produce spores. The bacteria are covered with a waxy coat that is predominantly composed of a fatty acid called mycolic acid. This mycolic acid layer (FIGURE 20.4) is part of the cell wall and limits the permeability of Gram stain and, thus, the bacteria are weakly Gram positive. Mycobacteria are often detected by acid-fast staining of sputum samples and, thus, are also called acid-fast bacilli (FIGURE 20.5). The decreased permeability also affects the uptake of nutrients, and the bacteria grow rather slowly with a doubling time of 12–24 hours.

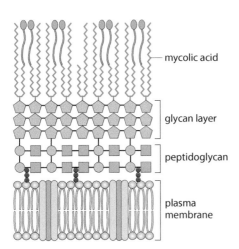

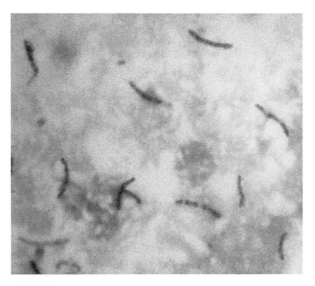

FIGURE 20.4 **Structure of *Mycobacterium* cell wall.** The cell wall of *Mycobacterium* consists of a peptidoglycan layer embedded into the plasma membrane and a glycan layer with mycolic acid on the outer surface.

FIGURE 20.5 **Acid-fast staining of *Mycobacterium tuberculosis*.** Staining of sputum samples with acid-fast stain. This can be used to diagnose tuberculosis bacteria, which appear as red rods against a blue background. [From Dr. George P. Kubica (CDC)].

The waxy coat does provide benefits to the bacteria such as tolerance to some antibiotics and disinfectants. More importantly, the waxy coat slows the desiccation process and increases its transmissibility. The waxy coat also makes the bacteria resistant to complement and destruction by phagocytosis. Therefore, the waxy coat is a virulence factor. Furthermore, the mycobacteria can escape from the phagosome and replicate within the cytoplasm of macrophages, and ultimately kill the macrophage. This ability to replicate in macrophages allows *M. tuberculosis* to establish long-term chronic infections.

Tuberculosis most often manifests as a latent asymptomatic infection

Tuberculosis is not extremely contagious and only 10–30% of people who are exposed to an infected person become infected. The primary mode of transmission is via droplet nuclei. These droplet nuclei develop from respiratory droplets and can be air-borne for several hours and reach the alveoli if inhaled. Because the disease typically causes a long-term chronic infection, an infected person with active tuberculosis may infect 12–15 other people over a year. In general, the factors involved in transmission are similar to other respiratory infections (see Table 20.1) and transmission is often among family members and coworkers.

Approximately 90% of people infected with *M. tuberculosis* develop latent tuberculosis (FIGURE 20.6). Latent tuberculosis is asymptomatic and not infectious to other people. Only individuals with active tuberculosis can transmit the disease. Approximately half of the people who develop active tuberculosis have symptomatic disease within 2 years of the initial infection and this is called primary tuberculosis. Tuberculosis that develops more than 2 years after the initial infection is called secondary tuberculosis and is due to an activation of the latent infection. Tuberculosis is treatable and can be cured. If tuberculosis is not treated, approximately half of the people with active tuberculosis die.

FIGURE 20.6 Natural history of tuberculosis. People infected with *M. tuberculosis* (Mtb) develop either an asymptomatic latent disease or an active disease with high mortality if not treated.

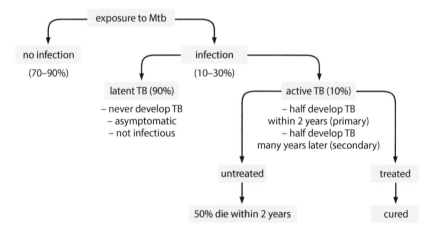

The immune system can contain but cannot eliminate a *M. tuberculosis* infection

After infection, as with most pathogens, the bacteria replicate and increase in numbers. As the adaptive immune response is activated the immune system begins to eliminate the bacteria and the number of bacteria decreases (FIGURE 20.7). However, the immune system does not completely eliminate all bacteria, and the infected individual develops a lifelong chronic infection called **latent tuberculosis infection** (LTBI). There is no definitive demonstration of anyone curing themselves without treatment of a tuberculosis infection, although this cannot be completely ruled out.

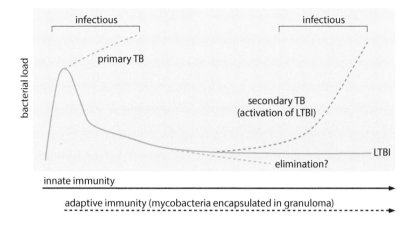

FIGURE 20.7 Bacterial growth curve and the progressions of tuberculosis. After infection, the bacteria increase in number until the activation of the adaptive immune response. The adaptive immune response contains the infection in cellular masses called granulomas (Figure 20.8) as a lifelong latent tuberculosis infection (LTBI). There is no evidence that mycobacteria are eliminated by the immune system. In approximately 5% of infected people the bacteria continue to replicate after the initial infection leading to primary tuberculosis (TB). Immune suppression later in life may lead to an activation of LTBI that causes secondary tuberculosis. Infected people are only infectious during active primary or secondary tuberculosis.

In approximately 5% of infected people the immune system does not contain the initial infection, and the bacteria continue to increase and cause primary tuberculosis. Generally, people who develop primary tuberculosis are immunocompromised in some fashion. For example, children, older people, and debilitated people are more likely to develop primary tuberculosis. Some individuals with latent tuberculosis may develop secondary active tuberculosis years or decades later. Secondary tuberculosis is an activation of the latent infection. Like primary tuberculosis, secondary tuberculosis is also associated with immune suppression. Secondary tuberculosis is a common opportunistic disease associated with AIDS (Chapter 22).

Granulomas contain the *M. tuberculosis* infection

Upon initial infection, macrophages are activated and try to eliminate the bacteria. However, the bacteria can survive within the macrophages and continue to replicate. Activation of the adaptive immune response involves the recruitment of lymphocytes to the site of infection. These various lymphocytes, and especially the T cells, help control the infection. Fibroblasts are also recruited to the site of infection and encapsulate the bacteria and infected macrophages to form a nodule called a **granuloma** (FIGURE 20.8). These granulomas are typically 0.5–3 mm in diameter and were previously called tubercles, hence the name tuberculosis for the disease. The granuloma contains the bacteria and prevents their spread and slows their replication. However, some of the bacteria are not killed, and the surviving bacteria maintain a lifelong latent infection.

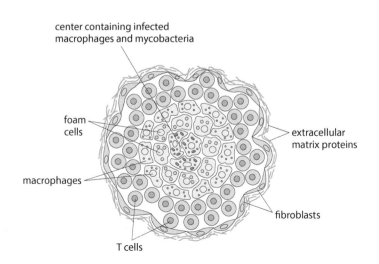

FIGURE 20.8 Granuloma structure. Infected macrophages and mycobacteria are surrounded by macrophages and lymphocytes. Macrophages that ingest mycobacteria become foamy in appearance and are called foam cells. T cells are also incorporated into the mass and assist the macrophages in killing the mycobacteria. Some of the mycobacteria survive and maintain long-term latent infection. This mass of cells and mycobacteria is encapsulated by fibroblast cells, which secrete extracellular matrix proteins to form the granuloma. Encapsulation of the mycobacteria prevents active disease with symptoms. Furthermore, encapsulation of the mycobacteria within the granuloma precludes the spread of infection to other individuals.

A person with latent tuberculosis has an approximately 10% risk during their lifetime of developing active tuberculosis. Factors that increase this risk are advancing age, diabetes, steroid use, chronic poor health, and especially AIDS, all of which lower immunity. People with latent tuberculosis who are co-infected with HIV have an approximately 10% risk of developing active tuberculosis per year. During activation of latent tuberculosis, the center of the granuloma becomes necrotic and soft and the granuloma ruptures. The rupture of the granuloma releases the bacteria, and the suppressed immune system cannot re-contain the bacteria into granulomas. Furthermore, these bacteria are released into the airways and the infected person is now capable of transmitting the infection. As is the case in primary tuberculosis, the bacteria can also enter the blood and spread throughout the body and cause extrapulmonary tuberculosis.

Pathophysiology and clinical manifestations of pulmonary tuberculosis are primarily associated with intense inflammation

Most clinical tuberculosis is due to secondary tuberculosis and manifestations are generally more severe in secondary tuberculosis than in primary tuberculosis. The cell wall of *M. tuberculosis* is highly immunogenic and induces a high level of inflammation. During secondary tuberculosis, T cells that were already sensitized during the initial exposure exhibit a more intense response following the rupture of the granuloma and release of bacteria. This results in extensive tissue damage surrounding the ruptured granuloma and can create cavities in the lungs. These cavities can be up to 15 cm in size and are often visible on chest X-rays (FIGURE 20.9). There is also extensive fibrosis and scarring of the lungs during secondary tuberculosis.

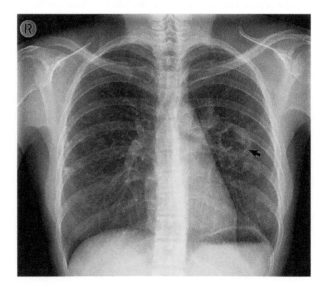

FIGURE 20.9 Chest X-ray demonstrating lung cavitation due to tuberculosis. The circular structure denoted by the arrow is a cavitating lesion. Patchy airspace opacities and nodules are also seen throughout both lungs that are also consistent with tuberculosis. (Courtesy of Dr. Henry Knipe, Radiopaedia.org, rID: 30280.)

Tuberculosis is a slowly progressing chronic infection and therefore the onset of clinical manifestations is usually subtle. Symptoms include coughing, low-grade fever, night sweats, fatigue, weight loss, and generalized wasting. Initially, the cough is non-productive, and later becomes productive. The sputum may also contain blood. These clinical symptoms, combined with characteristic features on chest X-rays or other imagining techniques, are highly suggestive of tuberculosis. Positive diagnosis, however, relies upon the detection of the bacteria in sputum.

Methods to detect *M. tuberculosis* include microscopy, culture, and molecular methods (TABLE 20.7). Diagnosis by microscopy is somewhat insensitive and culture can take several weeks. Molecular methods based on the Xpert MTB/RIF assay greatly speed up diagnosis, increase sensitivity, and provide information about drug resistance (BOX 20.3).

Table 20.7	Diagnosis of Tuberculosis	
Method	**Description**	**Limitation**
Microscopy	Examination of sputum for acid-fast bacilli (Figure 20.5)	Low sensitivity
Culture	In vitro cultivation of sputum sample	Can take 8–12 weeks
Xpert MTB/RIF	Sputum sample is analyzed by PCR in a special machine (Box 20.3)	Requires somewhat expensive equipment and reagents

BOX 20.3

Xpert MTB/RIF Assay

The Xpert MTB/RIF assay is a semi-automated procedure for the diagnosis of tuberculosis. The test is based on the amplification of bacterial DNA by the polymerase chain reaction (see Box 7.2). This method is more sensitive than conventional microscopy and more rapid than in vitro culture. The procedure is relatively simple to carry out. A sputum sample is mixed with reagents and placed into a cartridge (**BOX 20.3 FIGURE 1**), and the cartridge is then placed into the machine. From that stage, the process is fully automated and the results are printed out in approximately 90 minutes. In addition to detecting the presence of *Mycobacterium* the machine also detects a genetic marker for rifampicin resistance.

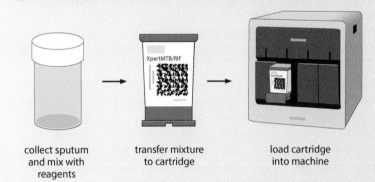

collect sputum and mix with reagents

transfer mixture to cartridge

load cartridge into machine

BOX 20.3 FIGURE 1 Xpert MTB/RIF procedure. Sputum samples are mixed with reagents, placed into a cartridge, and automatically analyzed by the Xpert machine.

Latent tuberculosis can be diagnosed with the tuberculin skin test

The methods used to diagnose active tuberculosis can not be used for the diagnosis of latent tuberculosis since bacteria are not present in the sputum. Diagnosis of latent tuberculosis relies on detecting sensitized T cells that were generated at the beginning of the infection. A preparation from *M. tuberculosis*, called purified protein derivative (PPD), is injected intradermally (FIGURE 20.10). A type 4 hypersensitivity reaction, also called delayed-type hypersensitivity, occurs at the injection site in individuals who have latent tuberculosis. Sensitized T cells migrate to the injection site and cause a localized swelling called an induration. The size of the induration, combined with the person's risk factors for having tuberculosis, is used to make the diagnosis. A major limitation of the skin test is a high false-positive rate and, therefore, individuals with a low risk for having tuberculosis need a larger induration to be considered a positive result.

Type 4 hypersensitivity reactions are described in Chapter 17, pages 321–323.

SIGNPOST

FIGURE 20.10 Mantoux tuberculin skin test.
An extract from *M. tuberculosis* called purified protein derivative (PPD) is injected intradermally. Sensitized T cells migrate to the injection site and cause localized inflammation called induration. The size of the induration associated with this delayed-type hypersensitivity reaction and the individual's risk of acquiring tuberculosis can be used to diagnose latent tuberculosis.

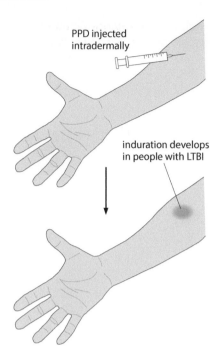

PPD injected intradermally

induration develops in people with LTBI

Most tuberculosis is curable

Tuberculosis is treated with a combination of drugs. A typical treatment protocol consists of a combination of isoniazid, rifampicin, pyrazinamide, and ethambutol for 2 months, followed by treatment with isoniazid and rifampicin for another 4 months. Latent tuberculosis is typically treated for 6–9 months with isoniazid alone or treated 3 months with combined isoniazid and rifampicin. These long treatments and multiple drugs are necessary to completely clear the bacteria and prevent the development of drug resistance.

Expectedly, compliance is a problem due to the long duration of treatment. Generally, after a few weeks the patient is no longer exhibiting symptoms, even though the infection persists. Discontinuation of the treatment may lead to a relapse and the potential for continued spread in the community. In addition, not completing the full course of treatment is a major factor in the development of drug resistance. Therefore, it is recommended that anti-tuberculosis drugs be delivered in a strategy called DOTS, which stands for directly observed therapy, short course. In this strategy a health-care professional or reliable family member directly observes the patient taking the drugs.

Multidrug resistance is a major problem in the management of tuberculosis

Isoniazid and rifampicin are two key drugs for the treatment of tuberculosis and resistance to these drugs is relatively common. Isoniazid is a prodrug that is activated by an enzyme encoded by the *kat G* gene of *Mycobacterium* (FIGURE 20.11). This activated isoniazid forms an adduct with nicotinamide adenine dinucleotide (NAD) and disrupts the synthesis of mycolic acids used to construct the cell wall by inhibiting an essential enzyme encoded by the *inh A* gene of *Mycobacterium*. Resistance is associated with point mutations in the *kat G* gene that block the activation of isoniazid and a mutation in the promoter region of the *inh A* gene that increases its expression. Rifampicin binds to bacterial RNA polymerase and inhibits its activity. A point mutation in the RNA polymerase gene lowers the binding affinity of rifampicin for RNA polymerase.

SIGNPOST

The development of drug resistance is described in Chapter 19, pages 379–382.

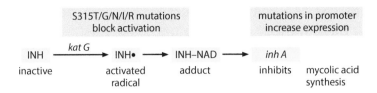

FIGURE 20.11 Mechanism of action of isoniazid. Isoniazid (INH) is a prodrug that needs to be activated by the *Mycobacterium*. A bifunctional catalase-peroxidase encoded by the *kat G* gene generates an INH radical that forms a chemical adduct with NAD. Several mutations in the *kat G* gene, and especially mutations in codon 315, block this activation. The IHN–NAD adduct blocks an enzyme in the mycolic acid synthesis pathway encoded by the *inh A* gene. Mutations in the promoter region of the *inh A* gene increase its expression.

A combination of resistance to at least both isoniazid and rifampicin is called **multidrug-resistant tuberculosis** (MDR-TB). MDR-TB can no longer be cured by the standard treatment and now requires other drugs called second-line drugs. The second-line drugs are more expensive, take longer to cure the patient (up to 2 years), and have more adverse side effects than the first-line drugs. In 2006, isolates resistant to second-line drugs were identified and called extensively drug-resistant tuberculosis (XDR-TB). There are few options for the treatment of XDR-TB, and it is often considered untreatable.

Drug resistance is compounded by the lack of a highly efficacious vaccine. There is a vaccine against tuberculosis called Bacillus Calmette-Guérin (BCG). BCG is an avirulent strain of *Mycobacterium bovis* from cows that was developed in 1921. However, the vaccine is of questionable efficacy, especially against pulmonary tuberculosis. The vaccine does appear to protect children against extrapulmonary tuberculosis and the WHO recommends vaccination of children in highly endemic countries.

THE COMMON COLD

The common cold, or simply a cold, is a viral infection of the upper respiratory tract that primarily affects the nose, throat, and sinuses. More than 200 different virus types have been implicated as causing colds with rhinovirus being the most common (TABLE 20.8). Infection with multiple cold viruses simultaneously is also common. Colds are the most frequent infectious disease, and on average, adults experience two to three colds per year and children may have six to eight episodes. Colds are more common in winter and the winter months are often referred to as cold and flu season. The reason for this seasonality has not been definitively determined. Perhaps people spending more time indoors and in close contact with others may increase transmission, or perhaps changes in the respiratory tract due to colder weather may increase infectivity.

Table 20.8 Most Frequent Common Cold Viruses
• Rhinovirus (30–80%)
• Coronaviruses (≈15%)
• Influenza viruses (10–15%)
• Adenoviruses (≈5%)
• Respiratory syncytial virus
• Enteroviruses
• Parainfluenza viruses
• Metapneumovirus

Symptoms of a cold are generally mild and resolve within a few days without extensive pathogenesis

The most common symptoms of colds are a runny nose, sneezing, nasal congestion, sore throat, and cough. These symptoms are sometimes accompanied by muscle aches, fatigue, headache, and loss of appetite. Fever may be present in infants and young children, but rarely in adults. Diagnosis is based on symptoms and there are no antiviral treatments. Treatment consists of rest and interventions, such as throat lozenges and decongestants, to relieve symptoms.

The clinical manifestations associated with a cold can be due to an immune response against the virus, damage to the epithelium, or a combination of both. A typical course of infection involves infection of epithelial cells (FIGURE 20.12). The initial host response is hypersecretion of mucus resulting in a runny nose. The irritation of the nasal mucosa can also provoke sneezing. Numerous virus particles are present in the expelled clear fluid of the runny nose and sneezes, thereby enhancing transmission. Damage to the epithelium caused by the virus may allow resident bacteria to partially invade the mucosa and cause a minimal inflammatory response. At this point, the phlegm, which was initially clear, becomes thicker and purulent due to cellular debris and dead bacteria. This acute inflammation only lasts a short time as the immune system eliminates the virus and secondary bacterial infections and the epithelium regenerates.

FIGURE 20.12 Pathogenesis of the common cold. Cold viruses infect epithelial cells and stimulate hypersecretion of mucus and sneezing. Disruption of the epithelium allows for commensal bacteria to cause a secondary infection. The immune response eliminates bacteria and viruses and promotes the regeneration of the epithelium.

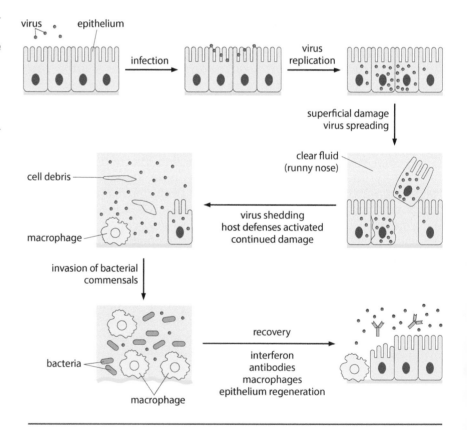

INFLUENZA

Among the viruses that infect the respiratory tract, the influenza virus is important in terms of both prevalence and disease severity. Up to a half million people may die each year of influenza. The clinical manifestations of influenza, commonly called the flu, are more severe and have a longer duration than the common cold. Furthermore, immunocompetent people can also experience severe disease and mortality with some frequency. This is distinct from most other respiratory viruses in which severe disease is generally limited to very young, very old, or immunocompromised individuals. Nonetheless, people often use cold and flu interchangeably despite the two being rather distinct diseases. Flu specifically refers to the disease caused by the influenza virus, and the flu is often accompanied by fever and more severe symptoms of longer duration.

Influenza virus is an enveloped negative-strand RNA virus. There are three types of influenza viruses, designated A, B, and C, as determined by their genetic structure, host range, and disease manifestations (TABLE 20.9). Influenza virus C causes mild disease and is grouped with the common cold viruses. Both influenza A and B can cause severe disease, but severe disease caused by influenza B is generally restricted to older individuals.

Table 20.9   Three Types of Influenza Virus

	Influenza A	Influenza B	Influenza C
Genetic structure	8 RNA segments	8 RNA segments	7 RNA segments
	10 viral proteins	11 viral proteins	9 viral proteins
	M2 protein unique	NB protein unique	HEF protein unique
Host range	Humans and several other animals and birds	Only humans	Humans and pigs
Variants	Numerous serotypes based on hemagglutinin and neuraminidase	Two: Yamagata, Victoria	Unknown
Epidemiology	Seasonal, occasional pandemic	Seasonal	No seasonality
Clinical features	Can cause significant mortality in young people	Severe disease generally confined to older individuals	Mild disease (common cold)

The viral surface proteins hemagglutinin and neuraminidase play key roles in the influenza virus life cycle

Influenza viruses have a segmented genome with each RNA fragment containing one or two genes. The capsid is a ribonucleoprotein complex, which consists of the viral RNA segments, nucleocapsid proteins, and RNA polymerase. The ribonucleoprotein complex is encased by matrix proteins. Two proteins are embedded in the outer envelope of the virus called hemagglutinin and neuraminidase (FIGURE 20.13). An ion channel, called M2, is also present in the envelope of influenza A.

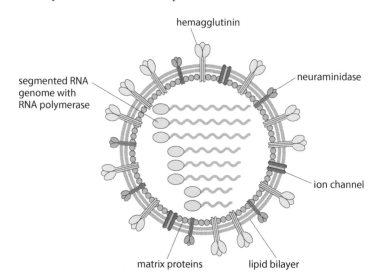

FIGURE 20.13 Structure of influenza A virus. The influenza virus genome consists of 8 RNA strands assembled into a nucleocapsid with nucleoproteins and RNA polymerase. The nucleocapsid is surrounded by matrix proteins. Hemagglutinin and neuraminidase are exposed on the surface of the viral envelope. An ion channel is also located on the envelope in influenza A.

The **hemagglutinin** protein on the virus surface recognizes and binds to sialic acid, which functions as the host cell receptor for the influenza virus. Sialic acids are a type of carbohydrate associated with glycoproteins and glycolipids on the surface of animal cells. After binding to the host cell, the virus is endocytosed and the nucleocapsid is released into the host cell cytoplasm following the fusion of the virus envelope with the membrane of the endosome (**FIGURE 20.14**). The viral ribonucleoprotein complex is released and translocated into the host nucleus. In the nucleus the viral RNA is transcribed into mRNA and copied into viral genomic RNA by the viral polymerase. The virus particles are assembled from the viral proteins and RNA genome on the host cell membrane. Progeny virus particles are released from the host cell by budding. During budding, the **neuraminidase** on the virus surface removes sialic acid from the surface of the host cell, which allows the virus to separate from the infected host cell.

FIGURE 20.14 Influenza virus life cycle. The hemagglutinin on the virus surface binds to the sialic acid associated with glycoprotein or glycolipid receptors on the host cell. The host cell endocytoses the virus and the acidic pH of the endosome results in the fusion of the virus envelope with the endosomal membrane. Viral RNA–protein complexes enter the nucleus and viral polymerases copy the RNA genome of the virus. Since influenza is a negative-stranded virus, the copied RNA molecules are positive-stranded RNA molecules that serve as mRNA, which is translated on host ribosomes in the cytoplasm. Some of the copied positive strands serve as templates (cRNA) to make negative-stranded RNA molecules which are the progeny viral genomes. The translated viral proteins and viral RNA (vRNA) genomes assemble into virus particles at the plasma membrane. Virus particles are released by a budding process, and the neuraminidase removes the sialic acid receptors on the host plasma membrane. This allows the virus to separate from the host cell and infect other cells.

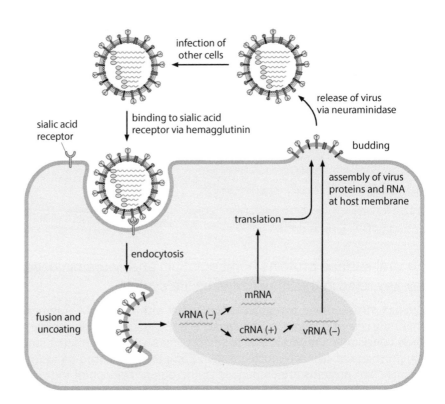

Uncomplicated influenza resolves in 1–2 weeks

Transmission of the influenza virus from person to person primarily involves respiratory droplets. The incubation period of influenza is typically 1–4 days. Adults can shed virus particles starting the day before the onset of symptoms through approximately 5 days after the onset of symptoms. Children and immunocompromised individuals might be infectious for a longer period. The initial symptoms of influenza are like other respiratory viral infections, and include runny nose, sore throat, fever, headache, muscle ache, malaise, and non-productive cough. However, influenza typically causes a higher fever that can last for up to 1 week. Weakness and fatigue, and sometimes an accompanying cough, can persist for an additional 1–2 weeks. Virus replication appears to be limited to epithelial cells of the respiratory tract and, therefore, severe disease manifestations are not likely due to systemic dissemination of the virus.

The symptoms that distinguish influenza from other respiratory viruses, such as high fever, headache, and muscle aches, are correlated with proinflammatory cytokines, primarily interleukin (IL)-6 and interferon-alpha (IFN-α) produced in response to virus replication and damage to epithelial cells (FIGURE 20.15). Some influenza strains are particularly pathogenic and are noted for inducing extremely high levels of cytokines called a **cytokine storm**. This cytokine storm can affect the airways or the alveoli and produce manifestations of bronchitis or pneumonia.

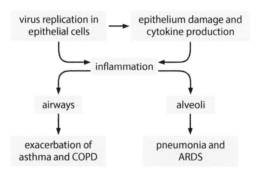

FIGURE 20.15 **Pathogenesis of influenza complications.** Replication of the influenza virus in epithelial cells causes inflammation associated with epithelium damage and cytokine production. Inflammation of the bronchi and bronchiole (airways) causes hypersecretion of mucus and constriction of the airways and leads to bronchitis-like symptoms. This bronchitis exacerbates other lung diseases such as asthma and chronic obstructive pulmonary disease (COPD). Inflammation of the air sacs causes pneumonia or acute respiratory distress syndrome (ARDS).

Pneumonia is a common complication associated with influenza

The risk of serious illness and death is highest among people over 65 years old, children less than 2 years old, and people who have medical conditions that place them at increased risk of developing complications from influenza. A common complication associated with influenza is pneumonia. This pneumonia can be primarily viral pneumonia due to the influenza virus reaching the alveoli, a combination of viral and bacterial pneumonia, or secondary bacterial pneumonia due to disruption of the epithelium. Primary pneumonia due to the influenza virus develops abruptly after the onset of symptoms and rapidly progresses to severe acute respiratory distress syndrome (ARDS). Hypotension and hypoxia (low levels of oxygen) are key features of ARDS and patients may exhibit cyanosis, which is characterized by a bluish pallor. Fatality rates associated with ARDS are as high as 50% and death can occur within 4 days without interventions such as ventilators.

Secondary bacterial pneumonia typically manifests after the patient has started to improve from the original influenza infection. *Streptococcus pneumoniae, Staphylococcus aureus,* and *Haemophilus influenzae* are the most common bacterial species associated with secondary bacterial pneumonia or combined viral–bacterial pneumonia. The manifestations of secondary bacterial pneumonia are similar to influenza viral pneumonia except for a slower onset and a lower case-fatality rate. Influenza infection can also exacerbate pre-existing chronic medical conditions, such as asthma (Chapter 17), chronic obstructive pulmonary disease (Chapter 18), and congestive heart failure (Chapter 14).

There are many serotypes of influenza virus A

Variants of influenza A can be distinguished by serotyping. This serotyping is based on antigenic variants of hemagglutinin and neuraminidase. Eighteen hemagglutinin antigenic variants and eleven neuraminidase antigenic variants have been described. These serotypes are designated as H#N# where H# refers to a hemagglutinin variant and N# refers to a neuraminidase variant. Among the 198 theoretical serotypes, 131 have been detected in nature. Most of these serotypes are found in aquatic birds, and aquatic birds represent the natural hosts for influenza virus A.

Serotyping is described in Chapter 9, pages 188–189.

SIGNPOST

In total, 14 serotypes representing 19 distinct genetic variants have been detected in humans. Five of these strains are efficiently transmitted human to human, 10 are predominantly zoonotic avian influenza viruses and 4 are predominantly zoonotic swine influenza viruses. Generally, zoonotic strains are not readily transmitted between humans and only cause mild disease in humans. One noted exception is H5N1, commonly called bird flu (BOX 20.4), which is highly pathogenic. The five influenza strains that are readily transmitted between humans likely emerged from animal strains and adapted to humans. Some of these influenza strains are also noted for major pandemics with substantial mortality. Therefore, there is always concern about the emergence and spread of a new highly pathogenic influenza virus from animal sources.

BOX 20.4

H5N1 Bird Flu

The H5N1 avian influenza virus primarily infects birds and is especially deadly for poultry. It is highly contagious and can quickly spread through a flock. Furthermore, farms often cull large numbers of birds to prevent virus spread, thereby adding to the economic impact of this virus. Several outbreaks of H5N1 have occurred – primarily in Asia – since its discovery in 1996. H5N1 not only is a concern of poultry farmers, but also can cause human infections during these outbreaks. Most human infections with H5N1 influenza are associated with prolonged and close contact with infected birds. H5N1 is rarely associated with human-to-human spread, and no community spread of this virus has ever been reported. Influenza H5N1 causes a severe disease in humans with a case-fatality rate that ranges from 14–33%. Not only is H5N1 capable of causing viral pneumonia and cytokine storms, but it can also disseminate beyond the respiratory tract and produce a systemic infection. Furthermore, the virus also impairs cytotoxic T-cell activity in vitro and therefore may suppress host immunity. There is some concern that mutations or reassortments could occur in H5N1 that would allow more efficient human-to-human transmission and possibly cause a deadly pandemic.

Influenza pandemics occur sporadically due to the introduction of a new strain of influenza A virus to humans

Four pandemics have occurred since the 20th century (TABLE 20.10). The most notable is the Spanish Flu pandemic that occurred near the end of World War I. This pandemic occurred in several waves and therefore the exact geographical origin is not known. The virus quickly spread around the world and possibly infected up to one-third of the world's population. Furthermore, the virus strain causing this pandemic had an exceptionally high case-fatality rate (CFR). Interestingly, the virus causing this pandemic has been reconstructed through genetic engineering and this virus strain causes an intense cytokine storm – more so than other influenza strains – in experimentally infected monkeys.

Table 20.10 Influenza Virus A Strains and Pandemics

Name (origin)	Pandemic years	Possible cases	CFR (%)	Sero-type	Circulation years	Origin
Spanish flu (?)	1918–1919	500 million	>2.5	H1N1	1918–1957	Avian-like influenza virus
Asian flu (China)	1957–1958	>500 million	0.67	H2N2	1957–1968	Reassortment between existing H1N1 and avian influenza virus
Hong Kong flu (China)	1968–1969	>500 million	0.2	H2N3	1968–present	Genetic shift from H2N2
Seasonal flu (?)	None	3–5 million per year	0.1	H1N1	1977–2009	Unknown
Swine flu or H1N1pdm (Mexico)	2009–2010	1.4 billion	0.02	H1N1	2009–present	Reassortment of human, swine, and bird viruses

The higher-than-normal levels of transmission and virulence associated with influenza pandemics are due largely to the introduction of an antigenically distinct virus. Therefore, there is little immunity against the new influenza strain in the human population. This lack of immunity in the population allows a large number of susceptible individuals to become infected, and those who are infected experience more severe disease. This sudden change in the antigenic properties of the virus is called **antigenic shift**. Influenza B is not capable of antigenic shift and thus is not associated with pandemics. As the pandemic progresses, more individuals become immune to the new virus and herd immunity develops. Herd immunity refers to the phenomenon in which disease transmission slows when a significant portion of the population becomes immune. As herd immunity develops, transmission decreases leading to the disease becoming endemic and seasonal.

Reassortment of viral genomes is associated with influenza pandemics

The segmented genome of the influenza virus may facilitate the generation of new virus types. If a cell is infected with two different virus types, the individual RNA molecules making the genomes could become shuffled, resulting in progeny that are hybrids of the two virus types. This reassortment of genomic RNA is analogous to the random sorting of chromosomes that occurs during meiosis and the production of gametes (Chapter 5). Progeny viruses with different genotypes and phenotypes are produced from this genome recombination. For example, a highly pathogenic avian virus that is not readily transmissible to humans could recombine with another virus that is capable of infecting humans to produce a new highly pathogenic human influenza virus (FIGURE 20.16).

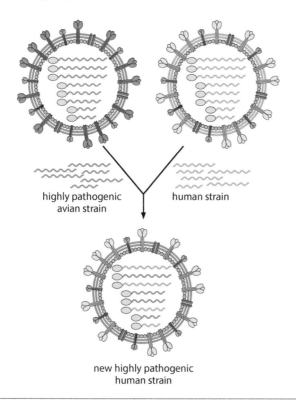

highly pathogenic
avian strain

human strain

new highly pathogenic
human strain

FIGURE 20.16 Generation of new influenza viruses by genomic reassortment. Infection of the same cell with two different types of influenza virus permits the production of progeny that are hybrids of the parental strains. This might result in the generation of highly pathogenic viruses that can infect humans.

Reassortments that have produced new human influenza viruses have involved avian viruses, swine viruses, or both. For example, the virus that was responsible for the 1957 pandemic was a recombination between the endemic H1N1 virus and a virus likely derived from geese (see Table 20.10). Similarly, the virus of the 2009 pandemic had genetic elements from human, swine, and avian viruses. Much attention has been paid to avian and swine viruses as a potential source of new human viruses since zoonotic transmission from either of these animals occurs with some frequency. In addition, the natural host of influenza viruses is wildfowl, and these viruses can potentially infect domestic fowl. Infections in domestic animals in a farm setting provide a scenario that increases the chances of co-infections with different virus types.

Following an influenza pandemic there is a tendency for the virus strain causing the pandemic to replace the previously circulating strain that existed before the pandemic (FIGURE 20.17). However, in the mid-1970s a new H1N1 influenza A emerged. It was first described in Russia and thus has sometimes been called the Russian flu. Subsequently, it was realized the virus had been circulating in China before appearing in Russia. The source of this virus was never positively identified and there was speculation that it was due to a release from a laboratory. However, that was never proven. Nonetheless, this H1N1 strain co-circulated with H3N2 until the 2009 pandemic, when it was replaced with another distinct H1N1 strain.

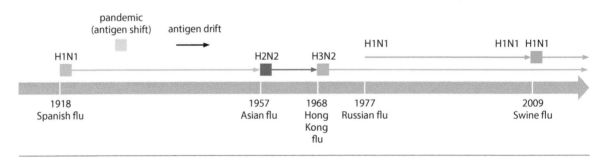

FIGURE 20.17 Human influenza strains. Timeline showing the appearance of new influenza strains associated with pandemics (boxes) followed by antigenic drift (arrows).

The influenza virus is continuously changing

Between pandemics, the circulating influenza strains undergo **antigenic drift**. Antigenic drift is a slow change in the antigenicity of the virus, in contrast with antigenic shift, which happens suddenly due to a recombination event. Mutations accumulate during the replication of the viral genome and some of these mutations affect the structure of epitopes recognized by the immune system (FIGURE 20.18). This means that a person that had previously been infected with the influenza virus may only be partially immune to a subsequent infection with a variant form of the virus. Subsequent infections are generally less severe due to partial immunity.

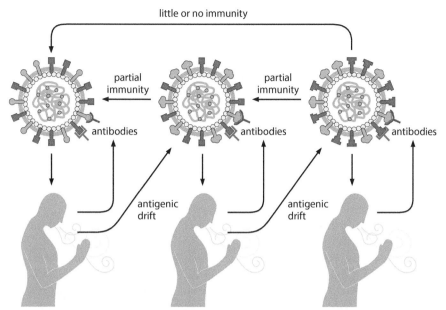

FIGURE 20.18 Antigenic drift. As viruses replicate, they mutate to produce forms with slightly different antigens. Therefore, as the virus spreads from person to person the repertoire of antibodies produced by individuals is slightly different. The continued spread of the virus within a population leads to more antigenic drift and now some individuals who had been previously infected may become susceptible to a new variant since their antibodies may no longer recognize that variant.

Antigenic drift is a phenomenon exhibited by most pathogens. By continuously changing epitopes, the pathogen can maintain itself in a host population. Without antigenic drift the percentage of susceptible individuals in the host population would decrease as the infection spread. At some point the only susceptible hosts without prior exposure would be infants and small children. For this reason, children under 2 years old are more susceptible to influenza infections, as well as most other infectious diseases. Antigenic drift also contributes to the variability of the influenza virus from season to season. During a typical flu season many variants of the virus are produced. However, only a few of those variants survive the off-season and emerge the next season. This might account for some of the variability between flu seasons as the virulence and transmissibility of the predominant circulating strain(s) determine the prevalence and amount of severe disease in the following flu season.

Vaccination against the influenza virus can minimize severe disease and reduce transmission

Antigenic drift complicates vaccination against influenza. In addition, four strains of influenza virus are currently circulating. Therefore, influenza vaccines are either trivalent, containing one of the influenza B types (Yamagata or Victoria) and both influenza A H1N1 and H2N3, or quadrivalent with both types of influenza B and both types of influenza A. Antigenic drift variants among these four circulating viruses are continuously monitored, and the vaccine is changed annually to incorporate the new variants that have been generated by antigenic drift. For that reason, it is recommended that everyone get vaccinated every year at the beginning of flu season.

Each year the WHO Global Influenza Surveillance Network chooses the exact strains of the four virus types that are likely to cause the most human suffering in the coming year. Vaccines are then manufactured using these strains. Both inactivated and live-attenuated vaccines are manufactured. Because the strains are chosen at the height of the previous flu season, the vaccines do not always match the virus variants that emerge the following season. Therefore, the efficacy of the vaccine varies from year to year. Nonetheless, the vaccine does reduce transmission and prevents some severe diseases and mortality.

The various types of vaccines are described in Chapter 9, pages 195–197.

SIGNPOST

Treatment of influenza includes antiviral drugs that slow virus replication

There are numerous rapid antigen detection tests and molecular assays to specifically diagnose influenza from nasal or throat swabs. These tests often distinguish influenza A and B. In addition, presumptive diagnosis based on symptoms is often used, especially at the height of the flu season and community transmission.

Antiviral drugs to treat influenza fall into three major classes based on the drug target: inhibitors of the M2 ion channel, inhibitors of neuraminidase, and inhibitors of a component of the virus RNA polymerase (TABLE 20.11). All these targets play necessary roles in the virus replication cycle. M2 inhibitors are only effective against influenza A and are no longer used due to nearly complete drug resistance. Neuraminidase inhibitors and polymerase inhibitors are effective against both influenza A and B. There are currently four FDA-approved drugs for the treatment of influenza. The drugs are most effective if treatment is initiated within 1 or 2 days of the appearance of symptoms. In addition, oseltamivir or zanamivir can also be used for chemoprophylaxis if taken within the first 48 hours after exposure.

Table 20.11 Anti-influenza Drugs			
Target	**Drug**	**Trade name**	**Approved**
M2 ion channel	Rimantadine	Flumadine®	No, drug resistance
	Amantadine	GOCOVRI®	No, drug resistance
Neuraminidase	Oseltamivir	Tamiflu®	Yes, US FDA*
	Zanamivir	Relenza®	Yes, US FDA
	Peramivir	Rapivab®	Yes, US FDA
	Laninamivir	Inavir®	Yes, Japan
RNA polymerase	Baloxavir marboxil	XOFLUZA®	Yes, US FDA
*United States Food and Drug Administration (US FDA).			

CORONAVIRUSES AND COVID-19

Seven different coronaviruses infect humans. Four of these viruses cause mild diseases and are considered common cold viruses, whereas three cause severe diseases with significant mortality (TABLE 20.12). All three of these highly virulent viruses emerged in the 21st century and, most recently, one of these caused a worldwide pandemic that greatly affected societies and economies during 2020 and beyond. Regarding emergence, all three viruses originated in bats and likely passed through another intermediate host before adapting to humans. Finally, all three can cause ARDS with case-fatality rates significantly higher than influenza.

Table 20.12 Coronavirus Diseases of Humans			
	SARS	**MERS**	**COVID-19**
Case-fatality rate (%)	9–10	37	2–3
Total cases	8100	2500↑	6.6 million↑
First report–(end)	11/2002–(2004)	6/2012–	12/2019–
Original host	Bats	Bats (distant)	Bats
Intermediate host	Civet	Camel (reservoir)	?
Receptor	ACE2	DPP4	ACE2

Spike proteins on the coronavirus surface give it a distinct appearance and play a key role in entering host cells

The name coronavirus refers to the spikes on the virus surface seen with the electron microscope that have the appearance of a crown or the Sun's corona (FIGURE 20.19). These spikes are viral proteins embedded in the virus envelope and, fittingly, are named spike proteins. The spike proteins bind to host cell receptors and initiate a fusion between the virus envelope membrane and the plasma membrane of the host. Host proteases cleave the spike protein, and this cleavage facilitates virus entry by causing a conformational change in the spike protein that activates the membrane fusion machinery. Fusion of the viral and host membranes deposits the nucleocapsid into the host cytoplasm and the genomic RNA is dissociated from the nucleocapsid proteins.

The viral genome is a positive-sense RNA and serves as the first mRNA of the infection. This mRNA is translated into a replicase complex (i.e., RNA polymerase) that then transcribes the viral genome into negative-strand intermediates (FIGURE 20.20). These negative-strand intermediates serve as templates for the synthesis of mRNA encoding the viral proteins and generation of the positive-stranded viral genome. The membrane-associated viral structural proteins are inserted into the endoplasmic reticulum (ER) membrane and serve as a scaffold for virion assembly. Nucleocapsid proteins bind to progeny genomic RNA and assemble into virions within the lumen of the ER. Progeny virions are transported to the plasma membrane in vesicles and released from host cells by exocytosis.

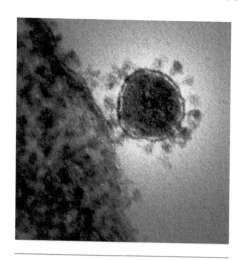

FIGURE 20.19 Coronavirus. An electron micrograph of coronavirus showing the spikes on the virus surface and its interaction with a host cell. [Courtesy of NIAID Rocky Mountain Laboratories (RML)].

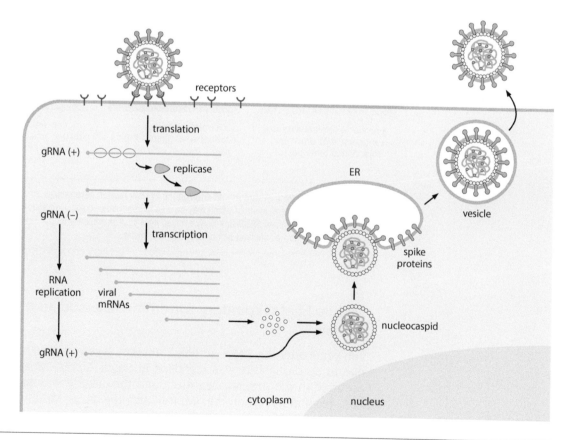

FIGURE 20.20 Coronavirus life cycle. Coronavirus binds to a host cell receptor and fuses with the plasma membrane. Upon entry into the host cell the positive-stranded genomic (g) RNA is immediately translated on host cell ribosomes to produce an RNA polymerase, called replicase. Replicase then transcribes the gRNA into a negative-stranded intermediate RNA that serves as a template to produce progeny gRNA and viral mRNA molecules, which are translated into viral proteins. Viral spike proteins are inserted into the endoplasmic reticulum (ER) membrane. Genomic RNA complexes with nucleocapsid proteins are assembled into virions within the ER. The newly formed virions are transported to the plasma membrane in vesicles and released by exocytosis.

The first newly emerging disease of the 21st century was due to a coronavirus

In November 2002, a severe influenza-like disease appeared in Guangdong Province, China and quickly started spreading, primarily in Asia. The increased severity and mortality associated with the disease suggested it was a newly emerged virus and the disease was named severe acute respiratory syndrome (SARS). The primary symptoms were high fever, headache, muscle ache, dry cough, and breathing difficulties. As the disease progresses there are decreasing levels of oxygen in the blood and patients require ventilators. In April 2003 the etiological agent of SARS was identified as a new human coronavirus that was named SARS-CoV (coronavirus). By July, the virus had disappeared from circulation. There were a few subsequent cases in late 2003 and early 2004 due to a probable accidental laboratory release. Containment of SARS was accomplished by isolation of cases and screening travelers for fever.

Molecular phylogenetics indicated the virus probably originated from horseshoe bats (*Rhinolophus affinis*). However, it is unlikely that the virus jumped directly from bats to humans, suggesting the involvement of an intermediate host. The initial focus was on animal markets since, at the beginning of the SARS epidemic, early index patients had animal exposure before developing the disease. Antibodies that recognize SARS-CoV were found in masked palm civets (*Paguma larvata*) and their animal handlers in the market, suggesting that civets were the intermediate hosts from which the human virus evolved (FIGURE 20.21).

SIGNPOST

Molecular phylogenetics is described in Chapter 5, pages 104–107.

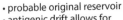
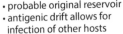
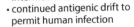

- probable original reservoir
- antigenic drift allows for infection of other hosts

- intermediate host
- continued antigenic drift to permit human infection

- human–animal contact
- further adaptation to humans
- sustained human-to-human transmission

FIGURE 20.21 Possible evolution and emergence of new human coronavirus. Antigenic drift produces a variety of progeny, some of which may be able to infect other hosts, leading to the possible creation of new viruses. Human SARS-CoV likely evolved from viruses originally found in bats that passed through civets as an intermediate host before adapting to humans.

Middle East respiratory syndrome is associated with camels

Ten years after the SARS outbreak another severe respiratory syndrome due to a coronavirus was identified in Saudi Arabia. Most cases have occurred in the Arabian Peninsula and the disease was named Middle East respiratory syndrome (MERS) and the virus MERS-CoV. Typical symptoms include fever, cough, diarrhea, and shortness of breath, and the disease is typically more severe in those with other health problems. MERS has a high case-fatality rate with more than one-third of infected individuals dying. A major risk factor for acquiring MERS is contact with camels and MERS is also known as camel flu. However, the exact factors associated with this camel-to-human transmission are not known. Person-to-person transmission is possible but is generally limited to health-care or family settings involving close contact with the infected person. No community transmission has been documented. Therefore, MERS is primarily a zoonotic infection with camels as a reservoir. Since 2013, typically 100–500 cases have been reported per year and there have been a few small outbreaks.

Molecular phylogenetic studies indicate that the isolates from humans and camels are greater than 99% identical. Furthermore, the isolates from humans and camels are contained within the same phylogenetic lineage (FIGURE 20.22) indicating multiple camel-to-human transmissions. In addition, neutralizing antibodies specific for human MERS-CoV are prevalent in camels, and antibodies against MERS-CoV have been detected in camel serum samples dating back to 1983. The combined data suggest that a MERS-CoV has been circulating in camels for at least several decades. RNA-sequence data suggest that the virus probably originated in bats before becoming endemic in camels.

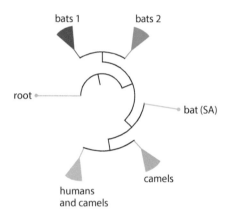

FIGURE 20.22 Molecular phylogenetics of MERS-CoV. A phylogenetic tree showing relationships between major lineages of viruses related to MERS-CoV. Human isolates of MERS-CoV are found within one of the two lineages of camel isolates. The lack of a lineage with only human isolates indicates a zoonotic transmission from camels to humans. The two camel lineages are derived from bats and there is a single bat isolate from South Africa (SA) that is the closest relative of the two camel lineages. [From Cui J, Li F & Shi Z-L (2019) Origin and evolution of pathogenic coronaviruses. *Nature Reviews Microbiology* 17: 181–192. doi: 10.1038/s41579-018-0118-9. With permission from Springer Nature.]

Another coronavirus emerged in 2019 and caused a worldwide pandemic

Late in 2019 another coronavirus causing a SARS-like disease was identified in Wuhan, China. The disease was named coronavirus disease 2019 (COVID-19) and the virus was named SARS-CoV-2. Unlike SARS and MERS, which are somewhat limited in prevalence, COVID-19 quickly spread around the world and, by June 2020, there were nearly seven million confirmed cases and over 400,000 deaths, and the disease is still ongoing. To control the spread most countries shut down their economies and had all but essential workers stay home or work from home. In addition, 'social distancing' was widely practiced in which people avoided large crowds, tried to stay 2 meters apart, and wore face coverings.

Like the first SARS-CoV, SARS-CoV-2 probably originated in bats and passed through an intermediate host before adapting to humans. However, the possible intermediate host has not yet been identified. There was focus on a market in Wuhan that sold live animals as the place of origin, since molecular epidemiological data indicated that the market was the epicenter of the pandemic. Many of the initial infections were among workers in the market or in people living close to the market. Furthermore, the market could have served as a super-spreader event in which one person potentially infected an atypically large number of other individuals. Analysis of the spike protein and focus on the receptor-binding domain demonstrated that SARS-CoV-2 and SARS-CoV evolved different strategies for binding to the same receptor (BOX 20.5).

BOX 20.5

Spike Protein Evolution in SARS-CoV-2

The spike protein on the surface of coronaviruses recognizes host cell receptors and plays a key role in virus entry into the host cell. Both SARS-CoV and SARS-CoV-2 utilize angiotensin-converting enzyme 2 (ACE2) as the host receptor. ACE2 converts angiotensin II into angiotensin and functions within the renin–angiotensin system (see Box 14.2). ACE2 is found on the surface of alveolar cells, epithelial cells of the small intestine, and endothelial cells in most organs. Even though the spike proteins from both SARS-CoV and SARS-CoV-2 bind to ACE2 with high affinity, only five out of the six key residues in the receptor-binding domain are the same between the two proteins (**BOX 20.5 FIGURE 1**). This indicates that the two viruses solved the same problem by different mechanisms and that the two viruses do not share a recent common ancestor. The bat virus that is most closely related to SARS-CoV-2 also only shares one of the six crucial binding site residues, suggesting the need for an intermediate host.

Another possibly significant change in the spike protein of SARS-CoV-2 is the insertion of a furin cleavage site not found in other coronaviruses. Furin is a host protease that cleaves proteins at polybasic sites and the insertion adds two arginine residues near another arginine residue (see Box 20.5 Figure 1). This insertion allows effective cleavage by furin and has a role in viral infectivity and host range. Cleavage of the spike protein into the S1 and S2 subunits facilitates viral entry into the host cell. Specifically, the cleavage of the spike protein activates the fusion complex and virus entry. Interestingly, hemagglutinin of the highly pathogenic influenza virus H5N1 (see Box 20.4) also has a polybasic furin cleavage site that facilitates virus entry.

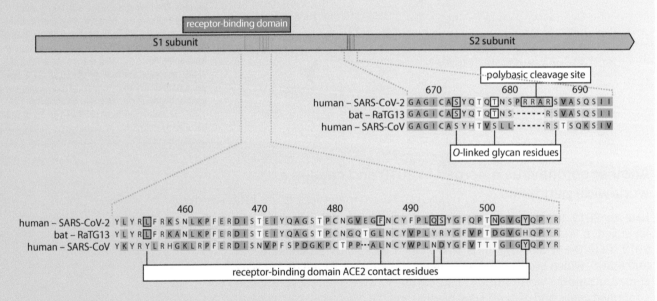

BOX 20.5 FIGURE 1 Receptor-binding domains and polybasic cleavage site of the spike protein. Alignment of the sequences of the spike protein from SARS-CoV-2, the bat virus with the closest sequence homology to SARS-CoV-2, and SARS-CoV reveals differences in the receptor-binding domain and an insertion of a polybasic site. Of the six residues known to be important in the interaction between spike protein and the ACE2 receptor, only one residue is identical (boxed Y) between the two human SARS viruses. Similarly, only one residue is identical between the bat virus and SARS-CoV-2 (boxed L). Insertion of three amino acids in the SARS-CoV-2 sequence creates a polybasic (furin) cleavage site. [Adapted from Andersen KG, Rambaut A, Lipkin WI, et al. (2020) The proximal origin of SARS-CoV-2. *Nature Medicine* 26: 450–452. doi: 10.1038/s41591-020-0820-9. With permission from Springer Nature.]

Comorbidities play a large role in the pathogenesis of COVID-19

Transmission of SARS-CoV-2 is primarily via respiratory droplets. However, air-borne transmission may also be possible. Approximately 80% of those infected exhibit mild symptoms that are similar to other respiratory tract infections and those individuals recover in a few days. Common symptoms include fever, cough, fatigue, shortness of breath, muscle pain, and loss of smell and taste. Complications associated with severe disease may include pneumonia, ARDS, multi-organ failure, and septic shock. Like severe influenza, severe COVID-19 is also associated with systemic hyperinflammation due to a cytokine storm. Most individuals who develop severe disease or die from COVID-19 have pre-existing conditions, such as COPD (Chapter 18), hypertension (Chapter 14), diabetes (Chapter 13), and cardiovascular disease (Chapter 14).

SUMMARY OF KEY CONCEPTS

- Numerous pathogens infect the respiratory tract and several of these pathogens cause common human diseases with substantial mortality

- Transmission is primarily via inhalation of respiratory droplets or droplet nuclei, as well as via contaminated hands or fomites

- Pathology associated with respiratory tract infections can range from superficial damage of the epithelium to intense inflammation and tissue damage

- Any portion of the respiratory tract can be affected with lower respiratory tract infections being more severe than upper respiratory tract infections

- Pneumonia and acute respiratory distress syndrome (ARDS) are the most severe manifestations of respiratory tract infections

- The more virulent bacterial diseases of the lungs are bacterial pneumonia and tuberculosis

- Pulmonary tuberculosis is a chronic infection of the lungs with high mortality if not treated

- Drug-resistant tuberculosis limits options for the management and control of the disease

- Among the numerous viral diseases of the respiratory tract, influenza and COVID-19 are the most morbid in addition to being relatively common

- Recombinations between influenza virus strains have produced notable pandemics

- Coronaviruses that originated in bats are agents of several severe human diseases known as SARS (severe acute respiratory syndrome), MERS (Middle East respiratory syndrome), and COVID-19

- MERS is primarily a zoonotic disease from camels that has a high mortality

- COVID-19 can manifest as SARS with substantial mortality and caused a major pandemic in 2020

KEY TERMS

acute respiratory distress syndrome (ARDS)	A life-threatening condition in which fluid builds up in the lungs leading to oxygen deprivation and often requiring mechanical ventilation.
alveoli	The small elastic air sacs that make up the lungs and where gas exchange occurs (singular = alveolus).
antigenic drift	Changes in the antigenic properties of a pathogen due to the constant and steady changes in its genome sequence that may affect antibody reactivity.
antigenic shift	A rapid and sudden change in antigenic or other properties of a pathogen due to genetic recombination between different strains of a pathogen.
coronavirus	A family of positive-strand RNA viruses responsible for severe respiratory diseases such as SARS, MERS, and COVID-19.
COVID-19	A respiratory disease caused by a coronavirus that emerged in 2019 and was named coronavirus disease 2019.

droplet nuclei	Particles less than 5 μm in diameter derived from respiratory droplets that facilitate air-borne transmission of infections.
granuloma	A tumor-like mass or nodule due to chronic inflammation that is associated with an infectious disease such as tuberculosis.
hemagglutinin	A protein found on the surface of the influenza virus that binds the sialic acid receptors on the host cell. It is also one of the proteins used in the serotyping of the influenza virus.
influenza	A potentially severe respiratory disease caused by the influenza virus and noted for several pandemics.
latent tuberculosis infections (LTBI)	An asymptomatic stage of tuberculosis in which the immune system contains the infection within granulomas and is not contagious.
multidrug-resistant tuberculosis (MDR-TB)	Tuberculosis caused by isolates of *Mycobacterium tuberculosis* that are resistant to at least isoniazid and rifampicin and may also be resistant to other drugs.
neuraminidase	A protein found on the surface of the influenza virus that removes sialic acid from the host cell and facilitates the release of virus particles. It is also one of the proteins used in the serotyping of the influenza virus.
Mycobacterium	The genus of bacteria that causes tuberculosis and has a waxy capsule composed of mycolic acid as a characteristic feature.
pneumonia	Inflammation of the alveoli characterized by the accumulation of fluid in the lungs.
tuberculosis (active) (TB)	A potentially mortal infection primarily of the lungs caused by *Mycobacterium tuberculosis* and exhibits an air-borne transmission.

REVIEW QUESTIONS

1. Describe the clinical manifestations associated with upper and lower respiratory tract infections and discuss why lower respiratory tract infections are more severe than upper respiratory tract infections.

2. What is the most common cause of bacterial pneumonia and describe the virulence factors associated with this pathogen?

3. Discuss the factors associated with the development of secondary active tuberculosis and its clinical manifestations.

4. Discuss the antigenic drift and antigenic shift of influenza virus in regard to pandemics and the use of vaccines.

5. What are three severe human diseases caused by coronaviruses? Discuss the possible emergence of each in regard to the relationship between human and animal viruses.

ADDITIONAL QUESTIONS

In addition to the Review Questions provided above, there is a range of free online questions designed to enable students to further test their understanding of the chapter material. In order to access these interactive questions, please visit the book's website: https://routledge.com/cw/wiser

Diarrheal Diseases

21

There are billions of cases of diarrheal diseases worldwide per year. Resource-poor countries and children are particularly affected and diarrhea is a major cause of death in children under 5 years old. Furthermore, diarrhea exacerbates malnutrition, which is a major problem in children of resource-poor countries. This exacerbation is due largely to the decreased absorption of nutrients associated with diarrhea. Decreased absorption of nutrients combined with malnutrition lowers immunity resulting in more frequent and more severe gastrointestinal (GI) infections, which then further lowers immunity. This creates a spiraling decline that can result in death (FIGURE 21.1).

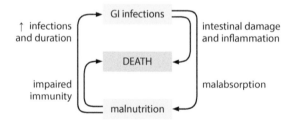

FIGURE 21.1 Vicious cycle of diarrhea and malnutrition. The intestinal damage and inflammation caused by gastrointestinal infections lead to decreased nutrient absorption and contribute to malnutrition, which impairs immunity. A lower immunity results in more gastrointestinal infections that last longer and are more severe, thus further exacerbating malnutrition and decreasing immunity. A high rate of childhood mortality is associated with a combination of malnutrition and gastrointestinal infections.

Most diarrhea is associated with GI infections. Non-infectious diarrhea can be caused by lactose intolerance, irritable bowel syndrome and related diseases, as well as some medications and toxins. Pathogens causing diarrhea are generally ingested with contaminated food or water and adapted to inhabiting some part of the GI tract. The pathology and inflammation associated with the infections manifest as diarrhea and other gastrointestinal symptoms. Numerous pathogens, ranging from viruses to multicellular parasites, have adapted to life in the GI tract.

TRANSMISSION

Acquisition of GI infections generally involves ingestion of the pathogen through contaminated food or water. The ingested pathogen then establishes an infection within the GI tract. Most pathogens have a tropism for a specific region of the GI tract, such as the stomach, small intestine, or large intestine. However, a few pathogens can inhabit multiple parts of the GI tract. The infection of the intestinal tract involves pathogen replication and the pathogen is excreted in the feces (FIGURE 21.2). Excreted pathogens then complete the cycle through subsequent ingestion by another host via the contamination of water, food, or hands.

FIGURE 21.2 Transmission of gastrointestinal infections. Gastrointestinal (GI) infections are most often acquired through the ingestion of pathogens that have contaminated water, food, or hands. The pathogen usually establishes a colony in the GI tract and replicates. A common manifestation of GI infections is diarrhea with pathogens excreted in the feces. These excreted pathogens can then be transmitted to other individuals. Some pathogens in the GI tract can cross the intestinal epithelium, spread to other parts of the body, and possibly cause enteric fevers. In some cases, disease pathology may be due to toxins associated with the pathogen.

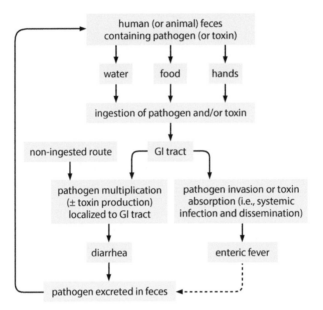

Personal hygiene plays a major role in the transmission of GI infections

The cycle of infectious pathogens being excreted in the feces and then establishing new infections after ingestion is called **fecal–oral transmission**. This does not mean that infected persons actively eat feces, but rather that fecal matter has contaminated something that is ingested. Several factors can play a role in fecal–oral transmission (TABLE 21.1). Among those, personal hygiene is probably the most important. For example, daycare centers are commonly the sites of outbreaks of GI infections due to fecal accidents. In addition, people living in close contact, such as prisons or hospitals, facilitate transmission. Another potential source of infection is food handlers who fail to wash their hands after using the restroom and then subsequently contaminate food. Another aspect of food-borne transmission is the growth of bacteria in food that has been left at room temperature for several hours.

Table 21.1	Fecal–Oral Transmission Factors
Factor	**Examples**
Poor personal hygiene	• Children (e.g., daycare centers) • Institutional living (e.g., prisons, hospitals) • Food handlers
Developing countries	• Poor infrastructure and sanitation (e.g., lack of plumbing) • Higher levels of endemicity • Travelers' diarrhea
Water-borne	• Contact with surface water (recreational or occupational) • Water treatment failures

The higher prevalence of diarrheal diseases in resource-poor areas is sometimes an infrastructure issue such as access to indoor plumbing, proper waste handling, and sufficient water treatment. Furthermore, a high prevalence of GI pathogens in a population increases the likelihood of transmission occurring. However, in such environments there is a higher immunity in the population against these pathogens and disease symptoms tend to be mild. Therefore, children, since they have not yet had multiple exposures, tend to be heavily affected. In addition, people without extensive prior exposure who travel to such areas often develop diarrheal diseases. This is often referred to as travelers' diarrhea.

Water-borne transmission is prone to outbreaks

Many intestinal pathogens can also be spread through contact with natural bodies of water via recreation or occupation. Water-borne transmission can be due to contamination of water with fecal matter containing intestinal pathogens or associated with a natural part of the pathogen's life cycle. In the case of the latter, some pathogens may have a life-cycle stage or an intermediate host that is normally associated with water. In addition, water-borne outbreaks of diarrheal disease are often associated with natural disasters such as civil unrest, earthquakes, or hurricanes. Often there is a breakdown in water treatment and the availability of clean water, and displaced persons are often forced into crowded refugee conditions that further facilitate fecal–oral transmission. Even in normal times, as well as in developed countries, there can be failures in water treatment that lead to outbreaks.

Prevention of diarrheal disease focuses on personal hygiene

Frequent handwashing is a major preventive measure to limit the spread of intestinal pathogens. This is especially true in communal living situations and for food preparers. Food preparation should be carried out under highly sanitary conditions. Furthermore, food should be properly stored to minimize bacterial growth once it is prepared. Any water that is questionable should be boiled. Treatment of asymptomatic carriers also minimizes transmission, especially to family members. On the societal level, special attention should focus on waste handling and water treatment to prevent outbreaks.

PATHOPHYSIOLOGY

The GI tract essentially consists of a long tube running from the mouth to the anus. This includes the esophagus, stomach, small intestine, large intestine, and rectum. The small intestine is divided into three sections: the duodenum (the first or proximal portion after the stomach); the jejunum (the middle portion); and the ileum (the distal or last portion before the large intestine). The symptoms associated with GI infections depend in part on the region of the intestinal tract that is infected (TABLE 21.2). For example, inflammation of the stomach is called gastritis, inflammation of small intestine is called **enteritis**, and inflammation of the large intestine is called **colitis**.

Table 21.2	Symptomatology of GI Infections
Gastroenteritis	Inflammation of the stomach and small intestine. Symptoms include nausea, vomiting, diarrhea, and abdominal discomfort
Colitis	Inflammation of the colon. Symptoms can include pain, abdominal cramps, and dysentery
Enterocolitis	Inflammation of the small and large intestines with a combination of gastroenteritis and colitis symptoms
Diarrhea	Frequent and/or watery stools usually resulting from disease of the small intestines
Dysentery	Blood and mucus in the feces and often associated with colitis. Generally, less voluminous than diarrhea

Diarrhea is the most common manifestation of GI intestinal infections

Inflammation of the GI tract typically results in diarrhea. Diarrhea is frequently defined as three or more watery or loose stools per day. Mild or transient diarrhea is generally not a problem. But the excretion of large volumes of water can result in dehydration and, if severe enough, cause shock and death. In addition, not unexpectedly, bloody diarrhea is a more serious condition. Generally, diarrhea is associated with inflammation of the stomach and small intestine. Inflammation of the large intestine is associated with **dysentery**. Dysentery is less voluminous than diarrhea and contains blood and mucus. Diarrhea is a relatively simple concept, but there are multiple mechanisms by which pathogens cause diarrhea and diarrhea can be classified into distinct types (TABLE 21.3). These types of diarrheas are not mutually exclusive and various mechanisms can simultaneously contribute to diarrhea.

Table 21.3	Types of Diarrheas
Osmotic	• Impaired absorption and enhanced secretion due to enterocyte malfunction • Excessive solutes in the intestinal lumen
Inflammatory	• Mucosal invasion or necrosis • Blood or leukocytes in stools
Secretory	• Enterotoxin mediated • Excessively watery diarrhea

Osmotic diarrhea is due to superficial damage to the intestinal epithelium

The intestinal tract consists of four basic layers: the mucosa, the submucosa, the muscle layer, and the serosa (see Figure 8.4). The mucosa is the innermost layer and consists of epithelial cells, sometimes called enterocytes, and a layer of connective tissue called the lamina propria. The epithelial cells are responsible for forming a barrier between the contents of the intestinal lumen and the underlying layers. In addition, the epithelial cells are also responsible for the uptake of nutrients and electrolytes via various membrane transporters located on the plasma membranes of epithelial cells. Thus, the mucosa must be able to efficiently take up nutrients and at the same time exclude toxins and microorganisms.

Damage to the intestinal epithelium results in villus blunting due to death or damage of the enterocytes at the distal ends of the villi (FIGURE 21.3). Most of the absorption is carried out by the enterocytes at the tips of the villi resulting in decreased absorption of sodium and other solutes. Coupled with the absorption of solutes is a cotransport of water molecules. In addition, when enterocytes are damaged or killed, the crypt cells – functioning as stem cells – replicate at higher rates to replace the damaged epithelium. Crypt cells normally secrete large volumes of water associated with the excretion of chloride ions and an increase in the crypt cells increases this excretion. This excreted water is normally reabsorbed in conjunction with the uptake of sodium ions and other solutes by mature enterocytes at the tips of the villi. However, since these enterocytes are damaged, there is now an impaired absorption of water and electrolytes and an increased secretion of water and electrolytes resulting in watery diarrhea. This is referred to as **osmotic diarrhea**. Osmotic diarrhea can also be caused by lactose intolerance and other conditions of excess solutes in the intestines (BOX 21.1).

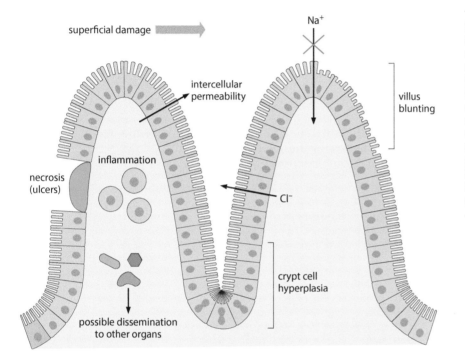

FIGURE 21.3 Intestinal pathogenesis. Superficial damage of intestinal epithelial cells can cause villus blunting and a loss of sodium absorption. This damage results in crypt cell hyperplasia and increased chloride secretion. Both the decreased absorption and increased secretion result in enhanced excretion of water and causes osmotic diarrhea. Disruption of the epithelial cell layer due to necrosis leads to submucosal inflammation and increased intercellular permeability as well as inflammatory diarrhea. Pathogens that have invaded the submucosa may be able to disseminate to other organs via the circulatory or lymphatic systems.

Lactose Intolerance

Osmotic diarrhea can result from an excessive concentration of solutes in the intestinal lumen or a dysfunction in the water and electrolyte transport in intestinal epithelial cells. High solute concentration in the intestinal lumen means that water flows from the epithelial cells of the small intestines into the intestinal lumen. Ingestion of poorly absorbed substrates can cause osmotic diarrhea by increasing the osmolality of the intestinal lumen. Some carbohydrates and some divalent cations are not absorbed by the intestines and, consequently, ingestion of large quantities of such solutes can result in osmotic diarrhea. The increased osmolarity of the intestinal lumen causes water to be secreted by the enterocytes and this water accumulates in the intestinal lumen.

Conditions leading to the accumulation of normally absorbed substrates can also cause osmotic diarrhea. One common example is lactose intolerance, which is caused by a deficiency in the enzyme lactase on the surface of intestinal epithelial cells. This lactase deficiency can be due to an infection, such as giardiasis, or a trait of the individual person. Lactase cleaves lactose, the principal sugar found in milk, into glucose and galactose, both of which are taken up by epithelial cells. Lactose, however, is not taken up by epithelial cells. Therefore, ingestion of dairy products by a lactase-deficient person leads to high levels of lactose in the intestinal lumen. This is called lactose intolerance and results in decreased absorption of water and consequently osmotic diarrhea.

BOX 21.1

Disruption of the epithelial cell barrier by pathogens can lead to ulcers and inflammatory diarrhea

Some pathogens cause extensive damage to the intestinal epithelium resulting in ulcers (see Figure 21.3). This destruction of the epithelial layer allows for pathogen invasion of the submucosal layer. The submucosa is made up of connective tissue, blood vessels, nerves, and immune effector cells. Pathogens in the submucosa induce an inflammatory immune response and the resulting inflammatory cytokines increase the permeability between the intestinal epithelial cells. This disruption of the epithelium can also lead to the exudation of serum and blood, as well as leukocytes, into the intestinal lumen. Thus, **inflammatory diarrhea** is often associated with leukocytes in the feces, bloody diarrhea, or dysentery.

There are several mechanisms by which a pathogen can disrupt the intestinal epithelium. For example, some pathogens invade enterocytes, and this invasion leads to cell death (**FIGURE 21.4**). The pathogen may then invade neighboring cells and, if this continued invasion is extensive enough, it will result in a large area of the intestinal epithelium being disrupted. Some bacteria, such as *Shigella* and enterohemorrhagic *Escherichia coli*, secrete cytotoxins that kill the enterocytes. This cytotoxic activity can also result in a loss of a large area of the intestinal mucosa. The protozoan pathogen *Entamoeba histolytica* kills enterocytes in a contact-dependent manner and disrupts the intestinal epithelium of the colon. Disruption of the intestinal epithelium allows for a subsequent invasion of the submucosa.

FIGURE 21.4 Pathogen-mediated necrosis of intestinal epithelium. Some pathogens can invade enterocytes and kill host epithelial cells. The extensive spread of the pathogen in the epithelium results in large areas of necrosis. Other pathogens directly kill enterocytes via direct contact or secretion of cytotoxins.

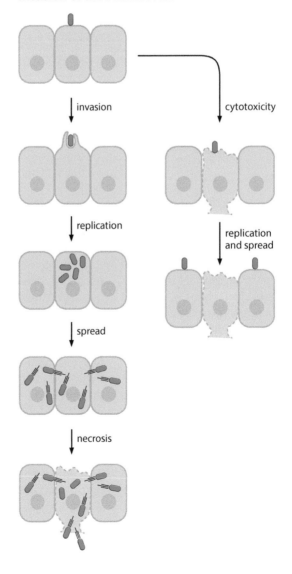

Dissemination of pathogens from the gastrointestinal tract results in a systemic disease

Most GI pathogens remain localized to the GI tract. However, following the invasion of the submucosa, some pathogens can possibly enter the circulatory system or lymphatic system and travel to other parts of the body (see Figure 21.3). To be able to disseminate, pathogens need mechanisms to evade the host's immune response. Some pathogens, such as some strains of *Salmonella*, can spread throughout the body and cause a generalized systemic infection. Whereas other pathogens, like the hepatitis A virus, may have a tropism for a specific organ or tissue. Movement of the pathogen out of the GI tract may reduce its transmissibility, unless some of the pathogens remain in the GI tract or the pathogen has the means to return to the GI tract.

Infections in the submucosa can also expand and penetrate the muscle layer and serosa. Penetration of the muscle layer and serosa can perforate the intestinal wall and connect the intestinal lumen to the peritoneal cavity. Bacteria and fecal matter can now seep into the peritoneal cavity and cause inflammation of the peritoneum. The peritoneum is the lining of the abdominal cavity and surrounds the internal organs. Inflammation of the peritoneum is called peritonitis and is characterized by abdominal pain and fever. If not treated, peritonitis has a high mortality rate. Peritonitis due to intestinal infections is rather rare, though.

VIRAL PATHOGENS

Several viruses infect the GI tract and cause gastroenteritis. These viruses exhibit a fecal–oral life cycle and the infectious virus is present in the feces. Specific diagnoses are difficult and viral gastroenteritis is generally diagnosed based on symptomology. Likewise, there are no specific antiviral treatments and the treatment is oral rehydration to reduce morbidity and mortality. Rotavirus and norovirus are common causes of gastroenteritis (FIGURE 21.5). Other viruses, such as poliovirus and hepatitis virus A, capitalize on the efficiency of fecal–oral transmission but cause systemic disease.

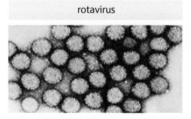

rotavirus

– extremely common in children
– significant mortality
– vaccine available

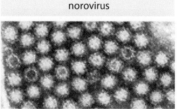

norovirus

– commonly associated with outbreaks
– seasonal transmission (winter vomiting bug)

FIGURE 21.5 Rotavirus and norovirus. The two most common causes of viral gastroenteritis are rotavirus and norovirus. [Courtesy of CDC Charles D. Humphrey (norovirus) and CDC/Erskine Palmer (rotavirus).]

Rotavirus and norovirus are highly transmissible and common viruses

Rotavirus is the most common cause of gastroenteritis in infants and children, with greater than 90% of children experiencing at least one rotavirus infection before the age of four and nearly all by the age of five. More than 10 trillion virus particles can be found per gram of feces resulting in a high degree of infectiousness, especially when considering that 100 virus particles may be sufficient to establish an infection. Subsequent infections elicit immunity and therefore adults rarely exhibit symptomatic disease. However, initial infections with rotavirus can produce severe watery osmotic diarrhea in children and there is a risk of mortality. Most rotavirus-associated mortality is in resource-poor countries. A vaccine is now available, and the incidence and severity of rotavirus infections have declined significantly in countries that have incorporated the rotavirus vaccine into their routine childhood immunization programs.

Norovirus is another common cause of viral gastroenteritis. It is particularly noted for outbreaks on cruise ships and in schools. It is found equally in resource-rich and resource-poor countries and, like most viruses, children are more severely affected. It is more common in the winter months and, thus, often referred to as the winter vomiting bug. Norovirus is an extremely contagious and infectious virus found in both vomit and feces. Flushing toilets can aerosolize the virus and allow for air-borne transmission (Chapter 20). This potential air-borne transmission combined with an infectious dose as low as 20 virus particles can quickly lead to outbreaks among people in enclosed spaces or in close contact.

Poliovirus can spread from the GI tract to the CNS and cause a paralytic disease

There are a few viruses that utilize the efficiency of fecal–oral transmission to pass from person to person, but cause disease in other organs. Notable among these are polio and hepatitis A. Upon ingestion, poliovirus initially infects gastrointestinal cells. Infection of the GI tract generally exhibits mild gastroenteritis or no symptoms. The virus can also infect the tonsils and other lymphatic tissues of the GI tract and continue to spread through the lymphatic system. The virus can also enter the blood, and this is called the viremic phase (TABLE 21.4). Patients with viremia may exhibit mild symptoms such as sore throat and fever. In 1–5% of infections the virus spreads to the central nervous system (CNS) and causes meningitis, which is an acute inflammation of the protective membranes covering the brain and spinal cord. Common symptoms of meningitis are fever, headache, and neck stiffness. Even less frequently the virus can spread to specific neurons and cause paralytic disease. Polio is now largely controlled with vaccination and is a rather rare infection.

Table 21.4 Phases of Poliovirus Infection			
Alimentary phase	**Lymphatic phase**	**Viremic phase**	**Neurological phase**
• Acquired by ingestion of virus • Virus replicates in the epithelium of throat and small intestine • No symptoms or mild gastroenteritis • Virus passed in feces	• Virus infects mucosa-associated lymphatic tissue (MALT) • Spreads to lymph nodes	• Virus enters blood • Mild fever or asymptomatic	• Virus spreads to the central nervous system • Meningitis • Can cause a paralytic disease (rare)

The liver is the primary site of replication for the hepatitis A virus

Viral hepatitis is caused by one of five distinct viruses (designated A–E). Among those, hepatitis A virus (HAV) is the most common and is transmitted by the fecal–oral route. Following ingestion, HAV enters the bloodstream and infects liver cells. Newly formed virus particles are returned to the intestines via the bile duct. Children are most often asymptomatic and among those with symptoms it is generally restricted to mild jaundice. A single exposure to HAV leads to a long-term, and perhaps a lifetime, immunity against re-infection. Acquisition of the infection later in life is generally associated with symptoms that can be severe, especially after the age of 50 or in patients with pre-existing liver disease such as alcoholic cirrhosis.

Alcoholic cirrhosis is described in Chapter 18, page 350.   SIGNPOST

INTESTINAL BACTERIA

Numerous bacterial species inhabit the GI tract and are part of the intestinal microbiota (see Box 6.2). Most species making up the microbiota do not cause disease and these organisms are often referred to as commensals since they do not harm the host. In fact, many of these commensal organisms are beneficial and have a mutualistic relationship with the human host. For example, commensal bacteria can protect from more pathogenic organisms by competing for resources and limiting the growth of harmful pathogens, or excreting substances that are harmful to pathogenic microorganisms. In other words, commensals function as a biological barrier against pathogenic microbes. The commensals also help train the immune system, leading to oral tolerance. In addition, some commensals play a nutritional role by secreting enzymes to digest complex carbohydrates, which we lack, or by synthesizing vitamins that are needed for proper nutrition.

Oral tolerance is described in Chapter 17, pages 331–333.   SIGNPOST

Although most bacteria are not harmful, there are many bacteria found in the GI tract that do cause human disease with disease manifestations ranging from mild to severe. There are numerous bacterial species or subtypes that are clearly pathogens and routinely cause human disease (TABLE 21.5). In addition, some normally non-pathogenic bacteria may cause inflammation and disease manifestations if the intestinal epithelium is disrupted.

Table 21.5   Examples of Pathogenic Bacteria in the Gastrointestinal Tract

Pathogen	Manifestation	Comments
Campylobacter	Gastroenteritis	One of the most common causes of diarrhea and gastroenteritis. Most often associated with poultry
Salmonella enterica	Gastroenteritis	Another common cause of food-borne illness with poultry and eggs most often implicated
Salmonella enterica Typhi	Typhoid (enteric fever)	A strictly human subtype that can survive in macrophages and cause a systemic disease called typhoid
Shigella dysenteriae	Bacillary dysentery	Secretes a potent cytotoxin called Shiga toxin that disrupts the intestinal epithelium
Escherichia coli 0157:H7	Hemorrhagic colitis	An enterohemorrhagic E. coli (EHEC) associated with undercooked ground beef (Box 21.2)
Enteropathogenic E. coli (EPEC)	Gastroenteritis	E. coli strains possessing an adhesin + receptor that mediates adherence and damage to enterocytes
Enterotoxigenic E. coli (ETEC)	Secretory diarrhea	E. coli strains with enterotoxins that are similar to cholera toxin and associated with a high level of childhood mortality
Vibrio cholera	Cholera	Causes severe toxin-mediated secretory diarrhea that is sometimes associated with water-borne transmission
Clostridioides difficile	Antibiotic-associated diarrhea	A common cause of nosocomial infections and often associated with the use of antibiotics
Helicobacter pylori	Stomach ulcers and cancer	Adaptations of the bacteria to survive in the acidic stomach can cause chronic gastritis

Bacterial gastroenteritis is often associated with food poisoning

Campylobacter species, *Salmonella enterica*, *Shigella* species, and some *Escherichia coli* strains are common causes of bacterial gastroenteritis. These are often acquired from ingesting food that has been contaminated. This can be part of a fecal–oral transmission cycle, or quite often is associated with the processing and preparation of food. Some of these bacteria are common in animals and cannot be completely removed during meat processing. For example, more than half of the gastroenteritis associated with *Campylobacter* species is associated with poultry. Similarly, *Salmonella* species is also often associated with poultry. Gastroenteritis associated with contaminated food is often called food poisoning. Food poisoning can also include food contaminated with abiotic toxins. However, the majority of food poisoning is due to bacteria.

Cooking food kills bacteria and other organisms and, for this reason, it is recommended to cook meat thoroughly. Since many of these bacteria are rather common, contamination after cooking is still possible. Therefore, humans are constantly exposed and can tolerate low amounts of bacterial contamination without developing gastroenteritis. A potential problem, though, is that in a nutrient-rich environment and warm temperatures bacteria rapidly replicate and can exponentially increase their numbers. Therefore, most food should not be left at room temperature for extended periods of time. Even with refrigeration, bacteria can still replicate, albeit at a much slower rate, and food does eventually go bad.

Bacterial virulence factors are associated with disease pathogenesis

One common feature of pathogenic intestinal bacteria is the expression of virulence factors (see Table 19.4). These various virulence factors facilitate the ability of the bacteria to colonize and replicate on the intestinal epithelium, to invade or destroy the intestinal epithelium, or to evade the immune system (FIGURE 21.6). Bacterial toxins are also virulence factors that can cause disease manifestations independent of the living bacterium.

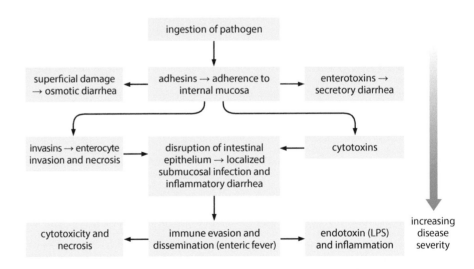

FIGURE 21.6 **Virulence factors and pathogenesis.** Diagram showing the relationship between bacterial virulence factors and the pathogenesis of the disease. LPS, lipopolysaccharide.

The nature of the virulence factor associated with a bacterial species determines its pathophysiology. For example, adhesins may facilitate the colonization of the intestinal epithelium. The resulting superficial damage of the intestinal epithelium may lead to osmotic diarrhea. Virulence factors that extensively disrupt the intestinal epithelium cause more serious clinical manifestations, such as bloody diarrhea or dysentery, and are associated with inflammatory diarrhea. Disruption of the epithelium can be mediated by cytotoxins or invasion of enterocytes. Invasins are virulence factors that allow bacteria to invade cells. This invasion ultimately results in the death of the host cell and the bacteria can continue to spread (see Figure 21.4). Other bacteria secrete cytotoxins that kill intestinal epithelial cells. For example, *Shigella dysenteriae* secretes a potent cytotoxin called Shiga toxin. *E. coli* O157:H7 (BOX 21.2) also secretes a similar toxin and is associated with outbreaks of bloody diarrhea and hemolytic uremic syndrome.

Escherichia coli O157:H7

The O157:H7 serotype of *E. coli* is occasionally in the news due to outbreaks of bloody diarrhea. The primary reservoir for O157:H7 is cattle and transmission to humans is primarily associated with the consumption of undercooked ground beef or unpasteurized milk. Outbreaks have also been traced to raw leafy vegetables when manure has been used as fertilizer. Person-to-person transmission can also occur during outbreaks. *E. coli* O157:H7 secretes the Shiga toxin, which is responsible for its increased virulence. This strain and other strains that express similar toxins are called enterohemorrhagic *Escherichia coli* (EHEC).

Shiga toxin is an AB toxin (see Table 19.6) that blocks eukaryotic protein translation. Cattle do not have the receptor for the binding (B) subunit and therefore do not exhibit disease manifestations. Humans, conversely, are sensitive to Shiga toxin and possibly experience disruption of the intestinal epithelium and inflammatory diarrhea. In 2–7% of the infections, particularly young children, older individuals, and immunocompromised people, the infection becomes systemic and can cause hemolytic uremic syndrome. This syndrome is characterized by the destruction of red blood cells and acute kidney failure.

BOX 21.2

E. coli and related bacteria constitute about 0.1% of the intestinal microbiota and, for the most part, are beneficial to human health. However, there are a few strains – often identified by serotype – that possess virulence factors that cause human disease (TABLE 21.6). One of these virulent types is called enteropathogenic *E. coli* (EPEC). EPEC (e.g., *E. coli* O127:H6) expresses an adhesin on its surface called intimin. Intimin binds to a receptor of bacterial origin called translocated intimin receptor (Tir). The Tir is secreted from the bacterium into the enterocyte membrane. Binding of the bacterium mediated by the intimin–Tir interaction causes a deformation of the enterocyte (FIGURE 21.7). This damage to the enterocyte affects the transport functions and causes osmotic diarrhea.

Table 21.6 Examples of Virulence Factors in Pathogenic *Escherichia coli*

Type	Diarrhea	Virulence factors
Enterotoxigenic *E. coli* (ETEC)	Secretory	Labile (LT) and stable (ST) enterotoxins
Enteropathogenic *E. coli* (EPEC)	Osmotic	Intimin and its receptor
Enterohemorrhagic *E. coli* (EHEC)	Inflammatory	Shiga toxin (cytotoxic)

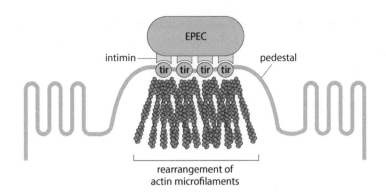

FIGURE 21.7 Adherence to enterocytes via intimin and translocated intimin receptor. Enteropathogenic *E. coli* (EPEC) secretes the translocated intimin receptor (Tir) into the membrane of the host enterocyte. Intimin on the surface of the bacterium binds to Tir and causes a rearrangement of the actin cytoskeleton (Chapter 2) associated with the microvilli of the enterocyte, and this rearrangement produces a pedestal. Deformation of the enterocyte affects absorption and contributes to osmotic diarrhea.

Salmonella Typhi can cause typhoid fever

Salmonella enterica is a common bacterial species found in all warm-blooded animals. There are two major types of disease caused by *S. enterica*: a usually self-limiting gastroenteritis called salmonellosis and a disseminated disease called **typhoid fever**. Salmonellosis is one of the most common causes of diarrhea globally and generally it is the second most common food-borne disease after campylobacteriosis. The non-typhoidal serotypes of *Salmonella* can originate from animals or humans and the infection is generally restricted to the intestinal epithelium, and usually only results in mild gastroenteritis. Very young, very old, and immunocompromised people are more likely to develop severe diarrhea or invasive disease.

Only certain strains of *Salmonella* can cause typhoid fever and these strains only infect humans. These typhoidal strains can survive in phagocytic cells such as macrophages and dendritic cells. Normally these phagocytic cells kill pathogens and present antigens to the immune system to activate a cellular immune response. However, the Typhi serotypes of *Salmonella* not only survive within these cells, but they also replicate. This ability of Typhi to survive and replicate in immune effector cells and to evade immunological destruction is a virulence factor. Furthermore, these strains of *Salmonella* exploit dendritic cells and intestinal epithelial cells to cross the epithelial cell layer (FIGURE 21.8).

FIGURE 21.8 Pathogenesis of *Salmonella* Typhi. *Salmonella* Typhi can invade enterocytes and survive within dendritic cells to be translocated across the intestinal epithelium. In the submucosa the bacteria are engulfed by macrophages, but the bacteria survive and replicate instead of being killed. The infected macrophages can remain in the submucosa or disseminate throughout the body and cause typhoid fever. Infected macrophages are killed by the replicating bacteria and the released bacteria maintain the infection.

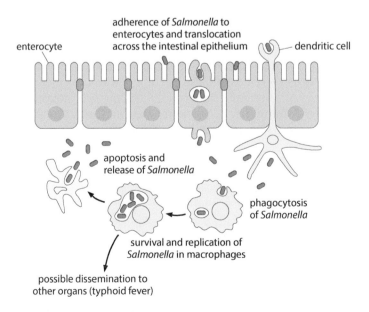

Normally bacteria crossing the intestinal epithelium would be eliminated by the macrophages in the submucosa. However, *Salmonella* Typhi is able to replicate within macrophages and eventually kills the macrophages. This ability to evade destruction by macrophages allows the bacteria to maintain an infection in the submucosa. Some of the bacteria infect enterocytes, and therefore infected individuals continue to shed bacteria into the intestinal lumen and, thus, maintain a fecal–oral transmission cycle. In addition, some of the infected macrophages disseminate throughout the body and infect other organs, especially the liver, spleen, and bone marrow. These organs are part of the reticuloendothelial system (RES) and have abundant macrophages to continue the infection. Typical symptoms associated with typhoid are fever, headache, and muscle aches. The symptoms are exacerbated by endotoxin (i.e., LPS) associated with the bacteria. Without treatment symptoms usually last for weeks or months.

Endotoxin is described in Chapter 19, pages 369–370.

SIGNPOST

Antibiotic-associated diarrhea is often due to an overgrowth of *Clostridioides difficile*

Antibiotic use can affect the intestinal microbiota. This change in microbiota affects carbohydrate metabolism and absorption, and potentially leads to osmotic diarrhea. Another consequence of antibiotic use is the potential overgrowth of pathogenic species in the intestinal tract. *Clostridioides difficile* (formerly *Clostridium difficile*) accounts for 10–20% of antibiotic-associated diarrhea cases. *C. difficile* is a spore-forming bacterium and spore formation contributes to its antibiotic resistance (see Box 19.3). The spores are more resistant to antibiotics and thus *C. difficile* has an advantage over non-spore-forming bacterial species in the presence of antibiotics. Furthermore, the metabolic changes associated with the disruption of the microbiota facilitate *C. difficile* spore germination. Disease pathology is due to exotoxins that disrupt the actin cytoskeleton of epithelial cells and lead to apoptosis, as well as other virulence factors that are associated with pathogenic strains of *C. difficile*.

Another notable feature of *C. difficile* infection is its transmission in hospitals, nursing homes, and other medical institutes. Pathogen transmission within medical facilities is known as nosocomial infection. Spore formation plays a large role in the increased risk of nosocomial spread of *C. difficile*. Spores passed in the feces can survive for long periods on surfaces and in the environment. Furthermore, the spores are heat resistant, and are not killed by alcohol-based sanitizers or routine surface cleaning.

Helicobacter pylori is adapted to survive in the stomach and is associated with ulcers and cancer

Helicobacter pylori is a common bacterial infection of the stomach and perhaps half of the world's population is infected. Most infected people do not develop disease and there is some discussion that *H. pylori* infection may have some benefits. For example, the decreasing prevalence of *H. pylori* has been correlated with an increase in allergies and autoimmune diseases, reminiscent of the hygiene hypothesis.

The hygiene hypothesis is described in Chapter 17, pages 333–335.

SIGNPOST

Comparatively few microorganisms colonize the stomach due to its extreme acidity. *H. pylori* has two major adaptations that allow it to survive in the stomach. For one, *H. pylori* is motile, and it uses its flagella to burrow through the mucous layer to reach the epithelial cells where the environment is less acidic. In addition, *H. pylori* secretes an enzyme that generates ammonia, which locally neutralizes acidic conditions. However, ammonia is toxic to epithelial cells and causes some pathology.

In addition, adhesins, cytotoxins, and other virulence factors further damage the epithelium resulting in inflammation and chronic gastritis. Chronic gastritis can lead to the formation of ulcers in the stomach and duodenum. In addition, chronic gastritis due to *H. pylori* infection is associated with an increased risk of developing stomach cancer.

ENTEROTOXIGENIC BACTERIA AND SECRETORY DIARRHEA

Several bacteria cause diarrhea through the expression of exotoxins, called **enterotoxins**, which stimulate the secretion of water from intestinal epithelial cells. Diarrhea mediated by enterotoxins is called **secretory diarrhea** (see Table 21.3). The most notable among these bacteria are *Vibrio cholera* and enterotoxigenic *E. coli* (ETEC). Enterotoxins cause diarrhea independent of living bacteria, and some toxins are even heat stable and thus are not inactivated by cooking.

Pathogenic *V. cholerae* has genes for cholera toxin and other virulence factors

V. cholerae is a bacterial species commonly associated with water, and especially salt or brackish water in coastal areas. Most *V. cholerae* serotypes are not pathogenic. Pathogenicity is restricted to the O1 and O139 serotypes of *V. cholerae*, and in these serotypes only strains infected with a bacteriophage called CTXφ are pathogenic (FIGURE 21.9). CTXφ contains the genes for the A and B subunits of cholera toxin (CT) and can convert a non-toxic *V. cholerae* to a toxic *V. cholerae*.

Another virulence factor associated with toxic *V. cholerae* is the toxin-coregulated pilus (TCP). Pili are hair-like structures on the surface of bacteria that facilitate adherence (Chapter 19). TCP is expressed in conjunction with cholera toxin and mediates adherence to the intestinal epithelium, thereby helping *V. cholerae* colonize the human intestine. In addition, TCP is encoded on a pathogenicity-associated genomic island and is amenable to horizontal gene transfer (see Box 19.4). A genomic island is a region of the genome that exhibits evidence of being transferred horizontally. In some pathogenic bacteria a genomic island may contain a block of genes related to virulence and pathogenicity.

FIGURE 21.9 Transformation of *Vibrio cholerae* into a toxigenic pathogen. Most *V. cholerae* in the environment are not pathogenic and do not infect humans. Pathogenic strains have acquired virulence factors via horizontal gene transfer that allow for human infection and cause secretory diarrhea. One of these virulence factors is cholera toxin (CT) which is acquired from a bacteriophage carrying the gene for the cholera toxin. Another virulence factor is toxin-coregulated pilus (TCP) which is acquired from a pathogenicity island. The pili help the bacteria to adhere to the intestinal epithelium and colonize the intestines.

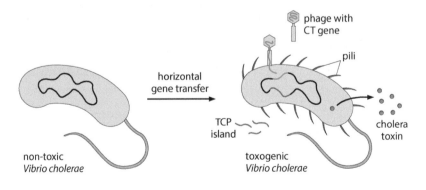

Cholera transmission is both endemic and subject to epidemics

Cholera infects over a million people per year worldwide and between 20,000 and 140,000 people die of cholera per year. Cholera is endemic in many resource-poor countries that lack sufficient waste disposal and water treatment. Cholera epidemics also occur with some frequency and are often associated with natural disasters and civil unrest. For example, a major cholera epidemic occurred in Haiti following the earthquake in 2010 (BOX 21.3). Transmission can occur via water contact (FIGURE 21.10). However, once established in humans, person-to-person transmission via fecal–oral transmission also occurs. The profuse watery diarrhea associated with cholera can introduce massive numbers of bacteria into the environment, which can facilitate a rapid spread of the infection through a population.

The 2010 Cholera Outbreak in Haiti

At 9 months after a catastrophic magnitude 7.0 earthquake struck Haiti in 2010, an outbreak of cholera began and quickly spread through the country. At 4 years after the earthquake 700,000 cholera cases had been reported, resulting in more than 8500 deaths. Prior to 2010, there had never been a reported case of cholera in Haiti, thus, raising questions about the origin of the outbreak. Initially there were two theories: (1) non-pathogenic *Vibrio cholerae*, indigenous in the coastal waters of Haiti, evolved into a pathogenic strain possibly due to environmental changes following the earthquake, or (2) cholera was introduced to Haiti by individuals from another country as part of the relief efforts. A combination of standard epidemiological outbreak tracing methods and molecular epidemiology clearly demonstrated that cholera was introduced by United Nations (UN) troops from Nepal.[1]

Early on it was suspected that the UN stabilization troops from Nepal may have been the source of the cholera outbreak. The first cholera case in Haiti occurred 4 days after the first Nepalese troops set up camp on a river in Haiti (**BOX 21.3 FIGURE 1**). This first case occurred two kilometers downstream of the camp. Furthermore, the dumping of untreated fecal waste into the river had been documented. The troops were asymptomatic for cholera before departing Nepal for Haiti, despite a cholera epidemic in Nepal shortly before their departure. Cholera is endemic in Nepal and therefore there is a high level of immunity, and it is likely that some of the troops were infected but asymptomatic.

After the first Haitian case the epidemic quickly spread downstream. The Haitians, never having experienced cholera before, had no immunity and infected individuals developed severe and life-threatening secretory diarrhea with some mortality. The massive amount of diarrhea excreted into the environment furthered the transmission. In addition, many Haitians were living in refugee camps due to the massive destruction caused by the earthquake. Such living conditions furthered the spread. Furthermore, the living conditions lowered immunity resulting in more severe disease in individuals and creating a catastrophic situation.

Molecular sequencing studies showed that the outbreak was clonal and due to a single introduction. A comparison of genomes demonstrated that isolates from Nepal and Haiti clustered together into a single monophyletic group indicating that these isolates shared a common ancestor. Furthermore, molecular clock calculations estimated the date of the most recent common ancestor was between July 23 and October 17, 2010, coincident with the epidemic in Nepal and the introduction of cholera to Haiti. This DNA-sequence data combined with standard epidemiological tracing convincingly demonstrated that cholera was introduced into Haiti from an external source and that the UN troops from Nepal were the most likely origin.

1. Orata FD, Keim PS and Boucher Y (2014). The 2010 cholera outbreak in Haiti: How science solved a controversy. *PLoS Pathogen* 10(4), e1003967.

BOX 21.3

23 SEPT	**10** OCT	**12** OCT	**20** OCT	**23** OCT
2010				
cholera outbreak near training facility of UN troops in Nepal	UN troops begin arriving in Haiti	first Haitian case of cholera 2 km downstream of UN camp	explosive cholera outbreak further downstream	first cholera outbreak in Haiti officially announced

BOX 21.3 FIGURE 1 Timeline of movement of United Nations troops and emergence of cholera in Haiti. An epidemic of cholera occurred in the vicinity of the UN troops in Nepal shortly before their assignment to Haiti. The first case in Haiti occurred 2 km downstream of the UN camp a few days after the troops arrived and then continued spreading downstream. [Modified from Orata FD, Keim PS & Boucher Y (2014) The 2010 cholera outbreak in Haiti: How science solved a controversy. *PLoS Pathogen* 10:e1003967. doi: 10.1371/journal. ppat.1003967. Published under CC by 4.0.]

FIGURE 21.10 Transmission of *Vibrio cholerae*. Cholera can be acquired from surface water contaminated with pathogenic strains of *V. cholerae*. *V. cholerae* can also be transmitted from person to person via fecal–oral transmission.

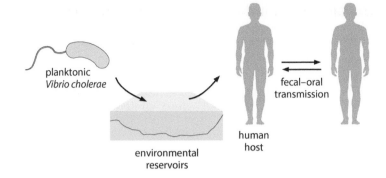

SIGNPOST

AB toxins are described in Chapter 19, pages 367–368.

FIGURE 21.11 Mechanism of action of cholera toxin. The B subunit of cholera toxin (CT) binds to gangliosides on the surface of enterocytes and facilitates the entry of the A subunit into the cell. The A subunit adds an ADP-ribose to the stimulatory G-protein (G) associated with adenylyl cyclase (AC). This activates adenylyl cyclase and elevates intracellular cAMP, which leads to an activation of protein kinase A (PKA). Protein kinase A phosphorylates the chloride transporter (CFTR) and the sodium/hydrogen exchanger (NHE). Phosphorylation of CFTR activates chloride secretion and phosphorylation of NHE blocks sodium absorption, leading to hypersecretion of water and electrolytes into the intestinal lumen. The secreted water and electrolytes are derived from the circulatory system which can lead to severe dehydration and possibly shock.

Cholera toxin activates a signal transduction pathway and turns on the chloride transporter

Cholera is characterized by profuse watery diarrhea. This diarrhea is due to a potent AB5 toxin in which a few micrograms can result in 20 liters of water being excreted with the feces. The B subunit binds with high affinity to GM1 ganglioside, a glycolipid found on the surface of enterocytes, and facilitates the entry of the A subunit into the cell (**FIGURE 21.11**). The A subunit activates the cAMP/protein kinase signaling pathway (see Box 3.1) leading to the phosphorylation of the chloride transporter. Phosphorylation of chloride channels activates chloride secretion by enterocytes. Simultaneously, protein kinase A also phosphorylates the sodium/hydrogen exchanger and, in this case, phosphorylation inhibits sodium absorption.

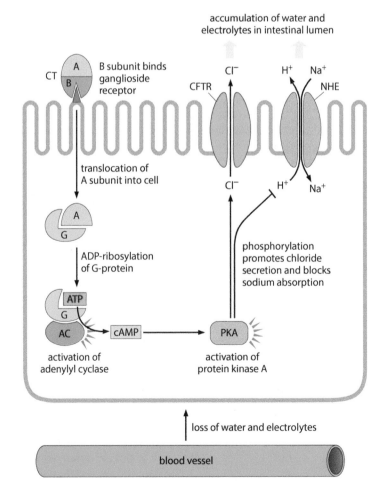

For every chloride ion expelled from the cell a water molecule is also expelled and, likewise, blocking sodium absorption blocks water absorption as well. Therefore, turning on the chloride transporter and simultaneously turning off the sodium transporter result in an efflux of water and electrolytes into the intestinal lumen, leading to severe secretory diarrhea. Patients with cholera, and especially children, might lose up to 80 liters of diarrhea. Often, this profuse diarrhea contains no fecal matter and is a cloudy liquid due to the high levels of bacteria. This massive diarrhea can quickly lead to dehydration and can cause death in a matter of hours if patients are not administered oral rehydration. Manifestations of the resulting electrolyte imbalance are muscle cramps and, if the dehydration is severe enough, there is a decrease in blood pressure that can lead to shock.

A hallmark of enterotoxigenic *E. coli* is the expression of one or more enterotoxins

E. coli is normally viewed as a commensal. However, pathogenic strains of *E. coli* have been described and enterotoxigenic *E. coli* (ETEC) is one such example. ETEC expresses one or more enterotoxins that cause secretory diarrhea. In addition, pili, which facilitate attachment to host intestinal cells, are present on the surface of ETEC. ETEC is a leading cause of bacterial diarrhea and more than 150,000 people – mostly children – may die per year.

Two distinct enterotoxins have been described in ETEC. One is heat sensitive and called labile toxin (LT), and the other is heat stable and called stable toxin (ST). Stable enterotoxins (STs) are small peptides that can withstand high temperatures. Therefore, cooking may not completely inactivate these toxins. ETEC strains possess either one of these toxins, or both. Labile toxin is a homolog of cholera toxin and exhibits the same mechanism of action (see Figure 21.11). ST has a similar effect on chloride secretion and sodium absorption as LT, but through a different signal transduction pathway. In addition, LT, ST, or both can be found on plasmids and can be transferred horizontally (see Box 19.4).

INTESTINAL PROTOZOA

Protozoa are eukaryotic microbes that cause several important human diseases. Several protozoan species infect the human gastrointestinal tract (TABLE 21.7). Many of these intestinal protozoa do not cause severe disease and some are no longer commonly found in human stools. In terms of prevalence and causing human disease, *Entamoeba histolytica*, *Giardia duodenalis*, and *Cryptosporidium* species are the most important human pathogens. Diarrhea is a major clinical manifestation for all three of these pathogens. Profuse watery diarrhea can be associated with giardiasis and cryptosporidiosis, and *E. histolytica* can produce an invasive disease with some risk of mortality.

Table 21.7 Protozoa Infecting Human Gastrointestinal Tract

Amoebas	Flagellates	Apicomplexa	Other
– *Entamoeba histolytica*	– *Giardia duodenalis*	– *Cryptosporidium hominis*	– *Blastocystis hominis* (stramenopile)
– *Entamoeba dispar*	– *Dientamoeba fragilis*	– *Cryptosporidium parvum*	– *Balantidium coli* (ciliate)
– *Entamoeba coli*	– *Chilomastix mesnili*	– *Cyclospora cayetanensis*	– Microsporidia (now considered highly derived fungi)
– *Entamoeba hartmanni*	– *Pentatrichomonas hominis*	– *Isospora belli*	
– *Endolimax nana*	– *Enteromonas hominis*		
– *Iodamoeba bütschlii*	– *Retortamonas intestinalis*		

Protozoan cysts play a key role in the transmission of infections

Although the protozoa that infect the human gastrointestinal tract are quite diverse, they all exhibit a similar life cycle that is predominantly fecal–oral transmission (FIGURE 21.12). The infection is acquired through the ingestion of food or water that has been contaminated with fecal matter. A specialized stage called the cyst initiates the infection. After ingestion, the cyst converts into a trophozoite. The trophozoite is often motile and exhibits an active metabolism. Most importantly, the trophozoite is the replicative form of the parasite and leads to an expansion of the pathogen population. Some of the trophozoites convert back into cyst stages, which are passed in the feces. The cyst stage is surrounded by a cyst wall that makes the parasite more resilient to environmental elements such as desiccation. Many protozoan cysts can survive for months outside of the body if kept moist and cool. Furthermore, the cysts are highly infectious and only a few cysts can initiate an infection.

FIGURE 21.12 Typical protozoan fecal–oral life cycle. Cysts are highly infectious forms that convert to trophozoites following ingestion via contaminated food or water. Trophozoites are feeding stages that replicate within the intestinal tract and are often motile. Some trophozoites convert back into cysts and are passed in the feces. Cysts have a thick wall and can survive outside of the host until ingested by another host.

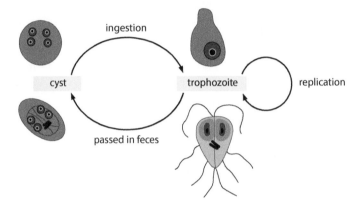

Amebiasis is associated with amoebic dysentery and other severe clinical manifestations

Disease manifestations associated with *E. histolytica* infections can range from asymptomatic to death. The disease is often characterized as being either non-invasive or invasive. During non-invasive disease the trophozoites establish a colony of amoeba in the colon and exhibit the typical fecal–oral life cycle. The trophozoites ingest bacteria and fecal matter as a food source and replicate by mitosis. The non-invasive disease is often asymptomatic or associated with diarrhea and other gastrointestinal symptoms such as cramping or pain. Most infections do not exhibit severe clinical manifestations and self-resolve in weeks to months.

However, the infection can progress to severe invasive disease (FIGURE 21.13). *E. histolytica* trophozoites can kill intestinal epithelial cells resulting in the formation of colonic ulcers that cause colitis and dysentery. The lesions can expand through the submucosa and occasionally through the colon wall leading to a perforation of the colon wall and peritonitis. In addition, the trophozoites in the submucosa can enter the circulatory system and metastasize throughout the body and cause extra-intestinal amebiasis. Since the mesentery blood vessels and portal vein system go directly from the intestines to the liver, the liver is the most common site of extra-intestinal amebiasis. From the liver the trophozoites can continue spreading throughout the body and affecting other organs. The pathology associated with amebiasis is due to a contact-dependent killing of host cells by the trophozoites. Severe diseases, such as fulminant amoebic colitis, peritonitis, or extra-intestinal amebiasis can be mortal if not treated.

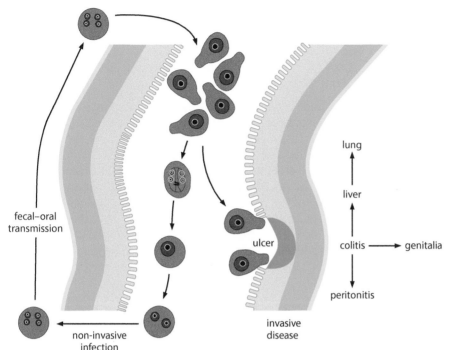

FIGURE 21.13 Non-invasive versus invasive amebiasis. During non-invasive disease the trophozoites of *E. histolytica* form a colony on the mucosa of the large intestine and exhibit a typical fecal–oral life cycle. Invasion of the epithelial layer of the colon by trophozoites results in ulcers and colitis. Trophozoites in the submucosa can enter the portal vein system and travel to the liver and other organs, such as the lung. The lesion can also expand and perforate the colon wall leading to peritonitis or trophozoites can spread to the genitalia. [Adapted from Wiser MF (2021) Nutrition and protozoan pathogens of humans: a primer. In Humphries DL, Scott ME and Vermund SH (eds), *Nutrition and infectious diseases. Nutrition and Health.* Humana Press, Cham. With permission from Springer Nature.]

Giardiasis is primarily associated with superficial damage to the intestinal epithelium and osmotic diarrhea

Giardia duodenalis, formerly known as *G. lamblia*, is a common protozoan of the intestinal tract. As implied by the name, the parasites have a strong tropism for the duodenum. *Giardia* species exhibits a typical fecal–oral life cycle and is periodically associated with water-borne outbreaks. *Giardia* trophozoites have a unique structure on their ventral surface called the adhesive disk (FIGURE 21.14). This disk is composed of cytoskeletal elements and contractile proteins and is involved in the attachment of the trophozoite to the epithelium of the small intestine. The attached trophozoites absorb nutrients from the intestinal milieu via pinocytosis.

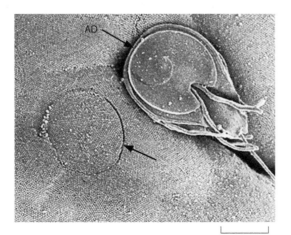

FIGURE 21.14 Adhesive disk of *Giardia* species. A scanning electron micrograph showing the ventral surface of a *Giardia* trophozoite. The suction-cup-like structure is the adhesive disk (AD) that attaches to the intestinal epithelium. The circle to the left (arrow) is an imprint left in the brush border (microvilli) of the intestinal epithelium following the detachment of the trophozoite. (Courtesy of the Public Health Image Library and attributed to CDC/Dr. Stan Erlandsen.)

Most infections with *Giardia* species are either asymptomatic or go undiagnosed. No mortality or severe pathology is associated with giardiasis. The most common symptom associated with giardiasis is acute diarrhea. Initially the stools are watery and profuse, but later may become greasy and foul smelling. Abdominal cramping, bloating, and flatulence are additional common symptoms. The symptoms usually clear spontaneously after 5–7 days. A few people develop chronic infections that can last for several months or even years. Chronic giardiasis is characterized by intermittent diarrheal episodes and progresses to weight loss and a failure to thrive.

Villus blunting and crypt cell hyperplasia are common pathological manifestations associated with giardiasis. This increased crypt cell proliferation leads to a repopulation of the intestinal epithelium by immature enterocytes and causes osmotic diarrhea characterized by increased chloride secretion and decreased sodium absorption (see Figure 21.3). There can also be damage to the brush border and microvilli and a reduction in enzymes, particularly disaccharidases such as lactase, on the surface of epithelial cells. In fact, some patients exhibit a lactose intolerance (see Box 21.1) that can persist for months after parasite clearance. This reduced digestion and absorption of metabolites and electrolytes increase fluid retention in the intestinal lumen and contribute to osmotic diarrhea.

Cryptosporidiosis is an opportunistic pathogen that causes severe disease in immunocompromised people

Cryptosporidium species is another common intestinal protozoan of humans. Two *Cryptosporidium* species infect humans: *C. hominis* is strictly a human pathogen and *C. parvum* is a parasite of cattle that infects a wide range of mammals including humans. *Cryptosporidium* species is a member of the Apicomplexa, which includes the malaria parasite (Chapter 23) and exhibits an apicomplexan life cycle. Although the details of the *Cryptosporidium* life cycle are more complex than other intestinal protozoal pathogens, it still exhibits a basic fecal–oral life cycle. The infection is initiated upon ingestion of the oocyst stage. Like the cyst, the oocyst is covered with a thick cyst wall and can survive for extended periods in the environment. These oocysts are quite robust and highly infectious. *Cryptosporidium* species is also noted for water-borne outbreaks with the most famous being in Milwaukee (BOX 21.4).

BOX 21.4

Massive Cryptosporidiosis Outbreak in Milwaukee

In the spring of 1993, an estimated 400,000 residents in the Milwaukee metropolitan area experienced gastroenteritis characterized by stomach cramps, fever, watery diarrhea, and dehydration due to cryptosporidiosis. *Cryptosporidium* oocysts were demonstrated to have passed through the filtration system of one of the city's two water treatment plants[1]. Initially the source of *Cryptosporidium* was believed to have been from agricultural runoff into lake Michigan during the spring snowmelt. However, later genotyping demonstrated that the causative agent was *C. hominis*, which only infects humans. It was believed, but never proven, that a broken pipe may have allowed for the contamination of clean water with untreated sewage. In the weeks before the outbreak the water treatment facility had noted higher-than-normal levels of turbidity in the treated water, but well within the prescribed limits.

This outbreak is usually cited as the largest water-borne disease outbreak in the history of the United States, and it changed our thinking about cryptosporidiosis. When human cryptosporidiosis was first described in 1976 it was believed to be a rare and exotic human infection. In the late 1970s and early 1980s cryptosporidiosis was the most common cause of AIDS-associated diarrhea, and therefore, *Cryptosporidium* species was designated as an opportunist that only produced clinical manifestations in immunocompromised individuals. Beginning in the early 1990s it was beginning to be realized that *Cryptosporidium* infections were rather common and could cause clinical manifestations in immunocompetent people. The Milwaukee outbreak solidified this belief. The commonness of cryptosporidiosis was likely missed because in immunocompetent people cryptosporidiosis is a self-limiting diarrhea that usually resolves in a few days. Many other gastrointestinal pathogens cause similar symptoms and specific diagnoses would rarely be prescribed for self-resolving diarrhea.

1. MacKenzie WR, Hoxie NJ, Proctor ME et al. (1994). A massive outbreak in Milwaukee of *Cryptosporidium* infection transmitted through the public water supply. *New England Journal of Medicine* 331: 161–167. doi: 10.1056/ NEJM199407213310304.

The most common clinical manifestation of cryptosporidiosis is mild-to-profuse watery diarrhea. Generally, the infection is self-resolving and the symptoms persist for a few days. However, recrudescence is common and the infection can persist for weeks to months. The symptoms are generally more severe in immunocompromised people and the disease is often life threatening in patients with AIDS. In patients with AIDS, diarrhea can be quite voluminous and is often described as cholera-like. However, diarrhea associated with cryptosporidiosis is osmotic due to superficial damage to the intestinal epithelium (see Figure 21.3). The infection is generally confined to the jejunum and ileum of the small intestine in immunocompetent individuals. However, in immunocompromised individuals, the infection spreads to the duodenum, biliary tract, colon, and stomach.

DIAGNOSIS AND TREATMENT

The diagnosis and treatment of intestinal infections are based largely on symptomatology. The primary symptom associated with intestinal infection is diarrhea accompanied by other gastrointestinal symptoms (see Table 21.2). The nature of diarrhea can vary depending on the nature of the pathogenesis and the part of the GI tract affected (FIGURE 21.15). In addition, most diarrhea is of short duration and resolves within a few days. Therefore, most intestinal infections go undiagnosed. Medical attention is usually sought only if the diarrhea is long-lasting or extremely profuse.

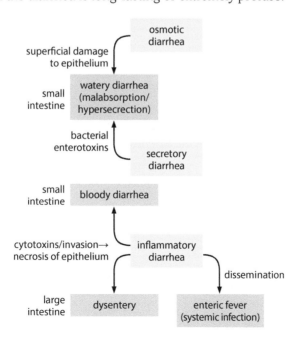

FIGURE 21.15 Clinical manifestations associated with the various mechanisms of pathogenesis. The primary clinical manifestations (purple boxes) of gastrointestinal infections are associated with the three types of diarrhea.

Laboratory diagnosis of specific pathogens can be difficult and is often not necessary

The sheer magnitude of organisms found in the gastrointestinal tract can make it rather difficult to identify a specific pathogen that is responsible for the disease manifestations. Microscopy of fecal samples can be used to diagnose protozoan pathogens, but microscopy is of little use for identifying viruses and bacteria. It is also possible to culture many of the pathogenic bacteria from stool samples. But, often, the symptoms have resolved by the time the culture results are available. Diagnostic tests based on antigen or nucleic acid detection are available for some of the pathogens. However, in the case of bacterial and viral pathogens specific diagnosis is often not necessary since the treatments are generally targeted at the symptoms, and not directed at a specific pathogen.

Treatment is generally restricted to small children or severe disease

Dehydration, especially in small children, is a potentially life-threatening consequence of severe diarrhea. **Oral rehydration therapy** (ORT) is the administration of water and electrolytes to replenish the loss of water and electrolytes caused by diarrhea. ORT does not remove the pathogen, but it does lower the risk of fatality until the pathogen can be eliminated by the immune system. Antibiotics are rarely used to treat diarrhea (TABLE 21.8). Antibiotics can disrupt the normal microbiota and, thus, might actually prolong the symptoms, or lead to antibiotic-associated diarrhea. Furthermore, the overuse of antibiotics can promote the development of drug resistance (Chapter 19). However, antibiotics are prescribed if there is evidence of disseminated disease. In addition, antibiotics may be beneficial for small children with severe diarrhea. Probiotics can also be used as an adjunct therapy to alter the microbiota with less harmful species. Probiotics are a mixture of live microorganisms that have health benefits when consumed or applied to the body.

Table 21.8 Treatment of Diarrhea	
Oral rehydration therapy (ORT)	• Administration of water and electrolytes • Used especially for small children and severe diarrhea, such as cholera

Antibiotics	• Rarely used except for:
	  ○ evidence of disseminated disease
	  ○ severe disease
	  ○ small children

Probiotics	• Adjunct therapy to replace microbiota • May shorten the duration of the illness
Anti-protozoal drugs	• Highly efficacious drugs are available for the treatment of giardiasis and amebiasis
Antimotility agents	• Antimotility agents slow intestinal movement • May improve symptoms in patients with watery diarrhea • Contraindicated in patients with bloody or inflammatory diarrhea

There are drugs that specifically target protozoan infections. Therefore, if the cause of the infection is suspected to be of protozoal origin, a laboratory test to identify protozoa in the feces is ordered. Metronidazole is a relatively non-toxic drug that is highly efficacious against both giardiasis and amebiasis. There are no highly effective drugs against cryptosporidiosis. Nitazoxanide does have some benefits in immunocompetent people, but nitazoxanide does not appear to be effective in patients with AIDS. Supportive care including rehydration and nutritional support, and possibly antimotility agents, is prescribed for patients with AIDS. Antimotility agents slow movement through the intestines and thus only treat the symptoms. However, antimotility agents are contraindicated for bloody diarrhea and other inflammatory diarrhea.

SUMMARY OF KEY CONCEPTS

- A wide range of pathogens infect the gastrointestinal tract and cause diarrheal disease
- Ingestion of a pathogen is the most common way to acquire a gastrointestinal infection, and many of the pathogens exhibit a fecal–oral life cycle
- Most intestinal pathogens establish a colony on the intestinal mucosa resulting in a superficial infection that causes osmotic diarrhea due to enterocyte dysfunction
- Invasion of the submucosa causes inflammatory diarrhea or dysentery characterized by mucus, blood, and white blood cells in the feces
- A few pathogens can disseminate to other organs and cause systemic infections after the invasion of the intestinal submucosa
- Norovirus and rotavirus are two common causes of viral gastroenteritis
- Numerous bacterial species can cause diarrheal disease and pathogenicity is usually associated with virulence factors such as adhesins, invasins, cytotoxins, or enterotoxins
- Enterotoxigenic bacteria, such as *Vibrio cholerae*, secrete exotoxins and cause secretory diarrhea characterized by the loss of large volumes of water in the feces
- *Entamoeba histolytica* causes amoebic dysentery after the invasion of the colonic mucosa
- *Giardia* species is a common protozoal infection that causes osmotic diarrhea
- Cryptosporidiosis causes self-limiting diarrhea in immunocompetent people, however, it can cause profuse and life-threatening diarrhea in patients with AIDS
- Diagnosis and treatment of diarrheal diseases largely focus on the symptoms
- Drug treatment is usually only prescribed for severe diarrhea in children, systemic infections, or protozoal infections

KEY TERMS

cholera	Acute infectious disease of the small intestine, caused by *Vibrio cholerae* and characterized by profuse watery diarrhea, vomiting, muscle cramps, severe dehydration, and depletion of electrolytes.
colitis	Inflammation of the mucosa of the colon and often associated with dysentery.
dysentery	A disease characterized by inflammation of the colon (i.e., colitis) and often manifests as diarrhea with blood or mucus in the feces.
enteritis	Inflammation of the mucosa of the small intestines and often associated with watery diarrhea.
enterotoxin	A bacterial exotoxin that specifically affects intestinal cells and generally causes secretory diarrhea.
fecal–oral transmission	A type of transmission in which infective stages of a pathogen are found in feces and infection is acquired by ingesting food or water contaminated with fecal matter.
inflammatory diarrhea	Diarrhea associated with pathogen invasion of the intestinal mucosa characterized by bloody diarrhea or dysentery.
oral rehydration therapy (ORT)	Treatment for diarrhea-related dehydration in which an electrolyte solution is administered by drinking.
osmotic diarrhea	Watery diarrhea due to increased excretion and decreased absorption of water and electrolytes by intestinal epithelial cells.
secretory diarrhea	Watery diarrhea caused by bacterial enterotoxins.
typhoid fever	An enteric fever caused by Typhi strains of *Salmonella enterica*.

REVIEW QUESTIONS

1. Describe the pathophysiologies associated with osmotic diarrhea, inflammatory diarrhea, and secretory diarrhea.

2. How do Typhi strains of *Salmonella enterica* differ from non-Typhi strains in regard to disease manifestations and what factors are involved in the increased virulence of *Salmonella* Typhi?

3. What are probiotics and oral rehydration therapy and how do they work?

4. What is cholera toxin and describe its mechanism of action?

5. Discuss the factors associated with the transmission of gastrointestinal pathogens.

ADDITIONAL QUESTIONS

In addition to the Review Questions provided above, there is a range of free online questions designed to enable students to further test their understanding of the chapter material. In order to access these interactive questions, please visit the book's website:
https://routledge.com/cw/wiser

HIV/AIDS and Other Sexually Transmitted Infections

22

Human immunodeficiency virus (HIV) causes acquired immune deficiency syndrome (AIDS). The disease was first recognized in the late 1970s, and the virus was discovered in the early 1980s. This virus is primarily transmitted via sexual contact. Pathogens that are transmitted via sexual contact are referred to as **sexually transmitted infections** (STIs). STIs have been called sexually transmitted diseases (STDs) or venereal disease in the past. The term STI is preferred since it includes people who are infected, but do not have symptoms. Indeed, most STIs are not extremely virulent and many STIs are more of a nuisance than a pathological condition. There are a few STIs that can be virulent, and pathology is generally an inflammation of the urogenital tract. However, except for AIDS, mortality is generally rather low.

HIV evolved from similar viruses in non-human primates called simian immunodeficiency virus (SIV). After jumping from primates to humans the virus spread around the world and created a pandemic that was recognized in the early 1980s. The virus remains endemic in the human population and is still a major cause of death. The virus infects helper T cells and other immune effector cells and establishes a long-term chronic and asymptomatic infection. During this latent period helper T cells are lost and the loss of these cells below a critical threshold marks the beginning of AIDS. Helper T cells play a crucial role in immunity and a hallmark feature of AIDS is infection with opportunistic pathogens. There are anti-retroviral drugs for the treatment and management of AIDS. However, AIDS cannot be cured and is a lifelong infection.

SEXUALLY TRANSMITTED INFECTIONS

Numerous pathogens including viruses, bacteria, protozoa, and fungi (TABLE 22.1) have adapted to person-to-person transmission via sexual activity, especially vaginal intercourse, anal intercourse, and oral sex. Like eating, drinking, and breathing, sex is a necessary and frequent activity in humans. Furthermore, sex involves direct contact and the exchange of body fluids, allowing for extremely efficient transmission. This efficiency is emphasized by the estimated one million people worldwide who acquire an STI every day. Generally, no special adaptations, such as resistance to desiccation or tolerance of stomach acid, are required for sexual transmission and most sexually transmitted pathogens have a simple life cycle that only involves the direct transmission of the pathogen from person to person.

Table 22.1 **Examples of Common Sexually Transmitted Pathogens**

Viruses	
Human immunodeficiency virus (HIV)	Infects helper T cells and causes acquired immune deficiency syndrome (AIDS)
Human papillomavirus (HPV)	Causes genital warts and some strains increase the risk of cervical cancer (Box 22.1)
Herpes simplex virus-2 (HSV-2)	An infection affecting >65% of adults characterized by sporadic flare-ups of genital blisters
Hepatitis B virus (HBV)	A blood-borne infection of the liver that can also be transmitted sexually
Bacteria	
Treponema pallidum	A spirochete bacterium that causes syphilis, which is characterized by a painless ulcer in the early stage of the disease. If not treated can affect the CNS, heart, and other organs years later
Neisseria gonorrhoeae	A Gram-negative bacterium that causes gonorrhea. Typical symptoms are painful urination and vaginal or penile discharge
Chlamydia trachomatis	Causes a common STI called chlamydia that has similar symptoms similar to gonorrhea
Protozoa	
Trichomonas vaginalis	A very common STI with low virulence and associated with vaginal discharge and painful urination
Fungi	
Candida species	Not always considered an STI since it is often believed to be part of the microbiota and is considered an opportunistic pathogen

Prevention of sexually transmitted infections focuses on safe sex

The primary mode of transmission of STIs involves sexual activity and, in particular, sexual intercourse. Many of the pathogens can also be transmitted from mother to child during childbirth due to the presence of the pathogens in the birth canal and the possible exchange of blood during delivery. Some sexually transmitted pathogens can also be found in the blood and blood-borne transmission from transfusions or sharing needles is also possible. Reduction in the number of sexual partners, mutual monogamy, and abstinence are strategies to reduce exposure to STIs. Reducing direct physical contact through the use of condoms also reduces the risk of transmission. These practices collectively are often referred to as **safe sex**. Sometimes these practices are referred to as safer sex or protected sex to emphasize that these practices do not eliminate the risk of acquiring an infection, but only lower the risk.

Pathophysiology of STIs typically involves the mucous membranes of the urogenital tract

Quite often STIs are asymptomatic. Symptoms for many STIs include vaginal discharge, penile discharge, ulcers on or around the genitals, and pelvic pain. Vaginal and penile discharge and ulcers around the genitalia are due to inflammation of the mucosa of the urogenital tract. Many of these sexually transmitted pathogens can invade the epithelium and establish a localized infection. However, most of these pathogens rarely cause systemic infections. Syphilis, caused by *Treponema pallidum*, is an exception in that if untreated it can spread to other parts of the body, including the heart and central nervous system. Another exception is HIV, which, after crossing the mucosa, infects immune effector cells and spreads throughout the body. Human papillomavirus (HPV) causes genital warts and generally is not virulent. However, HPV is noted for some strains increasing the risk of cervical cancer (BOX 22.1).

Human Papillomavirus and Cervical Cancer

Human papillomavirus (HPV) is the most common STI globally and most sexually active people will experience an HPV infection at least once in their life. Approximately 90% of HPV infections are asymptomatic and spontaneously resolve within 2 years. Symptoms are generally mild and the most common clinical manifestation is genital warts. Warts are non-cancerous growths of the outer layer of skin. Sometimes HPV infection persists and results in precancerous lesions that can develop into cancer of the cervix, anus, mouth, penis, and other regions of genitalia. It is now believed that nearly all cervical cancer is due to HPV infection. Vaccines against some of the more common and virulent strains of HPV are available and these vaccines reduce the incidence of both infection and cancer.

Among the common STIs, syphilis, gonorrhea, chlamydia, and trichomoniasis are curable with available drugs. However, resistance to some antibiotics has been noted in the case of gonorrhea. Antiviral drugs to treat herpes, hepatitis B (HBV), HIV, and HPV are available, but these drugs do not eliminate the infection and, therefore, these diseases are not curable. Vaccines against HBV and HPV are also available.

ORIGIN OF HIV/AIDS

In the late 1970s physicians began noting unusually high levels or two relatively uncommon syndromes. These were *Pneumocystis carinii* (now *jirovecii*) pneumonia (PCP) and Kaposi's sarcoma (TABLE 22.2). PCP is an opportunistic infection associated with immunodeficiency and, similarly, Kaposi's sarcoma is a rare cancer also associated with an immunocompromised state. These syndromes were observed in relatively diverse populations ranging from gay men in the USA and Sweden to heterosexual individuals in Haiti and Tanzania. The disease was also identified in people with hemophilia and intravenous drug users. In 1981 the disease was named **acquired immune deficiency syndrome** (AIDS).

Table 22.2 New Elements Leading to Description of AIDS	
Conditions	**Populations**
• *Pneumocystis jirovecii (carinii)* pneumonia (PCP) and other opportunistic infections • Kaposi's sarcoma	• Gay men in the USA and Sweden • Straight individuals in Tanzania and Haiti • Intravenous (IV) drug users • Hemophiliacs (blood transfusions)

One explanation to account for this rather heterogeneous population of affected people is that AIDS was an infectious disease that could be transmitted sexually, as well as via blood or blood products. In 1984 the causative agent of AIDS was identified as a virus and named **human immunodeficiency virus** (HIV) in 1986. HIV is in the **retrovirus** family and placed within a subgroup called lentivirus. Lentiviruses are known to cause chronic and deadly diseases characterized by long incubation periods.

HIV evolved from closely related viruses found in monkeys

The discovery of the virus explained the disease. However, it did not explain the origin of the disease. Molecular phylogenetics demonstrated that HIV was closely related to viruses isolated from monkeys called simian immunodeficiency virus (SIV). For the most part, each monkey species is infected with its own variant of SIV (**FIGURE 22.1**). In other words, SIV is composed of multiple subtypes and each subtype is restricted to a specific host monkey species. Two types of HIV were identified from molecular phylogenetics and are designated HIV-1 and HIV-2. HIV-1 is most closely related to SIV of chimpanzees and HIV-2 is the most closely related SIV of sooty mangabeys. HIV infections are not zoonoses, but, rather, an example of viruses that jumped from a simian species to humans and subsequently adapted to humans.

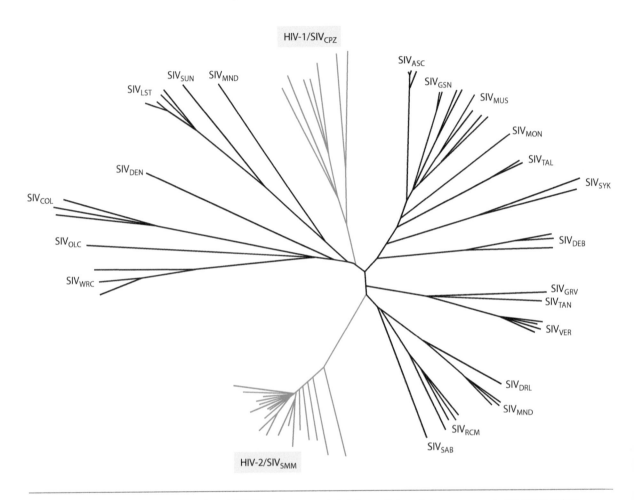

FIGURE 22.1 Relationship between HIV and SIV. Shown is an unrooted phylogenetic tree of virus sequences isolated from various primates including humans. The subscript of SIV is an abbreviation for the monkey species from which the isolates were obtained. Each monkey species is infected with its own subtype of SIV. Molecular phylogenetic analyses indicate that HIV-1 is derived from SIV of chimpanzees (SIV$_{CPZ}$) and HIV-2 is derived from SIV of sooty mangabeys (SIV$_{SMM}$) as denoted by the red branches.

A pathogen jumping from one host species to another host species implies close contact between the two host species. In the case of SIV jumping to humans, close contact most likely involves hunting or the preparation of meat. Monkeys and other wild animals are often hunted and sold in markets as bush meat. A cut or an animal bite could allow for the entry of animal blood or other bodily fluids into a human host. Most of the time such an occurrence is of little consequence since most pathogens have a relatively restricted host range. However, if a variant of SIV was able to infect human cells and replicate, this variant could evolve into HIV. Subsequent transmissions and further adaptation to the human host would result in a new human pathogen. Molecular analyses have detected several jumps of SIV to humans. Most of these cross-species transmissions have been inconsequential because relatively few people have been affected. However, some of the jumps have resulted in widespread human transmission and the origin of AIDS.

HIV-1 is more common and more virulent than HIV-2

HIV-1 causes more than 95% of AIDS and is found throughout the world. In contrast HIV-2 is found primarily in western Africa. However, some spread of HIV-2 to other parts of the world is occurring. Correlated with its increased prevalence, HIV-1 is more transmissible than HIV-2 (TABLE 22.3). Furthermore, HIV-2 is less likely to progress to AIDS and progresses to AIDS more slowly than HIV-1. The two viruses only exhibit 55% sequence identity, and this genetic difference may explain the lower infectivity and virulence of HIV-2. There are also some practical consequences of this genetic difference since not all diagnostic tests can detect both viruses and some of the efficacious drugs against HIV-1 do not work as well against HIV-2.

Table 22.3 HIV-1 vs. HIV-2			
	Virulence	**Infectivity**	**Prevalence**
HIV-1	High	High	Global
HIV-2	Lower	Low	West Africa

AIDS originated in central Africa approximately 100 years ago

Four types of HIV-1 have been described thus far and are designated as M, N, O, and P. Each type represents a distinct jump from chimpanzees or gorillas to humans. Group M (=main) is by far the most predominant type and causes more than 90% of AIDS worldwide. After appearing in humans, HIV-1 group M underwent massive divergence as measured by gene sequence diversity. At least nine subtypes of group M have been identified (FIGURE 22.2). These subtypes are generally denoted by letters (A, B, C, D, F, G, H, J, and K). In addition, recombinants between these subtypes have been identified and are called circulating recombinant forms (CRFs).

Through molecular phylogenetic analyses (Chapter 5) a hypothetical common ancestor of the group M HIV can be identified, and in this case that common ancestor represents the index case of the first human infected with the original SIV from a chimpanzee. The oldest known sample containing HIV is from a patient who died in 1959 of an AIDS-like illness. This retrospective sample can be used to adjust the molecular clock and geographical distributions of the subtypes can be used to predict a time and place of origin. The data suggest that the index case of HIV occurred around 1920 in the Democratic Republic of Congo. Furthermore, the genetic diversity of HIV is the greatest in central Africa, which is also consistent with the emergence of AIDS in central Africa. Since the emergence of AIDS in this hypothetical index case, more than 75 million people worldwide have developed AIDS.

HIV spread around the world causing a pandemic

After its origin, the migration of infected individuals from Africa spread the virus around the world. The spread of HIV was also facilitated by the lack of recognition of the disease for approximately 50 years after

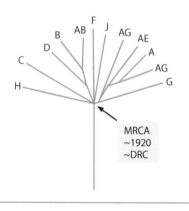

FIGURE 22.2 Radiation of HIV-1 group M. A phylogenetic tree showing group M subtypes and recombinant forms. The primary node of the branches represents the most recent common ancestor (MRCA). Using geographical and molecular clock data the MRCA, or index case, is estimated to have occurred around 1920 in the Democratic Republic of Congo (DRC).

its origin. This delayed recognition of the disease raises questions about why it was not recognized sooner. Several factors likely contributed to this lack of recognition of the newly emerged human disease (TABLE 22.4). The disease likely spread initially among rural low-income people who would not have frequently consulted physicians, and this is probably a major factor. Furthermore, physicians would not have seen the disease as something highly unusual against the background of the large number of tropical diseases that are prevalent in equatorial Africa. In the United States the disease was first recognized in the gay community of New York City and San Francisco. In contrast with the African population, this population was largely the upper-middle-class demographic and consulted physicians regularly. More importantly, the disease manifestations were rare opportunistic infections and cancers that stood out and were rather noteworthy.

Table 22.4   Factors Contributing to the Recognition of the AIDS Pandemic

Africa pre-1981	United States gay community pre-1981
• Individuals with HIV are likely to be poor and from rural communities • Rarely consulted physicians • Physicians would likely attribute disease symptoms (e.g., wasting) to many other tropical diseases	• Largely upper-middle-class Caucasians • Had health insurance and regularly consulted physicians • Increasing number of rare opportunistic infections and cancers

HIV TRANSMISSION

The predominant modes of HIV transmission are intimate sexual contact and mother-to-child transmission (TABLE 22.5). The virus is found in the blood and blood-borne transmission is also possible. In the early stages of the pandemic blood-borne transmission was relatively common. However, control measures have greatly reduced blood-borne transmission and currently blood-borne transmission is rather rare. For example, needle exchange programs have reduced transmission among intravenous drug users and testing of blood has nearly eliminated transmission from transfusion, transplantation, and blood products. Similarly, knowledge of the disease and extra precautions have greatly reduced transmission to health-care workers.

Table 22.5   HIV Transmission

- Intimate sexual contact (straight or gay)
- Mother-to-child transmission (MTCT) during gestation, labor, or postpartum (including breastfeeding)
- Needle stick or mucous membrane splash injuries (health-care workers)
- Shared needles (intravenous drug use)
- Blood or blood product transfusion and transplanted tissue (before routine testing)

Most HIV transmission is associated with sexual intercourse

For the most part, the risk factors for acquiring HIV infections are similar to the risk factors for acquiring other sexually transmitted infections, such as sexual promiscuity and unprotected sex. This virus is found in semen and vaginal secretions and transmission occurs primarily through vaginal or anal intercourse. In the early stages of the pandemic there was a focus on men who have sex with men. However, it is now recognized that heterosexuals are equally susceptible.

People with other STIs are at increased risk of becoming infected with HIV. This is due in part to similar risk factors. However, the prevalence of co-infections is greater than what could be simply attributed to similar risk factors. Some of the increased risks may be due to lesions or ulcers in the genital mucosa associated with many STIs. These lesions represent a breakdown of the barrier function of the mucosa, and this break in the epithelium facilitates the virus crossing the genital epithelium. In addition, there may also be inflammation of the genital mucosa associated with these other infections. This would result in more potential host cells in the area for the virus to infect and, thus, increase the chances of an infection becoming established. Furthermore, some studies have demonstrated that increased levels of HIV secretion are associated with co-infections by other STIs. Diagnosing and treating STIs in general have been demonstrated to reduce the incidence of HIV infection (BOX 22.2).

STI Treatment Reduces HIV Incidence

Having another sexually transmitted infection (STI) may increase the transmission of HIV due to increased inflammation in the genital mucosa. If this is true, then treatment of STIs may reduce the incidence of HIV. A study carried out in Mwanza, Tanzania by Grosskurth et al. (1995) demonstrated that diagnosing and treating STIs lowered the number of newly acquired HIV infections by approximately 40%. At 2 years of intervention, which consisted of establishing STI clinics, ensuring a regular supply of drugs and providing health education about STIs resulted in an incidence of 1.2% of newly acquired HIV infections. In the non-intervention comparison group the incidence was 1.9%.

Grosskurth H, Mosha F, Todd J et al. (1995) Impact of improved treatment of sexually transmitted diseases on HIV infection in rural Tanzania: randomized controlled trial. *Lancet* 346:530-536. doi: 0.1016/s0140-6736(95)91380-7.

BOX 22.2

Most children infected with HIV acquired the virus from their mothers

Mother-to-child transmission (MTCT) accounts for more than 90% of HIV infections in children under 15 years old. Without intervention 15–45% of children born to mothers infected with HIV acquire the infection. The infection can be transmitted from mother to child during gestation, during childbirth, or postpartum (TABLE 22.6). Approximately half of infected infants die within 2 years if not treated.

Table 22.6 MTCT of HIV	
Transmission	**Percentage infected mothers who transmit to the child**
During pregnancy	5–10
During labor/delivery	10–20
Through breastfeeding	5–20

VIRUS LIFE CYCLE

HIV is an enveloped double-stranded RNA virus and exhibits general features of other enveloped viruses (Chapter 19). A glycoprotein with a size of 120 kilodaltons, called gp120, is exposed on the outer surface of the envelope membrane of mature virus particles. Gp120 is a ligand that binds to a receptor protein called CD4 on the surface of host cells.

Binding of gp120 to CD4 initiates a fusion between the viral envelop membrane and the plasma membrane of the host cell. Fusion of these two membranes introduces the virus capsid into the cytoplasm of the host cell (FIGURE 22.3). The capsid then disassembles releasing the dsRNA viral genome.

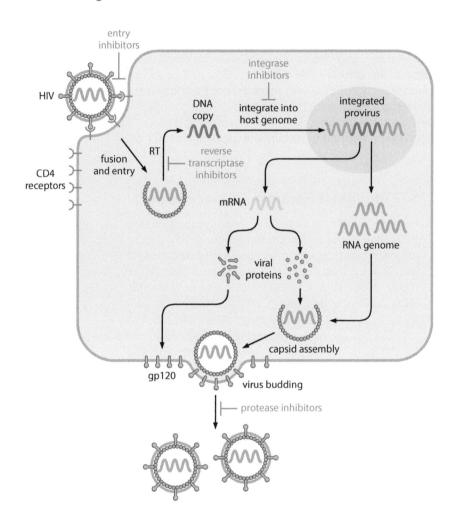

FIGURE 22.3 HIV life cycle. Binding of virus particles via receptor–ligand interactions between gp120 and the CD4 protein promotes the fusion of the virus envelope with the plasma membrane of the host cell. Following entry and disassembly of the capsid the viral RNA is converted by reverse transcriptase (RT) to a DNA copy that integrates into the host genome. Viral genes and the viral RNA genome are transcribed from the integrated provirus. Following translation, the viral proteins are assembled with the virus genome into capsids. Final viral assembly occurs at the host cell membrane and progeny viruses bud from the host cell. Chemotherapeutic targets are denoted by a red color.

Retroviruses generate a double-stranded DNA molecule that integrates into the host genome

HIV has a polymerase that converts the dsRNA genome into a dsDNA copy of the viral genome. This polymerase is best known as **reverse transcriptase** in reference to this synthesis being the opposite of transcription, which involves the generation of RNA from a DNA template (Chapter 4). In fact, the name retrovirus is in reference to this backward aspect of the normal molecular process of DNA serving as a template for the synthesis of RNA. Reverse transcriptase uses the viral RNA molecule as a template to synthesize a DNA copy of the viral genome. The dsDNA copy of the virus genome enters the nucleus and integrates into the host genome. A viral enzyme, called **integrase**, cuts the host DNA and splices the viral DNA into the cut site. This results in the viral DNA becoming a permanent part of the host genome. As the host cell replicates the viral DNA, called a **provirus**, is also replicated and thus included in all subsequent generations of the host cell.

Once integrated into the host genome, the viral DNA more or less functions like host cell genes. Viral genes are transcribed into viral mRNA molecules that are translocated to the host cell cytoplasm and translated via host cell ribosomes. The viral RNA genome is generated by transcription of the integrated provirus and the viral genome assembles with capsid proteins in the host cell cytoplasm (see Figure 22.3). Enveloped proteins, such as gp120, are transported to the host cell membrane via the secretory pathway. Virus assembly occurs on the host plasma membrane and the assembled virus forms by a process called budding, in which the host membrane with the embedded gp120 surrounds the capsid. Virus assembly also requires the action of a viral protease which generates the mature matrix proteins that make up the virus capsid. The newly formed virus particles are released from the host cell and are ready to infect other cells and repeat the replication cycle.

HIV infection results in a loss of helper T-cell function

CD4 protein, the receptor for HIV, is primarily found on helper T cells, dendritic cells, and macrophages. In fact, the CD4 protein is a marker for helper T cells and helper T cells are also called CD4+ (positive) cells. CD4+ cells play crucial roles in immunity. The first cells infected by HIV after it crosses the genital epithelium are likely macrophages and dendritic cells that are doing surveillance in the genital mucosa. These cells also serve as antigen-presenting cells and, as part of this function, migrate to the lymph nodes and other lymphatic tissues after encountering a pathogen (Chapter 9). There are numerous helper T cells in the lymphatics that can then subsequently become infected. As part of their normal function in eliminating pathogens, helper T cells also migrate throughout the body via the circulatory and lymphatic systems. Thus, the HIV infection spreads throughout the body, and does not remain localized in the genitalia.

This tropism for immune effector cells, and especially helper T cells, has pathological consequences in the development of immunity (**FIGURE 22.4**). Initially, the host makes a strong and effective immune response against the virus including antibodies and cytotoxic T cells. However, HIV infection interferes with the normal function of helper T cells, which is to stimulate other elements of the immune response such as antibody production and cytotoxic T cells. Normally helper T cells lead to pathogen elimination and protective immunity. Without the normal function of helper T cells, HIV is not eliminated and persists at a low level of viremia. This leads to chronic HIV infection. As the infection proceeds, the helper T cells die at a faster rate than they can be replenished, and therefore the number of helper T cells decreases. This decrease in helper T cells eventually leads to an immunodeficient state that progresses to AIDS.

PROGRESSION TO AIDS

The course of an HIV infection is generally divided into three phases (**TABLE 22.7**). The infection initially progresses through an acute stage in which viremia rapidly rises and subsequently falls (**FIGURE 22.5**). This is typical of many infections and represents the immune system controlling the infection. The symptoms during the acute phase may include fever, enlarged lymph nodes, muscle aches, rash, and tiredness. These are typical symptoms of many viral infections. Most people, however, do not develop symptoms during the acute phase and are unaware of being infected. Although the immune system controls the infection, it is unable to eliminate the infection and a low-level chronic infection ensues. The patient then enters into a long asymptomatic latent period before clinical AIDS develops.

Helper T cells are described in Chapter 9, pages 193–194.

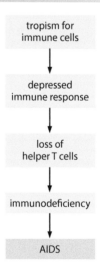

FIGURE 22.4 Progression of HIV infection to AIDS. HIV has a strong tropism for immune effector cells and especially helper T cells due to its affinity for the CD4 protein as a receptor. This eventually results in the loss of helper T cells and leads to immunodeficiency.

Table 22.7	Three Stages of HIV Infection
Acute infection	• Virus replication with viremia • Symptoms may include fever, enlarged lymph nodes, muscle aches, rash, tiredness
Asymptomatic latent phase	• Low viremia and decline in helper T cells (CD4⁺) • Seropositive (antibodies against HIV present) • Can last for years (average 8 years)
Symptomatic AIDS	• CD4⁺ cells fall below 200/μL of blood • Increased susceptibility to opportunistic pathogens

FIGURE 22.5 Natural course of HIV infection. After infection there is an acute phase characterized by a rise in viremia which is followed by a drop in viremia as the host mounts an immune response against the virus. Anti-HIV antibodies are present during the asymptomatic latent phase, but the virus is not eliminated. During the latent phase there is also a decrease in helper T cells. Once the helper T cells drop below a critical level, the viremia again increases, which marks the beginning of AIDS. The loss of helper T cells is also associated with increased susceptibility to opportunistic infections.

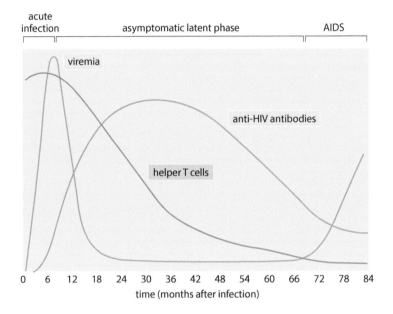

AIDS is characterized by a long latent period without clinical symptoms followed by immune deficiency

The latent phase can last for years with 8 years being the average. During the latent period patients are asymptomatic and viremia is difficult to detect. Viruses can only be detected by methods like PCR (see Box 7.2), which amplifies viral RNA and is very sensitive. The latent phase is also characterized by antibodies in the blood which can also be used to diagnose latent infections. As the latent phase progresses there is also a steady decline in the number of helper T cells (see Figure 22.5). If the helper T cells fall below a critical level a loss in immunity is observed marking the onset of clinical AIDS. One of the clinical criteria of AIDS is helper T cell levels below 200 cells per microliter (TABLE 22.8). Below this level helper T cells can no longer suppress the viremia and consequently viremia rises. Viremia refers to the number of virus particles found in the blood. In addition, the patient becomes more susceptible to other infectious diseases since helper T cells play a critical role in controlling most infections.

Table 22.8	Criteria for Diagnosing AIDS
	• Presence of opportunistic disease • CD4⁺ cell counts < 200/μl • Presence of HIV antibodies (latent) • Presence of viral RNA (detected by PCR)

A characteristic associated with AIDS is the presence of opportunistic infections

Opportunistic infections are caused by pathogens that normally do not cause disease or cause mild disease in immunocompetent people. However, in immunodeficient people these pathogens cause severe disease (BOX 22.3). In fact, before diagnostic tools such as antibody detection and PCR were available, opportunistic infections were the defining criteria for diagnosing AIDS. In addition, as AIDS progresses, there is a general decline in health and often a wasting syndrome develops. Major clinical manifestations include weight loss, chronic diarrhea, and prolonged fever (TABLE 22.9). Without treatment AIDS progresses to death.

Opportunistic Pathogens

A hallmark feature of AIDS is an opportunistic disease (**BOX 22.3 TABLE 1**). In general, opportunistic infections are normally mild or asymptomatic in immunocompetent people, but more severe or even fatal in immunodeficient people. The identification of AIDS was due in large part to an unusually high prevalence of *Pneumocystis jirovecii* pneumonia (PCP) among gay men. Until the AIDS pandemic, PCP was a relatively rare disease and generally associated with malnourished children or prolonged use of glucocorticoids. Like many other fungi (Chapter 19), *Pneumocystis jirovecii* (formerly *carinii*) is a commonly encountered fungus that is non-pathogenic in immunocompetent people. Many opportunistic pathogens also cause disease in immunocompetent individuals, but the disease is more severe or lasts longer in immunocompromised ones. Chronic infections are common in patients with AIDS. In addition, some pathogens caused localized infections in the immunocompetent host, while causing disseminated systemic infection in the immunocompromised host.

Box 22.3	Table 1 Common Opportunistic Pathogens and Tumors in AIDS
Viruses	Cytomegalovirus, herpes simplex, herpes zoster (shingles)
Bacteria	Tuberculosis, *Mycobacterium avium* complex (MAC), pneumonia, and septicemia
Protozoa	*Cryptosporidium, Toxoplasma*
Fungi	*Pneumocystis, Candida*, microsporidia, *Cryptococcus, Histoplasma*
Tumors	Kaposi's sarcoma, non-Hodgkin's lymphoma
Other	Wasting disease, HIV encephalopathy

BOX 22.3

Table 22.9	Clinical Features Associated with AIDS	
Major signs	**Minor signs**
• Weight loss > 10% • Chronic diarrhea > 1 month • Prolonged fever > 1 month	• Persistent cough > 1 month • Generalized dermatitis • Candidiasis of the mouth and throat • Disseminated herpes simplex • Generalized lymphadenopathy

ANTI-RETROVIRAL THERAPY

AIDS is treatable through chemotherapy and numerous drugs that slow virus replication are available. The life cycle of HIV provides several targets for the disruption of viral replication (see Figure 22.3). For example, reverse transcriptase is a unique polymerase that is not found in host cells and numerous drugs have been developed that inhibit reverse transcriptase. Similarly, integrase is another unique enzyme specific to HIV. Drugs that inhibit the protease involved in virus assembly are also available. Inhibiting any of these enzymes blocks virus replication and decreases viremia. The other antiviral strategy is to prevent viral entry through drugs that interfere with the virus binding to receptors or prevent the fusion of the viral envelope with the host plasma membrane.

Combinations of drugs are used to treat HIV infections

Treatment of HIV infections involves the use of drug combinations from the different classes of anti-retroviral drugs (TABLE 22.10). This is known as anti-retroviral therapy (ART). One advantage of simultaneously using multiple drugs against different targets is the synergistic effect. In simple terms, synergy means that the effect of the combined drugs is more than the sum of the effects of the individual drugs.

Table 22.10 Examples of Anti-retroviral Drugs	
Reverse transcriptase inhibitors (nucleotide)	Analogs of nucleotides that preferentially inhibit the viral polymerase
Reverse transcriptase inhibitors (non-nucleotide)	Inhibitors of the viral polymerase that are not analogs of nucleotides
Integrase inhibitors	Drugs that prevent the integration of the proviral DNA into the host genome
Protease inhibitors	Drugs that inhibit the virus protease that is important for virus assembly and budding from the host cell
Entry inhibitors	Drugs that either inhibit the fusion of the viral envelope with the host plasma membrane or drugs that block virus binding to the CD4 receptor

However, the major reason for drug combinations is to slow the development of drug resistance. HIV has a high mutation rate, and the virus rapidly generates resistant variants that can be selected under drug pressure. The high mutation rate is due to the high error rate of reverse transcriptase. This high error rate also accounts for the massive radiation observed during the evolution of HIV (see Figure 22.2). Using drug combinations against multiple targets, multiple mutations need to occur in a similar time frame to generate drug resistance. Therefore, drug resistance is slower to develop in the presence of drug combinations. The use of single drugs to treat HIV is essentially ineffective since resistance develops quicker than the drugs can substantially impact virus replication.

SIGNPOST

The development of drug resistance is discussed in detail in Chapter 19, pages 379–382.

ART does not cure HIV infections and treatment must continue for the life of the patient

ART can suppress blood levels of viral RNA to below the level of detection. Normally, this would be considered a cure. However, HIV can exist as a latent infection due to its integration into the host genome (BOX 22.4). Antiviral drugs are ineffective against this latent proviral form of HIV since the drugs target the production of progeny virus particles. This latent virus is passed to the progeny of the infected T cells as T cells replicate, and this passive viral replication is not affected by the drugs. The latent virus periodically reactivates and starts producing viral particles. Thus, if an infected person stops taking the drugs, reactivation of HIV replication produces viremia and the disease will continue to advance. Therefore, infected people must continue with ART for their entire lives to continuously suppress virus replication.

Proviruses and Latent HIV Infection

BOX 22.4

As part of its life cycle, HIV integrates its genome into the genome of the host cell. This integrated genome is called a provirus. The provirus is replicated with the host cell and passed to the progeny of the original infected cell. Therefore, the provirus maintains the infection even though viral particles are not being produced. In addition, anti-retroviral drugs are not effective against the provirus since the anti-retroviral drugs target various steps in the production of viral particles. Similarly, the immune system is unable to attack the provirus.

The primary reservoir of HIV-1 latency is probably long-lived memory helper T cells. Memory cells are generated after infection and remain in a quiescent state until stimulated by antigen (Chapter 9). Within these quiescent memory T cells, the HIV provirus is transcriptionally silent and undetectable by the immune system, and therefore the provirus provides a long-lived reservoir for the infection. The factors associated with the reactivation of provirus are not well understood but may be associated with the activation of the memory T cell by antigen. Activation of a productive viral infection means that the virus is once again susceptible to chemotherapeutic and immunological attack.

Even though ART does not cure HIV infection it has substantially increased the lifespan of infected individuals (FIGURE 22.6) and improved the quality of life. Before ART, patients diagnosed with AIDS generally died within a few years. Maintaining ART can lead to survival for decades. However, the efficaciousness of ART needs to be continually evaluated and the drug regimen may need to periodically change as drug resistance develops. Furthermore, some of the drugs have adverse side effects and this may lead to compliance problems.

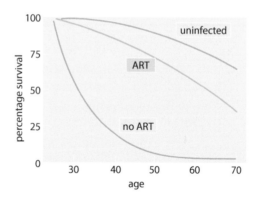

FIGURE 22.6 **Increased survival associated with anti-retroviral therapy (ART).** Treatment with ART increases the survival of individuals with HIV. However, survival is still less than in uninfected individuals.

Anti-retroviral drugs can also be used for prophylaxis

The availability of efficacious drugs that are relatively non-toxic has opened the possibility of chemoprophylaxis, in addition to safe sex as a means to prevent HIV infection. In the case of AIDS prevention, this is commonly referred to as **pre-exposure prophylaxis**, or PrEP. People at risk of becoming infected take the prescribed drugs in combination with other preventive measures for STIs. Similarly, mother-to-child transmission can be prevented with anti-retroviral drugs. Anti-retroviral drugs are administered to infected mothers during pregnancy and delivery. Newborn infants from infected mothers are also treated immediately after birth. Countries that have implemented chemoprophylaxis of infected mothers and their newborns have seen a substantial decrease in the number of children with HIV.

SUMMARY OF KEY CONCEPTS

- Sexually transmitted infections (STIs) refer to pathogens that are primarily transmitted via intimate sexual contact

- Prevention of STIs primarily focuses on practices that are collectively called safe sex

- The pathophysiology of STIs is primarily associated with the urogenital tract and most sexually transmitted pathogens do not exhibit a high level of mortality or morbidity

- Among the sexually transmitted pathogens, human immunodeficiency virus (HIV) has the highest rate of mortality

- HIV emerged from related retroviruses of non-human primates in Africa and spread around the world

- Modes of HIV transmission include sexual contact, mother-to-child transmission, and, to a lesser extent, blood-borne transmission

- HIV infects helper T cells and other immune effector cells leading to an immunodeficiency syndrome due to the loss of helper T cells

- Acquired immune deficiency syndrome (AIDS) is characterized by increased susceptibility to opportunistic infections and diseases

- Numerous drugs that target unique features of the retrovirus life cycle are available for the treatment and management of AIDS and HIV infections

- Anti-retroviral drugs do not cure HIV infections and need to be taken for the patients' entire lives to prevent the development of AIDS and early death

- The development of drug resistance is a major problem in the treatment of HIV infections and anti-retroviral drugs are always used in combinations

KEY TERMS

acquired immune deficiency syndrome (AIDS)	A sexually transmitted infection characterized by opportunistic infections that is caused by HIV.
anti-retroviral therapy (ART)	Combinations of drugs used to suppress the replication of HIV to prevent the development of AIDS.
CD4 protein	A protein expressed on the surface of helper T cells and other immune effector cells that also functions as the receptor for HIV.
human immunodeficiency virus (HIV)	The virus that causes AIDS in humans by infecting helper T cells and suppressing cell-mediated immunity.
integrase	An enzyme that incorporates the DNA copy of retroviral genomes into the genome of the host cell.
opportunistic infection	Microbes that normally do not cause disease except in individuals with a compromised immune system.
pre-exposure prophylaxis (PrEP)	The use of anti-retroviral drugs to prevent infection with HIV.
provirus	A viral genome integrated into the genome of the host cell.
retrovirus	A class of double-stranded RNA viruses that converts the RNA genome into a DNA molecule that is incorporated into the host genome.
reverse transcriptase (RT)	An RNA-dependent DNA polymerase found in retroviruses and that converts the RNA genome into a DNA copy of the genome.
safe sex	Practices such as limiting the number of sexual partners or barrier methods (e.g., condoms) to prevent STIs.
sexually transmitted infection (STI)	Infections that are primarily transmitted during sexual intercourse or other sexual activities.

REVIEW QUESTIONS

1. What are reverse transcriptase, integrase, and the HIV protease, and what roles do these proteins play in the life cycle of HIV and the treatment of HIV infections?

2. Describe the progression to AIDS following HIV infection to the appearance of clinical manifestations.

3. Why is it not possible to cure HIV infections and why do drug regimens need to be routinely changed in the management of HIV infections?

4. List and briefly describe some of the opportunistic infections that are often associated with AIDS.

ADDITIONAL QUESTIONS

In addition to the Review Questions provided above, there is a range of free online questions designed to enable students to further test their understanding of the chapter material. In order to access these interactive questions, please visit the book's website: https://routledge.com/cw/wiser

Malaria and Other Vector-transmitted Diseases

23

The transmission of most infectious diseases involves ingestion of contaminated food or water (Chapter 21), inhaling pathogens (Chapter 20), or intimate contact (Chapter 22). Thus, pathogens capitalize on activities necessary for the survival of individuals or species to ensure their own propagation. Some pathogens have adapted strategies involving another organism for transmission to humans. Intermediate hosts refer to another species infected by the pathogen as a necessary step in the pathogen's life cycle. Some parasite worms (Chapter 24) have life cycles involving intermediate hosts. **Vectors** are blood-feeding arthropods that transmit the pathogen from one host to another host (FIGURE 23.1). Mosquitoes and ticks are the most common vectors of human pathogens, but fleas, lice, and biting flies also transmit human diseases.

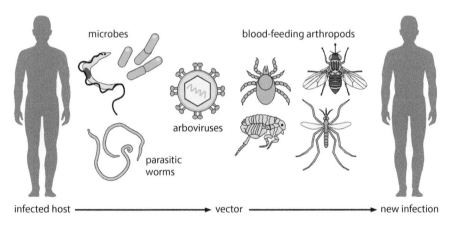

microbes   blood-feeding arthropods

arboviruses

parasitic worms

infected host ⟶ vector ⟶ new infection

FIGURE 23.1 Vector-transmitted diseases. A wide range of pathogens can be transmitted by a variety of blood-feeding arthropods. Blood-feeding arthropods serving as vectors acquire the infection by feeding on an infected person. The pathogen replicates and matures within the vector. A subsequent feeding on another person may initiate a new infection.

Approximately 17% of all infectious diseases are transmitted by arthropod vectors and these diseases result in more than 700,000 deaths annually. Among the vector-transmitted diseases malaria causes the most mortality and morbidity. The causative agent of malaria is a protozoan pathogen of the genus *Plasmodium*. Mosquitoes are the vector, and malaria is found in tropical areas throughout the world. In general, vector-borne diseases, and especially mosquito-transmitted diseases, are more prevalent in tropical climates.

VECTOR-BORNE TRANSMISSION

During vector transmission a blood-feeding **arthropod** becomes infected while feeding on an infected host. Thus, pathogens transmitted in this fashion need to be present in the blood at some point during their life cycle. This is one of several challenges for vector-transmitted pathogens (TABLE 23.1). The immune system is rather efficient at removing pathogens from the blood and vector-transmitted pathogens need mechanisms to avoid the immune system and to maintain sufficient levels of the pathogen in the blood. Many vector-transmitted pathogens establish long-term chronic infections in the host to increase the likelihood of transmission. In addition, there also needs to be enough vector present and, more importantly, sufficient vector–host contact to maintain the person-to-person transmission via the vector. For this reason, most vector-borne diseases have a defined geographical distribution as determined by the geographical distribution of the vector.

Table 23.1 **Challenges for Vector-transmitted Pathogens**
• Requires presence of vector and substantial vector–host contact
• Requires persistence of the pathogen in host blood or subcutaneous tissues
• The pathogen must be able to survive and proliferate in two relatively different hosts

Vector-transmitted pathogens generally have complex life cycles

Another challenge for vector-transmitted pathogens is the need to adapt to two rather different host organisms. The physiology of the arthropod vector is quite distinct from the physiology of humans or other vertebrates. Therefore, after being taken up by the vector, the pathogen needs to adapt quickly to this new environment and begin to proliferate. In addition, many of these pathogens undergo a developmental process or may need to migrate to different locations within the vector. There is a tendency for the vector to be viewed as a 'flying syringe' that transports the pathogen between hosts. However, quite often the life cycle of the pathogen within the vector is more complex than its life cycle in the host. It is possible for arthropods to become contaminated with a pathogen in their mouthparts, and then subsequently transmit the pathogen without development in the vector. This is called mechanical transmission and is not a true vector transmission.

After infecting the vector, the pathogen multiplies and matures within the vector. During this maturation period, the pathogen also needs to minimize the damage to the vector so that the vector survives long enough for the pathogen to mature. In addition, the maturation of the pathogen needs to be completed within the lifespan of the vector, which in some vectors is relatively short. Therefore, the pathogen is typically adapted to a specific vector species and vector-transmitted pathogens generally exhibit a high specificity for a single vector species or closely related vector species. If the pathogen sufficiently proliferates and matures to an infective stage, the vector is now capable of infecting another host.

During a subsequent bloodmeal the infected vector transfers the pathogen to a new host. The infective stage of the pathogen is often found in the saliva or associated with the mouth parts of the vector. Therefore, the pathogen usually gains entry to the new host via the bite wound. This is one advantage of vector transmission in that the vector helps the pathogen overcome the rather formidable barrier of the skin. After infecting the host, many vector-transmitted pathogens disseminate throughout the body and cause a systemic infection. Some pathogens, however, have a tropism for a particular tissue or organ. To complete the transmission cycle, infectious stages of the pathogen need to be present in the blood or subcutaneous tissues in sufficient numbers to infect a new vector.

Prevention and control of vector-borne disease involve vector control

Another advantage of vector transmission for the pathogen is that the vector potentially transmits the pathogen over comparatively long distances between hosts. Since many of the vectors for human diseases are flying insects, transmission can occur between hosts that are separated both physically and temporally. For example, a mosquito can become infected in one house and a few weeks later infect another person in a different house. This differs from the transmission of other infectious diseases that usually involves close and extended contact between the hosts. Therefore, the preventive measures used to control the transmission of other diseases often do not work well with vector-transmitted diseases.

Prevention of vector-transmitted diseases often focuses on the vector. There are three general strategies for the control of vector-transmitted diseases: (1) reduce human–vector contact, (2) reduce the vector, and (3) reduce the pathogen reservoir (TABLE 23.2). These approaches are not mutually exclusive and can be used in combinations. Obviously, methods that reduce vector–human contact such as repellents and screens on windows can lower the risk of acquiring a vector-transmitted disease. Similarly, widespread application of insecticides in a designated area can decrease the vector population, and that potentially lowers vector–human contact. Another approach is to decrease the number of infected people who serve as a source of infection for the vector. Diagnosis and treatment of infected people lower the pathogen reservoir, and this greatly reduces the chance of vector transmission.

Table 23.2 Control and Prevention Strategies	
Strategy	**Approaches**
Reduce human–vector contact	• Insecticide-treated bed nets (ITBNs) • Indoor residual spraying (IRS) • Screens on windows • Repellants • Protective clothing
Reduce vector	• Environmental modification • Larvicides/insecticides • Biological control
Reduce parasite reservoir	• Diagnosis and treatment of infected people

Vector-transmitted pathogens include a wide range of viruses, microbes, and parasites

Despite the challenges of vector transmission, many human pathogens are vector borne and some of them are quite prevalent. Numerous viruses are transmitted by vectors – primarily mosquitoes – and are collectively called **arboviruses** (arthropod-borne). This is not a phylogenetic group, but a convenience group since the various arboviruses are not necessarily related to each other. Several vector-borne diseases are caused by bacteria and protozoa as well. And somewhat unintuitively, there are even parasitic worms that are transmitted by vectors (Chapter 24). In biological terms, vector-transmitted diseases represent a wide range of vector species, including diverse ecologies, and a wide range of pathogen species. Furthermore, transmission involves complex interactions and interrelationships of the pathogen, the vector, and the host, and possibly a non-human reservoir in the case of zoonotic diseases. Ecology can also play a major role in vector transmission.

ARBOVIRUSES

There are several mosquito-transmitted viruses that infect humans. However, humans are not the natural hosts for most of these arboviruses and, thus, most arboviruses are zoonoses. In general, the incidences of these zoonotic arbovirus infections are low. However, sporadic outbreaks occasionally occur. Many of the viruses have geographically based names based on where they were initially described, such as the Japanese encephalitis virus, St. Louis encephalitis virus, West Nile virus, and Rift Valley fever virus. Some of these viruses do exhibit a defined geographical distribution, whereas others are more widespread. For example, West Nile virus is found in North America, Europe, Asia, Australia, and Africa. Birds being the natural reservoir for West Nile virus may account for this global spread. Most people infected with zoonotic arboviruses do not develop clinical disease. Mild symptoms generally include fever, headache, or malaise, and, with some arboviruses, rashes are common. As implied by many of the names, viral encephalitis is the most notable severe manifestation associated with arbovirus infections.

Among the anthroponotic arboviral diseases, which primarily involve human–vector–human transmission, yellow fever and dengue are the most prevalent and cause the most mortality (TABLE 23.3). Zika virus and chikungunya virus have garnered some attention lately due to their global spread and outbreaks during the early part of the 21st century. All four of these viruses can also infect non-human primates as well as humans. In the case of yellow fever, there are both sylvatic cycles in which non-human primates are the primary reservoir and urban cycles with humans as the reservoir. In general, the transmission of anthroponotic arboviruses is predominantly in urban or semi-urban areas of tropical and sub-tropical regions in many parts of the world.

Table 23.3	Anthroponotic Arboviral Diseases		
Disease	**Incidence per year**	**Deaths per year**	**Hallmark complication or severe manifestation(s)**
Yellow fever	200,000	30,000	Jaundice progressing to multi-organ failure
Dengue fever	500,000	20,000	Dengue hemorrhagic fever, dengue shock syndrome
Zika	Outbreaks	Rare	Microcephaly and other birth defects if infected in utero
Chikungunya	Outbreaks	Rare	Severe joint pain that may persist for months

Despite a vaccine and a good knowledge about transmission, yellow fever still causes significant mortality

Yellow fever was the first viral disease to be demonstrated to exhibit mosquito transmission, and it was the first human virus to be isolated. During the 18th and 19th centuries, yellow fever was considered one of the most dangerous infectious diseases. Approximately 15% of infected people develop severe disease initially characterized by jaundice, hence the name yellow fever. The disease can progress to severe hepatitis, renal failure, hemorrhage, shock, and multi-organ failure. A vaccine is available, and some endemic countries include the yellow fever vaccine as part of their routine immunizations or use the vaccine to control outbreaks. The vaccine is also recommended for people traveling to parts of Africa or South America where the disease is still endemic. Even with the vaccine, it will be difficult to eliminate yellow fever since the sylvatic cycle in non-human primates can serve as a source for the re-introduction of the virus into urban human transmission cycles.

Dengue virus occasionally causes a viral hemorrhagic fever

The most prevalent of the human arboviruses is dengue virus. It is estimated that there may be nearly 400 million infections per year. However, most of the infections are asymptomatic or mild. Among the symptomatic individuals, the most common symptoms of dengue are a high fever, headache, vomiting, muscle and joint pain, and a characteristic skin rash. This form of disease is called dengue fever. A small percentage of dengue fever cases develop into dengue hemorrhagic fever or dengue shock syndrome. Manifestations associated with severe dengue are vascular damage and plasma leakage, bleeding, and low levels of platelets. These are also typical symptoms of other viral hemorrhagic fevers. Continued plasma leakage can lead to a dangerous drop in blood pressure and the progression to dengue shock syndrome. Persons re-infected with a different serotype of dengue virus may be at increased risk of developing severe dengue due to antibody-dependent enhancement (BOX 23.1).

Severe Dengue Fever and Antibody-dependent Enhancement

BOX 23.1

There are four major serotypes of dengue virus designated 1–4. Infection with one serotype generally confers immunity against that serotype, but only confers partial immunity against the other three serotypes. This suggests that the antibodies produced against one serotype are suboptimal in clearing an infection caused by a different serotype. In addition, having a second dengue infection with a different serotype to the first infection may increase the risk of developing severe dengue hemorrhagic fever or dengue shock syndrome. It is believed that suboptimal antibodies from the primary infection, instead of a neutralizing virus from the secondary infection, actually increase the ability of the virus to invade host cells or to replicate. This phenomenon is called antibody-dependent enhancement (ADE).

The mechanism of ADE is not known and there are probably several different mechanisms. For example, the binding of a suboptimal antibody to the dengue virus may block the binding of neutralizing antibodies. Or memory cells from the primary infection may prejudice the immune response and highly effective antibodies against the secondary virus are not produced. Suboptimal antibodies may also enhance the ability of a virus to infect cells with Fc receptors, such as dendritic cells, macrophages, and monocytes (**BOX 23.1 FIGURE 1**). These suboptimal antibodies could also direct the virus to a compartment within the immune effector cells where the virus is not destroyed and is able to replicate.

Obviously, ADE needs to be considered during vaccine development since a vaccine can potentially worsen the disease via ADE. Dengue vaccines are complicated by the four serotypes and the potential for ADE. The only approved dengue vaccine is tetravalent and includes all four serotypes. Clinical trials indicated that the vaccine partially prevents infection. However, vaccinated individuals who have not been previously infected, as indicated by seronegativity, were at increased risk of developing severe disease following subsequent dengue infection. ADE is a possible explanation for more serious adverse outcomes in vaccinated seronegative people compared with vaccinated seropositive people. Therefore, the vaccine is only recommended for those individuals who have previously had dengue fever or for populations in which most people have been previously infected.

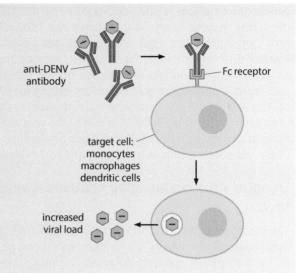

BOX 23.1 FIGURE 1 Possible mechanism of antibody-dependent enhancement of dengue virus. Anti-dengue virus (DENV) antibodies binding to the virus may allow the virus to utilize the Fc receptor as a mechanism to infect cells. This potentially increases the viral load if the virus is able to replicate within these Fc-receptor-bearing target cells. [Adapted from Laureti M, Narayanan D, Rodriguez-Andres J, et al. (2018) *Frontiers in Immunology* 9:2180. doi: 10.3389/fimmu.2018.02180. Published under CC BY license.]

ARTHROPOD-TRANSMITTED BACTERIAL DISEASES

Several bacteria are transmitted by blood-feeding arthropods (TABLE 23.4). Ticks, fleas, and lice are the most common vectors and most of these diseases are zoonoses. Currently, none of these diseases is highly prevalent and they do not account for much mortality. However, historically some of these diseases were major sources of mortality. For example, a pandemic of plague in the 14th century, called 'Black Death,' may have killed one-third of Europe's population. In the 19th century Napoleon lost more soldiers to typhus than to battles. Antibiotics, improved housing, and improved hygiene have greatly reduced bacterial diseases transmitted by arthropods.

Table 23.4	Examples of Arthropod-transmitted Bacteria			
Disease	**Pathogen**	**Vector**	**Reservoir**	**Epidemiology**
Plague	*Yersinia pestis*	Flea	Rats	600 cases/year worldwide, 30–90% mortality rate if untreated
Typhus	*Rickettsia prowazekii*	Body louse	Humans	Epidemics associated with refugee camps during civil unrest or natural disasters
Lyme disease	*Borrelia* species	Tick	Small mammals and birds	Most common tick-borne disease in the Northern Hemisphere
Human granulocytic anaplasmosis	*Anaplasma phagocytophilum*	Tick	Cattle and other ruminants	A few hundred cases per year
Carrión's disease	*Bartonella bacilliformis*	Sandfly	Humans	A rare infectious disease found in Peru, Ecuador, and Colombia
Cat scratch disease	*Bartonella henselae*	Flea	Cats	Transmitted between cats via fleas and humans become infected through scratches or bites
Trench fever	*Bartonella quintana*	Body louse	Humans	Originally described in soldiers during World War I. Currently found primarily in destitute people

ARTHROPOD-TRANSMITTED PROTOZOAL DISEASES

There are five vector-borne protozoal diseases of humans (TABLE 23.5). These are all rather distinct diseases in terms of the pathogens, vectors, and pathophysiology of the disease. The dynamics of transmission are complex with some being anthroponotic, whereas others are zoonotic.

Their geographical distribution tends to be focal depending on the presence of the vector and sufficient vector–host contact, as well as the presence of the non-human non-reservoir if needed. Except for babesiosis, vector-transmitted protozoal diseases are predominantly found in tropical or sub-tropical regions. In terms of prevalence and mortality, malaria is clearly the most important among these protozoan pathogens.

Table 23.5 Vector-transmitted Protozoa			
Disease	**Pathogen(s)**	**Vector**	**Features**
Malaria	*Plasmodium* species	Mosquitoes	A febrile disease that causes approximately 200 million clinical cases with 500,000 deaths per year
Cutaneous or visceral leishmaniasis	*Leishmania* species	Sandflies	A disease found in tropical and semi-tropical areas throughout the world that causes a range of disease manifestations depending on parasite species
Human African trypanosomiasis	*Trypanosoma gambiense* (anthroponotic), *T. rhodesiense* (zoonotic)	Tsetse flies	A disease, found in focal locations throughout equatorial Africa, that causes a fatal neurological disease if not treated, which is also known as African sleeping sickness
Chagas' disease	*Trypanosome cruzi*	Triatomine bugs	A major cause of cardiac disease in South and Central America primarily associated with living in houses infested with the vector
Babesiosis	*Babesia divergens* (Europe), *B. microti* (eastern USA), *B. duncani* (western USA)	Ticks	A relatively rare zoonotic infection in humans. The disease can be severe and fatal in immunocompromised people

Control efforts have significantly reduced the prevalence of African trypanosomiasis and Chagas' disease over the past several decades. The reduction in African trypanosomiasis has been primarily facilitated through case detection and treatment of infected people. This reduces the reservoir of *Trypanosoma gambiense* in humans and limits the spread of the infection. Reduction in Chagas' disease has primarily been accomplished through indoor residual spraying of houses infested with the vector *Triatoma infestans*. Removal of the vector from houses prevents transmission to humans. Control of leishmaniasis is substantially more complicated because more than 20 different species of *Leishmania* infect humans. Furthermore, most leishmaniasis is zoonotic and involves a wide range of reservoir hosts and numerous species of sandfly vectors. Therefore, the transmission dynamics of leishmaniasis depend on the specific parasite species, specific vector species, and specific reservoir host found in a particular location.

MALARIA

Malaria is a febrile disease that is found throughout tropical and sub-tropical regions. It is caused by protozoan parasites of the genus *Plasmodium*. There are five *Plasmodium* species that infect humans (TABLE 23.6). Four of these species are strictly human pathogens and one is largely zoonotic. Among these species, *P. falciparum*, causes the most morbidity and mortality. *P. vivax* is also common, but rarely results in death. About half of malaria outside of sub-Saharan Africa is due to *P. vivax*. The other species of human malaria parasites are less common. In addition, *P. ovale* and *P. malariae* cause a rather benign disease. *P. knowlesi* is a natural parasite of monkeys and can be mortal in humans. However, human infections with *P. knowlesi* are rather rare and primarily seen in Malaysia.

Table 23.6	*Plasmodium* Species Infecting Humans
Species	**Key features**
P. falciparum	• Found throughout the tropics and sub-tropics • Mortality associated with disease complications due to sequestration of infected erythrocytes in deep tissues
P. vivax	• Found in most tropical areas and can extend into temperate zones • The lower prevalence in Africa due to high prevalence of Duffy-negative phenotype in African populations (parasite uses Duffy antigen to invade erythrocytes) • It can cause severe febrile attacks, but very low mortality • Relapses from the liver stage are possible
P. malariae	• Spotty distribution throughout the tropics • Causes a rather benign disease • It can produce sub-patent infection lasting for decades
P. ovale	• Occurs primarily in tropical west Africa • Causes a rather benign disease • Relapses from the liver stage possible
P. knowlesi	• Natural parasite of macaque monkeys • Some human infections reported in Malaysia and southeast Asia • Some mortality associated with infection

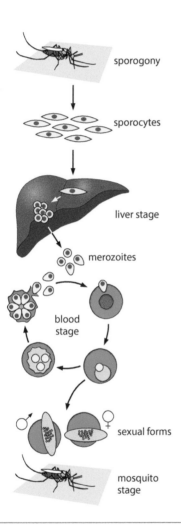

FIGURE 23.2 Malaria parasite life cycle.
Infection of humans is initiated by sporozoites that are injected into the blood during mosquito feeding. The sporozoites are carried to the liver and replicate within liver cells to produce merozoites. Merozoites are released from the infected liver cells and invade erythrocytes. Parasite replication in the erythrocytes produces more merozoites that are released from the infected erythrocytes and again invade erythrocytes. Repeated rounds of replication during the blood stage maintain the infection. Some merozoites develop into sexual forms called gametocytes which are infective for mosquitoes. After being taken up by a mosquito during feeding, the sexual forms fuse and undergo a process called sporogony to produce sporozoites and complete the life cycle.

The malaria parasite has a complex life cycle involving mosquitoes

Infection is initiated via the bite of a mosquito (**FIGURE 23.2**). Only mosquitoes of the genus *Anopheles* can transmit malaria to humans. **Sporozoite** stages in the mosquito saliva are injected as the mosquito takes a blood meal and these sporozoites enter the circulatory system. The sporozoites invade liver cells and undergo an asexual replication that results in the production of thousands of **merozoites**. This asexual replication is called merogony and during merogony the parasite undergoes multiple rounds of nuclear division without cytoplasmic division. Subsequently, the parasite undergoes a segmentation process to produce merozoites. The infected liver cell ruptures and releases the merozoites into the blood. The released merozoites invade erythrocytes and undergo another round of merogony and the newly produced merozoites are again released into the blood following the rupture of the infected erythrocyte. This invasion of erythrocytes and merogony is continuously repeated and maintains the infection. In addition, the blood stage of the infection is also responsible for the disease pathology.

As an alternative to this asexual replication, some of the merozoites differentiate into sexual forms – analogous to eggs and sperm that are called **gametocytes**. Gametocytes are infective to the mosquito and cause no pathology in the human host. When these sexual forms are taken up by a mosquito during feeding, they fuse (i.e., fertilization) and undergo a complex differentiation process involving meiosis called sporogony. Thousands of sporozoites are produced in the body cavity of the mosquito during sporogony and the sporozoites migrate to the salivary glands. The entire process of sporogony generally takes 9–21 days. After invading the salivary glands, the sporozoites are now ready to infect a new human host when the mosquito feeds again.

Periodic episodes of fever are a hallmark characteristic of malaria

A particularly characteristic manifestation of malaria is acute febrile attacks called paroxysms. These paroxysms have a defined periodicity of either 48 or 72 hours (FIGURE 23.3). A 48-hour periodicity means that an infected person has a paroxysm on one day – usually in the evening or night – and the next day feels fine before having another paroxysm the following day. *P. falciparum*, *P. vivax*, and *P. ovale* exhibit 48-hour periodicities between paroxysms. *P. malariae* exhibits a 72-hour periodicity and thus infected individuals have 2 days of feeling well between the paroxysms. Descriptions of these regular and periodic paroxysms have been noted in some of the earliest Indian and Chinese medical writings and were described by Hippocrates in 500 BC.

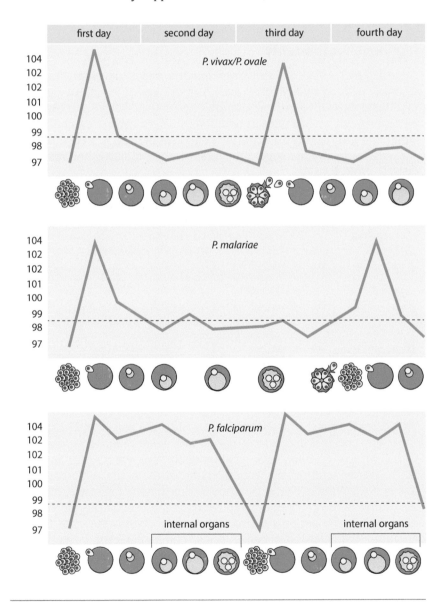

FIGURE 23.3 Correlation of malarial paroxysms to synchronous parasite replication.
During the blood stage of the infection, the replication cycles of the parasite are synchronized so that there is nearly simultaneous lysis of infected erythrocytes and release of merozoites. The release of immunogenic material associated with merozoite release causes periodic febrile attacks. Body temperature returns to normal while parasites are replicating within the infected erythrocytes. *P. falciparum* infections often exhibit continuous fever, and the infected erythrocytes sequester in the internal organs.

The periodicity of the paroxysm is due to the replication cycle of the parasite in the blood that takes either 48 or 72 hours depending on the parasite species. After a merozoite invades the host erythrocyte it undergoes a growth period and asexual replication to produce more merozoites. The host erythrocyte ruptures when the newly formed merozoites are mature and these merozoites invade new erythrocytes. The parasites within an infected individual are synchronous and therefore all the infected erythrocytes rupture at approximately the same time. Associated with the rupture of the infected erythrocytes is the release of a large amount of immunogenic material. This stimulates the production of inflammatory cytokines (Chapter 9) which cause periodic febrile attacks. *P. falciparum*, however, generally causes more severe disease manifestations characterized by continuous fever and does not always exhibit these periodic paroxysms.

P. falciparum can cause severe disease associated with a high level of mortality

Approximately 10% of *falciparum* malaria cases develop into severe disease with complications and a mortality rate of 10–50%. The manifestations of severe malaria are often organ specific and essentially all organs can be affected. In addition, multiple organs can be affected simultaneously. The most common complications associated with severe *falciparum* malaria are severe anemia, cerebral malaria, and respiratory distress (TABLE 23.7).

Table 23.7 Common Complications in *Falciparum* Malaria	
Severe anemia	• Especially prevalent in children • Due to both increased destruction and decreased production of blood cells • Sequestration in the bone marrow may affect the production of red blood cells
Cerebral malaria	• Neurological symptoms associated with sequestration in the brain • Consciousness ranges from stupor to coma • Convulsions frequently observed
Respiratory distress	• Associated with acidosis due to sequestration in the lungs • Can progress to respiratory failure • The best predictor of death

The increased severity associated with *falciparum* malaria compared with the other species is due in large part to the higher parasitemia associated with *P. falciparum* infections. Another distinctive feature of *P. falciparum* is the cytoadherence of infected erythrocytes to the endothelial cells lining the capillaries. Cytoadherence means that the infected erythrocytes are bound to the walls of the capillaries in the tissues and therefore are not found in the peripheral circulation. Only erythrocytes infected with immature parasites, often called 'rings,' can be detected in the blood. In other words, erythrocytes infected with mature parasites are sequestered in the tissues. This **sequestration** is mediated by modifications of the erythrocyte membrane by the parasite and parasite proteins exported to the host erythrocyte membrane (BOX 23.2).

BOX 23.2

Knobs and Cytoadherence

Cytoadherence is mediated by structures induced on the erythrocyte membrane by the parasite called knobs. These knobs are due to parasite proteins that are secreted by the parasite and associate with the erythrocyte membrane. Several parasite proteins are associated with the knobs and one of these is a transmembrane protein called PfEMP1 (*P. falciparum* erythrocyte membrane protein). PfEMP1 has a large extracellular domain and functions as a ligand that binds to receptors on the surface of endothelial cells. Although sequestration affords many advantages for the parasite, antigens on the surface of the infected erythrocyte provide a potential target for the immune system.

The parasite stays ahead of the immune system by having many different forms of PfEMP1. PfEMP1 is not a single protein, but rather is part of a gene family, called *var* genes for variable, and the parasite genome contains approximately 60 different alleles of PfEMP1. Sequential expression of the various *var* gene alleles results in antigenic variation (**BOX 23.1 FIGURE 1**) and allows the parasite to escape elimination by the immune system. This antigenic variation creates another potential problem for the parasite since a change in the protein sequence of PfEMP1 may affect its ability to bind to endothelial cell receptors. In fact, the various PfEMP1 proteins produced by the *var* gene family bind to different receptors found on endothelial cells. Therefore, the switch in PfEMP1 allele expression during antigenic variation also results in a change in the binding phenotype of the infected erythrocyte. This change in binding phenotype explains why multiple organs may be affected during severe *falciparum* malaria. A switch in the expression of a PfEMP1 allele may result in a different binding phenotype and sequestration in a different tissue.

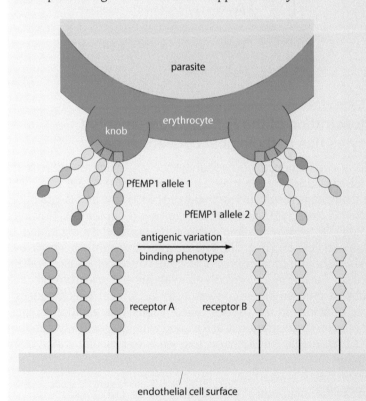

BOX 23.2 FIGURE 1 Antigenic variation of PfEMP1 alleles. *P. falciparum* parasites alter the erythrocyte membrane to produce knobs that have PfEMP1 expressed on the erythrocyte surface. PfEMP1 binds to a receptor on the surface of endothelial cells. Antigenic variation results in the expression of a different PfEMP1 allele on the surface of the erythrocyte. Accompanying this change in the PfEMP1 allele is a change in the binding phenotype of the infected erythrocyte due to the various PfEMP1 alleles binding to different endothelial cell receptors. [Adapted from Wiser MF (2011) *Protozoa and Human Disease*, Garland Science, Taylor & Francis.]

Sequestration has several benefits for the parasite. One major advantage is that infected erythrocytes that are adhered to the endothelial cells in the capillaries do not circulate through the spleen and, therefore, are not removed by the spleen (FIGURE 23.4). In addition, the low oxygen concentration in the deep tissues is better for the metabolism of the parasite and the parasite is subject to less oxidative stress. And, finally, the slow blood flow in the capillaries facilitates merozoite invasion. All these things increase parasite survival and result in higher parasitemia compared with the other malaria species. This greater parasite burden accounts for much of the increased disease severity.

FIGURE 23.4 Sequestration of *P. falciparum*-infected erythrocytes. Knobs appear on the surface of the infected erythrocyte as the intraerythrocytic parasite matures. These knobs mediate the adherence of the infected erythrocyte to endothelial cells lining the capillaries. This sequestration of the infected erythrocytes benefits the parasite in three ways: (1) the lower oxygen concentrations results in the formation of fewer oxygen radicals associated with parasite metabolism, (2) merozoite invasion is more efficient in the slowly moving blood of the capillaries, and (3) fewer infected erythrocytes are removed by the spleen.

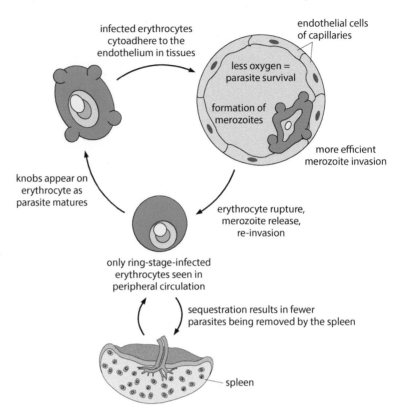

The sequestration of the *P. falciparum*-infected erythrocytes has pathological consequences

One obvious consequence of cytoadherence and sequestration is the blockage of capillaries by the infected erythrocytes (FIGURE 23.5). This mechanical blockage results in less localized blood flow and is analogous to the ischemia associated with atherosclerosis (Chapter 14). However, the pathology of severe malaria cannot be completely explained by decreased blood flow and hypoxia. The parasite also exhibits a high level of glycolysis without a tricarboxylic acid cycle which can cause localized hypoglycemia. Furthermore, glycolysis without the tricarboxylic acid cycle results in glucose being converted to lactic acid which causes a high degree of acidity in the tissues called acidosis. In addition to these metabolic effects, sequestration is associated with localized inflammation that affects endothelial cell function and leads to vascular leakage and hemorrhaging.

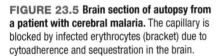

SIGNPOST   Glucose metabolism is discussed in greater detail in Chapter 3, pages 60–61.

FIGURE 23.5 Brain section of autopsy from a patient with cerebral malaria. The capillary is blocked by infected erythrocytes (bracket) due to cytoadherence and sequestration in the brain.

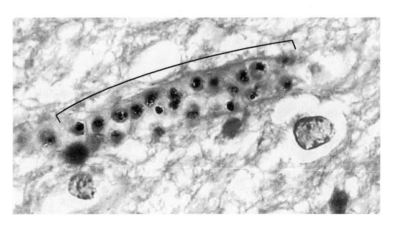

The ischemia, metabolic effects, vascular leakage, hemorrhaging, and inflammation all contribute to severe disease manifestations (FIGURE 23.6). For example, one potential complication of *falciparum* malaria is cerebral malaria, which is characterized by impaired consciousness and other neurological manifestations. In particular, hypoxia, hypoglycemia, acidosis, and vascular leakage directly interfere with neuron function and cause neurological manifestations. Similarly, respiratory distress is another complication associated with severe malaria and is a good predictor of death. Metabolic acidosis in the lungs is due to parasite sequestration and the associated high rate of glycolysis is the likely cause of lung injury, which can lead to respiratory failure. Likewise, sequestration in bone marrow suppresses hematopoiesis and contributes to the development of anemia. The organ-specific effects are due to the sequestration of infected erythrocytes in those particular tissues. Sequestration in a particular tissue is possibly due to the expression of a PfEMP1 allele that binds to a receptor found in the affected tissue (see Box 23.2).

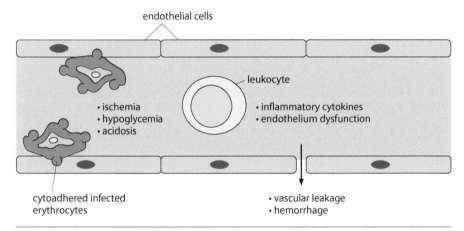

FIGURE 23.6 Pathology of severe *falciparum* malaria. Sequestration of infected erythrocytes leads to ischemia due to the cytoadherence of infected red blood cells which decrease blood flow. The high metabolism of the parasite produces localized hypoglycemia and acidosis, along with hypoxia due to the blockage of capillaries. Immune response to parasite antigens leads to high levels of inflammatory cytokines that disrupt normal endothelial cell function and cause vascular leakage and hemorrhage into the tissues.

Malaria is generally easy to diagnose and treat

Diagnosis of malaria relies heavily on detecting the parasite in the blood. Historically, malaria was diagnosed via microscopy of blood smears. Species can be distinguished based on morphological differences between species. Rapid diagnostic tests are probably used more frequently now than blood smears as the major means of diagnosing malaria. Many of the rapid diagnostic tests are specific for *P. falciparum* and some distinguish *P. falciparum* from the other species. As a general rule, all diagnosed malaria – even asymptomatic cases – should be treated since complications can develop. In addition, chemoprophylaxis is strongly encouraged for visitors to areas that are highly endemic for malaria.

Numerous antimalarial drugs have been developed and, except for drug resistance, the disease is generally easy to treat. The choice of drugs depends somewhat on the parasite species, the state of drug resistance, and the policy of the country or agency involved. Historically, chloroquine has been a widely used and efficacious drug for the treatment of malaria. Currently, though, *P. falciparum* is essentially resistant to chloroquine throughout the world. Chloroquine is still efficacious against the other

Plasmodium species, and still is often used to treat non-*falciparum* malaria. Since 2006, artemisinin combination therapy (ACT) has been recommended as the standard treatment for *falciparum* malaria (BOX 23.3). ACT refers to using artemisinin, or a derivative of artemisinin, combined with another antimalarial. The World Health Organization (WHO) strongly discourages using artemisinin as a monotherapy to slow the development of drug resistance that has plagued most of the previously developed antimalarial drugs.

BOX 23.3

Artemisinin

Artemisinin is a compound found in the sweet wormwood plant (*Artemisia annua*). This herb has long been used in Chinese traditional therapy for the treatment of fever. Research in the late 1960s led to the rediscovery of this antimalarial. Artemisinin is well tolerated with few side effects and the drug is fast acting with a short half-life. Several derivatives that also have antimalarial activity have been produced. The drug is only used in combination with another antimalarial. This promotes the efficient killing of the parasite and slows the development of drug resistance. The exact mechanism of drug action is not known but involves the cleavage of endoperoxide linkage of artemisinin (**BOX 23.3 FIGURE 1**). This cleavage occurs in the parasite's food vacuole in association with the digestion of

hemoglobin and thus is selectively toxic for the parasite. Reactive oxygen intermediates produced from the cleavage of the endoperoxide linkage damage parasite proteins and cause parasite death.

BOX 23.3 FIGURE 1 Chemical structure of artemisinin. A chemical feature of this compound is the endoperoxide (two oxygen atoms in red) linkage that spans the ring. Cleavage of the endoperoxide generates reactive oxygen intermediates that kill the parasite.

ANTIMALARIAL DRUG RESISTANCE

SIGNPOST

Mechanisms of drug resistance are described in Chapter 19, pages 379–382.

Drug resistance has been a problem in the treatment and control of malaria since the first antimalarial drugs were introduced. The parasite rapidly developed resistance to most drugs shortly after their introduction (FIGURE 23.7). Mechanisms associated with the development of drug resistance are generally similar to those of other pathogens. For example, point mutations in the genes of the drug target or the genes of transporters are prevalent mechanisms (TABLE 23.8). This includes ABC transporters that are often involved in multidrug resistance (see Box 19.5).

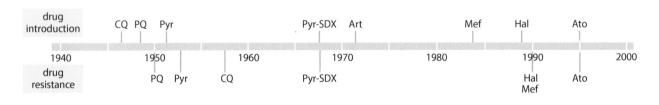

FIGURE 23.7 Timeline for the emergence of drug resistance against select antimalarial drugs. The text above the timeline indicates the year the drug was introduced and the text below the timeline indicates the year drug resistance was first observed. The drugs are chloroquine (CQ), proguanil (PQ), pyrimethamine (Pyr), sulfadoxine (SDX), artemisinin (Art), mefloquine (Mef), halofantrine (Hal), and atovaquone (Ato). The Pyr-SDX combination goes by the trade name Fansidar®.

Protein*	Location	Function	Drugs affected	Major polymorphisms**
CRT	Food vacuole	Transporter	Chloroquine	K76T
MDR1	Food vacuole	Transporter	Mefloquine	D86Y
DHFR	Cytoplasm	Folate metabolism	Pyrimethamine, proguanil	S108N, N51I, C59R, I164L
DHPS	Cytoplasm	Folate metabolism	Sulfadoxine, dapsone	A437G, K540E, A581G
Cytochrome *b*	Mitochondria	Electron transport	Atovaquone	Y268S/N/C

Table 23.8   Proteins and Mutations Involved in Drug Resistance

*CRT, chloroquine resistance transporter; DHFR, dihydrofolate reductase; DHPS, dihydropteroate synthetase; MDR1, multidrug resistance transporter.
**Polymorphisms associated with drug resistance in which the number refers to the amino acid (i.e., codon), and the first letter is the wild-type amino acid residue and the second letter is the polymorphism(s) associated with resistance.

Chloroquine resistance is due to a point mutation in a transporter gene

Chloroquine was first introduced as a treatment and prophylaxis against malaria in 1947. It rapidly became the drug of choice as an antimalarial due to its low toxicity and high efficaciousness. This high therapeutic index is due largely to the accumulation of chloroquine in the food vacuole of the parasite and the targeting of the drug to a unique parasite biochemical process (FIGURE 23.8). The food vacuole is a lysosome-like organelle that specializes in the digestion of hemoglobin. Digestion of hemoglobin releases free heme, which is highly toxic. The malaria parasite detoxifies the heme through a biocrystallization process that forms hemozoin, more commonly known as the malarial pigment. The host detoxifies heme by other mechanisms and therefore hemozoin formation is unique to the parasite.

Lysosomes are described in Chapter 2, page 37.

SIGNPOST

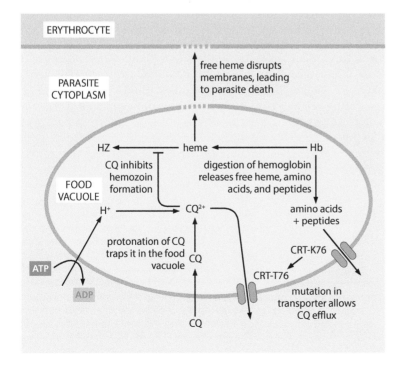

FIGURE 23.8 Mechanisms of chloroquine action and resistance. The site of action of chloroquine (CQ) is the food vacuole of the parasite which is an acidic lysosome-like organelle. The main activity of the food vacuole is the digestion of hemoglobin (Hb) into peptides and amino acids that are transported to the parasite cytoplasm. A by-product of Hb digestion is free heme that is detoxified in a biocrystallization process to form hemozoin (HZ). CQ enters the acidic food vacuole and is protonated to a positively charged form that cannot pass across the lysosomal membrane. Therefore, CQ is trapped in the food vacuole and reaches high concentrations. CQ blocks the formation of HZ, and the resulting free heme has detergent-like properties that disrupt membranes. Membrane disruption leads to parasite death. CQ resistance is associated with a mutation in the chloroquine resistance transporter (CRT) that converts the positively charged lysine (K) residue at position 76 to a threonine (T). The mutated CRT removes the protonated CQ from the food vacuole and the parasite now survives due to the efflux of CQ from the food vacuole.

Chloroquine is concentrated in the food vacuole and interferes with the formation of hemozoin. The free heme that is released during the breakdown of hemoglobin disrupts membranes and rapidly kills the parasite. The low toxicity and high efficacy of chloroquine resulted in its widespread use, including mass drug administration campaigns, and there was even talk of eliminating malaria. However, in the 1950s and 1960s chloroquine resistance emerged in southeast Asia and South America. From southeast Asia resistance spread to the Indian subcontinent and in the 1970s resistance was detected in Africa. The popularity of chloroquine and its use in mass drug administration likely contributed to the rapid spread of resistance around the world. Currently there is virtually no *P. falciparum* that is still sensitive to chloroquine.

The mechanism of chloroquine resistance was elucidated through gene-mapping studies of crosses between drug-resistant and drug-sensitive strains of *P. falciparum*. These gene-mapping studies identified a transporter gene that was subsequently named the chloroquine resistance transporter (CRT). The normal function of CRT is to transport peptides resulting from the digestion of hemoglobin from the food vacuole to the parasite cytoplasm. A point mutation in CRT confers the ability to transport chloroquine from the food vacuole (FIGURE 23.8). This efflux of chloroquine from the food vacuole means that chloroquine does not accumulate in the food vacuole and therefore does not block hemozoin formation.

Point mutations in drug targets often confer drug tolerance and resistance

Fansidar® is another common antimalarial and is a combination of pyrimethamine and sulfadoxine, which are inhibitors of dihydrofolate reductase (DHFR) and dihydropteroate synthetase (DHPS), respectively. Many drugs target folate metabolism and such drugs are powerful antimicrobials due to differences in folate metabolism between microbes and the human host (BOX 23.4).

BOX 23.4

Folate Metabolism and Mechanism of Action of Anti-Folates

Folates are co-factors that are needed for many biosynthetic processes. Humans cannot synthesize folate and must acquire folate from their diet. Microbes, conversely, can synthesize folate and often cannot utilize preformed folates. A major target of the de novo synthesis of folates is dihydropteroate synthetase (DHPS) and there are several antimicrobial drugs that inhibit this enzyme (BOX 23.4 FIGURE 1). These DHPS inhibitors provide selective toxicity toward the microbes since human cells do not have the pathway for de novo folate synthesis and the microbes are highly dependent on the synthesis of folate. Another target in folate metabolism is dihydrofolate reductase (DHFR). This enzyme plays a key role in the recycling of folates and is essential in all organisms. Drugs targeting DHFR distinguish host DHFR from pathogen DHFR and selectively inhibit the pathogen DHFR. The combination of DHPS and DHFR inhibitors acts synergistically.

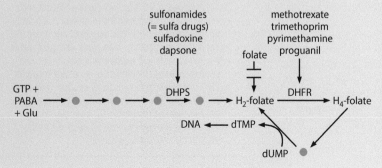

BOX 23.4 FIGURE 1 Drugs targeting folate metabolism. Microbes have the ability to synthesize folate from para-aminobenzoic acid (PABA), glutamate (Glu) and GTP. Several drugs inhibit DHPS, a key enzyme in this pathway. Sulfonamides, more commonly called sulfa drugs, are common anti-bacterial drugs and sulfadoxine and dapsone are used as antimalarial drugs. DHFR is another common drug target of folate metabolism. Pyrimethamine and proguanil are antimalarial drugs, trimethoprim is a common anti-bacterial drug, and methotrexate is used in cancer therapy as an anti-inflammatory drug. Inhibiting DHFR impacts the ability of the pathogen to synthesize DNA.

Several mutations in the *DHFR* and *DHPS* genes are associated with drug resistance (TABLE 23.9). The effects of the mutations are additive in that the first mutation decreases drug efficacy and subsequent mutations decrease the efficacy even further. The concentration of the drug that inhibits half of the enzyme activity is called the IC_{50} (for inhibitory concentration). The IC_{50} is a measure of drug efficacy and drugs with lower IC_{50} values are more effective than drugs with higher IC_{50} values. The introduction and accumulation of key mutations in both DHFR and DHPS raise the IC_{50} values of pyrimethamine and sulfadoxine. In the presence of drug pressure, mutations in target genes accumulate and over time the parasites become increasingly tolerant to the drug and are selected, and eventually the parasites become completely resistant to the drug.

Table 23.9 Polymorphisms and Drug Sensitivity

Pyrimethamine		Sulfadoxine	
DHFR haplotype	IC_{50}	DHPS haplotype	IC_{50}
S108/N51/C59 (wild-type)	3.7	A437/K540 (wild-type)	129
N108/N51/C59 (single mutation)	150.2	G437/K540 (single mutation)	614
N108/I51/R59 (triple mutation)	815.3	G437/E540 (double mutation)	1258

CONTROL AND PREVENTION OF MALARIA

The prevalence of malaria has been substantially reduced over the past few decades. Many approaches involving the various strategies for preventing and controlling vector-transmitted diseases have been utilized (see Table 23.2). Methods to reduce the number of mosquitoes primarily target the environment. For example, draining swamps and other water sources that are needed for mosquito breeding and larval development can decrease the abundance of mosquitoes. Spraying with insecticides or larvicides over large areas is also used to decrease mosquito abundance. Since insecticides and larvicides can have a negative environmental impact, biological control methods such as using fish that eat mosquito larvae or bacteria that are pathogenic to the larva are attractive alternatives.

Major methods to decrease mosquito-human contact include insecticide-treated bed nets, indoor residual spraying, screens on windows, mosquito repellants, and protective clothing. Most of these methods are aimed are reducing indoor mosquito biting since much of the transmission is due to mosquito biting inside houses during the night. In this regard, treated bed nets have proven to be efficacious, especially in tropical Africa. Controlling malaria in this region has been especially problematic due to limited resources and high transmission rates. The bed nets are relatively cheap, non-toxic, and easy to promote since they also reduce nuisance biting. A large portion of the reduction in malaria over the last two decades is believed to be associated with the use of bed nets and indoor residual spraying of insecticides (FIGURE 23.9). Case detection and treatment with artemisinin combination therapy have also contributed to this reduction in malaria prevalence. Since almost all malaria is restricted to humans, treatment of infected people reduces the parasite reservoir and decreases the risk of transmission.

FIGURE 23.9 Contributions of interventions to the decrease of malaria prevalence in tropical Africa. The prevalence of malaria has been decreasing due to control efforts. Insecticide-treated bed nets (ITNs) are believed to be a major factor in this reduction. Treatment of infected persons with artemisinin combination therapy (ACT) and indoor residual spraying (IRS) have also contributed to the reduction of *falciparum* malaria. [From Bhatt S, Weiss D, Cameron E et al. (2015) *Nature* 526: 207–211. doi:10.1038/nature15535. With permission from Springer Nature.]

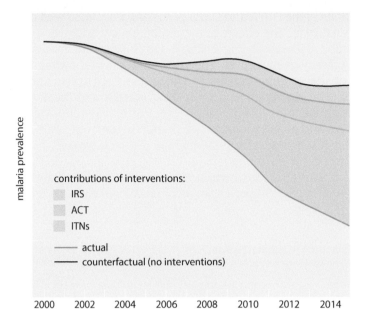

SUMMARY OF KEY CONCEPTS

- Numerous human pathogens, including viruses, bacteria, protozoa, and parasitic worms, are transmitted via blood-feeding arthropods

- Transmission dynamics depend on complex interactions of specific pathogens, specific vectors, possible non-human reservoirs, and humans

- Vector-transmitted viruses as a group are called arboviruses and dengue and yellow fever are the two most common arboviral diseases in humans

- Vector-borne bacterial diseases historically have been major human scourges, but, due to antibiotics and improved hygiene, their impact on human health has been substantially reduced

- Protozoal infections transmitted by vectors exhibit focal geographical distributions and are primarily found in tropical and sub-tropical regions

- Among the vector-transmitted diseases malaria causes the most mortality and morbidity

- The malaria parasite exhibits a complex life cycle involving both sexual and asexual replication and transmission by mosquitoes of the genus *Anopheles*

- Among the five species of *Plasmodium* that can infect humans, *P. falciparum* causes the most morbidity and mortality

- Severe *falciparum* malaria is associated with the sequestration of infected erythrocytes in the capillaries of organs

- Several efficacious drugs to treat malaria are available and the prognosis after treatment is good

- The malaria parasite rapidly develops resistance against antimalarial drugs and drug resistance is a major problem in the control of malaria

- Insecticide-treated bed nets, combined with indoor residual spraying, and treatment of infected perople have been effective in reducing the prevalence of malaria in tropical Africa

KEY TERMS

arbovirus	Any group of viruses that are transmitted by mosquitoes, ticks, or other arthropods (arthropod-borne virus).
arthropod	Invertebrate animals are characterized by an exoskeleton, a segmented body, and paired jointed appendages such as insects, arachnids, or crustaceans.
gametocyte	A stage in the malaria parasite life cycle that is transmitted from humans to mosquitoes.
malaria	A disease caused by protozoan parasites in the genus *Plasmodium* and transmitted by mosquitoes.
merozoite	A stage in the malaria parasite life cycle that invades red blood cells.
paroxysm	A sudden recurrence or intensification of a symptom. In malaria it refers to the periodic febrile attacks due to the synchronous replication and release of parasites into the blood.
Plasmodium falciparum	The species of malaria parasite that causes the most morbidity and mortality and is especially prevalent in tropical Africa.
sporozoite	The stage of malaria parasite that is found in the saliva of mosquitoes which is infectious to humans.
vector	An organism, generally an arthropod, that transfers (or carries) a pathogen from host to another host.

REVIEW QUESTIONS

1. Discuss the strategies used to control and prevent vector-transmitted diseases.

2. What are the two most common arboviral diseases in humans? Describe the disease manifestations of both infections.

3. What is meant by sequestration and what are the clinical manifestations that can result from the sequestration of the malarial parasite?

4. Discuss the diagnosis and treatment of malaria.

5. Describe the mechanism of action of chloroquine and the mechanism of chloroquine resistance.

ADDITIONAL QUESTIONS

In addition to the Review Questions provided above, there is a range of free online questions designed to enable students to further test their understanding of the chapter material. In order to access these interactive questions, please visit the book's website: https://routledge.com/cw/wiser

Parasitic Helminths

24

Most of the diseases covered in this book were chosen for their high mortalities. Infections with parasitic worms, also called helminths, are associated with a low rate of mortality. However, billions of people worldwide are infected with parasitic worms and **helminthiases** are a major problem in resource-poor countries. This affliction of primarily low-income people and the low mortalities mean that helminthiases do not receive a lot of attention and parasitic worm infections are often considered neglected tropical diseases.

A wide range of parasitic worms infects humans, including roundworms, flukes, and tapeworms. Major human diseases due to helminths are soil-transmitted helminthiases, filariasis, schistosomiasis, and neurocysticercosis. Various organs are affected by helminths and, similarly, there are numerous modes of transwmission. Depending on the species, an infection can involve ingestion of larvae or eggs, vector transmission, or direct penetration of the skin by larvae in association with soil or water contact. In natural human infections, adult worms reside in specific organs or tissues and produce eggs or larvae, which are key elements in the transmission of parasitic worm infections.

Infection with worms generally does not cause acute infection. Instead, human helminthiases are typically chronic infections that span years and sometimes decades. Thus, the morbidity associated with parasitic worms is usually subtle. For example, chronic infections are correlated with reduced birthweight, poor physical and cognitive development, or reduced productivity. These adverse manifestations promote the continuing cycle of poverty and poor health outcomes, especially in children. Severe pathology is typically due to inflammatory responses against the worm or damage caused by the migration of the worm through the body.

GENERAL HELMINTH MORPHOLOGY AND PHYSIOLOGY

SOIL-TRANSMITTED HELMINTHS

FILARIAL NEMATODES

SCHISTOSOMES

CESTODES

TREATMENT AND CONTROL OF HELMINTHIASES

GENERAL HELMINTH MORPHOLOGY AND PHYSIOLOGY

Worms are generally macroscopic organisms that are divided into four major taxonomic groups: platyhelminths (flatworms), acanthocephalans (thorny-headed worms), nematodes (roundworms), and annelids (segmented worms). Parasitic worms of humans are almost exclusively flatworms or roundworms (TABLE 24.1). There are parasitic thorny-headed worms; however, human infection is extremely rare. The only parasitic annelids are leeches, which are ectoparasites and not infectious disease. Parasitic nematodes of humans are divided into soil-transmitted helminths and filarial nematodes as convenience groups based on the mode of transmission. Trematodes (flukes) and cestodes (tapeworms) are two major types of parasitic flatworms.

Table 24.1 Major Helminthiases of Humans		
Class	**Major diseases**	**Prevalence**
Soil-transmitted helminths	Ascariasis, trichuriasis, hookworms	2 billion
Filarial nematodes	Lymphatic filariasis, river blindness, loiasis	200 million
Flukes (trematodes)	Schistosomiasis	250 million
Tapeworms (cestodes)	Cysticercosis	400,000

Roundworms have a complete digestive system and a body cavity

The roundworms have a long cylindrical shape and are characterized by a complete tubular digestive system consisting of a mouth on one end and an anus on the other end (FIGURE 24.1). Like other organisms, the digestive system of roundworms also has various regions along its length that specialize in different aspects of digestion. Roundworms also have a fluid-filled body cavity between the outer wall of the worm and the digestive tract. However, it is not considered a true body cavity since it is not lined with tissue. In addition, roundworms are covered with a waxy layer secreted by the epidermis, called the cuticle, which protects the worm. Gas exchange cannot occur across the cuticle and, thus, gas exchange occurs within the intestinal tract, as there is no respiratory system.

FIGURE 24.1 Basic roundworm body plan. Roundworms are characterized by a complete digestion system with a mouth and an anus. Distinct layers are seen in cross-section. Between the intestines and outer wall is a fluid-filled body cavity. However, it is not considered a true body cavity since it is not lined with a distinct tissue layer. Reproductive organs are also found in the body cavity (not shown).

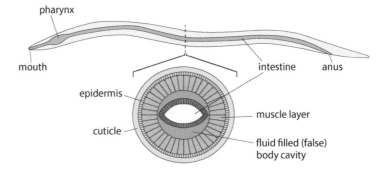

Flatworms have a simple body plan with a digestive cavity

As the name implies, flatworms have a flattened morphology. This morphology is due to the lack of a body cavity and the lack of specialized respiratory and circulatory organs. The flattened shape facilitates better nutrient diffusion and gas exchange to all the cells throughout the body. Flatworms have a digestive system with only a single opening into a digestive cavity (FIGURE 24.2). Food is taken up through this opening and digested within the cavity. Undigested material is expelled through this same opening. The space between the digestive cavity and the outer body wall is filled with a spongy connective tissue called parenchyma. The outer covering of flatworms is called the tegument. This tegument is a multinucleated tissue, called a syncytium, which serves as a protective covering, as well as permitting absorption and secretion. The outermost surface of the tegument is composed of glycolipids and glycoproteins and similar to the glycocalyx of microbes. This glycocalyx protects the parasite from host digestive enzymes, enhances nutrient absorption, and maintains the parasite's surface membrane.

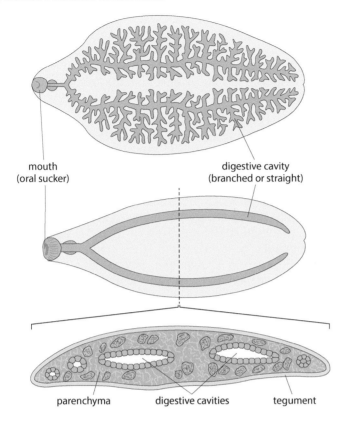

mouth
(oral sucker)

digestive cavity
(branched or straight)

parenchyma   digestive cavities   tegument

FIGURE 24.2 Basic flatworm body plan. The digestive cavity of flatworms has a single opening (mouth) for both ingress of food and egress of waste. In larger flatworms the digestive cavity is branched to provide more surface area. The outer surface of flatworms is called the tegument. Between the digestive cavity and tegument is connective tissue called parenchyma, which also contains the reproductive organs (not shown).

Helminth proliferation and transmission involve sexual reproduction

Helminths are metazoans with specialized reproductive systems involving testes and ovaries. Some species exhibit separate sexes (dioecious), and other species are hermaphroditic with an individual worm having both male and female sex organs. The gonads make up a large amount of body mass for most parasitic helminths and some worms can produce thousands of eggs per day. In general, the life cycle of parasitic worms involves the release of fertilized eggs and, after an embryonic period, **larvae** hatch from the eggs (FIGURE 24.3). A few worms release larval forms instead of eggs. In addition, some worms go through multiple larval stages before developing into adults. Sometimes the larval and adult stages of parasitic worms involve different hosts. Hosts of the larval stages are called **intermediate hosts** and hosts of the adult stages are called **definitive hosts**. For example, almost all trematodes have a vertebrate as the definitive host and a mollusk, usually a snail, as the intermediate host. Similarly, the life cycle of parasitic cestodes also involves an intermediate host and a definitive host with transmission involving ingestion of the intermediate host by the definite host.

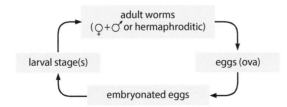

adult worms
(♀+♂ or hermaphroditic)

larval stage(s)

eggs (ova)

embryonated eggs

FIGURE 24.3 Generalized helminth life cycle. Parasitic helminths sexually reproduce via eggs. Some worms are hermaphroditic in that they contain both male and female reproductive organs and other worms have distinct sexes. Embryonated eggs hatch into larvae that develop into adults. Some worms undergo multiple larval stages before becoming egg-producing adults.

Humans are the definitive hosts of natural helminthiases

Depending on the worm species and its life cycle, humans are usually infected with either embryonated eggs or larvae. Larvae develop into adult worms, which produce eggs that are released into the environment or taken up by vectors to complete the life cycle. Thus, humans are the definitive host. Humans can sometimes become infected with larval forms of a helminth species not normally infecting humans. Since this represents an unnatural infection, the larvae do not develop into adults. Instead, the worms continuously migrate through the body and cause a disease known as visceral larva migrans.

In contrast with infectious diseases caused by microorganisms, which replicate within the host, adult worms do not proliferate within the human host. The adult worms survive for long periods in the human host – sometimes years or decades – and continuously release eggs or larvae. This lack of reproduction in the human host is one of the reasons helminth infections are not highly virulent, and quite often exhibit mild clinical manifestations. Pathogenesis associated with helminth infections is generally associated with the migration of the worms in various parts of the body or inflammatory reactions directed against the worm or eggs. In addition, parasitic worms are quite adept at modulating the host's immune response and suppressing inflammation (BOX 24.1). Furthermore, overt clinical manifestations are usually only associated with large numbers of worms. Hyperparasitism with numerous worms is referred to as a high worm burden.

BOX 24.1

Helminths as Masters of Immunomodulation

The establishment of long-term chronic infections by parasitic worms involves the secretion of products that downregulate both innate and adaptive immunity (Chapter 9). This immunomodulation prevents the elimination of the parasite by the host and, at the same time, minimizes severe pathology. For example, the migration of worms in tissues sets off alarm signals via damage-associated molecular patterns (DAMPs). Parasitic worms suppress these signals and the associated inflammation and possible tissue damage. Similarly, secretory products of parasitic worms suppress the activation of macrophages and dendritic cells. Macrophages and dendritic cells are phagocytic cells that also function as professional antigen-presenting cells. This suppression of professional antigen-presenting cells affects the development of T cells, and especially helper T cells. For example, it has long been known that helminth infections typically lead to a polarized type 2 helper T-cell response and a suppression of the type 1 helper T-cell response. Type 2 responses are associated with less inflammation than type 1 responses. This type 2 polarized response is also a key element of the hygiene hypothesis (Chapter 17). In addition, a large proportion of naive T cells develop into regulatory T cells (T_{reg} cells) during helminth infections. Regulatory T cells suppress immune responses in an antigen-specific manner. The suppression of inflammation by parasitic worms contributes to their ability to establish long-term chronic infections. Thus, inflammation and overt disease manifestations are not commonly associated with helminthiases.

SOIL-TRANSMITTED HELMINTHS

Soil-transmitted helminths (STHs), also called geohelminths, are a group of parasitic roundworms that inhabit the human intestines (TABLE 24.2). These infections are found worldwide, primarily in tropical and resource-poor areas. A common feature, as implied in the name, of these various worms is that the infective stages are found in soil and transmission involves contact with soil. Soil becomes contaminated when eggs or larvae are excreted in the feces (FIGURE 24.4). Generally, eggs mature in the environment and become infective after embryonation. The **embryonated egg** containing a larval form is the infective stage. In the case of *Ascaris* and *Trichuris* species, infection is initiated by the ingestion of embryonated eggs. Embryonated hookworm eggs hatch in the soil, and the resulting larvae initiate the infection by penetrating the skin of the human host. Eggs of *Strongyloides* species hatch in the intestines and the larvae are passed with the feces. Like hookworms, *Strongyloides* larvae in the soil penetrate the skin to initiate a new infection. In addition, *Strongyloides* larvae can penetrate the intestines and autoinfect the host. This autoinfection increases the worm burden in the already infected host.

Table 24.2	Soil-transmitted Helminths			
Species	**Common name**	**Infective stage**	**Larval stage**	**Major pathology**
Ascaris lumbricoides	Intestinal roundworms	Ingested embryonated egg	Migrate to lungs and swallowed	Bowel obstruction
Trichuris trichiura	Whipworms	Ingested embryonated egg	Remain in intestine	Dysentery and colitis
Ancylostoma duodenale	Hookworms	Larvae penetrate skin	Migrate to lungs and swallowed	Anemia due to intestinal bleeding
Necator americanus				
Strongyloides stercoralis	Threadworms	Larvae penetrate skin and autoinfection	Migrate directly to intestines	Disseminated disease in immunocompromised people
*Enterobius vermicularis**	Pinworms	Ingested eggs (autoinfection)	Remain in intestines	Itching around anus

*Not considered a soil-transmitted helminth but is an intestinal nematode with a life cycle similar to *Trichuris* species.

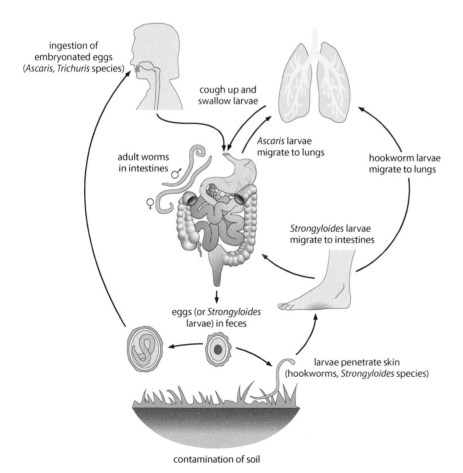

FIGURE 24.4 Life cycles of soil-transmitted helminths. One common feature of soil-transmitted helminths is the presence of adult worms in the intestines. Females lay eggs that are passed in the feces except for *Strongyloides* species. The eggs of *Strongyloides* species hatch within the intestines and the larvae are passed in the feces. Helminth eggs found in feces are not infective and must mature (embryonate) in the environment. Following embryonation, the eggs of *Ascaris* and *Trichuris* species become infective, and transmission is via ingestion of contaminated food or water. Ingested larval stages hatch in the intestines. *Trichuris* larvae mature to adults in the large intestine. *Ascaris* larvae migrate to the lungs via the circulatory system and are coughed up and swallowed. Following swallowing, the larvae mature into adults in the small intestine. Hookworm eggs hatch in the environment and the larvae penetrate the skin and migrate to the lungs. As with *Ascaris* species they are coughed up, swallowed, and develop into adults in the small intestine. The larvae of *Strongyloides* species also penetrate the skin, but these larvae can migrate directly to the intestines.

A common element in the life cycle of STH is the residence of adult worms in the intestines

Ultimately adult worms need to be in the intestines so that eggs or larvae can be passed in the feces. Different strategies are utilized by the various worms to reach the intestines. For example, *Trichuris* larvae hatch in the small intestines after ingestion and migrate to the colon where they develop into adults. The life cycle of *Ascaris* species is more complex in that the newly hatched larvae leave the intestines and migrate through the body until reaching the lungs. From the lungs they are coughed up and swallowed. The larvae from the lungs then develop into adults that reside in the intestinal tract. Similarly, hookworms, after penetrating the skin, migrate first to the lungs and then are coughed up and swallowed to reach the ultimate destination of the intestines. *Strongyloides* larvae can migrate directly to the intestines after penetrating the skin.

Contamination of soil with human feces is a key element of STH transmission

Several conditions are needed to maintain the transmission and endemicity of STH (TABLE 24.3). An important element of transmission is the presence of enough infected people living in the community to serve as reservoirs. Second, these individuals need to defecate outdoors and contaminate the ground. Thus, STH endemicity is associated with a lack of sanitation and generally afflicts a rural low-income population. In addition, the conditions must be appropriate for the eggs or larvae to mature and survive in the soil. Warm humid climates promote this survival and STH are particularly prevalent in the tropics. Under appropriate conditions eggs or larvae can survive years. Finally, there must be contact with the contaminated soil for infection to occur. In the case of *Ascaris* and *Trichuris* species, this involves the contamination of hands, food, or water that can lead to ingestion. Therefore, children at the crawling age and dirt-eating age are particularly prone to infections. For hookworms and *Strongyloides* species there must be sufficient contact of skin with soil, such as going barefoot.

Table 24.3 Conditions Promoting STH Transmission
• Infected people living in the community
• Defecation on the ground (i.e., deficiencies in sanitation)
• Appropriate conditions for eggs or larvae to mature and survive in the environment
• Contact with contaminated soil:
○ contamination of hands, food, or water → ingestion of eggs
○ direct skin contact with larvae (e.g., barefoot)

Soil-transmitted helminthiases are generally silent and insidious

Most infections with STH are asymptomatic. Except for the autoinfection associated with *Strongyloides* species, adult worms do not increase in number during infection. However, it is possible to be infected repeatedly and that leads to high worm burdens within infected hosts. Generally, clinical manifestations are only associated with high worm burdens. For example, high numbers of *Ascaris* species can result in blockage of the intestines. Heavy worm burden with *Trichuris* species is often associated with dysentery and chronic colitis. Heavy hookworm infections are especially prone to produce anemia. This is because hookworms attach to the intestinal mucosa via teeth-like structures (FIGURE 24.5) and ingest blood. The damage caused by the hookworm during feeding results in approximately 0.2 mL of blood loss per day per worm.

Although overt symptoms are not commonly seen with helminthiases, there is substantial morbidity – especially in children – that is subtle and insidious. Chronic helminth infection, combined with malnutrition that is prevalent in endemic areas, is correlated with poor birth outcome, poor physical (e.g., stunting) and cognitive development, poor school and work performance, and poor socioeconomic development. These silent clinical manifestations also contribute to the perpetuation of poverty.

Autoinfection by *Strongyloides* sp. can result in severe disease in immunocompromised individuals

Like other STH, *Strongyloides* spp often causes an asymptomatic infection or chronic disease with subtle clinical manifestations. However, *Strongyloides* spp. can also cause a hyperinfection syndrome that can lead to disseminated disease and possibly death if not treated. This hyperinfection is due in part to the ability of the newly hatched larvae in the intestines to penetrate the intestinal mucosa of the host. This is called autoinfection since it is not due to the normal route of infection involving contact with soil. This autoinfection can generate high worm burdens without re-infections, and also possibly results in a lifelong infection due to the continuous generation of adult worms. Infected individuals need to be treated to remove all adult worms to completely cure the infection. Disseminated strongyloidiasis caused by autoinfection is primarily associated with immune suppression. A wide range of clinical manifestations can be associated with disseminated disease due to multiple organs being affected.

Enterobiasis is the most common helminth infection in the United States and Europe

Most helminth infections are rare in developed countries and primarily found in resource-poor countries. One exception is *Enterobius vermicularis*, commonly known as pinworm, threadworm, or seatworm. The entire life cycle takes place in the intestines. Gravid females emerge from the anus and deposit eggs around the anus. This produces perianal itching, which is the primary clinical manifestation of enterobiasis. This itching results in scratching and contamination of the hands which can result in person-to-person transmission. Autoinfection in which individuals re-infect themselves by ingesting eggs is also possible. The eggs hatch in the small intestines and migrate to the colon where they mature into adults.

FILARIAL NEMATODES

Another group of roundworms that causes human disease is **filaria**. Filariae are distinguished from the STH by their mode of transmission. These parasites are transmitted from person to person via arthropod vectors (Chapter 23) or a copepod intermediate host (BOX 24.2). As with other parasitic helminths, adult worms inhabit the human host (FIGURE 24.6). The females, instead of laying eggs, release microscopic larval forms called **microfilariae**. Some view the microfilariae as actually being mobile eggs. The microfilariae are taken up by the vector as it feeds and the microfilariae develop into larvae. Mature larvae migrate to the head region of the vector and then infect a new host during a subsequent blood meal by the vector. Once within the host the larvae develop into adult forms to complete the life cycle.

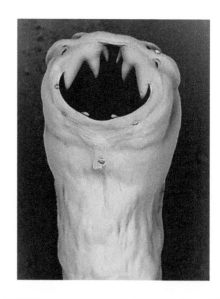

FIGURE 24.5 Scanning electron micrograph of a hookworm head. The mouths of hookworms have teeth-like structures that help the worm attach to the intestinal mucosa and result in damage and bleeding. The attached hookworms ingest blood from the host. [Modified from Loukas A, Hotez P, Diemert D et al. (2016) *Nature Reviews Disease Primers* 2: 16088. https://doi.org/10.1038/nrdp.2016.88. With permission from Springer Nature.]

BOX 24.2

Dracunculiasis

Another filaria that infects humans is *Dracunculus medinensis*, commonly known as the guinea worm. Transmission of guinea worm involves the ingestion of infected water fleas, which are the intermediate host. After ingestion, the larvae migrate out of the stomach into the abdominal cavity and seek out a mate. Following fertilization female worms migrate to the skin and cause a painful blister typically on the foot (**BOX 24.2 FIGURE 1**). The blister ruptures within a few days and results in an open sore associated with painful intense burning as the worm emerges. The sufferer typically submerges the foot in water to relieve the pain. When the worm contacts water, it releases hundreds of thousands of larvae, thus contaminating the water. The larvae then infect the water fleas and complete the life cycle.

Guinea worm has nearly been eradicated through control measures. Prevention relies on two basic complementary approaches. One measure is to prevent people from drinking water containing water fleas by providing safe water sources or by filtering water before drinking. The second approach is to prevent people with emerging guinea worms from entering water sources used for drinking. For example, the infected person places their foot with the lesion in a bucket of water and that water is poured on dry ground. The eradication program has eliminated dracunculiasis from many previously endemic countries and incidence has dropped from

3.5 million cases per year in the early 1980s to 24 cases in 2020.

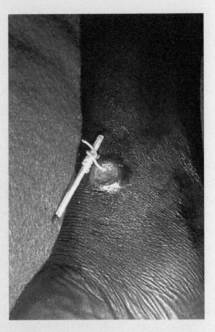

BOX 24.2 FIGURE 1 Extraction of emerging guinea worm. Shown is the lesion associated with an emerging guinea worm located near the ankle. One method to extract the worm is to slowly wrap it around a small stick. Care must be taken not to break the worm and since the worm can be up to one meter in length the extraction can take days to weeks. (From the Public Health Image Library of the Centers for Disease Control and Prevention.)

FIGURE 24.6 General filarial life cycle. Depending on the species, adult worms reside in the lymphatics or form subcutaneous nodules. Instead of laying eggs the female worm releases microfilariae that circulate in the blood, lymphatics, and skin. Microfilariae ingested by the vector during a blood meal undergo three larval stages. The final larval stage (L3) migrates to the head region of the vector. During a blood meal the L3 larva falls off the vector and penetrates the skin of the host or enters through the bite wound. Once within the host, the L3 larvae develop into adult forms to complete the life cycle.

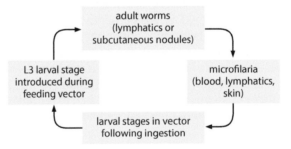

Disease manifestations of filariasis can be due to either the adult stages or the microfilarial stages

Five species of filarial nematodes commonly infect humans and cause two major types of diseases (TABLE 24.4). Lymphatic filariasis is mosquito transmitted and the disease manifestations are due to the adult worms living in lymphatic vessels. As with other parasitic worms, most lymphatic filariasis infections are asymptomatic. In some individuals, though, the adult worms or the damage to the lymph vessels caused by the worm blocks the flow of lymph, leading to swelling of the lower extremities and genitals. The accumulation of lymph in the lower extremities can be severe and the most severe manifestations are called elephantiasis (FIGURE 24.7).

Table 24.4	Human Filariasis		
Disease	**Species**	**Vector**	**Pathology**
Lymphatic filariasis	*Wuchereria bancrofti* *Brugia malayi* *Brugia timori*	Mosquitoes	Adult worms can block lymph vessels leading to swelling of lower extremities and in extreme cases cause elephantiasis
Subcutaneous filariasis	*Onchocerca volvulus* (river blindness)	*Simulium* spp. (blackflies)	Inflammatory responses against the microfilaria cause severe itching in the skin or blindness
	Loa loa	*Chrysops* spp. (deer flies)	Generally asymptomatic

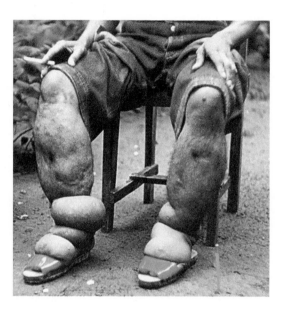

FIGURE 24.7 Severe manifestation of lymphatic filariasis. Blockage of lymph vessels can lead to massive accumulation of lymph in the lower extremities, leading to a syndrome known as elephantiasis. (From the Public Health Image Library of the Centers for Disease Control and Prevention.)

In the case of *Onchocerca volvulus* the adult worms are found in subcutaneous nodules that do not cause severe pathology. Instead, strong innate immune responses against the microfilariae, especially dying microfilariae, in the skin cause intense itching and other pathological effects in the skin. The cornea can be viewed as an extension of the skin, and onchocerciasis often results in blindness and visual defects as a consequence of corneal inflammation. Hence, onchocerciasis is also called river blindness due to the presence of the vector, blackflies of the genus *Simulium*, near rivers.

SCHISTOSOMES

Schistosomes, also referred to as blood flukes, are the most prevalent trematode infections in humans. However, food-borne trematode infections are increasingly being recognized as important neglected tropical diseases (BOX 24.3). Three *Schistosoma* species account for most human **schistosomiasis** (TABLE 24.5). Transmission involves snails as the intermediate hosts and each species of *Schistosoma* involves distinct genera of intermediate snail hosts, which also influences the geographical distributions. Although schistosomiasis does not exhibit a high level of mortality, it affects approximately 250 million people and has a large economic impact, primarily in resource-poor countries.

BOX 24.3

Food-borne Trematodiasis

Several trematode parasites of animals are capable of infecting and causing disease in humans. It is estimated that at least 40 million people may be infected with these various zoonotic trematodes. Most infections are acquired through ingesting intermediate hosts that are infected with larval stages of the parasites. Transmission is linked to methods of producing, processing, and preparing food and, thus, is often associated with cultural practices. In particular, the ingestion of raw or undercooked fish, crustaceans, or aquatic plants is a common source of food-borne trematodiasis.

The four most common genera causing food-borne trematodiasis are *Paragonimus, Clonorchis, Opisthorchis*, and *Fasciola* species (**BOX 24.3 TABLE 1**). Like other trematodes, snails are intermediate hosts for all four genera. However, unlike schistosomes, the cercariae do not directly infect the definitive host and transmission involves a second intermediate host, sometimes called a carrier.

Cercariae attach to crustaceans, fish, or plants and encyst into stages called metacercariae. Definitive hosts are infected by ingestion of the intermediate host or plant carrier containing metacercariae. Adult worms reside in the intestines (intestinal flukes), bile ducts (liver flukes), or lungs (lung flukes) and produce eggs, which are excreted with the feces (intestinal and liver flukes). Eggs of lung flukes are expelled with the sputum or coughed up and swallowed before being excreted in the feces.

As with most helminth infections, disease manifestations are usually observed only in cases of high worm burden. Pain in the upper right quadrant is a common symptom of liver flukes. Inflammation, obstruction of the bile ducts, or liver dysfunction is seen in chronic infections. A high rate of cholangiocarcinoma (cancer of the bile duct) is also observed in *Clonorchis* and *Opisthorchis* infections. Chronic cough and blood in the sputum are common symptoms of lung flukes.

Box 24.3	**Table 1 Common Food-borne Trematodes of Humans**		
Genus	**Carrier**	**Natural reservoir**	**Adult/Eggs**
Paragonimus	Crustaceans	Numerous crustacean-eating mammals	Lungs/coughed up and expelled or swallowed
Clonorchis	Fish	Dogs and other fish-eating carnivores	Bile ducts/passed in feces
Opisthorchis	Fish	Dogs, cats, and other fish-eating carnivores	Bile ducts/passed in feces
Fasciola	Aquatic, semi-aquatic plants	Numerous ungulates and herbivores	Bile ducts/passed in feces

Table 24.5	**Common *Schistosoma* Infections of Humans**			
Species	**Snail genera**	**Endemic areas**	**Eggs**	**Organs affected**
S. mansoni	*Biomphalaria*	Africa, South America, the Caribbean, the Middle East	Feces	Liver and intestines
S. japonicum	*Oncomelania*	China, East Asia, the Philippines	Feces	Liver and intestines
S. haematobium	*Bulinus*	Africa, the Middle East	Urine	Bladder

Contact with fresh water is the primary mode of transmission

A larval stage called the cercaria is released from snails into the water and cercariae penetrate the skin of the host (FIGURE 24.8). After gaining entry the cercaria develops into a stage called the schistosomula which enters the circulation. The schistosomula develops into either a male or a female adult in the liver. Paired adults migrate to veins surrounding the small intestines (*S. japonicum*), large intestines (*S. mansoni*), or the bladder (*S. haematobium*). Accordingly, eggs are shed in feces or urine. When eggs reach fresh water, a ciliated stage called the miracidium is released and the miracidium infects an appropriate snail intermediate host and develops into a sporocyst. The sporocyst replicates asexually and the daughter sporocysts develop into cercariae, thus completing the life cycle.

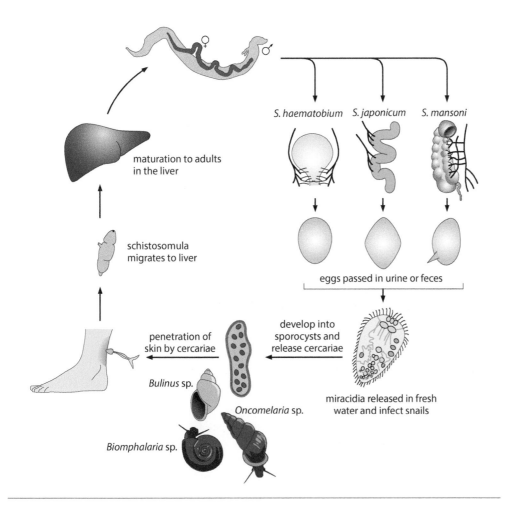

FIGURE 24.8 Schistosome life cycle. Adult worms are found in veins of the bladder or intestines resulting in eggs being passed in either the urine or the feces. Eggs hatch in fresh water and release miracidia that infect snails as the intermediate hosts. Within the snail the parasite develops into a sporocyst stage that replicates asexually to produce cercariae. Released cercariae penetrate the skin of the definitive host. After entry into the host, the cercariae lose their tails and develop into a schistosomula stage that enters the circulation. In the liver the schistosomula mature into either adult males or females, which pair and migrate to their final destinations of either the bladder or intestines and thus complete the life cycle.

Pathology associated with schistosomiasis is largely due to host inflammatory responses directed against the eggs trapped in tissues

As with many helminth infections little overt pathology is associated with the adult worms. Clinical manifestations associated with schistosomiasis are generally the consequence of progressive injury from chronic egg deposition in the tissues and the resulting inflammation. *S. mansoni* and *S. japonicum* primarily cause disease in the intestines and manifestations include abdominal pain and intermittent bloody stools. Eggs may also migrate back through the mesentery venules to the liver and a mass of immune cells, called a granuloma, surrounds the eggs. In heavy infections granulomas cause extensive fibrosis that can produce an enlarged liver over time. A clinical manifestation associated with *S. haematobium* is blood in the urine. Patients with high egg burdens may also experience dysuria due to obstruction of the urinary tract as a result of granuloma formation and fibrosis. Higher rates of bladder cancer are also observed in regions that are endemic for *S. haematobium*.

Granulomas are described in Chapter 20, pages 395–396.

SIGNPOST

CESTODES

Cestodes, commonly called tapeworms, are another type of parasitic flatworm that requires multiple hosts to complete its life cycle. Most tapeworms have a long ribbon-like structure, hence the name, with a head called the scolex and a long string of segments called proglottids (FIGURE 24.9). The scolex has hooks and suckers that attach to the intestinal epithelium. Each proglottid has its own ovary and testes that allow for both self-fertilization and cross-fertilization with nearby worms. The worms grow in length by adding proglottids to the end next to the scolex. Therefore, the more mature proglottids are found at the distal end of the worm. Gravid proglottids with embryonated eggs break off and are excreted with feces.

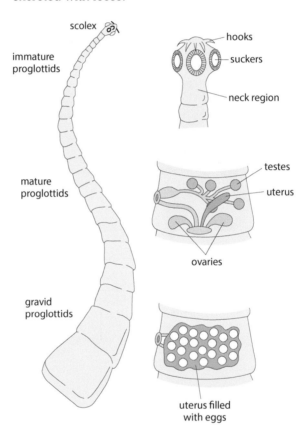

FIGURE 24.9 Tapeworm structure. The anterior end of the tapeworm has a head-like structure called the scolex. The scolex has suckers and hooks that attach it to the epithelium of the small intestine. The length of the worm is made of segments called proglottids. Newly formed proglottids are continuously produced at the neck below the scolex. Each proglottid functions independently and, as they mature, the proglottids develop both testes and ovaries. The most mature proglottids contain embryonated eggs and are found at the distal end. These gravid proglottids break off and are excreted with feces. The tapeworm has no digestive system and absorbs nutrients through the tegument on the surface of the worm.

Tapeworm infections usually involve pigs as the intermediate host

The most prevalent tapeworm infecting humans is *Taenia solium* of pigs. *T. saginata* and *T. asiatica* are acquired from cattle and are of less clinical significance. *Taenia* infections are acquired from the ingestion of undercooked meat containing larval stages of the worm called cysticerci. The larvae develop into adults in the small intestines and eggs are shed into the environment via the feces. Thus, humans are the definitive host. Intermediate hosts, such as pigs and cows, become infected by ingesting material contaminated with human feces. Following ingestion and hatching, the larvae migrate to the muscles or other tissues and form cysticerci. Cysticerci are dormant larval stages that are infectious to definitive hosts.

The most serious disease associated with cestodes is neurocysticercosis

Intestinal infections with adult tapeworms cause little if any disease manifestations – not even physical or cognitive impairment in children, as is often observed with nematode and trematode infections. Disease associated with tapeworm infections is primarily due to the cysticercus stage in tissues. In this case the infection was not acquired through the ingestion of undercooked meat, but, rather, through the ingestion of eggs from an infected person (FIGURE 24.10). In other words, humans can also be an intermediate host of tapeworms and develop cysticercosis. However, humans are considered a dead-end host in this situation since the infection cannot be passed on to other humans.

Cysticercosis in non-neural tissues is generally not problematic. However, neurocysticercosis due to cysticerci in the brain can result in seizures and other neurological symptoms. The manifestations of cysticercosis and neurocysticercosis depend largely on the anatomical location of the juvenile worm, and whether the worm is alive, dying, or dead. Disease manifestations are often associated with inflammatory responses directed at dying and dead worms.

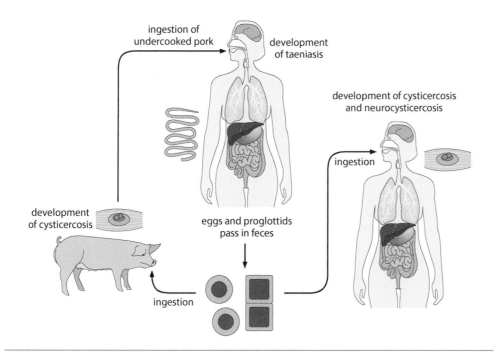

FIGURE 24.10 **Life cycle of *Taenia solium*.** In the natural life cycle, humans acquire *Taenia* infections through the ingestion of undercooked meat. Larval forms in the muscles of infected pigs develop into adult tapeworms in the small intestines causing taeniasis. Eggs and proglottids are excreted in the feces into the environment where they can survive for weeks. If ingested by a pig, the larval forms hatch from the eggs and migrate to tissues where they form cysticerci. These dormant cysticerci are infectious to humans. Humans can also become infected via the ingestion of eggs. In this aberrant life cycle the larvae migrate to the tissues and cause cysticercosis or to the brain and cause neurocysticercosis.

TREATMENT AND CONTROL OF HELMINTHIASES

Most helminth infections can be cleared through drug treatment (TABLE 24.6). Albendazole, praziquantel, and ivermectin generally target adult worms, whereas diethylcarbamazine targets the microfilaria stages. These drugs are relatively non-toxic and effective at clearing the worms. Since the pathology associated with many helminth infections is due to inflammation against dying worms, anti-helminthic treatment in some cases can exacerbate this inflammation. For example, the use of praziquantel to treat cysticercosis is often accompanied by an anti-inflammatory drug such as dexamethasone. Similarly, diethylcarbamazine is often contraindicated for onchocerciasis since the sudden death of the microfilaria can induce severe side effects.

Table 24.6 Common Anti-helminthic Drugs	
Drug	**Typical application**
Albendazole	Intestinal nematodes and filariasis
Praziquantel	Trematodes and cestodes
Ivermectin	Onchocerciasis
Diethylcarbamazine	Filariasis and loiasis

Control of helminthiases often involves mass drug administration

Mass drug administration (MDA) refers to treating a large population for a disease without first diagnosing the disease. This saves time and money since the diagnosis often costs more than the treatment. Furthermore, in many parts of the world the prevalence of parasitic worm infections is quite high and the majority of the population is infected. The low toxicity of anti-helminthic drugs, especially in non-infected persons, also makes mass drug administration possible. Mass drug administration is usually carried out in schools. This provides a convenient assembly of part of the population and children tend to be more affected by helminth infections through poor physical and cognitive development. In highly endemic areas, re-infection is common and mass drug administration is usually carried out annually or semi-annually. For this reason, some suggest that mass drug administration should strive to become community wide.

Prevention of infection involves sanitation, intermediate hosts, vectors, and food preparation

Since the transmission of many parasitic worms is largely due to contamination of the environment with helminth eggs, sanitation plays a key role in preventing and controlling helminth infections. In fact, development and the proper handling of human waste has led to the elimination of soil-transmitted helminths, schistosomiasis, and taeniasis in many parts of the world. Currently, these diseases primarily afflict the rural low-income population. Control of schistosomiasis could also include snail control and, similarly, the control of filarial nematodes could also incorporate vector control (Chapter 23). In addition, some helminthiases are associated with improper preparation of food, especially pork and fish. Thorough cooking of food can prevent these infections. Health education is also a key element in the prevention and control of helminthiases.

SUMMARY OF KEY CONCEPTS

- Parasitic worms affect billions of people, primarily in resource-poor countries or tropical areas, and are a major cause of human morbidity

- Important helminths infecting humans include intestinal nematodes, filarial nematodes, schistosomes and other trematodes, and cestodes

- Acquisition of infections can be via ingestion of eggs or larvae, ingestion of intermediate hosts infected with larvae, direct penetration of the skin by larvae, or arthropod vectors

- Many parasitic worms reside in the intestines and transmission involves the excretion of eggs or larvae in the feces

- Infection by soil-transmitted helminths, such as *Ascaris* and *Trichuris* or hookworms, involves contact with soil that has been contaminated by human feces

- Filariasis transmission involves arthropod vectors

- Schistosome infection involves contact with water and a snail as the intermediate host

- Normally helminth infections are not associated with severe clinical manifestations due to immunomodulation of the host by the parasite

- Overt clinical manifestations are generally associated with inflammation or tissue damage due to heavy worm burdens

- Subtle and insidious manifestations of helminthiases include suppressed physical and cognitive development, especially in children

- Control of helminthiases generally involves improved sanitation, health education about transmission, and possibly mass drug administration

KEY TERMS

cysticercosis	A disease caused by the larval stages (i.e., cysticerci) of tapeworms that was acquired from ingestion of matter contaminated with human feces.
definitive host	The host organism that harbors the adult stages of a parasite and where the parasite exhibits sexual reproduction.
embryonated eggs	Helminth eggs containing fully developed larvae.
filaria	A group of vector-transmitted roundworms characterized by a microfilaria stage that infects the vector.
helminthiasis	A disease caused by a parasitic worm.
intermediate host	A required host in the life cycle of a parasite that is essential for larval development.
larva	An immature stage (incapable of sexual reproduction) that hatches from an egg and develops into an adult form.
mass drug administration (MDA)	Administration of a drug to every member of a defined population at approximately the same time regardless of infection status.
microfilariae	Microscopic early larval forms of filaria parasites that are found in blood or subcutaneous tissue and that infect the vector.
schistosomiasis	A disease caused by a parasitic fluke of the genus *Schistosoma* and transmission involves an intermediate snail host and contact with water.
soil-transmitted helminth (STH)	Parasitic roundworms in which the eggs mature in the soil and then subsequently infect the human either via ingestion of embryonated eggs or penetration of the skin by larval forms.

REVIEW QUESTIONS

1. Discuss the transmission and life cycles of the various soil-transmitted helminths that infect humans.

2. Describe the pathogenesis of lymphatic filariasis and onchocerciasis and how the basis of the pathology differs between the two diseases.

3. What are the three most common schistosomes that infect humans and where are the adult worms and egg excretion typically found in the human body for each of them?

4. Describe the life cycle of *Taenia solium* and indicate the two ways in which humans can be infected and the possible disease manifestations in both situations.

ADDITIONAL QUESTIONS

In addition to the Review Questions provided above, there is a range of free online questions designed to enable students to further test their understanding of the chapter material. In order to access these interactive questions, please visit the book's website:
https://routledge.com/cw/wiser

Index